Nanobiotechnology for Sustainable Bioenergy and Biofuel Production

Editor

Madan L. Verma

Assistant Professor
Department of Biotechnology, School of Basic Sciences
Indian Institute of Information Technology Una
Himachal Pradesh, India

CRC Press is an imprint of the
Taylor & Francis Group, an **informa** business
A SCIENCE PUBLISHERS BOOK

Cover credit: Cover illustration reproduced from Trends in Biotechnology with permission of the publisher, Elsevier.

CRC Press
Taylor & Francis Group
6000 Broken Sound Parkway NW, Suite 300
Boca Raton, FL 33487-2742

Version Date: 20200320

International Standard Book Number-13: 978-0-367-08587-2 (Hardback)
International Standard Book Number-13: 978-0-367-54633-5 (Paperback)

Visit the Taylor & Francis Web site at
http://www.taylorandfrancis.com

and the CRC Press Web site at
http://www.crcpress.com

Preface

Presently, the prime focus of nanobiotechnology research in bioenergy sector is to improve the bioprocessing of biofuels production through the intervention of unique properties of nanomaterials. The nanomaterial is now documented as an excellent nanocarrier for diverse applications of biocatalyst through the structural-functional studies of nanomaterial immobilized enzyme (nanobiocatalyst). Through the amalgamation of nanotechnology and biotechnology fields, it is feasible to develop an efficient methodology for biofuel production.

Recently, researchers have employed novel nanobiotechnological approaches for the production of different forms of biofuels with easy and quick bioseparation of biocatalysts and products. Now methodologies for biofuels production with improved efficiency have revamped with the advancement in nanobiotechnology and analytical molecular biotechnology-based techniques. At the outset of new nanobiotechnology applications in the bioenergy sector, there is a pressing need to explore this area of biofuel production using the nanobiotechnological route and thus seeks special attention to publish the advancement in this area at this point of time. The present book is an attempt to bring together leading scientists to contribute review articles that cover the focal themes of all aspects of nanobiotechnology contributions in the bioenergy and biofuels production.

The first chapter summarizes various challenges and opportunities in the application of nanotechnology for sustainable biofuel production. The second chapter emphasizes on the recent advances in enzyme immobilization using nanomaterials and its applications for the production of biofuels. The third chapter discusses the contributions of carbon nanotubes-based enzyme immobilization for biofuel applications. The fourth chapter highlights the role of chitosan nanotechnology in biofuel production.

The fifth chapter focuses on the recent progress made in the bioenergy and biofuel production sector using nanobiocatalysis. The sixth chapter addresses the nanobiotechnological applications in biofuel production. The seventh chapter summarizes the various recent studies employed to enhance the bioethanol yield using ligninolytic and cellulolytic enzymes through nanotechnological intervention.

The eighth chapter provides an overview of the nanoparticles in methane production from anaerobic digesters with respect to its advantages, challenges and perspectives. The ninth chapter presents the contributions of nanomaterials in biohydrogen production.

The tenth chapter discusses nanobiotechnological solutions for sustainable bioenergy production. The eleventh chapter describes the potential of nanobiotechnology advances in bioreactors for biodiesel production. The twelfth chapter highlights the impact of nanotechnology in biorefineries.

As the editor, I sincerely thank all the authors for their outstanding efforts to provide the state-of-the-art information on the subject matters of their respective chapters. I thank several reviewers who evaluated the manuscripts and provided critical suggestions to improve these further. I hope this book will provide information about the latest research and advances, especially the innovations in nanobiotechnological solutions for bioenergy productions. I wish to thank everyone involved in making the possible publication of the book in a timely manner.

Most importantly, I am indebted to my parents (Sh. Rattan Chand Verma and Ms. Meera Kumari Verma) for inculcating values that has made me an established research academic. I am grateful to Dr. Aruna Verma (my wife and an academician) who has assisted in reviewing the contributed chapters. This endeavor would not have been possible without her motivation and constructive criticism as well as the cooperation extended by my sons Anmol Verma and Atharv Verma.

It is my hope that this book will be useful to the students and researchers from both academia and industry.

Madan L. Verma

Contents

Chapter 1

Challenges and Opportunities in the Application of Nanotechnology for Sustainable Biofuel Production

***Shailendra Kumar Singh,*[1,*] *Ajay Bansal,*[2] *Suman Kapur*[3] and *Shanthy Sundaram*[1]**

1. Introduction

To cope with the daunting energy challenges of day-to-day requirement, the focus of all nations has shifted toward the exploitation of alternative energies such as solar, wind, tidal, geothermal and bioenergy (Ahmadi et al. 2019). Among all alternatives, biofuels are one of the potent non-conventional alternatives that are comprised of diverse energy variables like biodiesel, biohydrogen, bioethanol, biogas, etc., and are obtained from living forms. In contrast to conventional fuel sources, the biofuels have not demonstrated any detrimental effects on the environment as they do not leave any toxic end-products after its processing (Cherubini 2010). They have gained huge attention in recent years due to their low GHG emission, reduced carbon footprint and, more importantly, their sustainable nature (Singh et al. 2012). However, due to the inadequate biochemical production techniques, there is still a very limited utilization of biological resources for biofuel application that may facilitate profitability.

Recently, nanotechnology has shown their significant potential on creating 'clean' and 'green' methods in biofuel production with substantial environmental remunerations. Though nanotechnology is an emerging science, it already has gained considerable attention in recent years due to its extremely fascinating and useful properties. At present, several engineered nanoparticles are being used for a

[1] Centre of Biotechnology, University of Allahabad, India - 211002.

[2] Department of Chemical Engineering, Dr B R Ambedkar National Institute of Technology, Jalandhar - 144011.

[3] Birla Institute of Technology & Science Pilani, Hyderabad Campus, Jawahar Nagar, Shamirpet Mandal, Ranga Reddy Dist., Telangana - 500078.

* Corresponding author: shailbiochem@gmail.com

variety of structural and non-structural applications in everyday life, ranging from consumer goods to medicine to improving the environment (Sahoo et al. 2017). Considering the environmental and economic issues, emerging nanotechnology could also offer potential solutions in enabling society to build and sustain a green economy. Nanoparticles have several interesting properties to provide faster and reliable methods of optimizing energy generation from the bio-resources. Nanotechnology targets entities' miniaturization to achieve unique characteristics like confining their electrons and producing quantum effects compared to the bulk material (Sudha et al. 2018). Thus, small nanoparticles are enough to enhance the performance of biofuel production through the improvement of its functionalities. Presently, exploitation of nanomaterials in the bioenergy area is under intense scientific study to generate products and processes that are energy efficient as well as economically and environmentally sustainable. The biggest advantage of coupling biofuel-nanotechnology is that it can facilitate innovations in material-science and engineering-technology to create energy-efficient routes and products for producing and storing energy.

In this context, the following sections of the chapter analyzes new advances and opportunities of nanotechnology to produce commercially viable biofuel. The review also examines the anticipated challenges related to nanotechnology for human and environmental safety.

2. Nanotechnology: Big Science in Small Packages

Nanotechnology is a molecular manipulation technique, i.e., it devises, synthesizes and manufactures matter on a nanoscale and applies structures or devices with the molecular precision scale. Today's researchers are finding a wide variety of ways to deliberately make materials ranging 1 to 100 nanometres (nm) or smaller such as nanoparticles, nanotubes, nanosheets, nanocomposites, nanocrystalline materials, metal-based nanomaterials, carbon-based nanomaterials, etc. (Faucher et al. 2019). A comparative scale of nanomaterials as to natural and man-made material is represented in Figure 1.

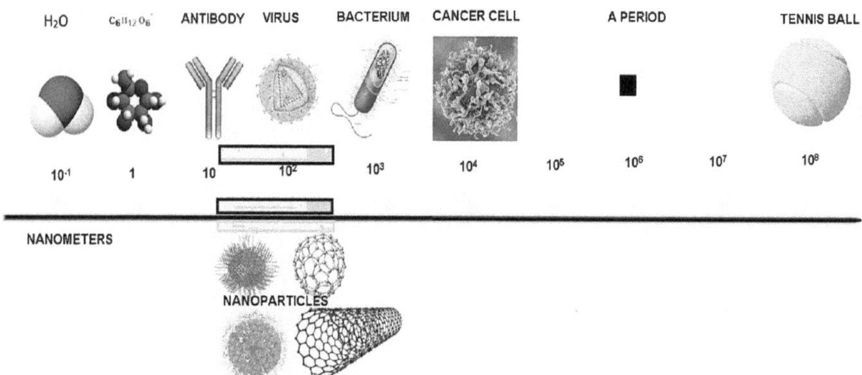

Figure 1. The relative size scale of macro, micro and nanomaterials in comparison to natural and man-made material.

In order to apply the amazing properties of nanomaterials to enhance biofuel production, it is first essential to understand the basic characteristics and production mechanisms of nanomaterials.

2.1 Methods of Nanomaterial Synthesis

Since the inception of nanomaterials (NMs) over half a century ago, scientists are continuously exploring advanced novel methods of synthesizing NMs with the optimal size and morphology that would be beneficial for various disciplines. NMs are materials with morphological features smaller than 1/10th of a micrometre in at least one dimension, and they display no less than one property unlike its bulk counterpart (Jeevanandam et al. 2018). This section overviews the commonly employed strategies for the fabrication of tailored nanostructured materials. Based on synthesis approach, NMs synthesis methods can be broadly divided into two categories (Figure 2):

(i) Top-down approach: Miniaturization of larger structures through etching from the bulk material, e.g., photolithography (Ginestra et al. 2019), electron beam lithography (Hong et al. 2018), milling techniques (Zhou et al. 2019), anodization (Gulati et al. 2018), ion and plasma etching (Pham 2018), etc.

(ii) Bottom-up approach: Growth and self-assembly to build up nanostructures from atomic or molecular precursors, e.g., self-assembly of monomer/polymer molecules, chemical or electrochemical (Ding et al. 2018), nanostructural precipitation (Kajita et al. 2018), sol-gel processing (Vela¡zquez et al. 2019), laser pyrolysis (Yeon et al. 2018), chemical vapor deposition (Dai et al. 2018), plasma or flame spraying synthesis (Zhou et al. 2018), bio-assisted synthesis (Nazir et al. 2018), etc.

The latter approach is said to be more difficult and expensive but is also known to contribute effectively to the sectors of energy development, transportation and electronics.

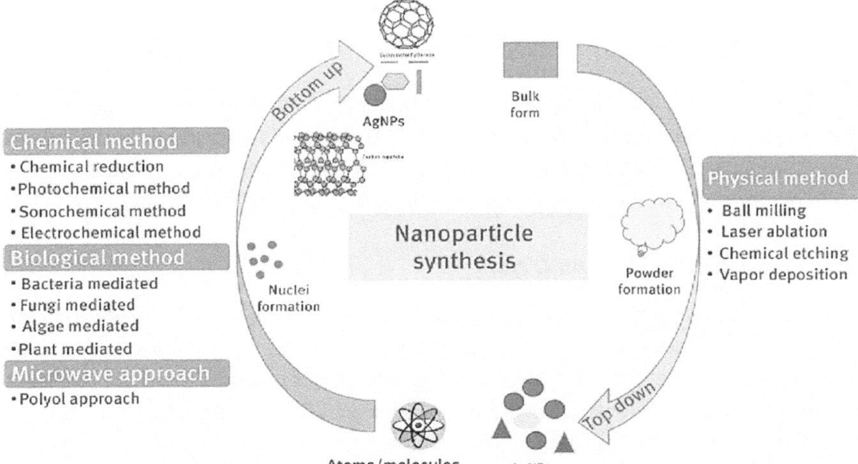

Figure 2. Top-down and bottom-up approach for nanomaterial synthesis.

However, broadly nanomaterial synthesis methods can be separated into three groups—physical methods, chemical methods and bio-assisted methods (summarized in Table 1).

Table 1. Nanomaterial synthesis methods.

	Physical Methods	**Chemical Methods**	**Bio-Assisted Methods**
Nanomaterial Synthesis	Mechanical pressure, high energy radiations, thermal energy or electrical energy to cause material abrasion, melting, evaporation or condensation.	Formation of atoms accomplished by using chemical reaction(s) under controlled but mild reaction conditions. Freshly formed atoms can then undergo elementary nucleation followed by growth processes leading to the formation of defined nanoparticles.	Employ biological systems like bacteria, fungi, viruses, yeast, actinomycetes, plant extracts, etc., for the synthesis of metal and metal oxide NMs.
Pros	Free of solvent contamination and produce uniform monodisperse NMs.	Produce NMs at the higher growth rate, leading to a reduction of the cost of the material.	Provides an environmentally benign, low-toxic, cost-effective and efficient protocol to synthesize and fabricate NMs.
Cons	Abundant waste produced during the synthesis makes physical processes less economical.	Solvent contamination and the use of additional step to remove organic contaminant after synthesis.	Large nanoparticles with broad particle size distribution. Low repeatability, organic contamination.
Examples	High energy ball milling, laser ablation, electrospraying, inert gas condensation, physical vapor deposition, laser pyrolysis, flash spray pyrolysis and melt mixing.	Sol-gel method, micro-emulsion technique, hydrothermal synthesis, polyol synthesis, chemical vapor synthesis and plasma enhanced chemical vapor deposition technique.	(i) Biogenic synthesis using microorganisms. (ii) Biogenic synthesis using biomolecules as the templates. (iii) Biogenic synthesis using plant extracts.

Generally, the nanomaterials—be it nanoparticles, nanofibers or nanotubes—are first produced as dry powders by either physical or chemical methods and then dispersed into a suitable fluid using either intense magnetic force agitation, high-shearing mixing, ultrasonic agitation or homogenizing and ball milling.

2.2 *Properties of Nanomaterials*

Prediction of the physical and chemical properties of NMs is very difficult and remains largely unknown. The principal parameters of nanoparticles are their shape, size, surface characteristics and inner structure. Recent studies have proven that at the nanoscale dimensions, NMs often have unique physical (e.g., electrical, optical, mechanical, topography, etc.) and chemical properties (e.g., magnetic, surface chemistry, hydrophilicity/lipophilicity, etc.) than their bulk macroscale counterparts

(Faucher et al. 2019). They can remain free or group together depending on the attractive or repulsive interaction forces between them. NMs can be encountered as aerosols (solids or liquids in the air), suspensions (solids in liquids) or as emulsions (liquids in liquids). It is well known that compared to their larger sized counterparts, the nanomaterials tend to have enhanced properties in terms of their electrical or heat-conducting capacity, surface area to volume ratio, reflecting light, rates of reactivity, elasticity, strength, etc. (Jeevanandam et al. 2018). The increase in the surface area results in an increase in surface atoms. Surface atoms in nanosize particles have lower coordination numbers compared to macro-particles.

Thus, nanomaterial offers an advantage for immobilization of several biological molecules and even a perfect scaffold to recruit functional molecules for multiple signal amplification due to high surface-to-volume ratio and surface activity. The activity is highly enhanced in nanoscale matters. Table 2 summarizes the transformed properties of nanomaterials from its bulk counterparts.

Table 2. Adjustable properties of nanomaterials.

Properties	**Examples**
Mechanical	Improved hardness and toughness of metals and alloys, ductility and super plasticity of ceramic. The intensity of the aggregates of carbon nanotubes is 100 times that of steel; the weight of the former is only one-sixth of the latter.
Magnetic	Increased magnetic coercivity up to a critical grain size, superparamagnetic behavior, e.g., silver is not magnetic in the bulk. It is nonmagnetic at macroscale but shows magnetic nature at nanoscale due to subtle electronic interactions.
Catalytic	Better catalytic efficiency through higher surface-to-volume ratio, e.g., inert materials like gold, silver and platinum serve as catalysts in their nanoscale.
Electrical	Increased electrical conductivity in ceramics and magnetic nanocomposition. Increased electric resistance in metals.
Optical	Spectral shift of optical absorption and fluorescence properties. Increased quantum efficiency of semiconductor crystals.
Sterical	Increased selectivity and hollow spheres for specific drug transportation and controlled release.
Biological	Increased permeability through biological barriers (membranes, blood-brain barrier, etc.), improved biocompability and antimicrobial properties.

3. From 1st Generation to Advanced 4th Generation Biofuels

Biofuel is a renewable substitute for fossil fuel, primarily derived from replenished biomass feedstock materials such as plant, bacteria, algae or animal waste. Due to inherent merits like better GHG emission abatement, renewability and long-term sustainability, biofuels have gained considerable attention (Singh et al. 2016a). Examples of biofuels include bioethanol (often made from corn and sugarcane), biodiesel (derived from vegetable oils, liquid animal fats algae and other plant sources), biogas (methane derived from animal manure and other digested organic material), biohydrogen (produced by biophotolysis of water by algae, dark and photo-fermentation of organic materials, usually carbohydrates by bacteria), bioelectricity

Table 3. Generations of biofuel.

	1st Generation	2nd Generation	3rd Generation	4th Generation
Feed Stocks	Synthesized from edible crops such as wheat, sugarcane, soya bean, rapeseed, etc., explicitly grown on arable land.	Derived from wide-ranging non-edible feedstocks biomass, such as agricultural residues, forestry waste and municipal and industrial wastes.	Derived from algae.	Photobiological solar fuels synthesized from genetically modified algae to enhance desired biofuel yields and electrofuels from microbial fuel cells.
Pros and Cons	Food crops grown for fuel production increases upwards pressure on food prices.	Made from lignocellulosic biomass or woody crops, agricultural residues or waste plant material; lower negative environmental impacts, overcome the shortcomings of the first-generation biofuel as the usage of arable land and food crops.	Algae do not require arable land for cultivation and is not an edible crop.	Creates an artificial carbon sink to make it a carbon-negative energy source. More CO_2 captured, higher production rate, high initial investment but profitable in the long run.
Methods	Biochemical conversion	Biochemical and thermochemical methods.	Biochemical and thermochemical methods.	Biochemical, thermochemical and electrochemical methods.
Examples	Biodiesel, bioethanol, bio-butanol, 2,5-dimethylfuran, etc.	Bioethanol, bio-methanol, FT-diesel, DME and bio-hydrogen.	Biodiesel, butanol, gasoline, methane, bioethanol and jet fuel.	Biohydrogen biodiesel, butanol, gasoline, methane, bioethanol, jet fuel and bioelectrivity.

(microbial fuel cells), etc. (Singh et al. 2016b). The generations of biofuel have been summarized in Table 3. During the coming 10–20 years, it is expected that various photobiological and electrochemical fuels are gradually entering into the market.

4. Strategic Role of Nanotechnology for Enhanced Biofuels Production

Technical advances of nanotechnology in the biofuel sector has lead to the production of more efficient, economically cheaper and environment-friendly energy production which has resulted in commercially competitive biofuel within existing forms. Currently, the application of nanotechnology in the biofuel industry mainly focuses on reducing the transportation cost of feedstock, breaking down the feedstock more efficiently and improving biofuel production efficiency (see Figure 3). Various metal oxide nanoparticles, such as iron oxide (Chang et al. 2010), titanium dioxide (Madhuvilakku and Piraman 2013), calcium oxide (Hu et al. 2011), magnesium

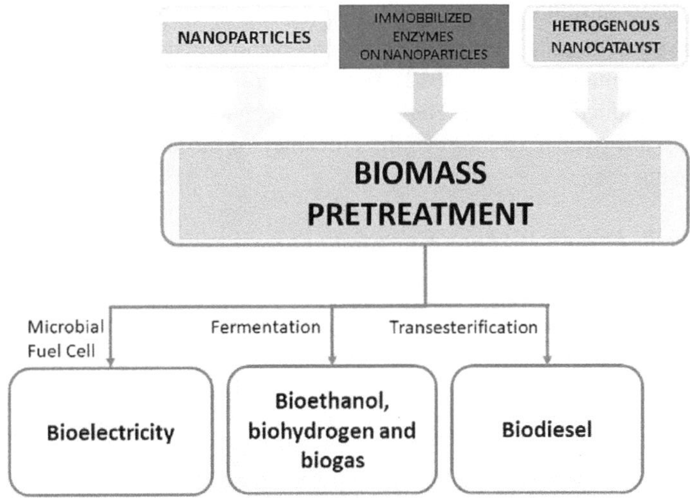

Figure 3. Strategic role of nanotechnology in biofuel production.

oxide (Verziu et al. 2008) and zirconium oxide (Qiu et al. 2011) have been developed with high catalytic performance. Magnetic nanoparticle-based catalysts are advantageous because of their ability to be easily separated from reaction media and reused (leading to more economical) industrial-scale biofuel production (Feyzi et al. 2013). Several mesoporous heterogenous nanocatalysts, with excellent structural properties for enhanced catalytic activities, have also been designed and successfully used in biodiesel production (Feyzi et al. 2013). Carbon-based nanocatalysts such as carbon nanotubes (Lee et al. 2010a), carbon nanofibers (Mohamad et al. 2015), etc., have also shown to hold great potentials for biofuel production from a wide range of feedstocks, especially from non-food ones.

Another promising application of nanotechnology in the biofuel industry is enzyme (biocatalysts) immobilization during lipase-catalyzed biodiesel and cellulosic ethanol production processes (Huang et al. 2015). These techniques could reduce the production cost at commercial-scale biofuel production.

4.1 Pre-Treatment of Biofuel Feedstock

On a large scale, biomass pre-treatment is a bottleneck and potentially costly step in biofuel production. Due to complex carbohydrates and glycoprotein, the multi-layered rigid cell wall of the biological organism provides a wide range of chemical resistance and high mechanical strength which makes pre-treatment more difficult. Therefore, to extract lipid and carbohydrate from biomass, the cell wall must be disrupted to the smallest pieces using pre-treatment process through solvents. Pre-treatment of biofuel feedstock involves the degradation or disruption of biomass to convert, accumulate and extract the carbohydrates and lipids which it contains. Methods of cell wall disruption and extracting solvents decide the efficiency of oil and carbohydrate extraction from biomass. Researchers have prescribed many different methods such as physical treatment including ultrasonication, thermal, microwave, bead milling and cryogrinding (Zheng et al. 2011), chemical methods including acid, alkaline, solvent soaking and osmotic shock (Ho et al. 2013) and enzymatic treatments (Li et al. 2012). But still there is a lack of feasible, economical and highly productive method. The use of nanotechnology and nanomaterials could be one possible avenue to improve the efficiency of biofuel production and reduce its processing cost.

4.1.1 Nanotechnology for Lipid Extraction

Oil extraction by cell-disruption from biomass is an expensive step for biodiesel production. Currently, most common methods for oil extraction are derived from conventional oil extraction methods for oil-bearing crop such as expelling, microwave/ultrasonic-assisted extraction, solvent extraction, supercritical carbon dioxide, etc., used either individually or in combination. However, extracted lipid yield significantly varies with the use of different solvent combinations and extraction conditions which affect the overall lipid yield. Thus, all these factors should be considered during optimization of extraction protocol with higher lipid yield. Application of nanotechnology in oil extraction offers a reliable, non-toxic, safe and cost-effective alternatives for lipid extraction. Furthermore, the solvent lipid separation step in the conventional extraction process, which is tedious and adds cost to the overall process, can also be eliminated with the use of nanotechnology. Many researchers have shown the tremendous potential of nanotechnology to achieve cost-effective and process-efficient lipid extraction. Lee et al. (2013a) utilized four amino-groups functionalized organic-nano clays, Mg-APTES clay, Al-APTES clay, Ca-APTES clay and Mg-N_3 clay for lipid extraction from wet microalgae *Chlorella* sp. biomass. The study shows that the presence of amino clay nanoparticles greatly enhanced the oil extraction efficiency along with the increase in microalgae harvesting efficiency. The cationic charged aminoclay nanoparticles weakened the algae cell wall by decreasing the water layer between cell wall and hydrophobic solvent, resulting in the higher amount of solvent to come in contact with the cell wall for the release of internal oil. Lee et al. (2013b) also disrupted the cells and effectively released internal lipids from wet microalgae biomass using OH^- free radicals produced through the activation of hydrogen peroxide (H_2O_2) by the Fe metal atoms embedded in the amino clay. Furthermore, Lee et al. (2014) demonstrated the application of aminoclay-TiO_2 nanocomposite in the presence of UV irradiation (365 nm) for simultaneous harvesting wet-microalgae biomass (~ 85% harvesting efficiency), thereby affecting the disruption of ~ 95% of cells. Apart from the metal embedded aminoclays, Razack et al. (2016) reported the increase in the oil content of *Chlorella vulgaris* from 8.44 to 17.68% with the increase in silver nanoparticles (AgNPs) concentration from 50 to 150 μg/g.

4.1.2 Nanotechnology for Carbohydrate Extraction

Cellulose and hemicellulose cover up to two-thirds of the lignocellulosic biomass besides lignin. Appropriate pre-treatment process can improve the bioconversion efficiency of cellulosic biomass by ease through release of fermentable sugars from the lignocellulosic material that leads to a high sugar yield and enhances bioethanol generation. Hydrolysis catalysts, such as cellulases and homogeneous acids, have many challenges like high cost, low efficiency, long reaction time, poor recyclability, reactor corrosion, etc. (Gumina et al. 2019). NMs can be a potent, efficient and environmentally friendly replacement for many conventional hydrolysis catalysts. They have many advantages over conventional catalyst, such as efficient activity, high selectivity, long catalyst life and ease of recovery and recycling.

4.1.2.1 Solid Acid NMs Catalysts for Lignocellulosic Pre-Treatment

Solid acid catalysts are the proton donators or electron acceptors during reactions. Recent studies reported that solid acid-mediated hydrolysis could provide ease in the conversion of lignocellulosic biomass into fermentable sugars. Solid acid nanomaterial catalysts are monodispersed nanoparticles in a solution form which help in easier access to oxygen atom in the ether linkage of cellulose as well as presenting good glucose and cello-oligomer yields (Eisa et al. 2018). They are accessible in many types such as metal oxides, functionalized silica, acid resins, supported metals, H-form zeolites, heteropoly acids, magnetic acids, immobilized ionic liquids, carbonaceous acids and hydrotalcite nanoparticle. Among all types, the magnetic acid nanoparticles have inherent promising potential to facilitate magnetic separation and are easily recycled (Mohanraj et al. 2014a). Some researchers have recently demonstrated that nanotechnology can offer easy, green, efficient and economically viable routes for lignocellulosic pre-treatment. Bootsma and Shanks (2007) demonstrated the catalytic activity of mesoporous silica functionalized with a sulfonic and carboxylic acid in the hydrolysis of cellobiose used as a model compound. In recent work, Zhong et al. (2015) produced a solid acid nanoscale catalyst (SO_4^{-2}/Fe_2O_3) with both Lewis and Bronsted acidity and were able to specifically hydrolyze hemicellulose fraction of wheat straw while maintaining the other cellulose and lignin fractions intact. They were able to achieve a hydrolysis yield of 63.5% in optimum conditions time (4.10 hours), temperature (141.97°C) and the ratio of wheat straw to catalyst (1.95:1;w:w). In addition, the catalyst could be recycled six times with high activity remaining. Although these researchers have opened a new possibility for the cost-effective lignocellulosic biomass pre-treatment, this area is still at the very early stage.

Considering the various advantages of magnetic nanocatalysts, like recovery after the completion of a reaction using an external magnetic field and reuse in subsequent cycles of pre-treatment, some researchers have given special focus on the application of acid-functionalized magnetic nanoparticles in the pre-treatment. Recently, Yang et al. (2015) reported a nanocomposite reduced graphene oxide functionalized with magnetic Fe_3O_4 nanoparticles and –$PhSO_3H$ groups (Fe_3O_4-RGO-SO_3H) for hydrolysis of cellulosic materials. Due to its crumpling feature with clear layers and the coexistence of –COOH and –OH groups, nanocomposite exhibited an outstanding catalytic performance for the hydrolysis of cellulosic materials. The unique structure of nanocomposite favors the accessibility of cellulose to the active sites of the material and facilitates the diffusion of the product molecules. Furthermore, nanocomposite can also be easily separated from the reaction residue with an extra magnetic force and can be further used for at least five times. Lai et al.

(2011) also demonstrated the synthesis and characterization of a sulfonated (sulfonic acid supported) mesoporous silica-magnetic nanocomposite, modified with iron oxide, as a promising hydrolysis catalyst. The catalyst shows improved performance for the hydrolysis of β-1,4-glucan and carbohydrate dehydration than a conventional solid acid and can be easily separated from the reaction residues with an external magnetic force. Similarly, some researchers synthesized magnetic nanoparticles like sulfonic acid functionalized silica-coated crystalline Fe/Fe_3O_4 core/shell (Wang et al. 2015), perfluoroalkyl-sulfonic (PFS) and alkylsulfonic (AS) acid-functionalized (Peña et al. 2012 and 2014) and demonstrated excellent stability and significant catalytic activity toward the hydrolysis of lignin cellulosic biomass.

4.1.2.2 NMs for Enzymatic Hydrolysis

Enzymatic hydrolysis has been recognized as an ideal alternative to alter the structure of lignocellulosic, cellulosic and starch materials. Cellulase mixture, which is composed of endoglucanases, cellobiohydrolyases and β-glucosidase, has been known to breakdown lignocellulosics (Baeyens et al. 2015). However, cellulases mixture for the hydrolysis of biomass is responsible for major portion of overall costs in the process of the production of bioethanol from lignocellulosic biomass. Thus, the immobilization of cellulases is important for the recovery and recycling of these biocatalysts and for cost reduction of the process. The main techniques for cellulase immobilization have been cross-linking, encapsulation and adsorption or covalent binding onto insoluble, reversibly soluble or soluble supports (Vaghari et al. 2016). Each of these techniques has advantages and disadvantages related to its toxicity, biodegradability, cost of materials and reagents used in immobilization, time spent and complexity in the preparation and performance of biocatalysts. Many researchers showed that the application of NMs allows simultaneous multiple enzyme co-immobilizations which increase the reaction yield, high specific surface area, great pH tolerability, enhance enzyme stability for long-term storage, low mass transfer resistance and high biocompatibility. Researchers have engineered composition and morphology of NMs as per required hydrolysis by coating the metal (Mg, Ni, Ti, Co, CoPt, Fe and FePt) or metal oxides (FeCO, Fe_3O_4, FeO, Fe_2O_3 and γ-Fe_2O_3) commonly with natural or synthetic polymeric material (Jordan et al. 2011). It allows the large-scale enzyme immobilization on magnetic nanoparticles without the use of synthetic surfactants and other toxic reagents.

Verma et al. (2013) attached a thermostable β-glucosidase from *Aspergillus niger* on a nanoscale magnetic NPs via glutaraldehyde activation, yielding an immobilization efficiency of about 93% by covalent binding. The cellobiose hydrolysis was significantly increased at 70°C in comparison to the free enzyme. The immobilized nanoparticle–enzyme conjugate retained more than 50% enzyme activity up to the 16th cycle. Maximum glucose synthesis from cellobiose hydrolysis by immobilized BGL was achieved by 16 hours. It seems that the success of an enzymatic process with nanoparticles also depends on the support used. For example, iron oxide nanoparticles are obtained by co-precipitation at alkaline pH (9–14) in which the morphology and particle composition depend on the salt used (chloride, sulfate or nitrate) and the molar ratio of Fe^{2+} and Fe^{3+}. Jordan et al. (2011) used

Fe_3O_4 nanoparticles activated with carbodiimide for enzyme immobilization, which was applied in the hydrolysis of microcrystalline cellulose, obtaining a thermostable enzyme with a remarkable maximum activity of 62.7 U/mg at 50°C and was able to be recycled for six times. The study revealed a thermal improvement by stabilizing the weak ionic forces and hydrogen bonds through nanoparticles activation using a carbodiimide cross-linker. Table 4 summarizes the immobilization of cellulase on different magnetic nanoparticles, methodology and main findings.

4.2 Improving Biofuel Production Efficiency

Nanotechnologies have significant potential to improve the biofuel production rate by enhancing reaction kinetics and may eventually reduce the production cost due to savings in material cost and processing time. The advantages of involving nanomaterials will certainly give extra value to the biofuels production process to become more sustainable by reducing the costs in addition to creating positive environmental impacts.

4.2.1 Nanomaterials for Enhancing Biodiesel Conversion

Natural oils can be effectively converted to biodiesel in four different ways, namely, directly blending with mineral diesel, microemulsion method, thermal cracking and transesterification (Noor et al. 2018). Among all, transesterification is the most commonly used process. During this process, oils derived from algae, plant, animal and other organism source react with alcohol to produce fatty acid methyl esters (FAME) which are also called as biodiesel (Singh et al. 2013a). Transesterification helps to bring down the inherent viscosity of oil by reducing fatty oils into FAME in the presence of a suitable alkaline catalyst, mostly NaOH. However, transesterification has its own set of problems that include the presence of free fatty acids, saponification, water content of oils or fats, deactivation of the catalyst, low reaction rate, reaction temperature and time and molar ratio of glycerides to alcohol (Meher et al. 2006). Currently, transesterification is carried out mainly through four types of catalysts, viz., acid, base, heterogeneous and enzyme catalysts (Obadiah et al. 2012, Cho et al. 2012, Abraham et al. 2014).

Homogeneous catalysts for transesterification, such as acid-catalyzed (HCL and H_2SO_4) and base-catalyzed (NaOH and KOH), act in the same liquid phase as the reaction mixture (Lam et al. 2010). However, both acid and base catalyze transesterification requires a higher amount of methanol and base, respectively, which makes the process costly. Furthermore, a series of issues are also found in this process, such as difficulty to recover the catalyst, soap formation and problems in the following downstream process. In order to overcome the mentioned hiccups, heterogeneous catalyzing transesterification with nanocatalyst is a promising green approach for high yields of methyl esters, i.e., biodiesel, that facilitate easy recovery and recycling of the catalysts. Many nanoscale catalysts such as carbonaceous acids, oxides, functionalized mesoporous silica, etc., have been explored to enhance the transesterification process and improve biodiesel yields (Han and Guan 2009, Qiu et al. 2011, Gao et al. 2015). Carbonaceous acids consist of stable sulfonated amorphous carbon which makes its catalytic activity more than half of the liquid sulfuric acid

Table 4. Immobilization of cellulase on different nanoparticles, methodology and main results.

Enzyme		Support Nanomaterial	Immobilization Methods	Results	Reference
	Source				
Natural Cellulase	*Aspergillus niger*	MNPs of TiO_2/ TiO_2 modified with aminopropyltriethoxysilane.	Adsorption and covalent methods.	Enzymes activity: Adsorbed (76%) and covalently (93%) when compared to free enzyme. Cellulase lost: After 60 minutes of incubation at 75°C; covalently immobilized (25%) and adsorbed (50%).	Ahmad and Sardar 2014
		β-cyclodextrin-conjugated magnetic particles.	Silanization and reductive amidation.	Cellulase activity: 10% decrease compared to free cellulose. Recyclability: Magnetized cellulase recycled by magnetic field with 85% of immobilized cellulase.	Huang et al. 2015
		Amine-functionalized Fe_3O_4 silica core-shell magnetic nanoparticles (MNPs).	Cross-linked cellulase aggregates (CLEA).	MNP retained 45% maximum activity at pH > 4.8 and 65% of its maximum activity at 80°C; compared to free cellulase which lost its activity sharply. 30% activity retained after six cycles.	Jafari Khorshidi et al. 2016
	Trichoderma viride	Superparamagnetic nanoparticles.	Physical Adsorption.	Enzyme stability increased even on higher pH.	Khoshnevisan et al. 2011
Commercial Cellulase		2D graphene with maghemite-magnetite nanoparticles.	Covalent bond by a combination of annealed polyelectrolyte brushes and zero-length spacer molecules.	Specific activity: Retained about 55% of the original even after four recycles. Highest cellulase activity reported for 2D flat carriers.	Gokhale et al. 2013
		Fe_3O_4 magnetic nanoparticles.	Covalent binding via carbodiimide activation.	Increase in stability over a wider range of temperatures. Enzyme activity was 30.2%; six recycles.	Jordan et al. 2011

		PVA/Fe_2O_3 nanoparticles.	Cyclically freezing and thawing procedure under microemulsion system.	Glucose yield: 1.89 $mg.ml^{-1}$ at least 3 times than sum of individual yield. Cellulose activity: 40% after four cycles of reuse.	Liao et al. 2010
		Magnetic Fe_3O_4 chitosan nanoparticles.	Covalent binding via glutaraldehyde activation method.	Cellulase activity: 112.3 mg/g, 5 IU/mg cellulase. Thermal stability: Cellulase retained 50% of its initial activity after 10 cycles.	Zang et al. 2014
Recombinant Cellulase	*Trichoderma reesei*	Activated magnetic support.	Covalent binding via glutaraldehyde activation method.	Immobilized enzyme retained 50% enzyme activity up to five cycles with thermostability at 80°C, superior to that of the free enzyme.	Abraham et al. 2014
	Co-immobilization of endoglucanase, exo-glucanase, and β-glucosidase)	AuNP (gold nanoparticles) and Au-MSNP (gold-doped magnetic silica nanoparticles).	Cystein residues of the enzyme were added to their C terminal to strongly co-immobilize these cellulases.	Exhibited higher pH stabilities; reused up to seven times with almost 60% and 90% of retained activity.	Cho et al. 2012

catalyst in the esterification of higher fatty acids. Many researchers have shown that different studies on the addition of large porous surface and recyclable catalyst-based nanoparticles increase during the contact between alcohol and oil which effectively improve the yield of products. Qiu et al. (2011) catalyzed the transesterification of soybean oil with methanol to biodiesel by nanocatalyst ZrO_2 loaded with $C_4H_4O_6HK$. A yield of 98.03% of biodiesel was obtained in 16:1 M ratio of methanol to oil, 6.0% catalyst, 60°C reaction temperature and 2.0 hours reaction time. Han and Guan (2009) prepared a nano-solid based syn-catalyst, i.e., $K_2O/\gamma\text{-}Al_2O_3$, and reported a yield of 94% in biodiesel production from rapeseed oil using 3% of catalyst, 12:1 M ratio of methanol to oil at 70°C and a reaction time of 3 hours. Furthermore, Gao et al. (2015) reported 96% biodiesel yield applying $KF/\gamma\text{-}Al_2O_3$/honeycomb ceramic (HC) monolithic catalyst from palm oil with the methanol; oil ratio 18:1 at 140°C and 33 minutes reaction time. Table 5 highlights the unseen potential of NMs for biodiesel production.

Furthermore, compared to acid, base and heterogeneous catalysts, transesterification using a biological catalyst or an enzyme seems to be a more efficient and environmentally friendly substitute. Enzymatic catalysts can be more resistant to interferences of free fatty acid and water associated with chemical catalysis. Thus, the use of lipase for transesterification would not only avoid product contamination but also favor an easier glycerol recovery (Antczak et al. 2009). However, the commercialization of lipase-catalyzed biodiesel production is limited due to higher cost and denaturation of enzyme by substrates and by-products. The biocatalytic activity of the enzyme is also decreased in the presence of methanol, which is used as an acyl acceptor (Ranganathan et al. 2008). Enzyme immobilization on nanocarriers and other porous materials can be a crucial way to solve abovementioned issues. Enzyme immobilization for biodiesel production consists of two important components: enzyme-like lipase and support materials. Commonly used lipases from bacteria and fungi include non-specific lipases (*Candida antarctica, Candida cylindracea, Candida rugosa, Pseudomonas cepacia* and *Pseudomonas fluorescens*) or specific lipases (*Rhizopus oryzae, Rhizopus delemar, Rhizomucor miehei, Thermomyces lanuginosus* and *Aspergillus niger*) for biodiesel production (Andualema and Gessesse 2012).

On the basis of magnetism property of support, nanoparticles have categorized into two types: non-magnetic (e.g., zirconia, silica, polystyrene, chitosan, polylactic acid, etc.) and magnetic ($\gamma\text{-}Fe_2O_3$, Fe_3O_4, Poly-(glycidyl methacrylate) grafted Fe_3O_4/SiO_x, etc.). The reuse of the non-magnetic nanoparticle-attached enzyme is often a difficult task as it could be well dispersed in the reaction solution. Thus, considering simplistic and fast separation of the immobilized enzyme from the reaction mixture, magnetic nanoparticles (MNPs) are routinely used as support for enzyme immobilization. Many researchers found that MNP-based lipase immobilization has excellent stability after repeated applications. Tran et al. (2012) showed that immobilized lipase on alkyl-functionalized $Fe_3O_4\text{-}SiO_2$ nanocomposites was 1.3-fold higher than non-functionalized $Fe_3O_4\text{-}SiO_2$. Nanocomposites were used in transesterification of olive oil for 10 cycles without significant loss in the activity. Tables 6 and 7 summarize the information about magnetic and non-magnetic investigations on nano-immobilized lipase.

Table 5. Potential of nanomaterials for biodiesel production.

Nano Catalyst	Size (nm)	Feedstock	Operation Condition					References
			Temperature (°C)	Alcohol: Oil Ratio	Catalyst (wt.%)	Reaction Time (minutes)	Yield (%)	
$Cs/Al/Fe_3O_4$	30–35	Sunflower oil	58	14:1	4	120	94.80	Feyzi et al. 2013
Hydrotalcite (Mg-Al)	4.66–21.1	Pongamia oil	65	6:1	1.5	240	90.8	Obadiah et al. 2012
MgO	50–200	Soybean oil	70–310	4:1	0.1–7	60	95	Mguni et al. 2012
		Rapeseed oil						
MgO supported on titania	–	Soybean oil	150–225	18:1	–	40–120	98	Verziu et al. 2007
ZrO_2 loaded with $C_4H_4O_6HK$	10–40	Soybean oil	60	16:1	6	120	98.03	Qiu et al. 2011
Lithium impregnated calcium oxide (Li-Cao)	40	Karanja oil	65	12:1	5	60	99	Kaur and Ali 2011
		Jatropha oil				120		
Magnetic solid base catalysts CaO/Fe_3O_4	49	Jatropha oil	70	15:1	2	80	95	Chang et al. 2010
KF/CaO	30–100	Chinese tallow seed oil	65	12:1	4	150	96	Wen et al. 2010
Hydrotalcite-derived particles with Mg/Al molar ratio of 3:1	7.3	Jatropha oil	45	4:1	1	90	95.2	Deng et al. 2011
Cao	20	Soybean oil	23–25	27:1	–	720	99	Reddy et al. 2006
Mgo	60	Soybean oil	200–260	6:1	0.5-3	12	99.04	Wang and Yang 2007
$KF/CaO–Fe_3O_4$	50	Stillingia oil	65	36:1	4	180	95	Hu et al. 2011
TiO_2-ZnO	34.2	Palm oil	60	12:1	–	300	92.2	Madhuvilakku and Piraman 2013
ZnO	28.4						83.2	
KF/Al_2O_3	50	Canola oil	65	6:1	3	480	97.7	Boz et al. 2009
CaO/MgO Ca	–	Jatropha oil	64.5	18:1	2	210	92	Chang et al. 2010
$Ca(OH)_2$-Fe_3O_4(Ca^{+2}: Fe_3O_4=7)	–	Jatropha oil	70	15:1	2	240	99	Chang et al. 2010

Table 6. List of various immobilized lipases on different nanomaterials.

Nanomaterials	Strain	Carrier	Binding Type	Effect of Nanomaterial	References
Nanoparticles	*Pseudomonas cepacia*	Zironia	Covalent	Increased activity and enantioselectivity.	Chen et al. 2009
	Mucor japonicus	Silica	Covalent	Enhanced enzyme loading and enzyme stability.	Kim et al. 2006
	Candida antarctica	Polystyrene	Adsorption	High hydrolytic activity.	Chronopoulou et al. 2011
	Candida rugosa	Chitosan	Covalent	High enzyme loading and activity retention.	Wu et al. 2010
	Candida rugosa	Polylactic acid	Adsorption	Enhanced activity and stability.	Chronopoulou et al. 2011
	Candida rugosa	γ-Fe_2O_3	Covalent	Enhanced stability.	Dyal et al. 2003
	Porcine pancreas	Magnetic	Adsorption	Good reusability.	Lee et al. 2009
Carbon nanotube	*Pseudomonas cepacia*	Single-walled carbon nanotube	Adsorption, covalent	Increased retention of enzyme activity.	Lee et al. 2010a
	Rhizopus arrhizus	Multiwalled carbon nanotube	Covalent	Enhanced resolution efficiency.	Ji et al. 2010
	Candida rugosa, C. antarctica B. Thermomyces lanuginosus		Adsorption	Enhanced stability.	Lee et al. 2010b
	Candida rugosa		Adsorption	High enzyme activity.	Shah et al. 2007
	Candida rugosa		Adsorption	Enhanced activity and thermal stability.	Mohamad et al. 2015
	Candida antarctica		Adsorption	Enhanced activity and stability.	Pavlidis et al. 2010
Nanofibers	*Candida antarctica*	Polyacrylnitrate	Covalent	High enzyme stability.	Li et al. 2007
	Candida rugosa	Poly-(acrylonitrile-comaleic acid)	Covalent	High activity and enzyme loading.	Ye et al. 2006
	Candida rugosa	Cellulose acetate	Covalent	Enhanced thermal stability.	Huang et al. 2011
	Burkholderia cepacia	Polycaprolactane	Covalent	Enhanced catalytic activity and reusability.	Song et al. 2012
	Candida rugosa	Polyvinyl alcohol (PVA)	Covalent	Equivalent esterification activity to that of Novozyme 435.	Nakane et al. 2007

Table 7. Biodiesel production using nano-immobilized lipase.

Strain	Nano-Support	Feedstock Oil	Biodiesel Coversion (%)	Reusability	References
Pseudomonas cepacia	Fe_3O_4	Soybean oil	88	10 days	Wang et al. 2011
		Rapeseed oil	94	20 days	Sakai et al. 2010
	PAN-nanofiber	Soybean oil	90	10 cycles	Li et al. 2011
Thermomyces lanuginosa	Amino-Fe_3O_4	Soybean oil	90	4 cycles	Xie and Ma 2009
		Palm oil	97	5 cycles	Raita et al. 2015
	Epoxy-silica	Canola oil	99	20 cycles	Babaki et al. 2016
Burkholderia sp.	Amino-Fe_3O_4-SiO_2	Waste cooking oil	91	3 cycles	Karimi 2016
		Olive oil	90	10 cycles	Tran et al. 2012
	Alkyl-Fe_3O_4-SiO_2	*Chlorella vulgaris*	90	2 cycles	Tran et al. 2013
Rhizomucor miehei	PAMAM-magnetic multiwalled carbon nanotube	Waste cooking oil	94	10 cycles	Fan et al. 2016
	Epoxy-silica	Canola oil	95	7 cycles	Babaki et al. 2016
Candida antarctica	Epoxy-Fe_3O_4-SiO_2	Waste cooking oil	100	6 cycles	Mehrasbi et al. 2017
	Epoxy-silica	Canola oil	59	15 cycles	Babaki et al. 2016

Thus, nanomaterials can be used as a good carrier for lipase immobilization. Immobilization helps in stabilization, reduces contamination issues, overcomes effluent problems and increases the reusability of the enzyme. It allows the design of continuous process and establishes an effective control over localized enzyme-substrate interaction and reaction parameters.

4.2.2 Nanomaterials for Biogas Production

Biogas is a flammable gas composed of a mixture of gases, mainly carbon dioxide and methane. It is usually produced by anaerobic digestion of animal manures, agricultural residues, mesophilic and thermophilic digestion of organic waste, sewage sludge and different energy crops (Luna-del Risco et al. 2011, Otero-González et al. 2014a, Mu and Chen 2011, Duc 2013). The effect of nano-additives on the anaerobic digestion process and consequently on the biogas yield is an active area of research. The nanomaterials for biogas production are classified into three categories: (1) metal oxides, (2) zero-valent metals and (3) nano-ash and carbon-based materials. The effect of nanomaterials on biogas production has been summarized in Table 8.

Table 8 showed the negative effects of some nanomaterials as ZnO, CuO, Mn_2O_3 and Al_2O_3 on the biogas production, attributed to the toxicity of the materials. However, the addition of nano-iron oxide (Fe_3O_4) enhanced the methane production by 234% due to the presence of the non-toxic Fe^{3+} and Fe^{2+} ions. Also, the addition of nano zero-valence iron (NZVI) is said to be a low release electron donor for the methanogenesis step, which resulted in a mixed effect on the methane production according to its concentration. Metal NPs encapsulated in porous SiO_2 showed a significant increase in methane production, especially in the case of nickel and cobalt. Micro/nano fly ash (MNFA) and micro/nano bottom ash (MNBA) were also found to influence the anaerobic digestion as they tend to provide more habitats for anaerobic organisms. From the above analysis, further research directions may include the use of bioactive nano-metal oxides to avoid the negative effect of currently used materials on the bacteria; use of mixtures of nano-metal oxides, nano zero-valence metals and ash and use of biocompatible photoactive nano-metal oxides to increase the amount of hydrogen produced and consequently the methane production.

4.2.3 Nanomaterials for Biohydrogen Production

Applications of nanomaterials have been much explored in the area of biohydrogen production among various existing biofuels. H_2 as a renewable source of energy is the cleanest and most energy-efficient (122 kJ/g) (Balat and Kırtay 2010). The biological H_2 production (BHP) at ambient physiological conditions is the most obvious and viable approach over energy-intensive conventional chemical or electrochemical processes. It is a known that biohydrogen production is a very complicated and microbial-mediated process. BHP can be mediated by either of two broad classes of enzymes, i.e., the hydrogenases or the nitrogenases (Singh et al. 2013b). Both classes are O_2 sensitive, requiring anoxic environments to function maximally. H_2 production efficiency or yield during the dark-fermentative and photo-fermentative depend on several physicochemical factors such as type of organisms, nature of substrates, inorganic nutrients including metal ions, operational

Table 8. Impact of nanomaterials on biogas production.

Categories	Nanomaterials Properties			Anaerobic Digestion Conditions			Impact on Production	References
	NMs	**Size**	**Conc.**	**Feedstock**	**Temperature (°C)**	**Incubation (Day)**		
Metal Oxides	CuO	37 nm	1.4 mg/l	AGS	30	83	A 15% decrease in methane production.	Otero-González et al. 2014a
		5 µm	15 mg/l	Cattle manure	36	14	No significant effect on biogas production.	Luna-delRisco et al. 2011
			120 mg/l				A 19% decrease in biogas production.	
			240 mg/l				A 60% decrease in biogas production.	
		30 nm	15 mg/l				A 30% decrease in biogas production.	
		40 nm	1,500 mg/l	Anaerobic granular sludge (AGS)	30	Theoretical maximum on methane production (TMMP)	Decrease of acetoclastic MA to 87% and no effect on hydrogentrophic MA.	Gonzalez-Estrella et al. 2013
	ZnO	15 mm	120 mg/l	Cattle manure	36	14	An 18% decrease in biogas production.	Luna-delRisco et al. 2011
			240 mg/l				A 72% decrease in biogas production.	
		50–70 nm	120 mg/l				A 43% decrease in biogas production.	
		140 nm	1 mg/g-TSS	WAS	35	105	No effect.	Mu and Chen 2011
			30 mg/g-TSS				An 18% decrease in methane production.	
			150 mg/g-TSS				A 75% decrease in methane production.	

Table 8 contd. ...

...Table 8 contd.

Categories	Nanomaterials Properties			Anaerobic Digestion Conditions			Impact on Production	References
	NMs	Size	Conc.	Feedstock	Temperature (°C)	Incubation (Day)		
			10 mg/g-TSS	AGS		8	No effect.	Mu et al. 2012
			50 mg/g-TSS				No effect.	
			100 mg/g-TSS				A 25% decrease in methane production.	
			200 mg/g-TSS				A 43% decrease in methane production.	
		< 100 nm	0.32 mg/l	Wastewater AGS	30	90 (Hydraulic retention time = 12 h)	Slight decrease in methane production.	Otero-González et al. 2014b
			34.5 mg/l				Complete inhibition after 1 week.	
		850 nm	10 mg/l	Sludge from USB	30	40	An 8% decrease in biogas production.	Duc 2013
			1000 mg/l	Reactor			A 65% decrease in biogas production.	
	ZnO	< 100 nm	6 mg/g-TSS	WAS	35	18	No effect.	Mu et al. 2011
			30 mg/g TSS				A 23% decrease in methane production.	
			150 mg/g TSS				An 81% decrease in methane production.	
		10–30 nm	1500 mg/l	AGS	30	TMMP	Decrease in acetoclastic MA to 53% and hydrogenic Mato 75%.	

	TiO_2	< 25 nm	6, 30, 150 mg/g TSS	WAS	35	Different fermentation times	No effect	Gonzalez-Estrella, et al. 2013
		25 nm	1500 mg/l	AGS	30	TMMP	No effect.	
		7.5 nm	1120 mg/l	Wastewater treatment sludge	37 55	50	A 10% increase in biogas production.	Garcia et al. 2012
	Al_2O_3	< 50 nm	6, 30, 150 mg/g TSS	WAS	35	Different fermentation times	No effect.	Mu et al. 2011
		< 50 nm	1500 mg/l	AGS	30	TMMP	No effect on acetoclastic MA and decrease in hydrogentrophic MA to 82%.	Gonzalez-Estrella et al. 2013
	Fe_3O_4	7 nm	100 mg/l	Wastewater sludge	37	60	An 180% increase in biogas production and 234% increase in methane production.	Casals et al. 2014
Zero-valent metals	NZVI	20 nm	0.1 wt%	WAS	37	17	Increase in biogas production by 30.4% and methane production 40.4%.	Su et al. 2013
		50 nm	1 g/l	TCE	22	21	Methane production increased from 58 µmol to 275 µmol.	Xiu et al. 2010
		55 nm	1, 10 mM	Wastewater	37	14	A 20% decrease in methane production.	Yang et al. 2013
			30 mM	Sludge			A 70% decrease in methane production.	
		ZVI < 212 µm	30mM				A 10% increase in methane production.	

Table 8 contd. ...

...Table 8 contd.

Categories	Nanomaterials Properties			Anaerobic Digestion Conditions			Impact on Production	References
	NMs	Size	Conc.	Feedstock	Temperature (°C)	Incubation (Day)		
Metal oxides encapsulated in porous SiO_2	Fe/SiO_2	–	10^{-5} mol/l	–	55	–	A 7% increase in methane production.	Al-Ahmad et al. 2014
	Pt/SiO_2	–	10^{-5} mol/l	–	55	–	A 7% increase in methane production.	
	Co/SiO_2	–	10^{-5} mol/l	–	55	–	A 48% increase in methane production.	
	Ni/SiO_2	–	10^{-5} mol/l	–	55	–	A 70% increase in methane production.	
Nano-ash and carbon-based materials	MNFA	0.4–10,000 nm	3 g/g VS	MSW	35	90	A 2.9 times increase in biogas production.	Lo et al. 2012
	MNBA	0.4–10,000 nm	36 g/g VS	MSW	35	90	A 3.5 times increase in biogas production.	

condition and the metabolites produced as end by-products. Pure sugars are widely used as feed for the H_2 production by different organisms, the well-known ones include *Bacillus, Caldicellulosiruptor, Citrobacter, Clostridium, Enterobacter, Klebsiella, Escherichia, Rhodobacter, Rhodopsedomonas, Thrmotoga,* etc., under either mesophilic or thermophilic conditions (Patel and Kalia 2013). Patel et al. (2012, 2014) showed that these organisms result in H_2 yields of 0.6–3.98 mol/mol of hexose. The complete utilization of hexose sugars results in maximum production of 4 and 8 mol via dark-fermentative followed with photo-fermentative processes, respectively, finally leading to 12 mol H_2/mol of hexose. Thus, the relative low energy conversion efficiency of the dark-fermentative process leads to practical limitations at a commercial scale. Furthermore, BHP through the dark-fermentative process showed only utilization of one-third of substrate utilization. However, NMs can play a vital role in the production of biohydrogen. They can significantly influence the microbial metabolic activity for H_2 production through similar phenomenon under aerobic conditions by efficient transfer of electrons. Thus, H_2 producing microorganisms can get the advantage of nanoparticles in anaerobic conditions due to the electron transfer which is more convenient to acceptors (Patel and Kalia 2013). Thus, a positive effect of various NPs, including metallic nanoparticles like silver (Ag), gold (Au), copper (Cu), iron (Fe), nickel (Ni), palladium (Pd); metal oxide nanoparticles (Fe_2O_3, F_3O_4, $NiCo_2O_4$, CuO, NiO, CoO, ZnO, SiO_2); titanium (Ti); activated carbon; carbon nano-tubes (CNTs) and nanocomposites ($SiCoFe_2O_4$, Fe_3O_4/alginate) were observed on BHP. Table 9 summarizes the influence of individual and mixtures of inorganic and organic NMs on the biological hydrogen production yield by the different organisms.

Briefly, these NMs might be stimulating BHP by their surface and quantum size effect. In the surface effect, if the size of NMs is smaller then it will give larger specific surface area, which will thus enable a strong ability to adsorb electrons. On the other hand, the extent of the quantum size is directly corelated with the rate of electron transfer between NMs and enzyme molecules such as hydrogenase, which is known to catalyze the conversion of H_2 to proton and vice versa either to act as electron sinks or deliver reducing power from H_2 oxidation. Though efforts have been made to improve the biological hydrogen production by adopting several approaches and a variety of nanomaterials, this area is still in its struggling phase at bench scale to initiate pilot plant study. Recent advancements and feasibility of nanomaterials as the potential solution for improved biohydrogen production process is likely to assist in developing an efficient, economical and sustainable biohydrogen production technology.

4.2.4 Biological Fuel Cells

A biological fuel cell (BFC) or microbial fuel cell (MFC) is a type of fuel cell that converts biochemical energy into electrical energy. Bioelectricity production relies on the capability of certain microbes for direct electron capture from electrodes (e.g., from solar cells or any renewable electricity source) to assimilate the reducing equivalents into metabolism along with CO_2 utilization. It uses an active microorganism as a biocatalyst in an anaerobic anode compartment and transfers

Table 9. Effect of individual and mixtures of inorganic and organic NMs on the biological hydrogen production.

Nanoparticles		Organisms	Feed	H_2 Yield efficiency (%)	References
Composition	**Conc. (mg/l)**				
Ag	20 nM	Mixed culture	Glucose	67.6	Zhao et al. 2013
Au	5.0 nM	Anaerobic culture	Wastewater	50.0	Zhang and Shen 2007
Cu	2.5	*Clostridium acetobutylicum* NCIM 2337	Glucose	3.5	Mohanraj et al. 2016
		Enterobacter cloacae 811101	Glucose	2.9	
Fe	5.0	Anaerobic sludge	Glucose	37.0	Taherdanak et al. 2016
	100	*Enterobacter cloacae* DH-89	Glucose	100	Nath et al. 2015
	400	Mixed bacterial consortium	Glucose	38	Zhang et al. 2015
	250	Mixed culture and *Clostridium butyricum* TISTR	Water hyacinth	55	Zada et al. 2013
	312	*Rhodobactersphaeroides* NMBL-02 + *Escherichia coli* NMBL-04	Malate	19.4	Dolly et al. 2015
Fe_2O_3	175	*C. acetobutylicum* NCIM2337	Glucose	33.9	Mohanraj et al. 2014a
	800	*Clostridium pasteurianum* CH5	Glucose	10.0	Hsieh et al. 2016
	200	*Enterobacter aerogenes* ATCC13408	Glucose	17.0	Lin et al. 2016
	125	*E. cloacae* 811101	Glucose	21.8	Mohanraj et al. 2014b
	50	Anaerobic sludge	Glucose	53.6	Engliman et al. 2017
	200	*E. cloacae* 811101	Sucrose	4.8	Mohanraj et al. 2014b
	200	Mixed culture	Sucrose	33.0	Han et al. 2011
	200	*E. aerogenes* ATCC13408	Cassava starch	63.1	Lin et al. 2016
	50.0	Anaerobic sludge	Dairy wastewater	24.0	Gandhe et al. 2015a
	200	Anaerobic sludge	Molasses wastewater	44.0	Gandhe et al. 2015b
	25 mg/g VSS	Anaerobic sludge	Starch wastewater	57.8	Nasr et al. 2015

Fe_3O_4	400	Anaerobic sludge	Glucose	26.4	Zhao et al. 2011
	50.0	Mixed culture	Wastewater	83.3	Malik et al. 2014
	200	Anaerobic sludge	Sugarcane bagasse	69.6	Reddy et al. 2017
SiO_2	40.0	*Chlamydomonas reinhardtii* CC124	Air:CO_2 (97:3)	45.2	Giannelli and Torzillo 2012
	5.1	*C. butyricum* CWBI1009	Glucose	4.3	Beckers et al. 2013
	120	Acidogenic mixed culture	Wastewater	666	Mohan et al. 2008
TiO_2	50	*C. pasteurianum* CH5	Glucose	5.0	Hsieh et al. 2016
	100	*Rhodopseudomonas palustris*	Waste sludge	46.1	Zhao and Chen 2011
	60	*R. sphaeroides* NMBL-02	Malate	69.9	Pandey et al. 2015
Fe + Ni	37.5 + 37.5	Anaerobic sludge	Starch	200	Taherdanak et al. 2015
Fe_2O_3 + NiO	200 + 5.0		Molasses wastewater	62.0	Gandhe et al. 2015b
	50.0 + 10.0		Dairy wastewater	27.0	Gandhe et al. 2015a
Granular activated carbon	10,000	Acidogenic mixed culture	Starch wastewater	94.5	Mohan et al. 2008
Powdered activated carbon	33.0	Anaerobic sludge	Sucrose	62.5	Wimonsong and Nitisoravut 2015
	33.3			73.0	Wimonsong and Nitisoravut 2014
	5000	Acidogenic mixed culture	Starch wastewater	44.0	Mohan et al. 2008

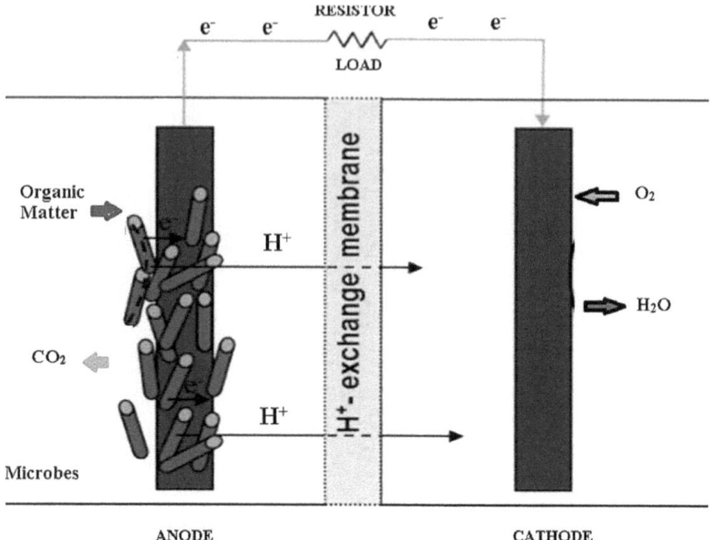

Figure 4. Principle of microbial fuel cells (MFCs).

electrons obtained from an electron donor (substrates glucose or acetic acid) to the anode electrode. Through BFC, energy from solar cells can be turned into convenient storable energy sources called electrofuels that are as efficient as a liquid fuel. Using nanomaterials to develop biofuel cells is a cutting-edge technique in the present day.

To improve MFCs' performance, researchers have conducted many studies on MFC technology coupled with nanotechnologies and have observed the potential of nanomaterials as both efficient catalysts and reaction media. Due to its high specific surface area, excellent stability, conductivity and uniform nanopore distribution, the nanocomposites-based electrodes have been used in MFC technology to enhance power density. Yan et al. (2008) produced a unique nanoscaled polyaniline (PANI)/ mesoporous TiO_2 composite to use it as an anode in *Escherichia coli* MFCs to investigate the catalytic behavior of nanostructured composite. The optimal performance of the composite was observed with 30wt% PANI. This nanostructured composite increased the 2-fold higher power density (1,495 mW/m^2), which was almost three times higher than that obtained with a conventional electrode in the MFC. Thus, nanocomposites-based electrodes have great potential to be used as the anode for a high-power MFC and may also provide a new universal approach for improving different types of MFCs. Similarly in another study, Sharma et al. (2008) constructed electrodes using novel electron mediators and carbon nanotubes (CNTs) and compared the performance with previously reported *E. coli* based MFCs with neutral red (NR) and methylene blue (MB) electron mediators and plain graphite electrode. CNT-based electrodes were showed as high as a ~ 6-fold increase in the power density (2,470 mW/m^2) compared to graphite electrodes (386 mW/m^2). Thus, nanocomposites-based electrodes have the potential for generating high energies from even simple bacteria like *E. coli*. Researchers also used nanoporous filters (e.g., nylon, cellulose, polycarbonate) to replace proton exchange membranes (PEMs) in MFCs to increase the output power in unique environments (Biffinger et al. 2007). As an additional benefit, the nanoporous membranes isolated the anode from invading natural bacteria, increasing the potential applications for MFCs beyond aquatic sediment environments. Thus, NMs applications in MFCs provide a practical solution for decreasing the cost of BFCs while incorporating new features for powering long-term autonomous devices. Moreover, the inherent unique electronic properties of metal nanoparticles, nanorods (or carbon nanotubes) and modified anode/cathode nanoelectrodes may also enhance the performance of enzyme-based biofuel cells. Enzymes like D-fructose dehydrogenase and laccases could be used to deposit onto the electrode surface to increase the rate of electron transfer to the anode without the need of a mediator. Stolarczyk et al. (2007) studied reduction of dioxygen catalyzed by laccase at the cathode coated with carbon nanotubes or carbon microcrystals, which was studied without any added mediators. They showed that dioxygen reduction started at 0.6 V versus Ag/AgCl, which is close to the formal potential of the laccase used. Thus, the use of enzyme coated nanoelectrodes may not only eliminate the need for mediators but also simplify the working of MFCs to operate under mild conditions.

5. Opportunities and Concerns Toward Application of Nanotechnology

Biofuel is considered the future fuel, and nanotechnology has interesting contributions to enable the production of next-generation biofuels. Various nanocatalysts, like magnetic nanoparticles, carbon nanotubes, metal oxide nanoparticles, engineered nanomaterials, etc., are promising to become an integral part of sustainable bioenergy production. The approach of using enzymes immobilized in nanoparticles is considered path-breaking research in the field of biofuel production and is favoring reuse or allowing continuous use of cellulases, lipases and other biocatalysts, including the whole-cell systems. The possibility of using nano-supports for biocatalysts can be a key technique to the economic feasibility of different processes, which is being studied nowadays, with low interference of nanosized particles in catalytic activity or may even enhance it. In addition to that, nanocatalysts can be used to chemically catalyze depolymerization of cellulose or transesterification of oils and fats with high efficiency.

However, despite their advantages, nanocatalysts also have several limitations. During the enzyme immobilization processes on nanomaterials, the size of some enzyme catalysts makes the coating process difficult. Due to similar or even bigger pore size than nanomaterials, catalysts are not embedded properly in the channels while others still stay on the surface and even hinder the diffusion of substrate/product to/off the enzyme. It can lower the enzyme activity or lose during the operation. Furthermore, selectivity problems are also associated with nanoparticles. Sometimes target reaction is enhanced but the affiliated reactions may also be accelerated.

In addition to the abovementioned technical challenges, there are environmental concerns about excessive and uncontrolled use of such heavy metal embedded nanoscale materials which could exert negative effects on human health (Singh et al. 2011). Nanoparticle surface has unique properties as a nanocatalyst can also cause oxidative reactions in cells. As they are small enough to enter human cells through inhalation, ingestion or dermal contact and can reach any organ in the human body to generate cytotoxic effects on various tissues. Metal nanoparticles are composed of toxic metals that are already known for biomagnification in tissues. Owing to the high metabolic rate of the kidneys, lungs and liver, these organs are generally at a higher risk to interact with nanosized particles. Many researchers have studied the harmful effects of the nanoparticles on the mentioned organs in animal models. Muller et al. (2005) showed that after being administered to Sprague-Dawley rats, multiwall carbon nanotubes (0.5, 2 or 5 mg) caused inflammatory and fibrotic reactions in the lungs. CNTs were able to agglomerate into the lungs and surrounding tissues to form collagen-rich granulomas protruding in the bronchial lumen and alveolitis. Furthermore, at the cellular level, CNTs induce TNF-α in macrophages, indicating that these particles present high inflammatory properties when administered. In another study, Tong et al. (2009) reported that an acid-functionalized single-walled carbon nanotube (AF-SWCNT), after being administered to mice through oropharyngeal aspiration at a dose of 10 or 40 μg, caused the development of sporadic clumps of particles. Also, the patches of cellular infiltration and edema in both small airways and interstitium were observed at higher doses. The study of Shukla et al. (2015) showed that the coating of iron oxide nanoparticles with chitosan nanoparticles can reduce its toxicity in various human cell lines. Table 10 summarizes the concerning *in vitro* and *in vivo* and shows the hazardous effects of different nanoparticles.

However, our understanding of the breadth, ecological significance and chemical interaction of the organism relationship are limited (Singh et al. 2018). The environmental toxicity of nanoparticles is not yet fully understood. The main concern derives from the fact that these particles can be easily accumulated by organisms in the food chain, causing major cellular damage as they are small enough to interact with biomolecules, such as proteins, lipids and DNA. The ecotoxicity of nanoparticles must be further studied in animal models, generating a better understanding of the precise damage that these particles can cause and the real danger of particle accumulation. Such studies will play a defining role in preventing or reducing the toxic effects of the nanoparticles to both humans exposed and the ecosystem as a whole.

6. Conclusion

This chapter gave a comprehensive overview of the application of nanotechnology and explored their potential to improve the overall process of biofuel production. Due to its nanoscale inherent property, possessing large surface-area-to-volume ratios, allows a greater number of reaction sites which increases the mass transfer during biofuel production. The recent development in nanotechnology has provided a wealth of diverse nanoscale scaffolds that could potentially be applied for immobilization enzymes like laccases, cellulases, etc. The nanomaterials loaded with enzymes demonstrated enhanced enzyme activity and stability, high enzyme

Table 10. Hazardous effects *in vitro* and *in vivo* of different nanoparticles.

Nanoparticles	Toxic Effects		References
	In Vivo	*In Vitro*	
Single-walled Carbon Nanotubes	Mice uphold allergic response genotoxicity and lung inflammation; 8-oxo-7, 8-dihydro-2'-deoxyguanosine (8-oxo-dG) increase in lung and liver of rat: Platelet activation and thrombus formation; toxicological responses in rat alveolar macrophages.	Inflammatory mediator response suppression in human lung epithelium; disruption of actin filament integrity and vein endothelial-cadherin (vascular endothelial-cadherin) distribution in human aortic endothelial cells; oxidative stress, DNA damage, micronuclei formation, oxidative stress, apoptosis and cytotoxicity in various cells.	Bihari et al. 2010, Folkmann et al. 2008, Yang et al. 2008, Haick et al. 2009, Nahle et al. 2019
Multiwalled Carbon Nanotubes	Mice induce inflammation but decrease production of reactive oxygen species in lung; induce apoptosis; pulmonary toxicity; granuloma formation; cytotoxicity and fibrosis in lungs.	Disruption of actin filament integrity and vein endothelial-cadherin distribution in human aortic endothelial cells; induce micronuclei and double strand breaks in DNA; induce apoptosis; oxidative stress and cytotoxicity.	Crouzier et al. 2010, Ma-Hock et al. 2013, Porter et al. 2010, Hirano et al. 2010, Poland et al. 2008, Walker et al. 2009, Kavosi et al. 2018, Duke et al. 2018, Knudsen et al. 2019, Gate et al. 2019
Gold NPs	Mice: Bioaccumulation in important body organs, cross-blood brain barrier; adverse effect on human sperm motility and apoptosis and acute inflammation in liver.	Autophagy, oxidative stress; mitochondrial damage which triggers necrosis and affects cellular motility.	Li et al. 2010, Tarantola et al. 2011, Taylor et al. 2014
Silver NPs	Rats: Blood-brain barrier destruction, astrocyte swelling and neuronal degeneration; brain edema formation. Mice: Expression of genes related to oxidative stress in brain.	Chromosome instability and mitotic arrest cytoskeleton deformations in human cells; mitochondrial disruption, decreased metabolism; apoptosis; DNA damage; cytotoxicity and INK activation in mammalian cells.	Asharani et al. 2009a, Foldbjerg et al. 2011, Miura and Shinhoray 2009, Tang et al. 2009, Rahman et al. 2009, Skalska et al. 2015, Huang et al. 2014, Manshian et al. 2015

Table 10 contd. ...

...Table 10 contd.

Nanoparticles	Toxic Effects		References
	In Vivo	*In Vitro*	
Fullerenes	Rats: Genotoxicity, elevated level of 8-oxo-dG in lung and liver. Increase in proinflammatory cytokines and ThI cytokines in BAL fluid.	Oxidative stress, DNA damage and cytotoxicity in mammalian cells.	Tokiwa et al. 2005, Folkmann et al. 2008, Jacobsen et al. 2008, Zhang et al. 2009, Sumi and Chitra 2019
Metal Oxide NPs, Copper Oxide, Zno, TiO_2 and Nickel oxide	Rats: Infiltration of macrophages, alveolities, inflammation of lung and collagen deposition.	Cytotoxicity and oxidative stress; alteration of gene expression and calcium homeostasis: DNA damage, apoptosis and disturbance of physiological functions and ionic homeostasis in rats' hippocampal CA3 pyramidal neurons by ZnO NPs.	Pan et al. 2009, Huang et al. 2010, Cho et al. 2010, Panas et al. 2012

loading capacity and recyclability for upstream and downstream biofuel production processes. Critical challenges related to the impact on health, and our environment is still to be considered for an effective sustainable strategy. However, with efforts from material scientists, bioprocessing engineers and biochemists, nanotechnology could be effectively worked to produce commercially viable and sustainable biofuel in the very near future.

References

Abraham, R.E., M.L. Verma, C.J. Barrow and M. Puri. 2014. Suitability of magnetic nanoparticle immobilised cellulases in enhancing enzymatic saccharification of pretreated hemp biomass. Biotechnology for Biofuels 7(1): 90.

Ahmad, R. and M. Sardar. 2014. Immobilization of cellulase on TiO_2 nanoparticles by physical and covalent methods: A comparative study. Indian Journal of Biochemistry and Biophysics 51: 314–320.

Ahmadi, M.H., M. Ghazvini, M. Alhuyi Nazari, M.A. Ahmadi, F. Pourfayaz, G. Lorenzini and T. Ming. 2019. Renewable energy harvesting with the application of nanotechnology: A review. International Journal of Energy Research 43(4): 1387–1410.

Al-Ahmad, A.E., S. Hiligsmann, S. Lambert, B. Heinrichs, W. Wannoussa, L. Tasseroul, F. Weekers and P. Thonart. 2014. Effect of encapsulated nanoparticles on thermophillic anaerobic digestion. In 19th National Symposium on Applied Biological Sciences. Gembloux, Belgium.

Andualema, B. and A. Gessesse. 2012. Microbial lipases and their industrial applications. Biotechnology 11(3): 100–118.

Antczak, M.S., A. Kubiak, T. Antczak and S. Bielecki. 2009. Enzymatic biodiesel synthesis: Key factors affecting efficiency of the process. Renewable Energy 34(5): 1185–1194.

Asharani, P.V., M.P. Hande and S. Valiyaveettil. 2009. Anti-proliferative activity of silver nanoparticles. BMC Cell Biology 10(1): 65.

Babaki, M., M. Yousefi, Z. Habibi, M. Mohammadi, P. Yousefi, J. Mohammadi and J. Brask. 2016. Enzymatic production of biodiesel using lipases immobilized on silica nanoparticles as highly reusable biocatalysts: effect of water, t-butanol and blue silica gel contents. Renewable Energy 91: 196–206.

Baeyens, J., Q. Kang, L. Appels, R. Dewil, Y. Lv and T. Tan. 2015. Challenges and opportunities in improving the production of bio-ethanol. Progress in Energy and Combustion Science 47: 60–88.

Balat, H. and E. Kırtay. 2010. Hydrogen from biomass: present scenario and future prospects. International Journal of Hydrogen Energy 35(14): 7416–7426.

Beckers, L., S. Hiligsmann, S.D. Lambert, B. Heinrichs and P. Thonart. 2013. Improving effect of metal and oxide nanoparticles encapsulated in porous silica on fermentative biohydrogen production by *Clostridium butyricum*. Bioresource Technology 133: 109–117.

Biffinger, J.C., R. Ray, B. Little and B.R. Ringeisen. 2007. Diversifying biological fuel cell designs by use of nanoporous filters. Environmental Science & Technology 41(4): 1444–1449.

Bihari, P., M. Holzer, M. Praetner, J. Fent, M. Lerchenberger, C.A. Reichel, M. Rehberg, S. Lakatos and F. Krombach. 2010. Single-walled carbon nanotubes activate platelets and accelerate thrombus formation in the microcirculation. Toxicology 269(2-3): 148–154.

Bootsma, J.A. and B.H. Shanks. 2007. Cellobiose hydrolysis using organic-inorganic hybrid mesoporous silica catalysts. Applied Catalysis A: General 327(1): 44–51.

Boz, N., N. Degirmenbasi and D.M. Kalyon. 2009. Conversion of biomass to fuel: Transesterification of vegetable oil to biodiesel using KF loaded nano-γAl_2O_3 as catalyst. Applied Catalysis B: Environmental 89: 590–596.

Casals, E., R. Barrena, A. Garca, E. GonzÃ¡lez, L. Delgado, M. Busquetsâ Fit, X. Font, J. Arbiol, P. Glatzel and K. Kvashnina. 2014. Programmed iron oxide nanoparticles disintegration in anaerobic digesters boosts biogas production. Small 10(14): 2801–2808.

Chang, A.C., R.F. Louh, D. Wong, J. Tseng and Y.S. Lee. 2011. Hydrogen production by aqueous-phase biomass reforming over carbon textile supported Pt-Ru bimetallic catalysts. International Journal of Hydrogen Energy 36: 8794–8799.

Chen, Y.Z., C.B. Ching and R. Xu. 2009. Lipase immobilization on modified zirconia nanoparticles: Studies on the effects of modifiers. Process Biochemistry 44(11): 1245–1251.

Cherubini, F. 2010. The biorefinery concept: using biomass instead of oil for producing energy and chemicals. Energy Conversion and Management 51(7): 1412–1421.

Cho, E.J., S. Jung, H.J. Kim, Y.G. Lee, K.C. Nam, H.-J. Lee and H.-J. Bae. 2012. Co-immobilization of three cellulases on Au-doped magnetic silica nanoparticles for the degradation of cellulose. Chemical Communications 48(6): 886–888.

Cho, W.-S., R. Duffin, C.A. Poland, S.E.M. Howie, W. MacNee, M. Bradley, I.L. Megson and K. Donaldson. 2010. Metal oxide nanoparticles induce unique inflammatory footprints in the lung: important implications for nanoparticle testing. Environmental Health Perspectives 118(12): 1699–1706.

Chronopoulou, L., G. Kamel, C. Sparago, F. Bordi, S. Lupi, M. Diociaiuti and C. Palocci. 2011. Structure-activity relationships of *Candida rugosa* lipase immobilized on polylactic acid nanoparticles. Soft Matter 7(6): 2653–2662.

Crouzier, D., S. Follot, E. Gentilhomme, E. Flahaut, R. Arnaud, V. Dabouis, C. Castellarin and J.-C. Debouzy. 2010. Carbon nanotubes induce inflammation but decrease the production of reactive oxygen species in lung. Toxicology 272(1-3): 39–45.

Dai, J., H. Chang, E. Maeda, S. Warisawa and R. Kometani. 2018. Approaching the resolution limit of WC nano-gaps using focused ion beam chemical vapour deposition. Applied Surface Science 427: 422–427.

Deng, X., Z. Fang, Y.H. Liu and C. Liu-Yu. 2011. Production of biodiesel from Jatropha oil catalyzed by nanosized solid basic catalyst. Energy 36: 777–784.

Ding, J., S. Ji, H. Wang, J. Key, D.J.L. Brett and R. Wang. 2018. Nano-engineered intrapores in nanoparticles of PtNi networks for increased oxygen reduction reaction activity. Journal of Power Sources 374: 48–54.

Dolly, S., A. Pandey, B.K. Pandey and R. Gopal. 2015. Process parameter optimization and enhancement of photo-biohydrogen production by mixed culture of *Rhodobacter sphaeroides* NMBL-02 and *Escherichia coli* NMBL-04 using Fe-nanoparticle. International Journal of Hydrogen Energy 40(46): 16010–16020.

Duc, N.M. 2013. Effects of CeO_2 and ZnO nanoparticles on anaerobic digestion and toxicity of digested sludge. Thailand: Asian Institute of Technology.

Duke, K.S., E.A. Thompson, M.D. Ihrie, A.J. Taylor-Just, E.A. Ash, K.A. Shipkowski, J.R. Hall, D.A. Tokarz, M.F. Cesta and A.F. Hubbs. 2018. Role of p53 in the chronic pulmonary immune response to tangled or rod-like multi-walled carbon nanotubes. Nanotoxicology 12(9): 975–991.

Dyal, A., K. Loos, M. Noto, S.W. Chang, C. Spagnoli, K.V.P.M. Shafi, A. Ulman, M. Cowman and R.A. Gross. 2003. Activity of *Candida rugosa* lipase immobilized on γ-Fe_2O_3 magnetic nanoparticles. Journal of the American Chemical Society 125(7): 1684–1685.

Eisa, W.H., A.M. Abdelgawad and O.J. Rojas. 2018. Solid-state synthesis of metal nanoparticles supported on cellulose nanocrystals and their catalytic activity. ACS Sustainable Chemistry & Engineering 6(3): 3974–3983.

Engliman, N.S., P.M. Abdul, S.-Y. Wu and J.M. Jahim. 2017. Influence of iron (II) oxide nanoparticle on biohydrogen production in thermophilic mixed fermentation. International Journal of Hydrogen Energy 42(45): 27482–27493.

Fan, Y., G. Wu, F. Su, K. Li, L. Xu, X. Han and Y. Yan. 2016. Lipase oriented-immobilized on dendrimer-coated magnetic multi-walled carbon nanotubes toward catalyzing biodiesel production from waste vegetable oil. Fuel 178: 172–178.

Faucher, S., P. Le Coustumer and G. Lespes. 2019. Nanoanalytics: history, concepts, and specificities. Environmental Science and Pollution Research 26(6): 5267–5281.

Feyzi, M., A. Hassankhani and H.R. Rafiee. 2013. Preparation and characterization of Cs/Al/Fe_3O_4 nanocatalysts for biodiesel production. Energy Conversion and Management 71: 62–68.

Foldbjerg, R., D.A. Dang and H. Autrup. 2011. Cytotoxicity and genotoxicity of silver nanoparticles in the human lung cancer cell line, A549. Archives of Toxicology 85(7): 743–750.

Folkmann, J.K., L. Risom, N.R. Jacobsen, H.k. Wallin, S. Loft and P. Maller. 2008. Oxidatively damaged DNA in rats exposed by oral gavage to C60 fullerenes and single-walled carbon nanotubes. Environmental Health Perspectives 117(5): 703–708.

Gadhe, A., S.S. Sonawane and M.N. Varma. 2015a. Enhancement effect of hematite and nickel nanoparticles on biohydrogen production from dairy wastewater. International Journal of Hydrogen Energy 40(13): 4502–4511.

Gadhe, A., S.S. Sonawane and M.N. Varma. 2015b. Influence of nickel and hematite nanoparticle powder on the production of biohydrogen from complex distillery wastewater in batch fermentation. International Journal of Hydrogen Energy 40(34): 10734–10743.

Gao, L., S. Wang, W. Xu and G. Xiao. 2015. Biodiesel production from palm oil over monolithic KF/γ-Al_2O_3/honeycomb ceramic catalyst. Applied Energy 146: 196–201.

Garcia, A., L. Delgado, J.A. Tora, E. Casals, E. Gonza¡lez, V. Puntes, X. Font, J.n. Carrera and A. Sa¡nchez. 2012. Effect of cerium dioxide, titanium dioxide, silver, and gold nanoparticles on the activity of microbial communities intended in wastewater treatment. Journal of Hazardous Materials 199: 64–72.

Gate, L., K.B. Knudsen, C. Seidel, T. Berthing, L. Chazeau, N.R. Jacobsen, S. Valentino, H. Wallin, S. Bau and H. Wolff. 2019. Pulmonary toxicity of two different multi-walled carbon nanotubes in rat: Comparison between intratracheal instillation and inhalation exposure. Toxicology and Applied Pharmacology 375: 17–31.

Giannelli, L. and G. Torzillo. 2012. Hydrogen production with the microalga *Chlamydomonas reinhardtii* grown in a compact tubular photobioreactor immersed in a scattering light nanoparticle suspension. International Journal of Hydrogen Energy 37(22): 16951–16961.

Ginestra, P.S., M. Madou and E. Ceretti. 2019. Production of carbonized micro-patterns by photolithography and pyrolysis. Precision Engineering 55: 137–143.

Gokhale, A.A., J. Lu and I. Lee. 2013. Immobilization of cellulase on magnetoresponsive graphene nano-supports. Journal of Molecular Catalysis B: Enzymatic 90: 76–86.

Gonzalez-Estrella, J., R. Sierra-Alvarez and J.A. Field. 2013. Toxicity assessment of inorganic nanoparticles to acetoclastic and hydrogenotrophic methanogenic activity in anaerobic granular sludge. Journal of Hazardous Materials 260: 278–285.

Gulati, K., S.M. Hamlet and S.o. Ivanovski. 2018. Tailoring the immuno-responsiveness of anodized nano-engineered titanium implants. Journal of Materials Chemistry B 6(18): 2677–2689.

Gumina, B., C. Espro, S. Galvagno, R. Pietropaolo and F. Mauriello. 2019. Bioethanol production from unpretreated cellulose under neutral selfsustainable hydrolysis/hydrogenolysis conditions promoted by the heterogeneous Pd/Fe_3O_4 catalyst. ACS Omega 4(1): 352–357.

Haick, H., M. Hakim, M. Patrascu, C. Levenberg, N. Shehada, F. Nakhoul and Z. Abassi. 2009. Sniffing chronic renal failure in rat model by an array of random networks of single-walled carbon nanotubes. ACS Nano 3(5): 1258–1266.

Han, H. and Y. Guan. 2009. Synthesis of biodiesel from rapeseed oil using K_2O/γ-Al_2O_3 as nano-solid-basecatalys. Wuhan University Journal of Natural Sciences 14(1): 75–79.

Han, H., M. Cui, L. Wei, H. Yang and J. Shen. 2011. Enhancement effect of hematite nanoparticles on fermentative hydrogen production. Bioresource Technology 102(17): 7903–7909.

Hirano, S., Y. Fujitani, A. Furuyama and S. Kanno. 2010. Uptake and cytotoxic effects of multi-walled carbon nanotubes in human bronchial epithelial cells. Toxicology and Applied Pharmacology 249(1): 8–15.

Ho, S.-H., S.-W. Huang, C.-Y. Chen, T. Hasunuma, A. Kondo and J.-S. Chang. 2013. Bioethanol production using carbohydrate-rich microalgae biomass as feedstock. Bioresource Technology 135: 191–198.

Hong, Y., D. Zhao, D. Liu, B. Ma, G. Yao, Q. Li, A. Han and M. Qiu. 2018. Three-dimensional *in situ* electron-beam lithography using water ice. Nano Letters 18(8): 5036–5041.

Hsieh, P.-H., Y.-C. Lai, K.-Y. Chen and C.-H. Hung. 2016. Explore the possible effect of TiO_2 and magnetic hematite nanoparticle addition on biohydrogen production by *Clostridium pasteurianum* based on gene expression measurements. International Journal of Hydrogen Energy 41(46): 21685–21691.

Hu, S., Y. Guan, Y. Wang and H. Han. 2011. Nano-magnetic catalyst KF/CaO–Fe_3O_4 for biodiesel production. Applied Energy 88: 2685–2690.

Huang, C.-C., R.S. Aronstam, D.-R. Chen and Y.-W. Huang. 2010. Oxidative stress, calcium homeostasis, and altered gene expression in human lung epithelial cells exposed to ZnO nanoparticles. Toxicology *In Vitro* 24(1): 45–55.

Huang, P.-J., K.-L. Chang, J.-F. Hsieh and S.-T. Chen. 2015. Catalysis of rice straw hydrolysis by the combination of immobilized cellulase from Aspergillus niger on Î²-cyclodextrin-Fe_3O_4 nanoparticles and ionic liquid. BioMed Research International 2015: 1–9.

Huang, X.-J., P.-C. Chen, F. Huang, Y. Ou, M.-R. Chen and Z.-K. Xu. 2011. Immobilization of *Candida rugosa* lipase on electrospun cellulose nanofiber membrane. Journal of Molecular Catalysis B: Enzymatic 70(3-4): 95–100.

Huang, Y., X. Li and J. Ma. 2014. Toxicity of silver nanoparticles to human dermal fibroblasts on microRNA level. Journal of Biomedical Nanotechnology 10(11): 3304–3317.

Jacobsen, N.R., G. Pojana, P. White, P. MÃ¸ller, C.A. Cohn, K. Smith Korsholm, U. Vogel, A. Marcomini, S. Loft and H.k. Wallin. 2008. Genotoxicity, cytotoxicity, and reactive oxygen species induced by single-walled carbon nanotubes and C_{60} fullerenes in the FE1-Mutatrade mark Mouse lung epithelial cells. Environmental and Molecular Mutagenesis 49(6): 476–487.

Jafari Khorshidi, K., H. Lenjannezhadian, M. Jamalan and M. Zeinali. 2014. Preparation and characterization of nanomagnetic cross-linked cellulase aggregates for cellulose bioconversion. Journal of Chemical Technology & Biotechnology 91(2): 539–546.

Jeevanandam, J., A. Barhoum, Y.S. Chan, A. Dufresne and M.K. Danquah. 2018. Review on nanoparticles and nanostructured materials: history, sources, toxicity and regulations. Beilstein Journal of Nanotechnology 9(1): 1050–1074.

Ji, P., H. Tan, X. Xu and W. Feng. 2010. Lipase covalently attached to multiwalled carbon nanotubes as an efficient catalyst in organic solvent. AIChE Journal 56(11): 3005–3011.

Jordan, J., C.S.S.R. Kumar and C. Theegala. 2011. Preparation and characterization of cellulase-bound magnetite nanoparticles. Journal of Molecular Catalysis B: Enzymatic 68(2): 139–146.

Kajita, S., S. Kawaguchi, N. Ohno and N. Yoshida. 2018. Enhanced growth of large-scale nanostructures with metallic ion precipitation in helium plasmas. Scientific Reports 8(1): 56.

Karimi, M. 2016. Immobilization of lipase onto mesoporous magnetic nanoparticles for enzymatic synthesis of biodiesel. Biocatalysis and Agricultural Biotechnology 8: 182–188.

Kaur, M. and A. Ali. 2011. Lithium ion impregnated calcium oxide as nano catalyst for the biodiesel production from karanja and jatropha oils. Renewable Energy 36: 2866–2871.

Kavosi, A., S.H.G. Noei, S. Madani, S. Khalighfard, S. Khodayari, H. Khodayari, M. Mirzaei, M.R. Kalhori, M. Yavarian and A.M. Alizadeh. 2018. The toxicity and therapeutic effects of single-and multi-wall carbon nanotubes on mice breast cancer. Scientific Reports 8(1): 8375.

Khoshnevisan, K., A.-K. Bordbar, D. Zare, D. Davoodi, M. Noruzi, M. Barkhi and M. Tabatabaei. 2011. Immobilization of cellulase enzyme on superparamagnetic nanoparticles and determination of its activity and stability. Chemical Engineering Journal 171(2): 669–673.

Kim, M.I., H.O. Ham, S.-D. Oh, H.G. Park, H.N. Chang and S.-H. Choi. 2006. Immobilization of *Mucor javanicus* lipase on effectively functionalized silica nanoparticles. Journal of Molecular Catalysis B: Enzymatic 39(1-4): 62–68.

Knudsen, K.B., T. Berthing, P. Jackson, S.S. Poulsen, A. Mortensen, N.R. Jacobsen, V. Skaug, J.z. Szarek, K.S. Hougaard and H. Wolff. 2019. Physicochemical predictors of multi-walled carbon nanotube: induced pulmonary histopathology and toxicity one year after pulmonary deposition of 11 different multi-walled carbon nanotubes in mice. Basic & Clinical Pharmacology & Toxicology 124(2): 211–227.

Lai, D.-m., L. Deng, Q.-x. Guo and Y. Fu. 2011. Hydrolysis of biomass by magnetic solid acid. Energy & Environmental Science 4(9): 3552–3557.

Lam, M.K., K.T. Lee and A.R. Mohamed. 2010. Homogeneous, heterogeneous and enzymatic catalysis for transesterification of high free fatty acid oil (waste cooking oil) to biodiesel: A review. Biotechnology Advances 28(4): 500–518.

Lee, D.-G., K.M. Ponvel, M. Kim, S. Hwang, I.-S. Ahn and C.-H. Lee. 2009. Immobilization of lipase on hydrophobic nano-sized magnetite particles. Journal of Molecular Catalysis B: Enzymatic 57(1-4): 62–66.

Lee, H.-K., J.-K. Lee, M.-J. Kim and C.-J. Lee. 2010a. Immobilization of lipase on single walled carbon nanotubes in ionic liquid. Bulletin of the Korean Chemical Society 31(3): 650–652.

Lee, S.H., T.T.N. Doan, K. Won, S.H. Ha and Y.-M. Koo. 2010b. Immobilization of lipase within carbon nanotube silica composites for non-aqueous reaction systems. Journal of Molecular Catalysis B: Enzymatic 62(2): 169–172.

Lee, Y.-C., Y.S. Huh, W. Farooq, J. Chung, J.-I. Han, H.-J. Shin, S.H. Jeong, J.-S. Lee, Y.-K. Oh and J.-Y. Park. 2013a. Lipid extractions from docosahexaenoic acid (DHA)-rich and oleaginous *Chlorella* sp. biomasses by organic-nanoclays. Bioresource Technology 137: 74–81.

Lee, Y.-C., Y.S. Huh, W. Farooq, J.-I. Han, Y.-K. Oh and J.-Y. Park. 2013b. Oil extraction by aminoparticle-based H_2O_2 activation via wet microalgae harvesting. RSC Advances 3(31): 12802–12809.

Lee, Y.-C., H.U. Lee, K. Lee, B. Kim, S.Y. Lee, M.-H. Choi, W. Farooq, J.S. Choi, J.-Y. Park, J. Lee, Y.-K. Oh and Y.S. Huh. 2014. Aminoclay-conjugated TiO_2 synthesis for simultaneous harvesting and wet-disruption of oleaginous *Chlorella* sp. Chemical Engineering Journal 245: 143–149.

Li, J.J., D. Hartono, C.-N. Ong, B.-H. Bay and L.-Y.L. Yung. 2010. Autophagy and oxidative stress associated with gold nanoparticles. Biomaterials 31(23): 5996–6003.

Li, S., D. Han and K.H. Row. 2012. Optimization of enzymatic extraction of polysaccharides from some marine algae by response surface methodology. Korean Journal of Chemical Engineering 29(5): 650–656.

Li, S.-F., J.-P. Chen and W.-T. Wu. 2007. Electrospun polyacrylonitrile nanofibrous membranes for lipase immobilization. Journal of Molecular Catalysis B: Enzymatic 47(3-4): 117–124.

Li, S.-F., Y.-H. Fan, R.-F. Hu and W.-T. Wu. 2011. *Pseudomonas cepacia* lipase immobilized onto the electrospun PAN nanofibrous membranes for biodiesel production from soybean oil. Journal of Molecular Catalysis B: Enzymatic 72(1-2): 40–45.

Liao, H., D. Chen, L. Yuan, M. Zheng, Y. Zhu and X. Liu. 2010. Immobilized cellulase by polyvinyl alcohol/Fe_2O_3 magnetic nanoparticle to degrade microcrystalline cellulose. Carbohydrate Polymers 82(3): 600–604.

Lin, R., J. Cheng, L. Ding, W. Song, M. Liu, J. Zhou and K. Cen. 2016. Enhanced dark hydrogen fermentation by addition of ferric oxide nanoparticles using *Enterobacter aerogenes*. Bioresource Technology 207: 213–219.

Lo, H.M., H.Y. Chiu, S.W. Lo and F.C. Lo. 2012. Effects of micro-nano and non micro-nano MSWI ashes addition on MSW anaerobic digestion. Bioresource Technology 114: 90–94.

Luna-delRisco, M., K. Orupaµld and H.-C. Dubourguier. 2011. Particle-size effect of CuO and ZnO on biogas and methane production during anaerobic digestion. Journal of Hazardous Materials 189(1-2): 603–608.

Madhuvilakku, R. and K. Piraman. 2013. Biodiesel synthesis by TiO_2–ZnO mixed oxide nanocatalyst catalyzed palm oil transesterification process. Bioresoure Technology 150: 55–59.

Ma-Hock, L., V. Strauss, S. Treumann, K. Kttler, W. Wohlleben, T. Hofmann, S. Grters, K. Wiench, B. van Ravenzwaay and R. Landsiedel. 2013. Comparative inhalation toxicity of multi-wall carbon nanotubes, graphene, graphite nanoplatelets and low surface carbon black. Particle and Fibre Toxicology 10(1): 23.

Malik, S.N., V. Pugalenthi, A.N. Vaidya, P.C. Ghosh and S.N. Mudliar. 2014. Kinetics of nano-catalysed dark fermentative hydrogen production from distillery wastewater. Energy Procedia 54: 417–430.

Manshian, B.B., C. Pfeiffer, B. Pelaz, T. Heimerl, M. Gallego, M. Moì^ller, P. Del Pino, U. Himmelreich, W.J. Parak and S.J. Soenen. 2015. High-content imaging and gene expression approaches to unravel the effect of surface functionality on cellular interactions of silver nanoparticles. ACS Nano 9(10): 10431–10444.

Meher, L.C., D.V. Sagar and S.N. Naik. 2006. Technical aspects of biodiesel production by transesterification: A review. Renewable and Sustainable Energy Reviews 10(3): 248–268.

Mehrasbi, M.R., J. Mohammadi, M. Peyda and M. Mohammadi. 2017. Covalent immobilization of *Candida antarctica* lipase on core-shell magnetic nanoparticles for production of biodiesel from waste cooking oil. Renewable Energy 101: 593–602.

Mguni, L.L., R. Meijboom and K. Jalama. 2012. Biodiesel production over nano-MgO supported on Titania. World academy of science, engineering and technology. International Journal of Chemical, Molecular, Nuclear, Materials and Metallurgical Engineering 6(4): 380–384.

Miletic, N., V. Abetz, K. Ebert and K. Loos. 2010. Immobilization of *Candida antarctica* lipase B on polystyrene nanoparticles. Macromolecular Rapid Communications 31(1): 71–74.

Miura, N. and Y. Shinohara. 2009. Cytotoxic effect and apoptosis induction by silver nanoparticles in HeLa cells. Biochemical and Biophysical Research Communications 390(3): 733–737.

Mohamad, N., N.A. Buang, N.A. Mahat, J. Jamalis, F. Huyop, H.Y. Aboul-Enein and R.A. Wahab. 2015. Simple adsorption of *Candida rugosa* lipase onto multi-walled carbon nanotubes for sustainable production of the flavor ester geranyl propionate. Journal of Industrial and Engineering Chemistry 32: 99–108.

Mohan, S.V., G. Mohanakrishna, S.S. Reddy, B.D. Raju, K.S.R. Rao and P.N. Sarma. 2008. Self-immobilization of acidogenic mixed consortia on mesoporous material (SBA-15) and activated carbon to enhance fermentative hydrogen production. International Journal of Hydrogen Energy 33(21): 6133–6142.

Mohanraj, S., S. Kodhaiyolii, M. Rengasamy and V. Pugalenthi. 2014a. Green synthesized iron oxide nanoparticles effect on fermentative hydrogen production by *Clostridium acetobutylicum*. Applied Biochemistry and Biotechnology 173(1): 318–331.

Mohanraj, S., S. Kodhaiyolii, M. Rengasamy and V. Pugalenthi. 2014b. Phytosynthesized iron oxide nanoparticles and ferrous iron on fermentative hydrogen production using *Enterobacter cloacae*: Evaluation and comparison of the effects. International Journal of Hydrogen Energy 39(23): 11920–11929.

Mohanraj, S., K. Anbalagan, P. Rajaguru and V. Pugalenthi. 2016. Effects of phytogenic copper nanoparticles on fermentative hydrogen production by *Enterobacter cloacae* and *Clostridium acetobutylicum*. International Journal of Hydrogen Energy 41(25): 10639–10645.

Mu, H. and Y. Chen. 2011. Long-term effect of ZnO nanoparticles on waste activated sludge anaerobic digestion. Water Research 45(17): 5612–5620.

Mu, H., Y. Chen and N. Xiao. 2011. Effects of metal oxide nanoparticles (TiO_2, Al_2O_3, SiO_2 and ZnO) on waste activated sludge anaerobic digestion. Bioresource Technology 102(22): 10305–10311.

Mu, H., X. Zheng, Y. Chen, H. Chen and K. Liu. 2012. Response of anaerobic granular sludge to a shock load of zinc oxide nanoparticles during biological wastewater treatment. Environmental Science & Technology 46(11): 5997–6003.

Muller, J., F.o. Huaux, N. Moreau, P. Misson, J.-F.o. Heilier, M. Delos, M. Arras, A. Fonseca, J.B. Nagy and D. Lison. 2005. Respiratory toxicity of multi-wall carbon nanotubes. Toxicology and Applied Pharmacology 207(3): 221–231.

Nahle, S., R. Safar, S.p. Grandemange, B. Foliguet, M.l. Lovera-Leroux, Z. Doumandji, A. Le Faou, O. Joubert, B. Rihn and L. Ferrari. 2019. Single wall and multiwall carbon nanotubes induce different toxicological responses in rat alveolar macrophages. Journal of Applied Toxicology 39(5): 764–772.

Nakane, K., T. Hotta, T. Ogihara, N. Ogata and S. Yamaguchi. 2007. Synthesis of (Z)-3-Hexen-1-yl acetate by lipase immobilised in polyvinylalcohol nanofibers. Journal of Applied Polymer Science 106(2): 863–867.

Nasr, M., A. Tawfik, S. Ookawara, M. Suzuki, S. Kumari and F. Bux. 2015. Continuous biohydrogen production from starch wastewater via sequential dark-photo fermentation with emphasize on maghemite nanoparticles. Journal of Industrial and Engineering Chemistry 21: 500–506.

Nath, D., A.K. Manhar, K. Gupta, D. Saikia, S.K. Das and M. Mandal. 2015. Phytosynthesized iron nanoparticles: effects on fermentative hydrogen production by *Enterobacter cloacae* DH-89. Bulletin of Materials Science 38(6): 1533–1538.

Nazir, S., M. Zaka, M. Adil, B.H. Abbasi and C. Hano. 2018. Synthesis, characterisation and bactericidal effect of ZnO nanoparticles via chemical and bio-assisted (Silybummarianum *in vitro* plantlets and callus extract) methods: A comparative study. IET Nanobiotechnology 12(5): 604–608.

Nguyen, M.D. 2013. Effects of CeO_2 and ZnO nanoparticles on anaerobic digestion and toxicity of digested sludge. Thailand: Asian Institute of Technology.

Noor, C.W.M., M.M. Noor and R. Mamat. 2018. Biodiesel as alternative fuel for marine diesel engine applications: A review. Renewable and Sustainable Energy Reviews 94: 127–142.

Obadiah, A., R. Kannan, P. Ravichandran, A. Ramasubbu and S.V. Kumar. 2012. Nano hydrotalcite as a novel catalyst for biodiesel conversion. Digest Journal of Nanomaterials and Biostructures 7(1): 321–327.

Otero-González, L., J.A. Field and R. Sierra-Alvarez. 2014a. Inhibition of anaerobic wastewater treatment after long-term exposure to low levels of CuO nanoparticles. Water Research 58: 160–168.

Otero-González, L., J.A. Field and R. Sierra-Alvarez. 2014b. Fate and long-term inhibitory impact of ZnO nanoparticles during high-rate anaerobic wastewater treatment. Journal of Environmental Management 135: 110–117.

Pan, Z., W. Lee, L. Slutsky, R.A.F. Clark, N. Pernodet and M.H. Rafailovich. 2009. Adverse effects of titanium dioxide nanoparticles on human dermal fibroblasts and how to protect cells. Small 5(4): 511–520.

Panas, A., C. Marquardt, O. Nalcaci, H. Bockhorn, W. Baumann, H.R. Paur, S. Malhopt, S. Diabat, and C. Weiss. 2012. Screening of different metal oxide nanoparticles reveals selective toxicity and inflammatory potential of silica nanoparticles in lung epithelial cells and macrophages. Nanotoxicology 7(3): 259–273.

Pandey, A., K. Gupta and A. Pandey. 2015. Effect of nanosized TiO_2 on photofermentation by *Rhodobacter sphaeroides* NMBL-02. Biomass and Bioenergy 72: 273–279.

Patel, S.K.S., H.J. Purohit and V.C. Kalia. 2010. Dark fermentative hydrogen production by defined mixed microbial cultures immobilized on ligno-cellulosic waste materials. International Journal of Hydrogen Energy 35(19): 10674–10681.

Patel, S.K.S., P. Kumar and V.C. Kalia. 2012. Enhancing biological hydrogen production through complementary microbial metabolisms. International Journal of Hydrogen Energy (14): 10590–10603.

Patel, S.K.S. and V.C. Kalia. 2013. Integrative biological hydrogen production: an overview. Indian Journal of Microbiology 53(1): 3–10.

Patel, S.K.S., P. Kumar, S. Mehariya, H.J. Purohit, J.-K. Lee and V.C. Kalia. 2014. Enhancement in hydrogen production by co-cultures of Bacillus and Enterobacter. International Journal of Hydrogen Energy 39(27): 14663–14668.

Pavlidis, I.V., T. Tsoufis, A. Enotiadis, D. Gournis and H. Stamatis. 2010. Functionalized multi-wall carbon nanotubes for lipase immobilization. Advanced Engineering Materials 12(5): B179–B183.

Peña, L., M. Ikenberry, K.L. Hohn and D. Wang. 2012. Acid-functionalized nanoparticles for pretreatment of wheat straw. Journal of Biomaterials and Nanobiotechnology 3(3): 342.

Peña, L., F. Xu, K.L. Hohn, J. Li and D. Wang. 2014. Propyl-sulfonic acid functionalized nanoparticles as catalyst for pretreatment of corn stover. Journal of Biomaterials and Nanobiotechnology 5(01): 8.

Pham, V.P. 2018. Plasma-related graphene etching: a mini review. Journal of Materials Science and Engineering with Advanced Technology 17: 91–106.

Poland, C.A., R. Duffin, I. Kinloch, A. Maynard, W.A.H. Wallace, A. Seaton, V. Stone, S. Brown, W. MacNee and K. Donaldson. 2008. Carbon nanotubes introduced into the abdominal cavity of mice show asbestos-like pathogenicity in a pilot study. Nature Nanotechnology 3(7): 423.

Porter, D.W., A.F. Hubbs, R.R. Mercer, N. Wu, M.G. Wolfarth, K. Sriram, S. Leonard, L. Battelli, D. Schwegler-Berry and S. Friend. 2010. Mouse pulmonary dose- and time course-responses induced by exposure to multi-walled carbon nanotubes. Toxicology 269(2-3): 136–147.

Qiao, Y., S.-J. Bao, C.M. Li, X.-Q. Cui, Z.-S. Lu and J. Guo. 2007. Nanostructured polyaniline/titanium dioxide composite anode for microbial fuel cells. ACS Nano 2(1): 113–119.

Qiu, F., Y. Li, D. Yang, X. Li and P. Sun. 2011. Heterogeneous solid base nanocatalyst: Preparation, characterization and application in biodiesel production. Bioresource Technology 102: 4150–4156.

Rahman, M.F., J. Wang, T.A. Patterson, U.T. Saini, B.L. Robinson, G.D. Newport, R.C. Murdock, J.J. Schlager, S.M. Hussain and S.F. Ali. 2009. Expression of genes related to oxidative stress in the mouse brain after exposure to silver-25 nanoparticles. Toxicology Letters 187(1): 15–21.

Raita, M., J. Arnthong, V. Champreda and N. Laosiripojana. 2015. Modification of magnetic nanoparticle lipase designs for biodiesel production from palm oil. Fuel Processing Technology 134: 189–197.

Ranganathan, S.V., S.L. Narasimhan and K. Muthukumar. 2008. An overview of enzymatic production of biodiesel. Bioresource Technology 99(10): 3975–3981.

Razack, S.A., S. Duraiarasan and V. Mani. 2016. Biosynthesis of silver nanoparticle and its application in cell wall disruption to release carbohydrate and lipid from *C. vulgaris* for biofuel production. Biotechnology Reports 11: 70–76.

Reddy, C.R.V., R. Oshel and J.G. Verkade. 2006. Room-temperature conversion of soybean oil and poultry fat to biodiesel catalyzed by nanocrystalline calcium oxides. Energy & Fuels 20: 1310–1314.

Reddy, K., M. Nasr, S. Kumari, S. Kumar, S.K. Gupta, A.M. Enitan and F. Bux. 2017. Biohydrogen production from sugarcane bagasse hydrolysate: effects of pH, S/X, Fe^{2+}, and magnetite nanoparticles. Environmental Science and Pollution Research 24(9): 8790–8804.

Sahoo, S.K., R. Misra and S. Parveen. 2017. Nanoparticles: a boon to drug delivery, therapeutics, diagnostics and imaging. *In*: Nanomedicine in Cancer: Pan Stanford, 73–124.

Sakai, S., Y. Liu, T. Yamaguchi, R. Watanabe, M. Kawabe and K. Kawakami. 2010. Production of butyl-biodiesel using lipase physically-adsorbed onto electrospun polyacrylonitrile fibers. Bioresource Technology 101(19): 7344–7349.

Shah, S., K. Solanki and M.N. Gupta. 2007. Enhancement of lipase activity in non-aqueous media upon immobilization on multi-walled carbon nanotubes. Chemistry Central Journal 1(1): 30.

Sharma, T., A.L.M. Reddy, T.S. Chandra and S. Ramaprabhu. 2008. Development of carbon nanotubes and nanofluids based microbial fuel cell. International Journal of Hydrogen Energy 33(22): 6749–6754.

Shukla, S., A. Jadaun, V. Arora, R.K. Sinha, N. Biyani and V.K. Jain. 2015. *In vitro* toxicity assessment of chitosan oligosaccharide coated iron oxide nanoparticles. Toxicology Reports 2: 27–39.

Singh, S.K., A. Bansal, M.K. Jha and A. Dey. 2011. Comparative studies on uptake of wastewater nutrients by immobilized cells of *Chlorella minutissima* and dairy waste isolated algae. Indian Chemical Engineer 53(4): 211–219.

Singh, S.K., A. Bansal, M.K. Jha and A. Dey. 2012. An integrated approach to remove Cr (VI) using immobilized *Chlorella minutissima* grown in nutrient rich sewage wastewater. Bioresource Technology 104: 257–265.

Singh, S.K., A. Bansal, M.K. Jha and R. Jain. 2013a. Production of biodiesel from wastewater grown *Chlorella minutissima*. Indian Journal of Chemical Technology 20: 341–345.

Singh, S.K., K. Dixit and S. Sundaram. 2013b. Bioengineering of biochemical pathways for enhanced photobiological hydrogen production in algae and cyanobacteria. International Journal of Biotechnology and Bioengineering 4(5): 511–518.

Singh, S.K., A. Rahman, K. Dixit, A. Nath and S. Sundaram. 2016a. Evaluation of promising algal strains for sustainable exploitation coupled with CO_2 fixation. Environmental Technology 37(5): 613–622.

Singh, S.K., S. Sundaram, S. Sinha, M.A. Rahman and S. Kapur. 2016b. Recent advances in CO_2 uptake and fixation mechanism of cyanobacteria and microalgae. Critical Reviews in Environmental Science and Technology 46(16): 1297–1323.

Singh, S.K., S.R. Major, H. Cai, F. Chen, R.T. Hill and Y. Li. 2018. Draft genome sequences of *Cloacibacterium normanense* IMET F, a microalgal growth-promoting bacterium, and *Aeromonas jandaei* IMET J, a microalgal growth-inhibiting bacterium. Genome Announcements 6(24): e00503–00518.

Skalska, J., M.g. Frontczak-Baniewicz and L. Strużyńska. 2015. Synaptic degeneration in rat brain after prolonged oral exposure to silver nanoparticles. Neurotoxicology 46: 145–154.

Song, J., D. Kahveci, M. Chen, Z. Guo, E. Xie, X. Xu, F. Besenbacher and M. Dong. 2012. Enhanced catalytic activity of lipase encapsulated in PCL nanofibers. Langmuir 28(14): 6157–6162.

Stolarczyk, K., E. Nazaruk, J. Rogalski and R. Bilewicz. 2008. Nanostructured carbon electrodes for laccase-catalyzed oxygen reduction without added mediators. Electrochimica Acta 53(11): 3983–3990.

Su, L., X. Shi, G. Guo, A. Zhao and Y. Zhao. 2013. Stabilization of sewage sludge in the presence of nanoscale zero-valent iron (nZVI): abatement of odor and improvement of biogas production. Journal of Material Cycles and Waste Management 15(4): 461–468.

Sudha, P.N., K. Sangeetha, K. Vijayalakshmi and A. Barhoum. 2018. Nanomaterials history, classification, unique properties, production and market. In Emerging Applications of Nanoparticles and Architecture Nanostructures: Elsevier, 341–384.

Sumi, N. and K.C. Chitra. 2019. Fullerene C60 nanomaterial induced oxidative imbalance in gonads of the freshwater fish, Anabas testudineus (Bloch, 1792). Aquatic Toxicology 210:196–206.

Taherdanak, M., H. Zilouei and K. Karimi. 2015. Investigating the effects of iron and nickel nanoparticles on dark hydrogen fermentation from starch using central composite design. International Journal of Hydrogen Energy 40(38): 12956–12963.

Taherdanak, M., H. Zilouei and K. Karimi. 2016. The effects of Fe^0 and Ni^0 nanoparticles versus Fe^{2+} and Ni^{2+} ions on dark hydrogen fermentation. International Journal of Hydrogen Energy 41(1): 167–173.

Tang, J., L. Xiong, S. Wang, J. Wang, L. Liu, J. Li, F. Yuan and T. Xi. 2009. Distribution, translocation and accumulation of silver nanoparticles in rats. Journal of Nanoscience and Nanotechnology 9(8): 4924–4932.

Tarantola, M., A. Pietuch, D. Schneider, J. Rother, E. Sunnick, C. Rosman, S. Pierrat, C. Sannichsen, J. Wegener and A. Janshoff. 2011. Toxicity of gold-nanoparticles: synergistic effects of shape and surface functionalization on micromotility of epithelial cells. Nanotoxicology 5(2): 254–268.

Taylor, U., A. Barchanski, S. Petersen, W.A. Kues, U. Baulain, L. Gamrad, L. Sajti, S. Barcikowski and D. Rath. 2014. Gold nanoparticles interfere with sperm functionality by membrane adsorption without penetration. Nanotoxicology 8(supl): 118–127.

Tokiwa, H., N. Sera and Y. Nakanishi. 2005. Involvement of alveolar macrophages in the formation of 8-oxodeoxyguanosine associated with exogenous particles in human lungs. Inhalation Toxicology 17(11): 577–585.

Tong, H., J.K. McGee, R.K. Saxena, U.P. Kodavanti, R.B. Devlin and M.I. Gilmour. 2009. Influence of acid functionalization on the cardiopulmonary toxicity of carbon nanotubes and carbon black particles in mice. Toxicology and Applied Pharmacology 239(3): 224–232.

Tran, D.-T., C.-L. Chen and J.-S. Chang. 2012. Immobilization of Burkholderia sp. lipase on a ferric silica nanocomposite for biodiesel production. Journal of Biotechnology 158(3): 112–119.

Tran, D.-T., C.-L. Chen and J.-S. Chang. 2013. Effect of solvents and oil content on direct transesterification of wet oil-bearing microalgal biomass of *Chlorella vulgaris* ESP-31 for biodiesel synthesis using immobilized lipase as the biocatalyst. Bioresource Technology 135: 213–221.

Vaghari, H., H. Jafarizadeh-Malmiri, M. Mohammadlou, A. Berenjian, N. Anarjan, N. Jafari and S. Nasiri. 2016. Application of magnetic nanoparticles in smart enzyme immobilization. Biotechnology Letters 38(2): 223–233.

Vela¡zquez, J.J., J. Mosa, G. Gorni, R. Balda, J. Ferna¡ndez, L. Pascual, A. DurÃ¡n and Y. Castro. 2019. Transparent SiO_2-GdF_3 sol-gel nano-glass ceramics for optical applications. Journal of Sol-Gel Science and Technology 89(1): 322–332.

Verma, M.L., R. Chaudhary, T. Tsuzuki, C.J. Barrow and M. Puri. 2013. Immobilization of β-glucosidase on a magnetic nanoparticle improves thermostability: Application in cellobiose hydrolysis. Bioresource Technology 135: 2–6.

Verziu, M., B. Cojocaru, J. Hu, R. Richards, C. Ciuculescu, P. Filip and V.I. Parvulescu. 2008. Sunflower and rapeseed oil transesterification to biodiesel over different nanocrystalline MgO catalysts. Green Chemistry 10: 373–378.

Walker, V.G., Z. Li, T. Hulderman, D. Schwegler-Berry, M.L. Kashon and P.P. Simeonova. 2009. Potential *in vitro* effects of carbon nanotubes on human aortic endothelial cells. Toxicology and Applied Pharmacology 236(3): 319–328.

Wang, H., J. Covarrubias, H. Prock, X. Wu, D. Wang and S.H. Bossmann. 2015. Acid-functionalized magnetic nanoparticle as heterogeneous catalysts for biodiesel synthesis. The Journal of Physical Chemistry C 119(46): 26020–26028.

Wang, L. and J. Yang. 2007. Transesterification of soybean oil with nanoMgO or not in supercritical and subcritical methanol. Fuel 86(328–333).

Wang, X., X. Liu, C. Zhao, Y. Ding and P. Xu. 2011. Biodiesel production in packed-bed reactors using lipase nanoparticle biocomposite. Bioresource Technology 102(10): 6352–6355.

Wen, L., Y. Wang, D. Lu, S. Hu and H. Han. 2010. Preparation of KF/CaOnanocatalyst and its application in biodiesel production from Chinese tallow seed oil. Fuel 89: 2267–2271.

Wimonsong, P. and R. Nitisoravut. 2014. Biohydrogen enhancement using highly porous activated carbon. Energy & Fuels 28(7): 4554–4559.

Wimonsong, P. and R. Nitisoravut. 2015. Comparison of different catalyst for fermentative hydrogen production. Journal of Clean Energy Technologies 3: 128–131.

Wu, Y., Y. Wang, G. Luo and Y. Dai. 2010. Effect of solvents and precipitant on the properties of chitosan nanoparticles in a water-in-oil microemulsion and its lipase immobilization performance. Bioresource Technology 101(3): 841–844.

Xie, W. and N. Ma. 2009. Immobilized lipase on Fe_3O_4 nanoparticles as biocatalyst for biodiesel production. Energy & Fuels 23(3): 1347–1353.

Xiu, Z.-m., Z.-h. Jin, T.-l. Li, S. Mahendra, G.V. Lowry and P.J.J. Alvarez. 2010. Effects of nano-scale zero-valent iron particles on a mixed culture dechlorinating trichloroethylene. Bioresource Technology 101(4): 1141–1146.

Yan, Q., B. Shu-Juan, M.L. Chang, C. Xiao-Qiang, L. Zhi-Song and G. Jun. 2008. Nanostructured polyaniline/titanium dioxide composite anode for microbial fuel cells. American Chemical Society (ACS) Nano 2(1): 113–119.

Yang, S.-T., X. Wang, G. Jia, Y. Gu, T. Wang, H. Nie, C. Ge, H. Wang and Y. Liu. 2008. Long-term accumulation and low toxicity of single-walled carbon nanotubes in intravenously exposed mice. Toxicology Letters 181(3): 182–189.

Yang, Y., J. Guo and Z. Hu. 2013. Impact of nano zero valent iron (NZVI) on methanogenic activity and population dynamics in anaerobic digestion. Water Research 47(17): 6790–6800.

Yang, Z., R. Huang, W. Qi, L. Tong, R. Su and Z. He. 2015. Hydrolysis of cellulose by sulfonated magnetic reduced graphene oxide. Chemical Engineering Journal 280: 90–98.

Ye, P., Z.-K. Xu, J. Wu, C. Innocent and P. Seta. 2006. Nanofibrous membranes containing reactive groups: electrospinning from poly(acrylonitrile-co-maleic acid) for lipase immobilization. Macromolecules 39(3): 1041–1045.

Yeon, J., I. Choi and M. Choi. 2018. Carbon nano-onions synthesized by laser pyrolysis as catalyst support in proton exchange membrane fuel cells. Paper Read at Meeting Abstracts.

Zada, B., T. Mahmood and M. SA. 2013. Effect of iron nanoparticles on hyacinths fermentation. International Journal of Science 2: 106–121.

Zang, L., J. Qiu, X. Wu, W. Zhang, E. Sakai and Y. Wei. 2014. Preparation of magnetic chitosan nanoparticles as support for cellulase immobilization. Industrial & Engineering Chemistry Research 53(9): 3448–3454.

Zhang, L., L. Zhang and D. Li. 2015. Enhanced dark fermentative hydrogen production by zero-valent iron activated carbon micro-electrolysis. International Journal of Hydrogen Energy 40(36): 12201–12208.

Zhang, Q., W. Yang, N. Man, F. Zheng, Y. Shen, K. Sun, Y. Li and L.-P. Wen. 2009. Autophagy-mediated chemosensitization in cancer cells by fullerene C60 nanocrystal. Autophagy 5(8): 1107–1117.

Zhang, Y. and J. Shen. 2007. Enhancement effect of gold nanoparticles on biohydrogen production from artificial wastewater. International Journal of Hydrogen Energy 32(1): 17–23.

Zhao, W., J. Zhao, G.D. Chen, R. Feng, J. Yang, Y.F. Zhao, Q. Wei, B. Du and Y.F. Zhang. 2011. Anaerobic biohydrogen production by the mixed culture with mesoporous Fe_3O_4 nanoparticles activation. Paper Read at Advanced Materials Research.

Zhao, W., Y. Zhang, B. Du, D. Wei, Q. Wei and Y. Zhao. 2013. Enhancement effect of silver nanoparticles on fermentative biohydrogen production using mixed bacteria. Bioresource Technology 142: 240–245.

Zhao, Y. and Y. Chen. 2011. Nano-TiO_2 enhanced photofermentative hydrogen produced from the dark fermentation liquid of waste activated sludge. Environmental Science & Technology 45(19): 8589–8595.

Zheng, H., J. Yin, Z. Gao, H. Huang, X. Ji and C. Dou. 2011. Disruption of *Chlorella vulgaris* cells for the release of biodiesel-producing lipids: a comparison of grinding, ultrasonication, bead milling, enzymatic lysis, and microwaves. Applied Biochemistry and Biotechnology 164(7): 1215–1224.

Zhong, C., C. Wang, F. Huang, F. Wang, H. Jia, H. Zhou and P. Wei. 2015. Selective hydrolysis of hemicellulose from wheat straw by a nanoscale solid acid catalyst. Carbohydrate Polymers 131: 384–391.

Zhou, F., Y. Wang, L. Wang, Y. Wang, W. Chen, C. Huang and M. Liu. 2018. Synthesis and characterization of nanostructured t′-YSZ spherical feedstocks for atmospheric plasma spraying. Journal of Alloys and Compounds 740: 610–616.

Zhou, F., L. Ouyang, M. Zeng, J. Liu, H. Wang, H. Shao and M. Zhu. 2019. Growth mechanism of black phosphorus synthesized by different ball milling techniques. Journal of Alloys and Compounds 784: 339–346.

Chapter 2

Recent Advances in Enzyme Immobilization Using Nanomaterials and its Applications for the Production of Biofuels

Sujit Sadashiv Jagtap[1,2,*] and *Ashwini Ashok Bedekar*[1]

1. Introduction

Enzymes play a vital role as the catalyst to accelerate the rate of reaction with remarkable specificity and efficiency in mild to harsh conditions of temperature, pH and solvents (Blanco and Blanco 2017). Enzymes, as catalysts, have applications in industrial, biomedical and analytical processes. The applications of enzymes at the commercial level are limited because of high cost and low reusability. In addition to that, many enzymes are active in mild conditions, unstable at high temperatures and require an aqueous medium for their function, which diminishes their applications in various fields. On the other hand, in the immobilized state, activity, specificity and stability of the enzymes have found to be improved.

Immobilization of enzymes offers multiple advantages over free enzymes, including improved stability against solvents and temperature, increased functional efficiency, enhanced reproducibility, continuous use, reusability, lower reaction time, fewer chances of contamination in a product, a higher enzyme to substrate ratio and also use of the multi-enzyme system to obtain the desired product is possible. In addition to that, this process is cheap, easy to use and environment-friendly (Ansari and Husain 2012, Mohamad et al. 2015, Sirisha et al. 2016). On the other side, it has disadvantages including high cost for isolation and purification of active

[1] Department of Chemical and Biomolecular Engineering, University of Illinois at Urbana-Champaign, 600 S. Mathews Ave., Urbana, IL 61801, USA.

[2] DOE Center for Advanced Bioenergy and Bioproducts Innovation, University of Illinois at Urbana-Champaign, 600 S. Mathews Ave., Urbana, IL 61801, USA.

* Corresponding author: jagtap@illinois.edu

enzyme, lower or loss of activity of some enzymes due to conformational changes after binding to support, low stability and alteration in their properties and reduced efficiency against insoluble substrates such as cellulose (Singh et al. 2013, Singh et al. 2011).

The four main enzyme immobilization methods including adsorption or carrier binding, entrapment, covalent binding or cross-linking and membrane confinement have been used successfully in the last two decades (Sirisha et al. 2016). Almost all enzymes are proteins, and proteins are composed of amino acids. The enzyme immobilization methods take advantage of functional groups present in the side chain of amino acid. These amino acids bind with a carrier through various types of interactions including the weak bonds (ionic interaction, hydrogen bonds and Van der Waal forces) and covalent bonding. In the covalent bonding method, the reactive functional group of carriers forms covalent bonds with functional groups of protein (Brena and Batista-Viera 2006, Sheldon 2007a).

Protein engineering has been widely used to overcome the limitations of natural enzymes. It has been used to optimize enzyme traits such as chemoselectivity, regioselectivity, stereoselectivity, thermostability and tolerance toward organic solvents. The rational protein design, semi-rational protein design and site-directed mutagenesis are used to improve the trait of the enzyme. The recent advancement in understanding the protein structure coupled with a good screening system has been used to change the enzyme property of interest. The choice of enzyme immobilization method depends upon desired enzyme property, existing structural and mechanistic information of protein and high throughput screening method (Singh et al. 2013).

The characteristics of materials used for immobilization of enzymes have the utmost importance. The ideal support or carrier used for enzyme immobilization should be cost-effective, eco-friendly, inert, highly stable, thermally and mechanically resistant, should be able to regenerate, be resistant to microorganisms and be enzyme-specific (Sirisha et al. 2016). The support or matrix used for the enzyme immobilization allows the exchange of medium containing substrate, cofactor or product molecules.

The advancement in nanotechnology over the last few decades led to the development of various nanomaterials that have important applications in different areas. The nanostructured materials including nanoparticles, nanofibers, nanocomposites and nanotubes are favored as excellent support for enzyme immobilization (Asuri et al. 2006, Kim et al. 2006a, Mohamad et al. 2015). They are expensive but still applicable as they offer ideal characteristics that determine the efficiency of enzymes. They inherently have a large surface area and high mechanical properties, which allows high enzyme loading with minimum diffusion limitation. The nanomaterials-based immobilization have multiple advantages as compared to conventional immobilization techniques, such as easy synthesis in high solid loading, control over synthesis of particle size and homogeneity (Ansari and Husain 2012, Saifuddin et al. 2012, Verma et al. 2013b, Wang et al. 2010, Wang et al. 2009a). In addition to that, the nanoparticles have optical, electrical, thermal, chemical, mechanical and catalytic properties (Ansari and Husain 2012, Kadam et al. 2018, Kadam et al. 2016, Kadam et al. 2017, Liberman et al. 2014, Verma et al. 2013b).

Enzymatic cascade reactions catalyzed by immobilized multi-enzymatic systems have been emerging technology of great interest (Xia et al. 2015, Xue and Woodley 2012). Multistep cascade reactions have many advantages over conventional single enzyme immobilization techniques. Multi-enzymatic immobilized systems have advantages, such as easy operation, high catalytic rate, cofactor regeneration, synergistic actions and less substrate and product inhibition (Ji et al. 2016).

Not surprisingly, the synthesis of suitable nanocarrier for single or multi-enzyme immobilization and engineering of proteins for improved function has attracted significant attention. There are multiple good review papers and books providing detailed information about enzyme immobilization techniques, carriers and their applications (Al-Lazikani et al. 2001, Asuri et al. 2006, Bernal et al. 2018, Blanco and Blanco 2017, Celinska and Grajek 2009, Hanefeld et al. 2009, Hedstrom 2010, Kim et al. 2008, Liese and Hilterhaus 2013, Puri et al. 2013, Sheldon and van Pelt 2013, Verma et al. 2013a, White and White 1997, Wong et al. 2009). We are mainly focusing on the recent developments in nanocarrier, the combination of protein engineering and immobilization methods for improved enzyme function and advancement in multi-enzymatic immobilization systems for biofuels production.

2. Techniques of Enzyme Immobilization

The selection of appropriate immobilization method is the most important part of the immobilization process. The enzyme properties such as catalytic activity, enzyme deactivation, reusability, the toxicity of immobilization materials and the cost of the immobilization process are considered while choosing an enzyme carrier for a particular reaction. There are four principal techniques used for immobilization of enzymes including entrapment, adsorption, covalent linking and crosslinking (Figure 1) (Adlercreutz 2013, Hanefeld et al. 2009, McMorn and Hutchings 2004).

2.1 Entrapment

The pore size of the matrix has an important role in entrapment. The porous structure offers a high surface area for enzyme immobilization and holds the enzyme in a place for a longer time. The enzymes are entrapped in support, inside of fibers or in membranes which allow the transfer of substrate and product passing through it but restrict the enzymes in a confined place (Datta et al. 2013). Entrapment also provides the microenvironment required for the enzyme and helps to lower the enzyme leaching. Also, in enzyme immobilization by entrapment method, there is no need for any bonding between enzyme and carrier matrix. The disadvantages of enzyme immobilization by entrapment include mass transfer limitations, enzyme leakage and low loading capacity. The most used entrapment method is the gelation of poly-cationic or poly-anionic polymers by adding multivalent counterions. The alginate, polyacrylamide, gelatin and collagen are few examples of polymers that are used as a matrix (Won et al. 2005).

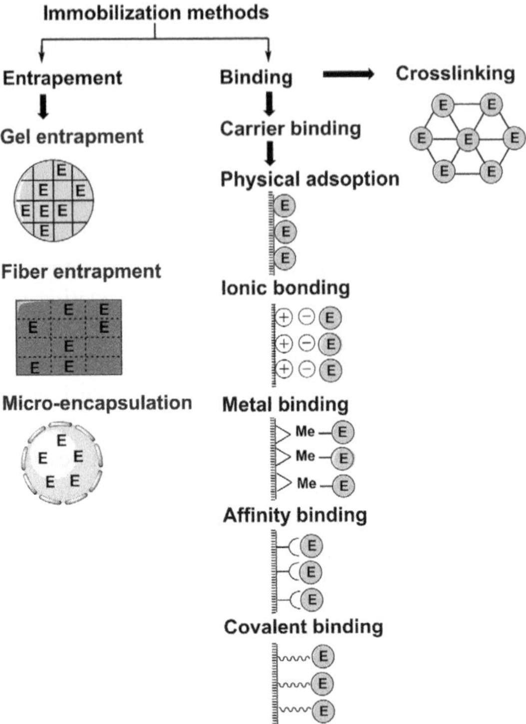

Figure 1. The enzyme immobilization methods.

2.2 Cross-linking

It is an irreversible method of immobilization and is also termed as carrier-free immobilization (Subramanian et al. 1999). The intermolecular cross-linkages between the enzyme and multifunctional reagents form a cross-linking. The most commonly used cross-linking reagent is glutaraldehyde which cross-links proteins via a surface amino group of lysine residues of the enzyme (Hanefeld et al. 2009). The cross-linking agent has low molecular weight compared with that of the enzyme which offers high enzyme activity and low production cost.

2.3 Carrier Binding

This is a simple immobilization method based on adsorption or ionic bonding. In the physical adsorption method, the enzyme is attached to the carrier through hydrogen bonding, van der Waals forces or hydrophobic interactions (Datta et al. 2013). The advantage of physical adsorption is the immobilization of enzymes takes place externally and does not have any diffusion limitations. Salt linkages are used for the binding of enzymes through ionic bonding. This method is easy to perform and can be reversed by changing the conditions that influence the strength of the interaction. However, the optimum conditions of pH and optimum temperature of enzymes can be changed because of partition or diffusion (Sirisha et al. 2016).

In chelation or metal-binding method, heating or neutralization is used for precipitation of the transition metal salts or hydroxides onto the carrier. This method is quite simple and high specific activities of the immobilized enzyme can be obtained; however, the operational stabilities are highly variable (Kennedy and Cabral 1995). In the disulfide bond method, enzyme bearing exposed non-essential thiol groups is immobilized onto thiol-reactive carriers under mild conditions. The reactivity of thiol groups can be modulated, and activity yield is usually high considering the thiol-reactive adsorbent is used (Ovsejevi et al. 2013).

2.4 Covalent Bonding

Immobilization of enzymes to an insoluble polymer by a covalent bond is the most widely used method for irreversible enzyme immobilization. The enzyme properties of the covalently bonded enzyme depend on several factors, such as the size and shape of a carrier, the composition of carrier and conditions used during the coupling method (Hanefeld et al. 2009). The direction of the enzyme binding determines the enzyme activity, and high activity is achieved when the catalytic amino acid residues are not involved in the binding with the support. The coupling of support depends on active groups present in the enzyme. The carrier is modified to generate activated groups or by adding reactive functional groups without modifications. The nucleophiles of protein react with generated electrophilic groups of the carrier. The functional group of enzymes that provide chemical groups to form covalent bonds are α-amino group of N terminal and α-carboxyl group of C terminal, ε-amino groups of lysine and arginine, α- and β-carboxyl groups of aspartate and glutamate, phenol ring of tyrosine, thiol group of cysteine, hydroxyl group of serine and threonine, indole ring of tryptophan and imidazole group of histidine (Brena and Batista-Viera 2006, Sheldon 2007a).

3. Surface Analysis Techniques to Study Enzyme-Carrier Interaction

The enzymes undergo conformational changes following an immobilization on a carrier. Therefore, it is necessary to understand the orientation of enzymes on the surface, functionality and homogeneity of surface coverage to ensure high enzyme stability and activity. Multiple techniques are available to obtain information about enzyme interaction with support surface and interaction between enzymes. These techniques provide crucial qualitative and quantitative information to check the efficacy of enzyme immobilization strategies.

Thermal gravimetric analysis (TGA) has been used to get information about the thermal stability of the carrier used in immobilization by measuring the decomposition of the sample. It is used to estimate a new structure of the carrier, reaction rates and the number of functional groups attached to support (Huang and Cheng 2008, Rajan et al. 2008). The surface information on nanomaterials in nanometer (nm) scale resolution is obtained by electron microscopy techniques such as scanning electron microscopy (SEM) or field emission scanning electron microscopy (FESEM). SEM observes the morphological changes to confirm enzyme immobilization. FESEM visualizes morphology changes in the entire surface or fraction of the surface. Transmission electron microscopy (TEM) obtains structural information, morphology and particle size. It has been widely used to check the size of supports, including carbon nanotubes and nanoparticles (Bai et al. 2006, Liu et al. 2005, Song et al. 2013). Surface plasmon resonance (SPR) method is a ruler of distance for dynamic biological processes, and it is used to measure the maximal binding and surface coverage as the concentration of enzyme in solution is stepped up and down (Schuck et al. 1998).

Circular dichroism (CD) spectroscopy is another commonly used surface analysis technique used to examine the conformation and stability of proteins in different conditions, like temperature or ionic strength. In addition to that, it also provides information on protein-ligand interaction, protein-protein interaction and kinetic and thermodynamic information (Greenfield 2006, Ribeiro and Ramos 2005). It analyzes secondary structure components such as α-helix, β-turns, parallel sheets, antiparallel sheets and random coils of immobilized enzyme compared to a free enzyme (Kumar et al. 2018). Atomic force microscopy (AFM) is a microscopic technique used to visualize biomolecules at the single-molecule level in liquid. It checks the surface topography of enzyme molecules on the surface of the support (Binnig et al. 1986, Liese and Hilterhaus 2013). Microcalorimetry can provide direct information on the stability and biological activity of immobilized proteins (Battistel and Rialdi, 2006). Isothermal batch and flow calorimetry observe the effects of the immobilization and the type of entrapment on the active site of the enzyme (Koenigbauer 1994).

Scanning electrochemical microscopy (SECM) detects redox reaction catalyzed by an immobilized enzyme. It measures the kinetic parameters of the electron transfer reaction at the surface of a carrier (Liu and Mirkin 2001). The time of flight secondary ion mass spectroscopy (TOF-SIMS) has been used for mapping the surface from protein-adsorbed materials to analyze the distribution pattern of a selective protein. It is also used to get surface information including composition, structure, orientation and distribution of molecules. It is used in the analysis of steric hindrance, lipid-lipid and lipid-protein interaction and enzyme immobilization (Belu et al. 2003, Zheng et al. 2008). Altogether, these techniques provide useful information about protein-nanomaterial interactions to optimize the overall enzyme immobilization system.

4. Protein Engineering for Enzyme Modifications

The amino acid sequence information of protein is enough to predict the secondary and tertiary structure using various algorithms (Al-Lazikani et al. 2001, Zhang 2008). This information advancement is crucial for the rational design of proteins with unknown structures. The advancement in the molecular docking and simulation of small molecules facilitated the study of protein-ligand interactions for the rational design of proteins for a property of interest (Figure 2) (Meng et al. 2011).

Protein engineering has been generating the enzymes with improved properties at specific conditions by directed evolution, rational design and genetic modifications. The combination of protein engineering and enzyme immobilization has been used for enhancing the performance of the enzyme (Sheldon and Pereira 2017, Sheldon and

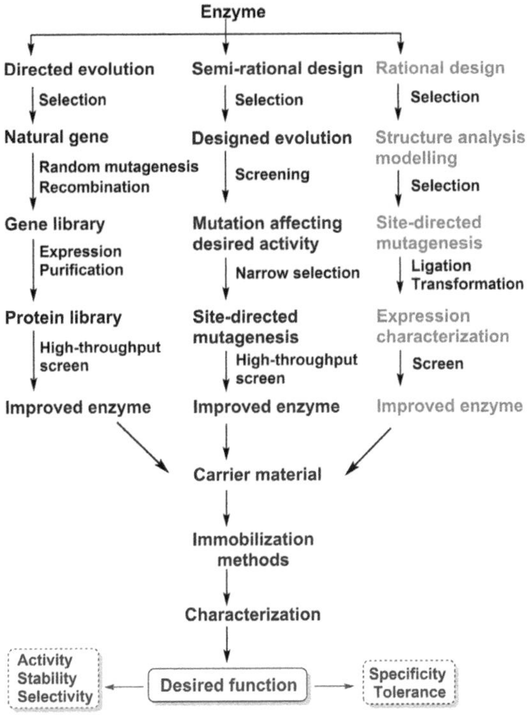

Figure 2. Integration of protein engineering and enzyme immobilization for improvement in properties of the enzyme.

van Pelt 2013, Sheldon and Woodley 2018). Both fields are developing in parallel for tailoring traits of the enzyme and optimizing binding chemistry for immobilization. The enzymes from natural sources are efficient in native conditions, but they become inefficient when exposed to industrial bioprocess in harsh conditions. Therefore, modifications of the enzymes by protein engineering through changes in amino acid residues improve the enzyme functions (Sheldon and Woodley 2018). The disadvantage of protein engineering is the requirement of structural and functional information.

Directed evolution method mimics natural evolution and it is independent of structural information of the protein. It works as the iterative cycle for gene diversity generation, screening or selecting of desired variants and analysis of desired variants. The better performing variant gets selected for a new round of iterative cycles (Farinas et al. 2001). Ration design is a more classical approach in protein engineering that relies on structural and functional information of the existing protein. It limits the screening of the mutagenesis library. The combination of directed evolution and protein engineering has demonstrated to improve enzyme function, including thermal stability, altered pH and altered tolerance toward solvents (Bernal et al. 2018).

Thermomyces lanuginosus lipase (TLL) was engineered to improve the hydrolytic activity, enantioselectivity and organic solvent tolerance using site-saturation mutagenesis coupled with high-throughput screening method (Skjold-

Jørgensen et al. 2017). The active site of lipase was covered with a lid and activated when the enzyme was in contact with the hydrophobic surface. This activation mechanism of TLL had been altered through site-directed mutagenesis and resulted in lipase with increased activation in contact with hydrophobic surface (Skjold-Jørgensen et al. 2014). Directed evolution method was used to increase the tolerance of lipase toward methanol for biodiesel production. Iterative saturation mutagenesis targeted high B-factor amino acids to enhance methanol tolerance in TLL. The new hydrogen bond formed by double mutant showed 30% more activity than TLL and retained 70% of activity after incubation with methanol (Tian et al. 2017).

5. Improvement in Enzyme Properties

5.1 Enzyme Activity

Immobilized enzymes exhibit higher activity than the free enzymes. The improvement in enzyme activity depends on several factors including microenvironment (pH and temperature), diffusion effect, enzyme orientation, conformational change in the enzyme after binding to a carrier and tolerance toward substrate and product (Hanefeld et al. 2009). The correct orientation of enzymes, where catalytic residues are not exposed to support, exhibits higher activity and stability as compared to free enzyme. The pore volume is available for the diffusion of substrate and the enzyme is reduced at high enzyme loading. The addition of a spacer to the carrier for enzyme immobilization has resulted in improved activity due to the conformational flexibility of the enzyme (Shim et al. 2017). The conformation changes in enzyme led to a decrease in affinity to the substrate.

The higher enzyme activity was achieved for allosteric enzymes, such as lipase and cellulase (Adlercreutz 2013, Hanefeld et al. 2009). When the active site of lipase was covered with a lid in closed confirmation led to inactivation of lipase, and in an open confirmation, the catalytic site was exposed to reaction medium (Adlercreutz 2013, Hanefeld et al. 2009, McMorn and Hutchings 2004). The lipase from *Mucor risopus* immobilized in organic solvent was more active than the immobilized inactive lipase in an aqueous medium. The position of binding of enzymes to support is different in the organic solvent and an aqueous medium (Stark and Holmberg 1989).

Lipase and cellulase are active on the insoluble substrate that is commonly used in the industry. The mobility of these enzymes at the surface of the substrate is limited by diffusion and enzyme desorption. The conditions in the interface between the enzyme and substrate and the surface properties of enzymes determine the substrate turnover. The substitution of Glu87 and Trp89 in the lid region of the lipase changed the hydrolytic activity. The hydrolytic activity at the water/lipid interface was substrate-dependent, and amino acid replacement led to an altered chain length specificity (Martinelle et al. 1996). *T. reesei* cellobiohydrolase II is exoglucanase cleaving cellobiose units from the non-reducing end of cellulose. In addition to catalytic residue Asp221, Tyr169 is involved in the distortion of glucose ring into a more reactive conformation (Koivula et al. 1996).

5.2 Thermal Stability

The stability of immobilized enzyme depends on multiple factors including the structure of carrier, interaction with the support, properties of a spacer, pH, temperature, solvent, the presence of surfactants, the orientation of enzyme, conformational flexibility and conditions used for the immobilization (Singh et al. 2013). Many enzymes get unfolded or inactivated at a higher temperature. In the cross-linking method, the activity of an immobilized enzyme is decreased to compensate for increased stability. The binding of the enzyme to carrier introduces additional covalent or non-covalent interactions. The number of bonds in multipoint attachment between carrier and enzyme determines the extent of enzyme stabilization. The additionally introduced bonds decreased the structural flexibility and liability for enzyme denaturation and increased the rigidity of the enzyme. The multiple covalent bonds between carrier and enzyme reduced the conformational flexibility and thermal vibrations and thus protecting the enzyme from denaturation (Wong et al. 2009).

The combination of directed evolution and site-directed mutagenesis has been frequently used to engineer the enzyme to improve thermostability (Packer and Liu 2015). It is impossible to cover the entire mutational library of a typical protein. The crucial part is the screening method used to identify library members with desired properties. Random mutagenesis provides a high chance of functional library members screening in the absence of enzyme structure information. The chemical and physical agents have been used to damage DNA (Packer and Liu 2015). Defined methods have been used for mutagenesis of structurally characterized protein. Recently, recombination-based methods are used for protein evolution (Packer and Liu 2015).

The thermal stability of the enzyme includes three different types including thermodynamic, process stability and kinetic stability (Kumar et al. 2000). Multiple factors including hydrophobic interactions and hydrogen bonds affect the thermostability of enzymes. Several methods have been used to improve the thermal stability of the enzyme, including the introduction of disulfide bonds, salt bridges, hydrogen bonds, optimizing surface charge of the protein, optimizing the free energy of protein unfolding and chemical cross-links (Verma et al. 2012, Yang et al. 2015). The single or few point mutations also largely changed its thermal stability (Packer and Liu 2015).

The shorter spacers are preferred for lipase immobilization to increase thermal stability because they restrict the lipase mobility and prevent the unfolding of lipase. The B-factor strategy was used to improve the thermostability of *Candida antarctica* lipase. The half-life of Ala162 mutation was increased by 4.5 fold (Le et al. 2012). The β-1,4-glucosidase from *Agaricus arvensis* covalently immobilized onto functionalized silicon oxide nanoparticles that showed higher optimum temperature and even enhanced thermostability. The enzyme half-life improved 288 fold over the free enzyme (Singh et al. 2011).

5.3 Solvent Stability

The use of enzymes in the organic solvent system, instead of aqueous media for enzymatic reactions, has increasingly drawn the attention of researchers. It offers several advantages such as the increase in solubility of hydrophobic substrates like lipids and phospholipids, change in substrate specificity, easy product recovery, reduced microbial contamination in reaction and use of novel chemistry for difficult reactions (Arnold 1990, Khmelnitsky and Rich 1999, Singh et al. 2013). In addition to that, it offers an advantage through limitations on water-dependent side reactions. However, the majority of available enzymes get inactivated in non-aqueous solvents. Solvents such as acetone or dimethylformamide reduce the catalytic activity of the enzyme by interacting with the enzyme and water molecules.

Multiple strategies have been used to increase the activity and lifetime of enzymes in the presence of aqueous media. The three main effectively used strategies include screening and isolation of novel enzymes functioning in non-aqueous solvents, engineering of enzymes to improve tolerance toward non-aqueous solvents and modification of solvent-enzyme reaction conditions to decrease the denaturing effect on enzymes (Stepankova et al. 2013). The aggregated enzymes in organic solvents are less accessible to the substrate. The immobilized enzyme is in limited contact with molecules that are soluble in the aqueous phase but not in contact with the organic phase. The immobilized enzyme inside the porous solid carrier provides the operational stability to the enzyme (Singh et al. 2013).

Enzymes have been redesigned by engineering their amino acid sequences to improve catalysis in non-aqueous solvents. Physical and chemical properties of engineered enzymes suit the new non-aqueous environment. Directed evolution and rational designs are widely used strategies to engineer proteins that are suitable for catalysis in the non-aqueous environment. The amino acid sequence modifications lowered the free energy of the active form of the enzyme or raised the free energy of inactive forms in the presence of solvents (Arnold 1990). The compatibility of the protein surface with a solvent has changed by removing surface charges and hydrogen bonding sites. The conformational stability of protein has improved by introducing new disulfide bonds, hydrogen bonds and electrostatic interactions (Arnold 1990).

Rhizopus oryzae lipase showed good thermal stability, increased activity in acetone and high solvent tolerance when immobilized onto graphene oxide nanocarrier. The covalently immobilized lipase exhibited better resistance to heat inactivation in comparison with the free enzyme. In addition to that, the immobilized enzyme retained high activity over 100% of initial activity in multiple polar solvents (Hermanova et al. 2015).

5.4 Selectivity and Specificity

Enzymes are the highly selective catalysts that are capable of choosing a single substrate from multiple closely related substrates (Hedstrom 2010). Specificity is the most important characteristic of an enzyme. The enzyme specificity is expressed in the rate at which the substrate is converted to product, and it arises from the three-dimensional structure of an active site. A good enzyme can induce an active conformation that cannot be accessed by a secondary substrate. The enzyme restricts its conformation so that the reaction is channeled in one direction to yield a single product. The additional active site of the enzyme that catalyzes the proofreading reaction enhances the specificity (Hedstrom 2010). The selectivity of enzyme includes the substrate selectivity, stereoselectivity and regioselectivity. The difference in activation energy forms an unequal mixture of stereoisomers from the single reactant. The enzymes also show regioselectivity as they give predominance to one product between two products, major product and minor product.

The modulation of the enzyme-substrate selectivity has been achieved using immobilization techniques. The enantioselectivity of *candida* lipase in transesterification in organic solvents has been modulated using the diffusion-controlled immobilization technique (Palomo et al. 2007, Palomo et al. 2004). The enantioselectivity of lipase is different when different support is used for immobilization (Mateo et al. 2007).

6. Recent Advances in Nanomaterials for Enzyme Immobilization

The recent advances in nanotechnology have made enzyme immobilization affordable (Puri et al. 2013, Verma 2017). These nanomaterials have advantages over conventional carriers including large surface areas, control on pore size, a balanced ratio of hydrophobicity and hydrophilicity and surface chemistry (Kim et al. 2008). The larger surface area and controlled pore size of nanomaterial allow high enzyme loading which increases the enzyme activity per unit mass. The shape and size of nanomaterials increase enzyme immobilization efficiency and recycling stability (Hwang et al. 2011). Moreover, these nanomaterials provide good features such as plasticity, an increase in strength and chemical bonding. There are several organic and inorganic nanomaterials that have been used for enzyme immobilization.

The major advantages of using immobilized enzymes over soluble enzymes are the recycling and stability of immobilized enzymes (Diego et al. 2017). Also, immobilization adapts enzymes to exploit their properties such as selectivity and specificity which are industrially important (Diego et al. 2017). Different techniques of enzyme immobilization such as physical adsorption, hydrophobic interactions, covalent binding, hydrogen bonding and encapsulation or cross-linking—on or indifferent nanocarrier including polymer, silica, carbon and metal-based—have been developed magnificently over the time to develop efficient nanocatalysts (Figure 3) (Table 1). Physical and chemical properties of a carrier and the enzyme play an important role in the development of nanobiocatalyst.

6.1 Polymer Nanocarriers

There are many important applications of polymer-carriers such as easy production in higher quantities, high stability and functionality of nanobiocatalyst assembly (Lee et al. 2012, Xu et al. 2013). Because of the large surface area, nanopolymers can accommodate high enzyme loading, and their distinctive properties help enzyme to remain stable, active and enhance their biological activity (Neri et al. 2011). But the pore size, porosity and molecular structure of conventional polymeric carriers impact negatively on their immobilization efficiency (Misson et al. 2015). Also, the major drawbacks of polymer-supported immobilization include high cost, resistance to diffusion because of size, infrastructure, the time required for the overall process

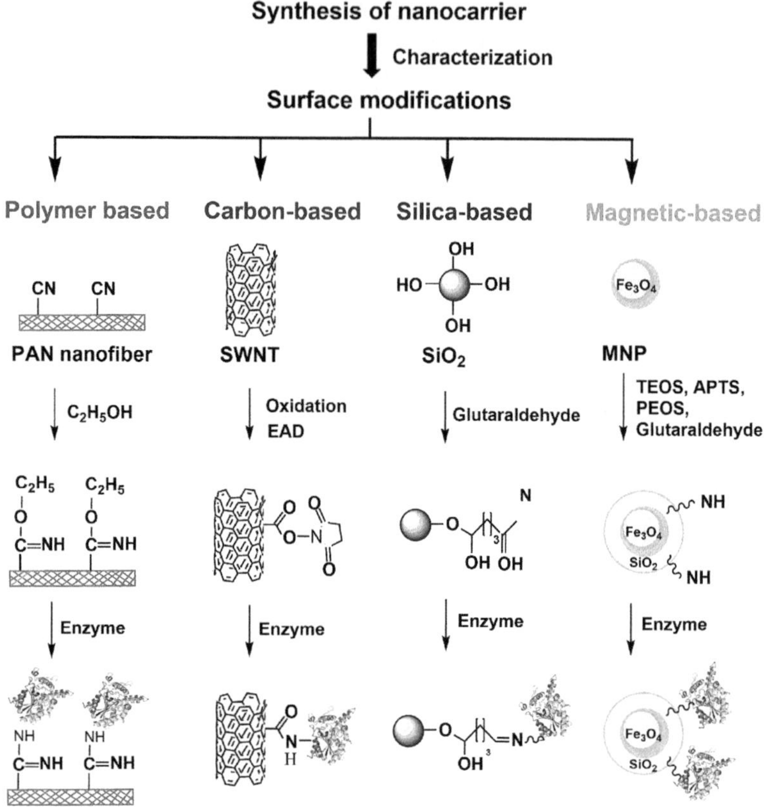

Figure 3. Preparation of nanocarriers for enzyme immobilization.

and the expertise required to assemble these polymers (Wu et al. 2005). The various methods like solvent evaporation, dialysis, salting-out and supercritical fluid technologies are used for the preparation of polymer nanocarriers.

Nanofibers have been used previously as enzyme carriers for the production of biofuels and bioproducts. The distinctive properties such as interconnectivity, porosity and easy separation of nanofibers make them good hosts for enzyme immobilization (Wang et al. 2009b). Physical adsorption is the most common method used for the immobilization of enzymes onto nanofibers. The main drawback of the physical adsorption method is the leaching of the enzyme. Enzyme immobilization on nanofibers by covalent binding is a successful and efficient way of immobilization (Wu et al. 2005).

6.2 Carbon Nanocarriers

Carbon-based nanomaterials have been widely used for enzyme immobilization due to their biocompatibility, inertness and thermal stability (Gao and Kyratzis 2008). Currently, carbon-based nanomaterials are considered as one of the most fascinating nanomaterials having different attractive forms, including single and multi-walled

Table 1. The list of nanocarriers used for the immobilization of enzymes.

Nanomaterial		**Enzyme Source**	**Immobilization Method**	**Modified Feature**	**Product**	**Reusability**	**Reference**
Polymer-based nanocarrier	Onto PAN nanofibers	Lipase from *Candida rugosa*	Covalent linkage	High operational stability	90% of biodiesel conversion from soybean oil	91% of initial activity retained after 10 repetitive cycles	(Li et al. 2007)
	Cellulose nanofiber	Lipase from *Candida rugosa*	Covalent linkage	Improved stability	ND	30% activity remained after 8 cycles	(Huang et al. 2011)
Carbon-based nanocarrier	Multi-walled carbon nanotube	Lipase from *Candida rugosa*	Adsorption method	High retention of lipase activity	ND	ND	(Domínguez de María et al. 2006)
Silica-based nanocarrier	Silica nanoparticles	Lipase from *Mucor javanicus*	Covalent linkage	High stability and lipase loading	ND	ND	(Kim et al. 2006)
	Silica nanoparticles	Cellulase	Physical adsorption	Improved stability	Ethanol	ND	(Lupoi and Smith 2011)
Magnetic or composite nanocarrier	Manganese oxide nanoparticles	Cellulase from *Aspergillus fumigatus* JCF	Covalent linkage	Higher thermal, pH, temperature stability	Ethanol	60% of initial activity was retained after 5 cycles	(Cherian et al. 2015)
	Ferric oxide magnetic nanoparticles	APTES-SiO_2-Fe_3O_4 and MPTMS-Fe_3O_4 lipase	Covalent linkage	Faster reaction rate, easy separation	89% and 81% of biodiesel yield	Activity lost during repetitive use	(Thangaraj et al. 2016b)
	Ferric oxide magnetic nanoparticles	APTES- SiO_2-Fe_3O_4 and MPTMS-Fe_3O_4 lipase	Covalent linkage	Faster reaction rate, easy separation	84% and 83% of biodiesel yield respectively	Activity lost during repetitive use	(Thangaraj et al. 2016a)

Table 1 contd. ...

...Table 1 contd.

Nanomaterial		**Enzyme Source**	**Immobilization Method**	**Modified Feature**	**Product**	**Reusability**	**Reference**
	Magnetic nanoparticles	Lipase from *Thermomyces lanuginosus*	Covalent linkage	High operational stability, simple separation	97.2% FAME yield	80% of initial activity was retained after 5 cycles	(Raita et al. 2015)
	Magnetic multiwalled carbon nanotubes	Lipase from *R. miehei*	Covalent linkage	Higher stability	Up to 94% biodiesel conversion	94% of initial activity was retained after 10 cycles	(Fan et al. 2016)
	Supermagnetic iron oxide nanoparticles	Lipase from *B. cepacia*	Covalent linkage	Higher activity	Up to 91% biodiesel conversion	Activity was slightly lost during reuse	(Karimi 2016)
	Ferric oxide magnetic nanoparticles	Lipase from isolated strain *Burkholderia* sp. C20	Covalent linkage	Higher conversion and reusability	90% FAME yield	Up to10 cycles without any loss in activity	(Tran et al. 2012)

PAN – Polyacrylonitrile, ND – not reported, ATPES – (3-Aminopropyl)triethyoxysilane, MPTMS – (3-Mercaptopropyl)trimethoxysilane.

carbon nanotubes, fullerenes, nanofibers and carbon nanoparticles (Verma et al. 2013b). Among them, carbon nanotubes are widely used for the immobilization of biofuel producing enzymes. A number of covalent binding and physical adsorption methods have been studied for enzyme immobilization using carbon nanotubes for biofuel production. Single-walled carbon nanotubes (SWCNTs) and multi-walled carbon nanotubes (MWCNTs) have been used for the immobilization of enzymes (Figure 3). The MWCNTs are composed of a central tubule surrounded by several graphite layers, while in SWCNTs the central tubule is present without any graphitic layer. The SWCNTs have been known for their higher surface area, while MWCNTs are easily dispersible and inexpensive. In carbon nanotubes, both covalent and non-covalent type of immobilization has been reported earlier (Gao and Kyratzis 2008).

6.3 Silica Nanocarriers

Porous materials are considered good support carriers for entrapment or covalent binding of enzymes because of their natural porosity, pore geometry, thermal stability, biocompatibility, stability in organic solvents and inertness. The designing and manufacturing a desired porous structure with suitable properties and biocompatibility is a challenging task (Liberman et al. 2014). Silica carriers are good porous materials used for enzyme immobilization. The main advantages of using silica nanocarriers are fine pore size, shape and crystallinity adjustment that is possible along with their scalable synthesis (Liberman et al. 2014). Silica nanocarriers are robust and resistant to breakage due to their high tensile strength which makes them suitable for multiple reuses.

The various configuration of silica, such as nanoparticles, nanosheets and nanowires, has been reported for lipase immobilization (Macario et al. 2013, Shang et al. 2015). The surface properties of mesoporous silica nanoparticles including surface charge density, pore structure and functional groups play an important role in the adsorption of enzymes (Du et al. 2013, Takahashi et al. 2000). The hydrophobic/hydrophilic and electrostatic attractions are mainly responsible for the physical adsorption of enzymes on silica nanoparticles. The altered pore size of mesoporous silica creates a suitable microenvironment and improves the enzyme stability in aqueous and organic media.

Ethanol production was found to be increased when silica nanoparticles were used for physical adsorption of cellulase (Lupoi and Smith 2011). When cellulase was physically adsorbed on silica nanoparticles, ethanol yields were 2.1 fold higher compared to the free enzyme. Mesoporous silica generates a suitable environment for immobilization and improves the stability of the enzyme in both the aqueous and organic media, which results in maximum reaction efficiency (Cherian et al. 2015).

6.4 Magnetic or Composite Nanoparticles

The metal-based nanomaterials have been widely used for nanobiocatalysts production. The surfaces of metal nanoparticles are prepared by adding different functional groups, such as carboxylate, amino, phosphate or thiolate, that form a strong binding with enzyme molecules. In some cases, a stable immobilization is achieved by coating a thin layer of polymers onto the surface of metal nanoparticles and then the functional groups interact with enzyme molecules (Figure 3). The magnetic nanoparticles are non-porous, and hence they could cause erosional damage to naked nanoparticles (Sen et al. 2006). The coating of layers of gold/silica/polymer has been used to avoid this reaction. In addition to that, the materials such as PEG, chitosan, polyethyleneimine and polyvinyl alcohol have been used for the coating purpose (Crespilho et al. 2009, Xu et al. 2007). Coating/encapsulation of nanoparticles is useful in nanosupport applications. Magnetic nanoparticles are well known as nanobiocatalysts due to their high reusability and enzyme immobilization capacity.

7. Multi-Enzyme Immobilization

Enzymatic cascade reactions have multiple advantages over traditional stepwise synthesis methods, including short reaction time, high efficiency, less requirement of chemicals and solvents, intrinsically green as it reduces waste, less product inhibition and regeneration of cofactors (Sheldon 2007b, Xue and Woodley 2012). In addition to that, the production of a targeted product using an immobilized multi-enzymatic system was higher because of theproximity of enzymes and reduced diffusion rates.

7.1 Factors Influencing the Selection of a Multi-Enzymatic System

The immobilized multi-enzymatic systems have been influenced by multiple factors, including optimum conditions such as pH and temperature of the coupling reaction, a proportion of each enzyme considering the turnover number, substrate specificity, product inhibition, the synergy of enzymes and the ability of cofactor regeneration (Figure 4) (Ji et al. 2016).

The optimum pH of each enzyme has to be taken into consideration while developing a multi-enzyme system because the enzyme activity is highly influenced by pH. The right combination of close optimum conditions such as acidic-acidic, basic-basic or neutral-neutral is a better performing combination than that of acidic-basic. The change in optimum pH of the enzyme after immobilization has been reported due to alteration in ionization of charged side chains of proteins (Hanefeld et al. 2009).

The immobilization offers a microenvironment to protect the enzyme from thermal inactivation. Immobilized enzymes in the multi-enzyme system can have different optimum temperatures. The right combination of close optimum temperature conditions of the enzymes immobilized on carrier improves the reaction efficiency. The half-life of the enzyme at optimum temperature is also an important parameter which affects reaction performance. Therefore, variable temperature control for a defined period is necessary for the high performing immobilized multi-enzymatic system.

The turnover number is defined as the number of times each enzyme site converts substrate to product per unit time (Eisenthal et al. 2007). The different enzymes show different turnover numbers after immobilization. The quantity of each enzyme can be controlled in the immobilized multi-enzymatic system. The substrate specificity

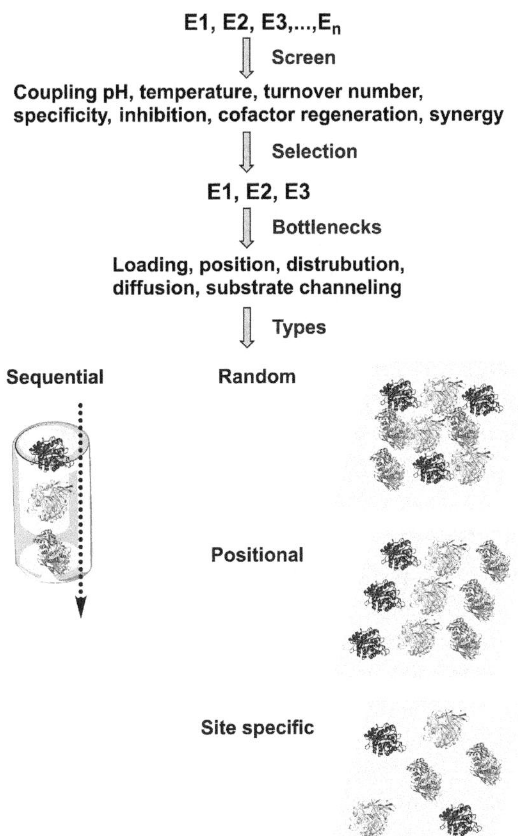

Figure 4. Strategies used for immobilization of multi-enzyme system.

and product inhibition of the enzyme are important parameters for the immobilized multi-enzymatic system. The enzyme should be substrate selective and unaffected by the products of other immobilized enzymes. The use of cofactor-regenerating enzymes reduces the cost of the immobilized multi-enzymatic system (Wichmann and Vasic-Racki 2005).

The synergistic action of the immobilized multi-enzymatic system is a favorable solution for improving the efficiency of a system. The best example is the cellulase system that acts together for the conversion of cellulose to glucose (Cho et al. 2012). The different hydrolase has different specificities, and their co-immobilization enhances the synergistic effect.

7.2 Bottlenecks in Immobilized Multi-Enzymatic System

The overall rate of reaction is influenced by the reaction rate of each enzyme in the multi-enzymatic system. The slow rate of the enzyme is compensated by higher loading; otherwise, it affects the overall rate of reaction. The main bottlenecks in the immobilized multi-enzymatic system are loading of each enzyme, optimum conditions, distribution of enzymes, position and orientation of enzymes and diffusion of substrate and product (Ji et al. 2016).

The addition of the right proportion of the rate-limiting enzyme decides the reaction rate of the immobilized multi-enzymatic system. The optimum conditions are necessary to increase the reaction rate, including pH and temperature, and a suitable enzyme carrier is also necessary (Kuo et al. 2012, Luo et al. 2015). The uniform distribution of enzyme on carriers is preferred for high immobilization yield. The rate-limiting enzymes positioned closer to an external surface of a carrier were found to be less affected by the diffusion rate of substrates and products. The reaction intermediates or the product is transferred from the catalytic site of one enzyme to another one. The orientation of the catalytic site of the enzyme and inter-enzyme distance affects the channeling of substrates. The distribution of substrate and product is uneven in the multi-enzymatic system. The mass transport limitation arises due to lower diffusion rates of substrate and product than the enzyme catalytic rates, and thus lead to the substrate or product inhibition (Fu et al. 2012).

7.3 Types of Immobilized Multi-Enzymatic System

The stepwise/sequential system and co-immobilization enzyme system are the two main types of immobilized multi-enzymatic systems (Figure 3) (Ji et al. 2016). In a sequential system, multiple reactions are catalyzed in a stepwise manner by multiple enzymes. Each enzyme catalyzes one reaction in a separate module. This system offers the flexibility of operation and detection of activity and the stability of each enzyme. In a mixed immobilization system, the loading of each enzyme is controlled and there is less product inhibition on other enzymes. In a co-immobilization multi-enzyme system, the multiple enzymes are immobilized on the same support to increase the reaction rate and decrease the lag time. The detection of stability and activity of each enzyme is difficult in this system. Co-immobilization uses different approaches such as random, positioned and site-specific co-immobilization (Schoffelen and van Hest 2013).

The immobilized multi-enzymatic system is more applicable to redox reaction and hydrolysis reaction. The oxidoreductase requires nicotinamide cofactors such as flavin adenine dinucleotide (FAD), NAD(H) or NAD(P)H or both. The regeneration of cofactors was achieved using multi-enzymatic systems to improve the reaction rate of the system. The combination of hydrolases is more efficient than individual ones (Schrittwieser et al. 2011).

8. Production of Bioproducts and Biofuels

8.1 2,3-Butanediol and Acetoin Production

2,3-Butanediol (2,3-BDO) is a high-value chemical with a heating value of 27.2 kJ g^{-1}. It can be used to produce cosmetics, fumigants, antifreeze agents, transport fuels, polymers, acetoin, diacetyl and solvents such as 1,3-butadiene, methyl ethyl ketone and gamma-butyrolactone (Celinska and Grajek 2009, Ji et al. 2011, Syu 2001). 2-butanol is a stereoisomer of butanol. It has a higher energy density and hygroscopicity as compared to bioethanol. It is used with gasoline without any change to the vehicle system (Nigam and Singh 2011). It is also used as a paint thinner, base for perfumes and component of brake fluids (Nigam and Singh, 2011). Acetoin is used in the food industry as a flavor enhancer and as a building block for the synthesis of chemicals such as pyrazines, diacetyl and acetylbutanediol (Yang et al. 2017).

2,3-BDO and acetoin can be efficiently produced from ethanol using enzymes immobilized on silicon oxide nanoparticles (Figure 5). The immobilized enzymes on silicon oxide nanoparticles are inexpugnable because of the covalent linkages between the enzymes and activated silicon oxide nanoparticles (Singh et al. 2014, Singh et al. 2011). Purified enzymes such as ethanol dehydrogenase (EtDH), formolase (FLS), 2,3-butanediol dehydrogenase (BDH), diol dehydratase (DDH) and NADH oxidase (NOX) were immobilized on glutaraldehyde activated silicone oxide nanoparticles (Zhang et al. 2018). Ethanol dehydrogenase was used to convert ethanol to acetaldehyde which was condensed to yield acetoin by formolase. 2, 3-BDO was produced by the reduction of acetoin by the NADPH dependent 2,3-butanediol dehydrogenase. Diol dehydratase was used for conversion of 2,3-butanediol to 2-butanone which was again converted to 2-butanol by 2,3-butanediol dehydrogenase.

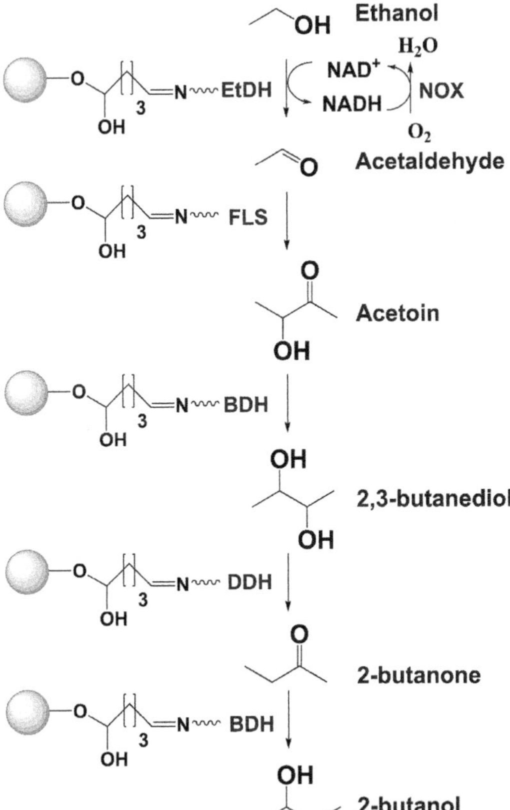

Figure 5. The conversion of ethanol to 2,3-butanediol using the immobilized multi-enzyme system.

NADH oxidase was used to recycle NAD^+. 2-butanol production was hampered by a low activity of diol dehydratase.

Immobilization of enzymes on glutaraldehyde-functionalized silicon oxide nanoparticles enabled acetoin and 2,3-BDO production. Immobilized enzymes exhibited more than 90% immobilization efficiency. The immobilized catalytic system showed 94% of acetoin production even after 10 cycles (Table 2). Likewise, immobilized enzymes showed 73% of 2,3-BDO production after 10 cycles. The recyclability of the multi-enzyme immobilized catalytic system is an important parameter, and this immobilized system can be used for industrial production of acetoin and 2,3-butanediol.

8.2 Biodiesel

Biodiesel is a renewable fuel produced from vegetable oils and fats. Fatty acids and alcohol are transesterified to produce biodiesel (Jagtap and Rao 2018a, Jagtap and Rao 2018b, Zhang et al. 2016). Biodiesel is a clean-burning renewable fuel, blended with petroleum diesel fuel in any proportion. Compared to petroleum fuel, biodiesel has lower toxicity, reduced exhaust emissions and it is safer to handle. Lipid production using microorganisms is advantageous over the animal or plant-based lipids because microbial lipids are produced in a short time, are less labor-intensive and are less affected by location or climate (Jagtap and Rao 2018a, Jagtap and Rao 2018b, Zhang et al. 2016). Oleaginous microorganisms can accumulate 10–70% lipid of their dry cell mass.

Lipases are classified as a subclass of esterase used to catalyze the hydrolysis of fats or lipids. Lipases act mainly on a specific position on the glycerol backbone of lipids (Kim et al. 2018). Lipases are stable in non-aqueous media used for the transesterification process during biodiesel production. The immobilization methods including adsorption, covalent bonding and cross-linking are used for biodiesel production using nano-immobilized lipases (Table 2) (Puri et al. 2013, Verma 2017, Verma et al. 2011, Verma et al. 2013a). Nanoparticles including Fe_2O_3, Fe_3O_4 and silica nanoparticles have been used for the immobilization of lipases (Figure 6) (Dyal et al. 2003, Wang et al. 2009a). The immobilization is based on an amine group of the enzyme with surface acetyl or amine groups of a nanoparticle. Amino-functionalized Fe_3O_4 nanoparticles loaded with lipase showed a broad particle size distribution owing to a large surface area, stability and paramagnetic attraction (Kim et al. 2018). The covalently immobilized lipase was stable and reactive even after 30 days (Dyal et al. 2003).

In one study, a packed bed reactor containing Fe_3O_4 nanoparticles-lipase was developed for biodiesel production from soybean oil (Wang et al. 2009a). Emulsification of soybean oil before methanolysis was important to increase the rate of the reaction. The conversion rate and stability of the four-packed bed reactor system were higher as compared to the single-packed bed reactor system. The conversion rate was 75% after 240 hours of reaction (Wang et al. 2009a). In another study, lipase immobilized on Fe_3O_4 modified by (3-aminopropyl)triethyoxysilane (APTES) used for transesterification of palm oil. The optimized conditions such as 23.2% w/w enzyme loading, 4.7:1 methanol to free fatty acids (FFAs) molar ratio,

Table 2. The list of biofuels and bioproducts produced using immobilized enzymes on nanocarriers.

Product	Carrier	Substrate	Enzyme Source/Strain	Product (g/L)	Conversion (%)	Reusability (Cycles)	Reference
Acetoin	Glutaraldehyde-functionalized silicon oxide nanoparticles	Ethanol (100 mM)	EtDH, FLS:L482S, and NOX	44.39	88.78	0	(Zhang et al. 2018)
	Glutaraldehyde-functionalized silicon oxide nanoparticles	Ethanol (100 mM)	EtDH, FLS:L482S, and NOX	41.72	83.44	10	(Zhang et al. 2018)
2,3-butanediol	Glutaraldehyde-functionalized silicon oxide nanoparticles	Ethanol (100 mM)	EtDH:D46G, FLS:L482S, BDH:S199A, and NOX	44.14	88.28	0	(Zhang et al. 2018)
	Glutaraldehyde-functionalized silicon oxide nanoparticles	Ethanol (100 mM)	EtDH:D46G, FLS:L482S, BDH:S199A, and NOX	32.22	64.44	10	(Zhang et al. 2018)
Biodiesel	Fe_3O_4	Soybean oil	*Pseudomonas cepacia* lipase	–	88	10	(Wang et al. 2009)
	Amino-Fe_3O_4	Palm oil	*Thermomyces lanuginose* lipase	–	97	5	(Raita et al. 2015)
	Amino-Fe_3O_4-SiO_2	Waste cooking oil	*Burkholderia cepacia* lipase	–	91	3	(Karimi 2016)
	Amino-Fe_3O_4-SiO_2	Waste cooking oil	*Burkholderia cepacia* lipase	–	54	5	(Karimi 2016)
		Waste cooking oil	*Rhizomucor cepacia* lipase	–	94	10	(Fan et al. 2016)

Ethanol dehydrogenase (EtDH) from *C. necator*, formolase (FLS) from *P. fluorescens*, 2,3-butanediol dehydrogenase (BDH) from *C. autoethanogenum*, diol dehydratase (DDH) from *L. brevis*, and NADH oxidase (NOX) from *L. rhamnosus*.

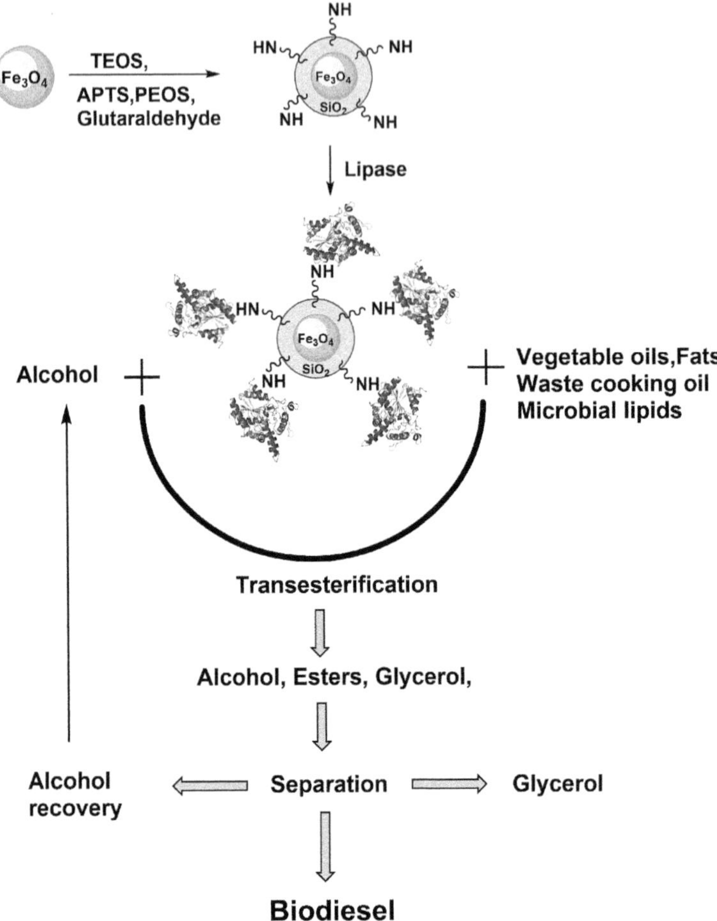

Figure 6. Lipase immobilization on nanocarriers for biodiesel production.

3.4% water content and 1:1 (v/v) *tert*-butanol to oil resulted in 97% FAME yield after 24 hours at 50°C (Raita et al. 2015). The immobilized system showed higher than 80% activity even after recirculation for five batches.

The superparamagnetic iron oxide nanoparticles covered with silica and functionalized by aldehyde groups have been used to immobilize *Burkholderia cepacia* lipase for transesterification of a waste cooking oil with methanol for biodiesel production (Karimi 2016). The conversion of waste cooking oil to biodiesel was 91% under optimized conditions. The conversion efficiency was 54% after five cycles (Karimi 2016, Kim et al. 2018). The dendrimer-coated magnetic multi-walled carbon nanotubes exhibited magnetic properties with amino groups functionalized large surface. It was used to immobilize *Rhizomucor miehei* lipase for biodiesel production using waste vegetable oil (Fan et al. 2016). The transesterification activity was found to be 27 fold higher than the free enzyme. The biodiesel conversion reached 94% under optimal conditions from waste vegetable oil in a *tert*-butanol solvent system. The conversion rate was above 90% after 10 cycles of reuse in a *tert*-butanol system (Fan et al. 2016, Kim et al. 2018).

Candida rugosa lipase was immobilized by covalent method onto polyacrylonitrile (PAN) nanofibers by amidination (Li et al. 2007). The nitrile groups of PAN nanofibers were activated by amidination followed by reaction with lipase solution in phosphate buffer solution. After cutting absolute ethanol and hydrogen chloride treatment, the membrane was washed with distilled water and kept in 1 ml of lipase solution (5 mg/ml) followed by agitation. The unbound enzyme from the membrane was removed by washing with phosphate buffer solution several times. FESEM and FTIR analysis were carried out to confirm aggregation of enzyme molecules on to nanofiber surface and covalent bond formation between enzyme and nanofiber, respectively. The immobilized lipase showed up to 70% of its specific activity even after 10 repeated batch cycles. *Pseudomonas cepacia* lipase was used for biodiesel production using PAN nanofiber (Noureddini et al. 2005). Under the optimal reaction conditions, 90% of the biodiesel conversion of soybean oil was obtained after 24 hours reaction. The initial conversion capacity of the immobilized *P. cepacia* lipase was still retained by 91% even after 10 repetitive cycles.

Cellulose has been used as a nanofiber carrier for enzyme immobilization. *Candida rugosa* lipase was immobilized by covalent immobilization method on cellulose nanofiber membrane (Huang et al. 2011). At optimum conditions ($NaIO_4$ content 2–10 mg/mL, reaction time 2–10 hours, reaction temperature 25–35°C and reaction pH 5.5–6.5), the highest enzymatic activity was 29.6 U/g of the immobilized lipase. In addition to that, it also showed significantly higher thermal stability and durability than the free enzyme.

There are many reports on enzyme immobilization on carbon-based nanomaterials for biofuel production (Tan et al. 2010). Immobilization of *Candida rugosa* lipases on the MWCNTs was done by the adsorption method (Che Marzuki et al. 2015). For adsorption, lipase solution at varying concentrations was added to the MWCNTs after sonication. After incubation, centrifugation and washing, the activity of immobilized lipase was checked. The 2.2-fold and a 14-fold increase in transesterification activity were observed after the immobilization of lipase and then the lyophilized powdered enzyme. The conversion was 64% in immobilized lipases at 24 hours as compared to free enzyme 14%.

The highly porous multi-walled carbon nanotube silicon oxide (MWCNT-SiO_2) particles were synthesized by spray pyrolysis (Kumar et al. 2018). The synthesized particles functionalized with glutaraldehyde and were used for immobilization of lipase from *Thermomyces lanuginosus*. Mesoporous SiO_2 showed higher protein loading, maximum velocity and catalytic efficiency than dense SiO_2 particles. Dense SiO_2-bound lipase and mesoporous SiO_2-bound lipase retained 74% and 92% of its initial activities after 12 cycles of reuse.

Mucor javanicus lipase was immobilized on silica nanoparticles (Kim et al. 2006b). A reactive epoxide group bearing glycidyl methacrylate (GMA) was embedded onto the nanoparticle surfaces and directly used for multipoint enzyme coupling. Amine residues were also introduced by coupling ethylene diamine (EDA) to the epoxide group. The covalent immobilization of *M. javanicus* lipase onto the amine-activated silica nanoparticles was done by using coupling agents, glutaraldehyde (GA) or 1,4phenylene diisothiocyanate (NCS). It was found that after modifications, the lipase loading capacities became much higher (EDA-GA 81.3 mg g^{-1} and EDA-NCS 60.9 mg g^{-1}) compared to unmodified GMA nanoparticles (18.9 mg g^{-1}). In addition to that, the relative hydrolytic activities were high and almost in the same range as the free enzyme. The immobilized lipases were found to be more temperature resistant than the free form (Kim et al. 2006b). Also, thermal stability and pH range were considerably increased after modified immobilization. Hence, the size-controlled silica nanoparticles can be used as efficient hosts for enzyme immobilization resulting in high activity and stability of the immobilized enzymes.

Ferric oxide magnetic nanoparticles showed integrated advantages of traditionally immobilized lipase as well as free lipase. This Fe_3O_4 immobilized lipase showed a faster reaction rate along with easy separation. In this report, organosilane compounds (3-aminopropyl)triethyloxysilane (APTES) and (3-mercaptopropyl) trimethoxysilane (MPTMS) along with Fe_3O_4 magnetic nanoparticles were used as carriers for lipase immobilization (Thangaraj et al. 2016b). Under optimal conditions, 89% and 81% of biodiesel yield were achieved by lipase immobilized on APTES-Fe_3O_4 and MPTMS-Fe_3O_4 magnetic nanoparticles, respectively. The recovery was easy by applying an external magnetic field for further use.

The organosilane-modified Fe_3O_4-SiO_2 core magnetic nanocomposites were used for lipase immobilization (Thangaraj et al. 2016a). The maximum 84% and 83% activity was recovered by APTES and MPTMS, respectively, and > 90% biodiesel was produced by Fe_3O_4-SiO_2 immobilized lipase (Thangaraj et al. 2016a). *Thermomyces lanuginosus* lipase was immobilized on magnetic nanoparticles for biodiesel production using palm oil (Raita et al. 2015). The 97.2% of biodiesel production yield was observed under optimum conditions (enzyme loading –23.2 wt%, methanol to oil molar ratio –4.7:1 and water content 3.4%). The immobilized lipase retained more than 80% of the activity even after five recycles.

Grafting of a polyamidoamine (PAMAM) dendrimer was performed onto magnetic multi-walled carbon nanotubes (m-MWCNTs) to express magnetic properties with functionalized amino groups (Fan et al. 2016). Immobilized lipase showed a 27-fold higher relative esterification activity than free enzyme. Also, biodiesel production from waste vegetable oil using immobilized lipase has been tested using waste vegetable oil in a *tert*-butanol solvent system. Under optimal conditions (lipase loading of 0.48 mL, pH of 6.86 and temperature of 46.25°C) biodiesel conversion reached up to 94% and even after 10 cycles of reuse. The synthesis of superparamagnetic iron oxide nanoparticles coated by silica and grafted by aldehyde groups used in the immobilization of *B. cepacia* lipase for the conversion of waste cooking oil to biodiesel (Karimi 2016). The total biodiesel produced from waste cooking oil was 91% along with the easy recovery of immobilized lipase and its repetitive use.

Kinetic modeling of mixed immobilized lipase and co-immobilized lipase has investigated for lipid-catalyzed transesterification of soybean oil and methanol (Lee et al. 2013). The effect of substrate inhibition by methanol was lower in the mixed-immobilized lipase system and the initial reaction rate was higher in the co-immobilized lipase system.

8.3 Ethanol

Ethanol is simple alcohol with chemical formula C_2H_5OH. It is used in antiseptic liquid, an ingredient in cosmetic products, and is a solvent used to dissolve water-insoluble medications. It has been widely used as an engine fuel, fuel blend and feedstock for the synthesis of organic compounds (Mohd Azhar et al. 2017). It has contributed to the reduction of crude oil consumption and pollution. Ethanol is less toxic, readily biodegradable and has a higher octane number (Balat and Balat 2009, John et al. 2011).

Lignocellulosic biomass is composed of cellulose, hemicellulose and lignin (Dhiman et al. 2012, Dhiman et al. 2013, Jagtap et al. 2014a, Jagtap et al. 2014b, Rubin 2008). The synergistic action of cellulase including endoglucanase, exoglucanase and β-glucosidase is necessary for the conversion of cellulose to glucose (Himmel et al. 2007, Jagtap et al. 2012, Jagtap et al. 2013a, Jagtap et al. 2013b, Rubin 2008). The monomers of glucose are used for ethanol production by microbes. The use of cellulolytic enzymes for bioethanol production from renewable biomass has developed sustainable technology.

The cellulose used for hydrolysis is a large insoluble molecule and it creates steric hindrances in the binding with the active site of enzymes. Immobilization of cellulase on nanoparticles can avoid the chances of a steric hindrance (Estell et al. 1986). Cellulolytic enzymes immobilized on nanomaterials have shown high catalytic efficiency, high yield of immobilization and more stability against pH, temperature and inhibitory products. The list of nanoparticles used for the immobilization of cellulolytic enzymes has been reviewed recently (Qayyum 2017).

Gold nanoparticles (AuNP), iron oxide magnetic nanoparticles (Fe_3O_4MNP) and silicon dioxide nanoparticles (SiO_2NP) have been used for the immobilization of cellulolytic enzymes (Jordan et al. 2011, Phadtare et al. 2004). The immobilized enzymes retained more than 50% of initial activity and reused six times. Fe_3O_4MNPs functionalized with amine and aldehyde and gold functionalized radial mesoporous silica nanoparticles (Au-MsNP) were used for immobilization cellulase (Alahakoon et al. 2012, Cheng and Chang 2013). Titanium dioxide nanoparticles (TiO_2) were used for the immobilization of cellulase from *Aspergillus niger*. The catalytic efficiency of immobilized cellulase increased by a 10-fold (Ahmad and Sardar 2014). *A. niger* cellulase immobilized on β-cyclodextrin coupled magnetic nanoparticles showed 90% of initial activity, reused 16 times with a negligible change in initial activity and produced 20-fold higher glucose than free enzyme (Zhang et al. 2015). The immobilized exoglucanase and β-glucosidase have been used for hydrolysis of lignocellulosic biomass (Das et al. 2011, Zheng et al. 2013). In one study, cellulase from *Trichoderma reesei* was immobilized on an activated magnetic nanoparticle and used in hydrolysis of microcrystalline cellulose (CMC) and hemp hurds (Abraham et al. 2014). The immobilized cellulase hydrolyzed 83% CMC and 93% hemp hurd biomass. It showed 50% of its initial activity after five cycles of reuse.

Aspergillus fumigatus JCF cellulase was immobilized on to MnO_2 nanoparticles (Cherian et al. 2015). The maximum cellulase-binding efficacy was found to be 75%. The immobilized cellulase showed higher activity, thermal, pH, temperature stability and more reusability than free enzyme. The results confirmed that the efficiency of cellulase immobilized on MnO_2 nanoparticles was remarkable in terms of cellulolytic activity and ethanol production. The thermostability (70°C for 2 hours) and reusability (60% activity was retained after five cycles) of the enzyme were found to increase after immobilization than that of free enzyme. Also, immobilized cellulase in combination with yeast for simultaneous saccharification and fermentation produced about 21.96 g/L of ethanol (Cherian et al. 2015).

Simultaneous saccharification and fermentation (SSF) have been preferred over separate hydrolysis and fermentation (SHF) because SSF shows lower product inhibition of the enzyme, higher ethanol yield and lower production of ethanol (Beldman et al. 1985). On the other side, it has disadvantages as well such as the requirement of compatibility of optimal conditions used for hydrolysis of cellulose and microbes used for fermentation. For example, *Trichoderma viride* cellulase hydrolyzes the cellulose at pH 4–5 and 50°C (Lupoi and Smith 2011, White and White 1997). In contrast, typical fermentation temperatures for ideal ethanol-producing yeasts are 28–38°C (Beldman et al. 1985, Szczodrak and Fiedurek 1996). In addition to that, the ethanol produced during the fermentation can lead to the inactivation of cellulase.

The immobilization of multiple enzymes on the same support can help other enzymes to stay in close proximity to increase its efficiency. In one study, cellulase from *T. viride* is immobilized on non-porous silica nanoparticles (Lupoi and Smith 2011). The optimum conditions were optimized for the hydrolysis of microcrystalline cellulose using immobilized cellulase on silica nanoparticles. Immobilized cellulase produced 1.6-fold higher glucose than the free enzyme. In SSF, ethanol yields were 2.1-fold higher compared to the free enzyme reaction. Immobilization of cellulase on silica nanoparticles stabilized the enzyme and promoted the cellulase activity at non-optimal conditions (Table 2) (Lupoi and Smith 2011).

The three cysteine-tagged cellulases including endo-glucanase, exo-glucanase and β-glucosidase co-immobilized on gold nanoparticles and gold-doped magnetic silica nanoparticles (Au-MSNP). These cellulases were involved in the breaking down of cellulose into glucose used for bioethanol production (Figure 7) (Cho et al. 2012). The co-immobilization of cellulase on Au-MSNP enhanced the yields of cellobiose and glucose by 158 and 179%, respectively. In addition to that, it retained more than 80% of initial activity after 7 cycles of reuse (Cho et al. 2012).

9. Conclusion

We have discussed the recent immobilization techniques using advanced nanomaterials for the production of biofuels. In recent years, advances in nanocarrier and protein engineering techniques have combined to improve the enzyme trait and performance of the immobilized system in aqueous and non-aqueous media. There are multiple methods for immobilization and the choice of carriers that changes from enzyme to enzyme. The selection of right support and enzyme is a crucial part of enzyme immobilization. The engineered superior enzymes using protein-engineering techniques combined with the best support can produce advanced biofuels. The multi-enzyme co-immobilization on nanomaterials is an emerging field in enzyme immobilization. The selection of the right enzyme, carrier, optimum conditions,

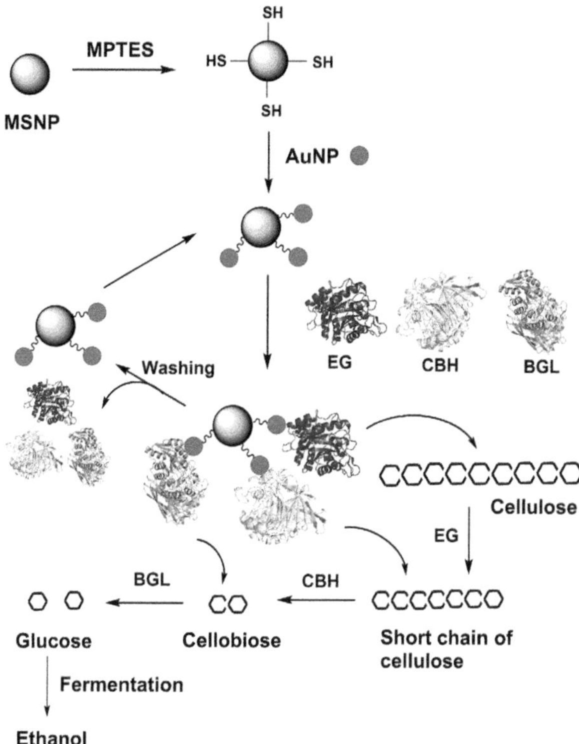

Figure 7. The conversion of cellulose to ethanol by the immobilized multi-enzyme system.

cascade synergy and co-immobilization method selection are necessary criteria for the efficient co-immobilized multi-enzymatic system. The immobilized enzyme system used for the production of biofuels.

References

Abraham, R.E., M.L. Verma, C.J. Barrow and M. Puri. 2014. Suitability of magnetic nanoparticle immobilised cellulases in enhancing enzymatic saccharification of pretreated hemp biomass. Biotechnology for Biofuels 7: 90–90.

Adlercreutz, P. 2013. Immobilisation and application of lipases in organic media. Chemical Society Reviews 42(15): 6406–36.

Ahmad, R. and M. Sardar. 2014. Immobilization of cellulase on TiO_2 nanoparticles by physical and covalent methods: A comparative study. Indian Journal of Biochemistry and Biophysics 51(4): 314–20.

Al-Lazikani, B., J. Jung, Z. Xiang and B. Honig. 2001. Protein structure prediction. Current Opinion in Chemical Biology 5(1): 51–6.

Alahakoon, T., J.W. Koh, X.W. Chong and W.T. Lim. 2012. Immobilization of cellulases on amine and aldehyde functionalized Fe_2O_3 magnetic nanoparticles. Preparative Biochemistry and Biotechnology 42(3): 234–48.

Ansari, S.A. and Q. Husain. 2012. Potential applications of enzymes immobilized on/in nano materials: A review. Biotechnology Advances 30(3): 512–523.

Arnold, F.H. 1990. Engineering enzymes for non-aqueous solvents. Trends in Biotechnology 8: 244–249.

Asuri, P., S.S. Karajanagi, E. Sellitto, D.Y. Kim, R.S. Kane and J.S. Dordick. 2006. Water-soluble carbon nanotube-enzyme conjugates as functional biocatalytic formulations. Biotechnology and Bioengineering 95(5): 804–811.

Bai, S., Z. Guo, W. Liu and Y. Sun. 2006. Resolution of (±)-menthol by immobilized *Candida rugosa* lipase on superparamagnetic nanoparticles. Food Chemistry 96(1): 1–7.

Balat, M. and H. Balat. 2009. Recent trends in global production and utilization of bio-ethanol fuel. Applied Energy 86(11): 2273–2282.

Battistel, E. and G. Rialdi. 2006. Characterization of immobilized enzymes by microcalorimetry. pp. 295–310. *In*: Guisan, J.M. (ed.). Immobilization of Enzymes and Cells. Humana Press. Totowa, NJ.

Beldman, G., M.F. Searle-Van Leeuwen, F.M. Rombouts and F.G. Voragen. 1985. The cellulase of Trichoderma viride. Purification, characterization and comparison of all detectable endoglucanases, exoglucanases and beta-glucosidases. European Journal of Biochemistry 146(2): 301–8.

Belu, A.M., D.J. Graham and D.G. Castner. 2003. Time-of-flight secondary ion mass spectrometry: techniques and applications for the characterization of biomaterial surfaces. Biomaterials 24(21): 3635–53.

Bernal, C., K. Rodríguez and R. Martínez. 2018. Integrating enzyme immobilization and protein engineering: An alternative path for the development of novel and improved industrial biocatalysts. Biotechnology Advances 36(5): 1470–1480.

Binnig, G., C.F. Quate and C. Gerber. 1986. Atomic force microscope. Physical Review Letters 56(9): 930–933.

Blanco, A. and G. Blanco. 2017. Chapter 8—Enzymes. pp. 153–175. *In*: Blanco, A. and G. Blanco (eds.). Medical Biochemistry, Academic Press.

Brena, B.M. and F. Batista-Viera. 2006. Immobilization of enzymes. pp. 15–30. *In*: Guisan, J.M. (ed.). Immobilization of Enzymes and Cells. Humana Press. Totowa, NJ.

Celinska, E. and W. Grajek. 2009. Biotechnological production of 2,3-butanediol—current state and prospects. Biotechnology Advances 27(6): 715–25.

Che Marzuki, N.H., N.A. Mahat, F. Huyop, N.A. Buang and R.A. Wahab. 2015. *Candida rugosa* lipase immobilized onto acid-functionalized multi-walled carbon nanotubes for sustainable production of methyl oleate. Applied Biochemistry and Biotechnology 177(4): 967–84.

Cheng, C. and K.C. Chang. 2013. Development of immobilized cellulase through functionalized gold nano-particles for glucose production by continuous hydrolysis of waste bamboo chopsticks. Enzyme and Microbial Technology 53(6-7): 444–51.

Cherian, E., M. Dharmendirakumar and G. Baskar. 2015. Immobilization of cellulase onto MnO_2 nanoparticles for bioethanol production by enhanced hydrolysis of agricultural waste. Chinese Journal of Catalysis 36(8): 1223–1229.

Cho, E.J., S. Jung, H.J. Kim, Y.G. Lee, K.C. Nam, H.-J. Lee and H.-J. Bae. 2012. Co-immobilization of three cellulases on Au-doped magnetic silica nanoparticles for the degradation of cellulose. Chemical Communications 48(6): 886–888.

Crespilho, F.N., R.M. Iost, S.A. Travain, O.N. Oliveira and V. Zucolotto. 2009. Enzyme immobilization on Ag nanoparticles/polyaniline nanocomposites. Biosensors and Bioelectronics 24(10): 3073–3077.

Das, S., D. Berke-Schlessel, H.-F. Ji, J. McDonough and Y. Wei. 2011. Enzymatic hydrolysis of biomass with recyclable use of cellobiase enzyme immobilized in sol–gel routed mesoporous silica. Journal of Molecular Catalysis B: Enzymatic 70(1): 49–54.

Datta, S., L.R. Christena and Y.R.S. Rajaram. 2013. Enzyme immobilization: an overview on techniques and support materials. 3 Biotech. 3(1): 1–9.

Dhiman, S.S., S.S. Jagtap, M. Jeya, J.-R. Haw, Y.C. Kang and J.-K. Lee. 2012. Immobilization of Pholiota adiposa xylanase onto SiO_2 nanoparticles and its application for production of xylooligosaccharides. Biotechnology Letters 34(7): 1307–1313.

Dhiman, S.S., D. Kalyani, S.S. Jagtap, J.-R. Haw, Y.C. Kang and J.-K. Lee. 2013. Characterization of a novel xylanase from Armillaria gemina and its immobilization onto SiO_2 nanoparticles. Applied Microbiology and Biotechnology 97(3): 1081–1091.

Diego, C., W. Lorena and B. Lorena. 2017. Lipase immobilization on siliceous supports: Application to synthetic reactions. Current Organic Chemistry 21(2): 96–103.

Du, X., B. Shi, J. Liang, J. Bi, S. Dai and S.Z. Qiao. 2013. Developing functionalized dendrimer-like silica nanoparticles with hierarchical pores as advanced delivery nanocarriers. Advanced Materials 25(41): 5981–5985.

Dyal, A., K. Loos, M. Noto, S.W. Chang, C. Spagnoli, K.V. Shafi, A. Ulman, M. Cowman and R.A. Gross. 2003. Activity of *Candida rugosa* lipase immobilized on gamma-Fe_2O_3 magnetic nanoparticles. Journal of the American Chemical Society 125(7): 1684–5.

Eisenthal, R., M.J. Danson and D.W. Hough. 2007. Catalytic efficiency and kcat/KM: a useful comparator? Trends in Biotechnology 25(6): 247–249.

Estell, D.A., T.P. Graycar, J.V. Miller, D.B. Powers, J.A. Wells, J.P. Burnier and P.G. Ng. 1986. Probing steric and hydrophobic effects on enzyme-substrate interactions by protein engineering. Science 233(4764): 659–63.

Fan, Y., G. Wu, F. Su, K. Li, L. Xu, X. Han and Y. Yan. 2016. Lipase oriented-immobilized on dendrimer-coated magnetic multi-walled carbon nanotubes toward catalyzing biodiesel production from waste vegetable oil. Fuel 178: 172–178.

Farinas, E.T., T. Bulter and F.H. Arnold. 2001. Directed enzyme evolution. Current Opinion in Biotechnology 12(6): 545–551.

Fu, J., M. Liu, Y. Liu, N.W. Woodbury and H. Yan. 2012. Interenzyme substrate diffusion for an enzyme cascade organized on spatially addressable DNA nanostructures. Journal of the American Chemical Society 134(12): 5516–5519.

Gao, Y. and I. Kyratzis. 2008. Covalent immobilization of proteins on carbon nanotubes using the cross-linker 1-ethyl-3-(3-dimethylaminopropyl)carbodiimide—a critical assessment. Bioconjugate Chemistry 19(10): 1945–1950.

Greenfield, N.J. 2006. Analysis of the kinetics of folding of proteins and peptides using circular dichroism. Nature Protocols 1(6): 2891–2899.

Hanefeld, U., L. Gardossi and E. Magner. 2009. Understanding enzyme immobilisation. Chemical Society Reviews 38(2): 453–68.

Hedstrom, L. 2010. Enzyme Specificity and Selectivity. Encyclopedia of Life Sciences. London: Nature: 1–7.

Hermanova, S., M. Zarevucka, D. Bousa, M. Pumera and Z. Sofer. 2015. Graphene oxide immobilized enzymes show high thermal and solvent stability. Nanoscale 7(13): 5852–8.

Himmel, M.E., S.Y. Ding, D.K. Johnson, W.S. Adney, M.R. Nimlos, J.W. Brady and T.D. Foust. 2007. Biomass recalcitrance: Engineering plants and enzymes for biofuels production. Science 315(5813): 804–7.

Huang, L. and Z.-M. Cheng. 2008. Immobilization of lipase on chemically modified bimodal ceramic foams for olive oil hydrolysis. Chemical Engineering Journal 144(1): 103–109.

Huang, X.-J., P.-C. Chen, F. Huang, Y. Ou, M.-R. Chen and Z.-K. Xu. 2011. Immobilization of *Candida rugosa* lipase on electrospun cellulose nanofiber membrane. Journal of Molecular Catalysis B: Enzymatic 70(3): 95–100.

Hwang, E.T., R. Tatavarty, H. Lee, J. Kim and M.B. Gu. 2011. Shape reformable polymeric nanofibers entrapped with QDs as a scaffold for enzyme stabilization. Journal of Materials Chemistry 21(14): 5215–5218.

Jagtap, S.S., S.S. Dhiman, M. Jeya, Y.C. Kang, J.-H. Choi and J.-K. Lee. 2012. Saccharification of poplar biomass by using lignocellulases from Pholiota adiposa. Bioresource Technology 120(Supplement C): 264–272.

Jagtap, S.S., S.S. Dhiman, T.-S. Kim, J. Li, Y. Chan Kang and J.-K. Lee. 2013a. Characterization of a β-1,4-glucosidase from a newly isolated strain of Pholiota adiposa and its application to the hydrolysis of biomass. Biomass and Bioenergy 54: 181–190.

Jagtap, S.S., S.S. Dhiman, T.-S. Kim, J. Li, J.-K. Lee and Y.C. Kang. 2013b. Enzymatic hydrolysis of aspen biomass into fermentable sugars by using lignocellulases from Armillaria gemina. Bioresource Technology 133(Supplement C): 307–314.

Jagtap, S.S., S.S. Dhiman, T.-S. Kim, I.-W. Kim and J.-K. Lee. 2014a. Characterization of a novel endo-β-1,4-glucanase from Armillaria gemina and its application in biomass hydrolysis. Applied Microbiology and Biotechnology 98(2): 661–669.

Jagtap, S.S., S.M. Woo, T.-S. Kim, S.S. Dhiman, D. Kim and J.-K. Lee. 2014b. Phytoremediation of diesel-contaminated soil and saccharification of the resulting biomass. Fuel 116: 292–298.

Jagtap, S.S. and C.V. Rao. 2018a. Microbial conversion of xylose into useful bioproducts. Applied Microbiology and Biotechnology 102(21): 9015–9036.

Jagtap, S.S. and C.V. Rao. 2018b. Production of d-arabitol from d-xylose by the oleaginous yeast Rhodosporidium toruloides IFO0880. Applied Microbiology and Biotechnology 102(1): 143–151.

Ji, Q., B. Wang, J. Tan, L. Zhu and L. Li. 2016. Immobilized multienzymatic systems for catalysis of cascade reactions. Process Biochemistry 51(9): 1193–1203.

Ji, X.-J., H. Huang and P.-K. Ouyang. 2011. Microbial 2,3-butanediol production: A state-of-the-art review. Biotechnology Advances 29(3): 351–364.

John, R.P., G.S. Anisha, K.M. Nampoothiri and A. Pandey. 2011. Micro and macroalgal biomass: A renewable source for bioethanol. Bioresource Technology 102(1): 186–193.

Jordan, J., C.S.S.R. Kumar and C. Theegala. 2011. Preparation and characterization of cellulase-bound magnetite nanoparticles. Journal of Molecular Catalysis B: Enzymatic 68(2): 139–146.

Kadam, A.A., J. Jang and D.S. Lee. 2016. Facile synthesis of pectin-stabilized magnetic graphene oxide Prussian blue nanocomposites for selective cesium removal from aqueous solution. Bioresource Technology 216: 391–398.

Kadam, A.A., J. Jang and D.S. Lee. 2017. Supermagnetically tuned halloysite nanotubes functionalized with aminosilane for covalent laccase immobilization. ACS Applied Materials & Interfaces 9(18): 15492–15501.

Kadam, A.A., J. Jang, S.C. Jee, J.-S. Sung and D.S. Lee. 2018. Chitosan-functionalized supermagnetic halloysite nanotubes for covalent laccase immobilization. Carbohydrate Polymers 194: 208–216.

Karimi, M. 2016. Immobilization of lipase onto mesoporous magnetic nanoparticles for enzymatic synthesis of biodiesel. Biocatalysis and Agricultural Biotechnology 8: 182–188.

Kennedy, J.F. and J.M. Cabral. 1995. Immobilisation of biocatalysts by metal-link/chelation processes. Artificial Cells Blood Substitutes and Biotechnology 23(2): 231–52.

Khmelnitsky, Y.L. and J.O. Rich. 1999. Biocatalysis in nonaqueous solvents. Current Opinion in Chemical Biology 3(1): 47–53.

Kim, J., H. Jia and P. Wang. 2006a. Challenges in biocatalysis for enzyme-based biofuel cells. Biotechnology Advances 24(3): 296–308.

Kim, J., J.W. Grate and P. Wang. 2008. Nanobiocatalysis and its potential applications. Trends in Biotechnology 26(11): 639–646.

Kim, K., O. Lee and E. Lee. 2018. Nano-immobilized biocatalysts for biodiesel production from renewable and sustainable resources. Catalysts 8(2): 68.

Kim, M.I., H.O. Ham, S.-D. Oh, H.G. Park, H.N. Chang and S.-H. Choi. 2006b. Immobilization of Mucor javanicus lipase on effectively functionalized silica nanoparticles. Journal of Molecular Catalysis B: Enzymatic 39(1): 62–68.

Koenigbauer, M.J. 1994. Pharmaceutical applications of microcalorimetry. Pharmaceutical Research 11(6): 777–783.

Koivula, A., T. Reinikainen, L. Ruohonen, A. Valkeajärvi, M. Claeyssens, O. Teleman, G.J. Kleywegt, M. Szardenings, J. Rouvinen, T.A. Jones and T.T. Teeri. 1996. The active site of Trichoderma reesei cellobiohydrolase II: the role of tyrosine 169. Protein Engineering, Design and Selection 9(8): 691–699.

Kumar, A., G.D. Park, S.K.S. Patel, S. Kondaveeti, S. Otari, M.Z. Anwar, V.C. Kalia, Y. Singh, S.C. Kim, B.-K. Cho, J.-H. Sohn, D.R. Kim, Y.C. Kang and J.-K. Lee. 2019. SiO_2 microparticles with carbon nanotube-derived mesopores as an efficient support for enzyme immobilization. Chemical Engineering Journal 359: 1252–1264.

Kumar, S., C.J. Tsai and R. Nussinov. 2000. Factors enhancing protein thermostability. Protein Engineering 13(3): 179–91.

Kuo, C.-H., Y.-C. Liu, C.-M.J. Chang, J.-H. Chen, C. Chang and C.-J. Shieh. 2012. Optimum conditions for lipase immobilization on chitosan-coated Fe_3O_4 nanoparticles. Carbohydrate Polymers 87(4): 2538–2545.

Le, Q.A.T., J.C. Joo, Y.J. Yoo and Y.H. Kim. 2012. Development of thermostable *Candida antarctica* lipase B through novel *in silico* design of disulfide bridge. Biotechnology and Bioengineering 109(4): 867–876.

Lee, J.H., S.B. Kim, H.Y. Yoo, J.H. Lee, C. Park, S.O. Han and S.W. Kim. 2013. Kinetic modeling of biodiesel production by mixed immobilized and co-immobilized lipase systems under two pressure conditions. Korean Journal of Chemical Engineering 30(6): 1272–1276.

Lee, S.J., R. Tatavarty and M.B. Gu. 2012. Electrospun polystyrene-poly(styrene-co-maleic anhydride) nanofiber as a new aptasensor platform. Biosensors and Bioelectronics 38(1): 302–7.

Li, S.-F., J.-P. Chen and W.-T. Wu. 2007. Electrospun polyacrylonitrile nanofibrous membranes for lipase immobilization. Journal of Molecular Catalysis B: Enzymatic 47(3): 117–124.

Liberman, A., N. Mendez, W.C. Trogler and A.C. Kummel. 2014. Synthesis and surface functionalization of silica nanoparticles for nanomedicine. Surface Science Reports 69(2): 132–158.

Liese, A. and L. Hilterhaus. 2013. Evaluation of immobilized enzymes for industrial applications. Chemical Society Reviews 42(15): 6236–49.

Liu, B. and M.V. Mirkin. 2001. Charge transfer reactions at the liquid/liquid interface. Analytical Chemistry 73(23): 670a–677a.

Liu, X., Y. Guan, R. Shen and H. Liu. 2005. Immobilization of lipase onto micron-size magnetic beads. Journal of Chromatography B 822(1): 91–97.

Luo, J., A.S. Meyer, R.V. Mateiu and M. Pinelo. 2015. Cascade catalysis in membranes with enzyme immobilization for multi-enzymatic conversion of CO_2 to methanol. New Biotechnology 32(3): 319–327.

Lupoi, J.S. and E.A. Smith. 2011. Evaluation of nanoparticle-immobilized cellulase for improved ethanol yield in simultaneous saccharification and fermentation reactions. Biotechnology and Bioengineering 108(12): 2835–43.

Macario, A., F. Verri, U. Diaz, A. Corma and G. Giordano. 2013. Pure silica nanoparticles for liposome/lipase system encapsulation: Application in biodiesel production. Catalysis Today 204: 148–155.

Martinelle, M., M. Holmquist, I.G. Clausen, S. Patkar, A. Svendsen and K. Hult. 1996. The role of Glu87 and Trp89 in the lid of Humicola lanuginosa lipase. Protein Engineering, Design and Selection 9(6): 519–524.

Mateo, C., J.M. Palomo, G. Fernandez-Lorente, J.M. Guisan and R. Fernandez-Lafuente. 2007. Improvement of enzyme activity, stability and selectivity via immobilization techniques. Enzyme and Microbial Technology 40(6): 1451–1463.

McMorn, P. and G.J. Hutchings. 2004. Heterogeneous enantioselective catalysts: strategies for the immobilisation of homogeneous catalysts. Chemical Society Reviews 33(2): 108–22.

Meng, X.-Y., H.-X. Zhang, M. Mezei and M. Cui. 2011. Molecular docking: a powerful approach for structure-based drug discovery. Current Computer-aided Drug Design 7(2): 146–157.

Misson, M., H. Zhang and B. Jin. 2015. Nanobiocatalyst advancements and bioprocessing applications. Journal of the Royal Society Interface 12(102).

Mohamad, N.R., N.H.C. Marzuki, N.A. Buang, F. Huyop and R.A. Wahab. 2015. An overview of technologies for immobilization of enzymes and surface analysis techniques for immobilized enzymes. Biotechnology & Biotechnological Equipment 29(2): 205–220.

Mohd Azhar, S.H., R. Abdulla, S.A. Jambo, H. Marbawi, J.A. Gansau, A.A. Mohd Faik and K.F. Rodrigues. 2017. Yeasts in sustainable bioethanol production: A review. Biochemistry and Biophysics Reports 10: 52–61.

Neri, D.F.M., V.M. Balcão, F.O.Q. Dourado, J.M.B. Oliveira, L.B. Carvalho and J.A. Teixeira. 2011. Immobilized β-galactosidase onto magnetic particles coated with polyaniline: Support characterization and galactooligosaccharides production. Journal of Molecular Catalysis B: Enzymatic 70(1): 74–80.

Nigam, P.S. and A. Singh. 2011. Production of liquid biofuels from renewable resources. Progress in Energy and Combustion Science 37(1): 52–68.

Noureddini, H., X. Gao and R.S. Philkana. 2005. Immobilized Pseudomonas cepacia lipase for biodiesel fuel production from soybean oil. Bioresource Technology 96(7): 769–777.

Ovsejevi, K., C. Manta and F. Batista-Viera. 2013. Reversible covalent immobilization of enzymes via disulfide bonds. Methods in Molecular Biology 1051: 89–116.

Packer, M.S. and D.R. Liu. 2015. Methods for the directed evolution of proteins. Nature Reviews Genetics 16: 379.

Palomo, J.M., R.L. Segura, G. Fernandez-Lorente, J.M. Guisán and R. Fernandez-Lafuente. 2004. Enzymatic resolution of (±)-glycidyl butyrate in aqueous media. Strong modulation of the properties of the lipase from Rhizopus oryzae via immobilization techniques. Tetrahedron: Asymmetry 15(7): 1157–1161.

Palomo, J.M., R.L. Segura, G. Fernandez-Lorente, R. Fernandez-Lafuente and J.M. Guisán. 2007. Glutaraldehyde modification of lipases adsorbed on aminated supports: A simple way to improve their behaviour as enantioselective biocatalyst. Enzyme and Microbial Technology 40(4): 704–707.

Phadtare, S., S. Vyas, D.V. Palaskar, A. Lachke, P.G. Shukla, S. Sivaram and M. Sastry. 2004. Enhancing the reusability of endoglucanase-gold nanoparticle bioconjugates by tethering to polyurethane microspheres. Biotechnology Progress 20(6): 1840–1846.

Puri, M., C.J. Barrow and M.L. Verma. 2013. Enzyme immobilization on nanomaterials for biofuel production. Trends in Biotechnology 31(4): 215–216.

Qayyum, H. 2017. Nanomaterials immobilized cellulolytic enzymes and their industrial applications: a literature review. JSM Biochemistry & Molecular Biology 4(3): 1019.

Raita, M., J. Arnthong, V. Champreda and N. Laosiripojana. 2015. Modification of magnetic nanoparticle lipase designs for biodiesel production from palm oil. Fuel Processing Technology 134: 189–197.

Rajan, A., J.D. Sudha and T.E. Abraham. 2008. Enzymatic modification of cassava starch by fungal lipase. Industrial Crops and Products 27(1): 50–59.

Ribeiro, E.A., Jr. and C.H. Ramos. 2005. Circular permutation and deletion studies of myoglobin indicate that the correct position of its N-terminus is required for native stability and solubility but not for native-like heme binding and folding. Biochemistry 44(12): 4699–709.

Rubin, E.M. 2008. Genomics of cellulosic biofuels. Nature 454: 841.

Saifuddin, N., A. Raziah and A. Junizah. 2012. Carbon nanotubes: a review on structure and their interaction with proteins. Journal of Chemistry 2013.

Schoffelen, S. and J.C.M. van Hest. 2013. Chemical approaches for the construction of multi-enzyme reaction systems. Current Opinion in Structural Biology 23(4): 613–621.

Schrittwieser, J.H., J. Sattler, V. Resch, F.G. Mutti and W. Kroutil. 2011. Recent biocatalytic oxidation–reduction cascades. Current Opinion in Chemical Biology 15(2): 249–256.

Schuck, P., C.E. MacPhee and G.J. Howlett. 1998. Determination of sedimentation coefficients for small peptides. Biophysical Journal 74(1): 466–474.

Sen, T., A. Sebastianelli and I.J. Bruce. 2006. Mesoporous Silica−Magnetite Nanocomposite: Fabrication and applications in magnetic bioseparations. Journal of the American Chemical Society 128(22): 7130–7131.

Shang, C.-Y., W.-X. Li and R.-F. Zhang. 2015. Immobilization of *Candida rugosa* lipase on ZnO nanowires/macroporous silica composites for biocatalytic synthesis of phytosterol esters. Materials Research Bulletin 68: 336–342.

Sheldon, R.A. 2007a. Cross-linked enzyme aggregates (CLEAs): stable and recyclable biocatalysts. Biochemical Society Transactions 35(Pt 6): 1583–7.

Sheldon, R.A. 2007b. Enzyme immobilization: The quest for optimum performance. Advanced Synthesis and Catalysis 349(8-9): 1289–1307.

Sheldon, R.A. and S. van Pelt. 2013. Enzyme immobilisation in biocatalysis: why, what and how. Chemical Society Reviews 42(15): 6223–6235.

Sheldon, R.A. and P.C. Pereira. 2017. Biocatalysis engineering: the big picture. Chemical Society Reviews 46(10): 2678–2691.

Sheldon, R.A. and J.M. Woodley. 2018. Role of biocatalysis in sustainable chemistry. Chemical Reviews 118(2): 801–838.

Shim, E.J., S.H. Lee, W.S. Song and H.R. Kim. 2017. Development of an enzyme-immobilized support using a polyester woven fabric. Textile Research Journal 87(1): 3–14.

Singh, R.K., Y.-W. Zhang, N.-P.-T. Nguyen, M. Jeya and J.-K. Lee. 2011. Covalent immobilization of β-1,4-glucosidase from Agaricus arvensis onto functionalized silicon oxide nanoparticles. Applied Microbiology and Biotechnology 89(2): 337–344.

Singh, R.K., M.K. Tiwari, R. Singh and L.K. Lee. 2013. From protein engineering to immobilization: promising strategies for the upgrade of industrial enzymes. International Journal of Molecular Sciences 14(1): 1232–77.

Singh, R.K., M.K. Tiwari, R. Singh, J.-R. Haw and J.-K. Lee. 2014. Immobilization of l-arabinitol dehydrogenase on aldehyde-functionalized silicon oxide nanoparticles for l-xylulose production. Applied Microbiology and Biotechnology 98(3): 1095–1104.

Sirisha, V.L., A. Jain and A. Jain. 2016. Chapter Nine—Enzyme immobilization: An overview on methods, support material, and applications of immobilized enzymes. pp. 179–211. *In*: Kim, S.-K. and F. Toldrá (eds.). Advances in Food and Nutrition Research, Vol. 79, Academic Press.

Skjold-Jørgensen, J., J. Vind, A. Svendsen and M.J. Bjerrum. 2014. Altering the activation mechanism in Thermomyces lanuginosus lipase. Biochemistry 53(25): 4152–4160.

Skjold-Jørgensen, J., J. Vind, O.V. Moroz, E. Blagova, V.K. Bhatia, A. Svendsen, K.S. Wilson and M.J. Bjerrum. 2017. Controlled lid-opening in Thermomyces lanuginosus lipase–An engineered switch for studying lipase function. Biochimica et Biophysica Acta (BBA)—Proteins and Proteomics 1865(1): 20–27.

Song, C., L. Sheng and X. Zhang. 2013. Immobilization and characterization of a thermostable lipase. Marine Biotechnology (NY) 15(6): 659–67.

Stark, M.-B. and K. Holmberg. 1989. Covalent immobilization of lipase in organic solvents. Biotechnology and Bioengineering 34(7): 942–950.

Stepankova, V., S. Bidmanova, T. Koudelakova, Z. Prokop, R. Chaloupkova and J. Damborsky. 2013. Strategies for stabilization of enzymes in organic solvents. ACS Catalysis 3(12): 2823–2836.

Subramanian, A., S.J. Kennel, P.I. Oden, K.B. Jacobson, J. Woodward and M.J. Doktycz. 1999. Comparison of techniques for enzyme immobilization on silicon supports. Enzyme and Microbial Technology 24(1-2): 26–34.

Syu, M.J. 2001. Biological production of 2,3-butanediol. Applied Microbiology and Biotechnology 55(1): 10–8.

Szczodrak, J. and J. Fiedurek. 1996. Technology for conversion of lignocellulosic biomass to ethanol. Biomass and Bioenergy 10(5-6): 367–375.

Takahashi, H., B. Li, T. Sasaki, C. Miyazaki, T. Kajino and S. Inagaki. 2000. Catalytic activity in organic solvents and stability of immobilized enzymes depend on the pore size and surface characteristics of mesoporous silica. Chemistry of Materials 12(11): 3301–3305.

Tan, T., J. Lu, K. Nie, L. Deng and F. Wang. 2010. Biodiesel production with immobilized lipase: A review. Biotechnology Advances 28(5): 628–34.

Thangaraj, B., Z. Jia, L. Dai, D. Liu and W. Du. 2016a. Effect of silica coating on Fe_3O_4 magnetic nanoparticles for lipase immobilization and their application for biodiesel production. Arabian Journal of Chemistry 12(8): 4694–4706.

Thangaraj, B., Z. Jia, L. Dai, D. Liu and W. Du. 2016b. Lipase NS81006 immobilized on Fe_3O_4 magnetic nanoparticles for biodiesel production. *In*: Ovidius University Annals of Chemistry 27: 13.

Tian, K., K. Tai, B.J.W. Chua and Z. Li. 2017. Directed evolution of Thermomyces lanuginosus lipase to enhance methanol tolerance for efficient production of biodiesel from waste grease. Bioresource Technology 245: 1491–1497.

Verma, M.L., W. Azmi and S.S. Kanwar. 2011. Enzymatic synthesis of isopropyl acetate by immobilized Bacillus cereus lipase in organic medium. Enzyme Research 2011: 7.

Verma, M.L., C.J. Barrow and M. Puri. 2013a. Nanobiotechnology as a novel paradigm for enzyme immobilisation and stabilisation with potential applications in biodiesel production. Applied Microbiology and Biotechnology 97(1): 23–39.

Verma, M.L., M. Naebe, C.J. Barrow and M. Puri. 2013b. Enzyme immobilisation on amino-functionalised multi-walled carbon nanotubes: structural and biocatalytic characterisation. PloS One 8(9): e73642.

Verma, M.L. 2017. Nanobiotechnology advances in enzymatic biosensors for the agri-food industry. Environmental Chemistry Letters 15(4): 555–560.

Verma, R., U. Schwaneberg and D. Roccatano. 2012. Computer-Aided protein directed evolution: A review of web servers, databases and other computational tools for protein engineering. Computational and Structural Biotechnology Journal 2(3): e201209008.

Wang, L., L. Wei, Y. Chen and R. Jiang. 2010. Specific and reversible immobilization of NADH oxidase on functionalized carbon nanotubes. Journal of Biotechnology 150(1): 57–63.

Wang, X., P. Dou, P. Zhao, C. Zhao, Y. Ding and P. Xu. 2009a. Immobilization of lipases onto magnetic Fe_3O_4 nanoparticles for application in biodiesel production. Journal of Chemistry and Sustainability, Energy and Materials 2(10): 947–950.

Wang, Z.-G., L.-S. Wan, Z.-M. Liu, X.-J. Huang and Z.-K. Xu. 2009b. Enzyme immobilization on electrospun polymer nanofibers: An overview. Journal of Molecular Catalysis B: Enzymatic (4): 189–195.

White, J.S. and D.C. White. 1997. Source Book of Enzymes. CRC Press.

Wichmann, R. and D. Vasic-Racki. 2005. Cofactor regeneration at the lab scale. Advances in Biochemical Engineering-Biotechnology 92: 225–60.

Won, K., S. Kim, K.-J. Kim, H.W. Park and S.-J. Moon. 2005. Optimization of lipase entrapment in Ca-alginate gel beads. Process Biochemistry 40(6): 2149–2154.

Wong, L.S., F. Khan and J. Micklefield. 2009. Selective covalent protein immobilization: Strategies and applications. Chemical Reviews 109(9): 4025–4053.

Wu, L., X. Yuan and J. Sheng. 2005. Immobilization of cellulase in nanofibrous PVA membranes by electrospinning. Journal of Membrane Science 250(1): 167–173.

Xia, S., X. Zhao, B. Frigo-Vaz, W. Zheng, J. Kim and P. Wang. 2015. Cascade enzymatic reactions for efficient carbon sequestration. Bioresource Technology 182: 368–372.

Xu, J., F. Zeng, S. Wu, X. Liu, C. Hou and Z. Tong. 2007. Gold nanoparticles bound on microgel particles and their application as an enzyme support. Nanotechnology 18(26): 265704.

Xu, R., Q. Zhou, F. Li and B. Zhang. 2013. Laccase immobilization on chitosan/poly(vinyl alcohol) composite nanofibrous membranes for 2,4-dichlorophenol removal. Chemical Engineering Journal 222: 321–329.

Xue, R. and J.M. Woodley. 2012. Process technology for multi-enzymatic reaction systems. Bioresource Technology 115: 183–195.

Yang, H., L. Liu, J. Li, J. Chen and G. Du. 2015. Rational design to improve protein thermostability: Recent advances and prospects. Chemical and Biochemical Engineering Reviews 2(2): 87–94.

Yang, T., Z. Rao, X. Zhang, M. Xu, Z. Xu and S.T. Yang. 2017. Metabolic engineering strategies for acetoin and 2,3-butanediol production: advances and prospects. Critical Reviews in Biotechnology 37(8): 990–1005.

Zhang, L., R.D.S. Singh, Z. Guo, J. Li, F. Chen, Y. He, X. Guan, Y.C. Kang and J.-K. Lee. 2018. An artificial synthetic pathway for acetoin, 2,3-butanediol, and 2-butanol production from ethanol using cell free multi-enzyme catalysis. Green Chemistry 20(1): 230–242.

Zhang, S., J.M. Skerker, C.D. Rutter, M.J. Maurer, A.P. Arkin and C.V. Rao. 2016. Engineering Rhodosporidium toruloides for increased lipid production. Biotechnology and Bioengineering 113(5): 1056–1066.

Zhang, W., J. Qiu, H. Feng, L. Zang and E. Sakai. 2015. Increase in stability of cellulase immobilized on functionalized magnetic nanospheres. Journal of Magnetism and Magnetic Materials 375: 117–123.

Zhang, Y. 2008. Progress and challenges in protein structure prediction. Current Opinion in Structural Biology 18(3): 342–348.

Zheng, L., C.M. McQuaw, M.J. Baker, N.P. Lockyer, J.C. Vickerman, A.G. Ewing and N. Winograd. 2008. Investigating lipid-lipid and lipid-protein interactions in model membranes by ToF-SIMS. Applied Surface Science 255(4): 1190–1192.

Zheng, P., J. Wang, C. Lu, Y. Xu and Z. Sun. 2013. Immobilized β-glucosidase on magnetic chitosan microspheres for hydrolysis of straw cellulose. Process Biochemistry 48(4): 683–687.

Chapter 3

Carbon Nanotubes (CNT) Based Enzyme Immobilisation for Biofuel Applications

Aravind Madhavan,[1] *Sharrel Rebello,*[2] *Raveendran Sindhu,*[3,*]
Parameswaran Binod[3] and *Ashok Pandey*[4]

1. Introduction

Biological catalysts possess distinct qualities like superior substrate preferences, high product turnover and high catalytic capability over other synthetic catalysts. Most of the industrial conversion processes are inhibited by the presence of different organic solvents, extremely high temperature, unavailability of appropriate storage conditions, etc. Enzyme immobilisation is one of the most important aspects of modern biology (Cang-Rong and Pastorin 2009, Feng and Ji 2011). This technology regularly overcomes the limitation of the sensitivity of the enzyme to harsh environments (like high temperature, pH, varying ionic concentration, fermentation inhibitors, presence of metal ions and mechanical agitation) and provides long-lasting strength/reusability of the biocatalyst and is also nature-friendly. So, the commercial applications of enzyme catalyst are highly linked to the establishment of productive immobilisation techniques (Laurent et al. 2008).

Nanotechnology driven biocatalytic techniques have emerged as a highly efficient, practically easy, appropriate and reproducible technique for the enzyme immobilisation. The very important and basic necessity for enzyme immobilization is that the immobilisation matrix should be inert with the immobilisation enzyme of interest, i.e., the supporting matrix should not react with the immobilised enzyme which thereby could copout its native bioactivity.

[1] Rajiv Gandhi Centre for Biotechnology, Jagathy, Trivandrum – 695 014, India.
[2] Communicable Disease Research Laboratory, St. Joseph's College, Irinjalakuda, India.
[3] Microbial Processes and Technology Division, CSIR-National Institute for Interdisciplinary Science and Technology (CSIR-NIIST), Trivandrum – 695 019, India.
[4] Center for Innovation and Translational Research, CSIR-Indian Institute of Toxicology Research (CSIR-IITR), 31 MG Marg, Lucknow – 226 001, India.
* Corresponding author: sindhurgcb@gmail.com; sindhufax@yahoo.co.in

Researchers were successful in developing matrix based on carbon nanotubes (CNTs) for enzyme immobilization (Zhang et al. 2009, Feng and Ji 2011, Mubarak et al. 2014). CNT-based matrix delivers significant qualities, including thermal, mechanical and electrical attributes, which make them a suitable candidate for a variety of functions (Mubarak et al. 2014). The enzymes immobilised on CNTs can withstand high temperature that results in enhanced and stable performance of the enzyme. This particular property allows for the extensive usage of CNTs in catalyst recycling and low mass transfer resistance (Asuri et al. 2006, Mubarak et al. 2014). This chapter will focus on CNT-based enzyme immobilisation and its implications in biofuel.

2. Chemical Modes of Carbon Nanotube Synthesis

CNT synthesis is done usually by multidisciplinary approaches, such as arc discharge, laser ablation, sono-chemical or hydrothermal, electrolysis and chemical vapour methods (Prasek et al. 2011). The arc discharge process involves the formation of carbon nanoparticles between two graphite electrodes in a chamber containing an inert gas along with plasma created at a temperature above 2,000°C (Herrera-Ramirez et al. 2019, Iijima and Ichihashi 1993). Laser beam induced synthesis of carbon nanoparticle can be carried out either at room temperature or as temperatures as high as 1,200°C (Arepalli 2004). Alternatively, room temperature based synthesis of CNT using graphene immersed in sulphuric acid, nitric acid and potassium chlorate, heating at 70°C and subsequent drying for three days have also been used as simple CNT synthetic steps (Lee and Seo 2010).

Catalyst-assisted chemical vapour deposition of hydrocarbons is the most popular method of CNT synthesis due to the high quantitative yield of the process and often this process is assisted by the presence of iron and cobalt catalysts. Advancements in the above process can be made by using silica encapsulated iron oxide (Fe_3O_4) nanoparticles yielding better activity and selectivity in CNT synthesis (Atchudan et al. 2019). Yet other attempts in this direction involve the catalyst-free chemical vaporisation of acetylene on aluminium oxide (Fleming et al. 2019). Synthesis of carbon nanorings using a single step with NiO/Al_2O_3 as catalyst material and acetylene as the precursor gas has also been attempted to avoid the need of multistage purification steps of nanotubes (Venkatesan et al. 2019). Variant types of chemical vaporisation technique following different methods such as hot filament method, water-assisted method, oxygen assisted method, microwave method, radio frequency assisted method and thermal as well as plasma-assisted techniques also contribute to its popular use in CNT synthetic techniques (Prasek et al. 2011).

The prospects of converting various plastics such as polyethylene, polypropylene, polyvinyl alcohol and PET into carbon nanotubes following the processes of catalytic pyrolysis and further degradation may also help in utilising the plastic waste to useful products (Mukherjee et al. 2019). More advancement in the field of CNT synthesis has resulted in patents for constructing devices to synthesise them by techniques involving deposition of catalyst on moving substrate and CNT formation on the substrate followed by its separation to yield purified CNT particles (Nguyen 2019).

3. Biogenic Synthesis of Carbon Nanotubes

Biogenic synthesis of nanoparticles using plants and bacteria serve as simple, easy and less-time consuming alternative to chemical synthetic methods (Durán et al. 2011). The concept of green synthesis of nanoparticles could help to reduce the use of harsh chemicals and in some instances, bioconversion of various wastes to useful nanoparticles is achieved (Ingale and Chaudhari 2013, Sharma et al. 2015). Various biological sources, such as plants, bacteria, algae, etc., serve as machinery for nanoparticle synthesis (Narayanan and Sakthivel 2011) with some exceptions were even dead biological matter could serve as tools of nanoparticle synthesis (Arya et al. 2018). Moreover, such biogenic methods would confer nanoparticles some additional physical and morphological properties, thereby, increasing the effective utility of nanoparticles (Lee et al. 2007). However, the feasibility of carbon CNT synthesis by bacteria has not been yet reported, except for various studies indicating that bacteria could serve as templates in developing CNT bacterial particle composites. In yet another study, the synthesis of As-S nanotubes rather than CNT were reported in metal dissimilatory bacteria *Shewanella* sp. that served as tools in the photo and electronic devices (Lee et al. 2007).

The use of bacteria, viz., *Magnetospirillum magneticum* as support to the synthesis of 3D structures of CNT has been attempted to modify its properties. Apart from increasing the porosity of CNT, the scaffold also provided better mechanical stability and increased surface area, thereby, contributing to the generation of lightweight three-dimensional CNT surfaces with better connectivity (Ozden et al. 2017). The choice of *Magnetospirillum magneticum* bacteria for the scaffold synthesis also possess an additional advantage as it is noted for the synthesis of bacterial magnetic nanoparticles and siderophores which would confer the resultant nanocomposites extra magnetic properties (Arakaki et al. 2003, Calugay et al. 2003).

Attempts to biosynthesise bacterial cellulose incorporated carboxylic multi-walled carbon nanotubes proved successful in generating biologically modified CNTs by simple processes (Lv et al. 2016). The use of plant materials such as coconut oil as substrates to synthesise multi-walled CNTs (MWCNTs) of 80–90 nm diameter served as a green strategy of nanoparticle synthesis following the principles of vapour deposition methods (Paul and Samdarshi 2011). The development of bionic composites of CNTs and most common baker's yeast, *Saccharomyces cerevisiae,* not only improved the tensile strength of the nanoparticles but also its electrical properties (Valentini et al. 2016).

4. Carbon Nanotubes and Enzyme Immobilisation

Carbon nanotubes are classified into two different types, single and multi-walled CNTs (SWNTs and MWNTs), which are used to immobilise different biocatalysts (Ong and Annuar 2018). A SWNT is composed of a single graphite layer, and a MWNT is composed of different graphite layers and a middle tubule. SWNTs have an advantage of larger area in the surface for immobilised catalyst interaction and multi-walled CNTs have higher dispersibility and low cost. Non-covalent and covalent

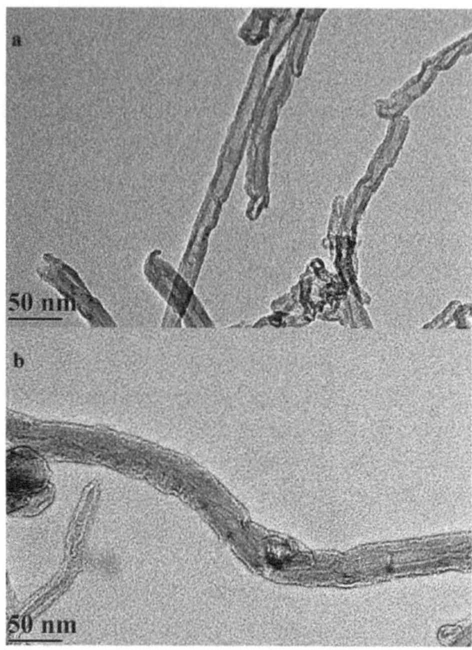

Figure 1. TEM images of (a) functionalised MWNT and (b) MWNT-bound lipase (adapted from Verma et al. (2013) doi:10.1371/journal.pone.0073642.

immobilisations have been performed for the immobilisation of different industrial catalysts (Gao and Kyratzis 2008). Covalent enzyme immobilisation preserves the attachment and it is more durable but the enzyme structure may be damaged. Non-covalent attachment, on the other hand, maintains the structure and function of enzyme and CNTs but the attachment is less durable (Gao and Kyratzis 2008, Fan et al. 2017). Functional activation of CNTs using different organic molecules, polymeric molecules and other biological molecules can develop its nanomaterials which are highly biocompatible with particular active functional groups. The performances of CNT immobilised enzyme complexes are influenced by chemical nature nanotube material and type of immobilisation method used (Verma et al. 2013). Figure 1 depicts TEM images of functionalised MWNT and MWNT bound lipase (adapted from Verma et al. 2013).

5. Non-Covalent Enzyme Immobilisation

Adsorption is the most commonly used non-covalent approach in which the biocatalyst is physically adsorbed onto the CNTs (Asuri et al. 2006). The process of adsorption involves the immersion of CNTs in a miscible solution of the required enzyme and subsequent mixing that allows the enzyme to adhere onto the CNT surface and wash off the excess of the enzyme that does not adhere. This method can also be performed by adsorption onto CNTs modified using polymers, surfactants and linking molecules (Feng and Ji 2011).

5.1 Physical Adsorption

The force involved in the immobilisation of physical adsorption is mainly hydrophobic (Gao and Kyratzis 2008). The surface hydrophobic region of immobilised enzyme interacts with walls of CNTs. The aromatic ring of enzymes and walls of CNTs are involved in the stacking interaction (π-π) which promotes adsorption (Matsuura et al. 2006). Another study reported the involvement of electrostatic interaction in the immobilisation of lysozyme (Nepal and Geckeler 2006). In glucose oxidase immobilisation, the adhesion onto CNT is strengthened by hydrogen bonding interaction which involves dihydroxybenzaldehyde and bovine serum albumin (Yu et al. 2010). The amount of enzyme attached to CNT in direct physical adsorption technique is influenced by various factors like the nature of the immobilised enzyme, nature of CNT material, nature of functional groups, etc. Different enzymes adsorbed onto CNT at a different rate. β-glucosidase adsorbed on CNTs 630 μg per mg of CNTs; whereas, in soybean peroxidase, only 575 μg protein per mg of CNT had been adsorbed (Karajanagi et al. 2004, Gao and Kyratzis 2008, Feng and Ji 2011). Shah et al. (2007) reported the immobilisation of CRL on to the MWCTs through direct physical adsorption that resulted in the high holding of its enzymatic activity. The entrapped enzyme displayed a 14-fold enhancement compared to the previous rate of transesterification.

5.2 Adsorption onto CNT Functionalised with Polymers, Biomolecules and Surfactants

Different types of biomolecules and polymers were used to functionalise CNTs. The functional group modified CNTs have superior water miscibility and the stability of the CNTs' immobilised enzyme is enhanced. Functionalised CNTs have many advantageous properties like hydrophobic interactions, electrostatic interactions, capacity to form hydrogen bonding, etc. (Mu et al. 2008). CNTs coated with polymers can provide different positive and negatively charged groups on its surface. For example, single-walled CNTs can be modified by poly(sodium 4-styrene sulphonate) which imparts positive charge (Wu et al. 2009). In another study, CNTs modified by chitosan enhanced the binding of glucose oxidase (Lee and Tsai 2009).

Biomolecules also gain attention for CNTs functionalisation. DNA-based electrochemical sensor has been constructed. The molecular recognition property of DNA has been applied to electrochemical sensors (Moghaddam et al. 2004). In this technique, single-stranded oligonucleotide chains have been attached to CNT electrode for hybridisation with redox-labelled complementary DNA. The attachment of oligonucleotide to CNT is by π-π stacking interaction with the oligonucleotide aromatic bases (Martin et al. 2008). The construction of bioelectrodes was achieved by the wrapping of DNA with one strand onto CNT and then the enzyme of interest was immobilised (Lee et al. 2010b). The bioelectrode showed the enhanced catalytic activity of laccase and glucose oxidase, and the output power generation was increased. The ideal role of oligonucleotide DNA is the specific binding of redox protein onto CNTs (Withey et al. 2008). The enzyme-CNT based transducers have shown enhanced electron transfer. The streptavidin and FAD cofactor (Yim et al.

2005, Patolsky et al. 2004, Feng and Ji 2011), were also used for attaching enzyme and CNTs. Surfactant-assisted adsorption can also be used for the immobilisation. CNTs functionalised using Triton X-100 could specifically attach to streptavidin (Chen et al. 2003). Enzymes (HRP and cytochrome *c*) and surfactants (SDS, CTAB and Triton X-100) have been co-assembled onto CNTs.

6. Covalent Linking

Several methods of covalent linking of biocatalyst on CNTs have been reported. For example, the interaction of the free amino acid group of the protein and carboxylic acid group of the CNTs generated by the sidewall oxidation and further activation by carbodiimide has been extensively adopted for the covalent immobilisation of biocatalysts in CNTs (Jiang et al. 2004, Gao and Kyratzis 2008). For the development of enzyme-based biofuel cells, different enzymes were immobilised onto CNTs (Li et al. 2011, Zhao et al. 2009). Covalently immobilised lipase on CNTs has shown clear supremacy compared to normal lipase (Ji et al. 2010). Several linkers are also used for covalent immobilisation of enzymes onto CNTs. Linking molecules exploit the advantage of hydrophobic and π-π interactions (Pang et al. 2010). For example, 1-pyrene butanoic acid succinimidyl ester were coated on CNTs (Kim et al. 2006) that enhanced the HRP enzyme immobilisation. Laccase immobilisation was enhanced by aminopyrene linking molecule (Pang et al. 2010).

7. Carbon Nanotubes and Enzymatic Fuel Cells

Compared to other nano-elements and materials, CNTs possess nanowire morphology, compatible nature and very good conductive nature (Smart et al. 2006). High specific area (> 1,000 CNTsm^2/g) is another advantage for the construction of a nanostructured CNT-based electrode (Peigney et al. 2001). These distinctive attributes make CNTs an interesting material for the construction of biofuel cells.

Cosnier et al. (2014) extensively reviewed the use of CNTs for the development of biofuel cells. The CNTs are capable of establishing an electrical conversation with enzymes through their conductive nature or by transfer of electrons to enzymes. The nanoscale size makes more direct contact with the prosthetic group of the enzyme and thus by establishes a successful contact between the enzyme and electrode. Thus, electrode modified by CNTs gains more attention for the construction of enzymatic biofuel cells (Holzinger et al. 2012). CNTs can also be used for the storage of energy in the form of electricity and thus it is employed as base nanostructured material for the development of capacitors with high-performance value, i.e., capacitors with less time for charging and enhanced capacitance. CNTs can act as a connecting link between batteries and capacitors due to their high power storage capacity, fast charge and discharge cycles. Hence, highly compatible material of enzymes-CNT complex has been used as highly active capacitors and bielectrodes in biofuel cells (Cosnier et al. 2014). A list of the enzyme-based enzymatic fuel cell is included in Table 1.

Table 1. Enzyme fuel cell developed based on CNTs.

Type of Carbon Nanotube	Enzyme System	Reference
Conductive multi-walled carbon nanotube (CNT)	Glucose oxidase and laccase	Chung et al. 2018
Single-walled carbon nanotubes (SWCNTs) functionalised with reduced graphene oxide	Alcohol dehydrogenase	Umasankar et al. 2017
Polyethyleneimine and carbon nanotube (CNT)	Glucose oxidase	Christwardana et al. 2017
Aniline monomers + functionalised carbon nanotubes (CNTs)	Glucose oxidase and laccase	Kang et al. 2018
Graphene-single walled carbon nanotubes	Glucose oxidase and laccase	Prasad et al. 2014
Carbon nanotube (CNT)	Glucose oxidase and laccase	Christwardana et al. 2016

8. Immobilisation of Enzyme for Biofuel Applications

Biofuels are highly advanced research themes that provide alternative methods of sustainable and renewable energy. Current developments in the area of CNT have gained much attention to increase the efficacy of bioenergy production and thus meet the energy crisis. CNTs are reported as a highly efficient material for the immobilisation of β-glucosidase which has a molecular weight of approximately 135 kDa. Immobilisation of around 630 mg per gram of support resulted within 12 hours. The interacting force between CNT and the enzyme was electrostatic due to the presence of surface charges of the enzyme and the support. The immobilised β-glucosidase gave an enzymatic unit of 400 U/g (Gomez et al. 2005).

Solid acid catalysts are important for the production of biodiesel by triglyceride trans-esterification. Oliveira et al. (2014) reported an efficient method for the synthesis of a highly active sulphonated MWCNTs catalyst for the transesterification of fatty acid ethyl ester generation. The use of sulphonated CNTs as biocatalysts for the conversion of levulinic acid into ethyl levulinate proved that levulinic acid firmly adsorbed in the active sites of the sulphonated CNTs. Recently, cellulase from *Aspergillus niger* was immobilised onto functionalised multi-walled CNTs (MWCNTs) via carbodiimide coupling. This bio-nanoconjugate can be used in saccharification of biomass and the making of platform chemicals (Ahmad et al. 2018). In a recent study, highly active sulphonated multi-walled CNTs catalyst was developed for the transesterification of trilaurin in ethanol solvent at temperatures of above 130°C. The different properties (physical and structural) of sulphonated multi-walled CNTs are characterised by TEM, FTIR, etc. (Guan et al. 2017). Another study reported the activation of multi-walled carbon nanotubes (MWNTs) and their functionalisation using ethylene diamine and evaluated for its rate of functionalisation. Lipase from *Thermomyces lanuginosus* has been then immobilised onto the glutaraldehyde cross-linker functionalised multi-walled CNTs. The enzymatic capability of the immobilised lipase especially in the form of temperature tolerance and reusability were also enhanced. This immobilisation resulted in overall yield, improved thermo-stability and reusability make this nanobiocatalyst a potent one (Verma et al. 2013).

Amyloglucosidase (AMG) was reported as an effective catalyst to effectively breakdown starch that was present in biomass, which is an initial step for the biomass to start energy development process. The enzyme amyloglucosidase was immobilised on CNT covered with magnetic iron oxide nanoparticles (Goh et al. 2012).

Given the recent world situation like energy crisis and focus on renewable sources of energy, oil transesterification by lipase is considered as one of the topmost important methods for the development of biofuel. The main disadvantage of lipase catalysed reaction is its expensive nature and that it cannot be recycled. Tan et al. (2011) reported the successful immobilisation of lipase on multi-walled magnetic CNTs which were covered with magnetic iron oxide. Zhao et al. (2011) and Fan et al. (2016) modified the magnetic SWCNTs and MWCNTs using dendrimers to further enhance the number of active CNT sites for enhanced immobilisation of the enzyme on the CNTs, and this resulted in the enhanced enzyme loading. Superparamagnetic multi-walled magnetic CNTs were developed by coating multi-walled CNTs with iron oxide and further functionalised by linking polyamidoamine dendrimers on the surface. Then, *Burkholderia cepacia* lipase (BCL) was successfully immobilised via covalent method. The activity of lipase was enhanced to 77.460 U/g-protein, which is > 10 fold increase than that of the normal enzyme. The immobilised lipase exhibited high thermo-stability and pH-tolerance and resulted in the effective transesterification to produce biodiesel with the conversion percentage of 92.8 (Fan et al. 2017). An updated list of immobilised enzymes is included in Tables 2, 3 and 4. Figure 2 depicts a schematic representation of the build-up of the enzyme on SWCNTs via π-π interactions (adapted from Poulpiquet et al. 2013).

9. Enzymatic Fuel Cells: Current Updates

Biofuel cells have been designed to produce energy from plants and animals as these are the richest source of mono/disaccharides. The bioanodes (lactate oxidase/glucose oxidase/glucose dehydrogenase) and biocathodes (bilirubin oxidase/laccase/

Table 2. Updated list of immobilised enzymes.

Enzyme	Type of Immobilisation	Reference
Lipase	Physical adsorption and Carboxylated MWCNTs	Deep et al. 2015
Cellulase	Physical adsorption and Functionalised MWCNTs	Mubarak et al. 2014
β-Glucosidase	Covalent Immobilisation	Celik et al. 2016
Lipase	Adsorption	Shah et al. 2007
Lipase	Adsorption	Pavlidis et al. 2010
Lipase	MWNT, Covalent	Ji et al. 2010
Lipase	MWNT, Adsorption	Lee et al. 2010a
Lipase	SWNT, Covalent	Lee et al. 2010b
Cellulase	MWCNTs Functionalised MWCNTs carbodiimide coupling	Ahmad et al. 2018

Table 3. Microbial lipases immobilised on CNTs.

Microbe	Binding	Carrier	Reference
Candida rugosa	Adsorption	MWNT	Shah et al. 2007
Rhizopus arrhizus	Adsorption	MWNT	Ji et al. 2010
Candida antartica	Adsorption	MWNT	Lee et al. 2010a
Pseudomonas cepacia	Covalent	SWNT	Lee et al. 2010b
Candida antartica	Adsorption	MWCNTs	Markiton et al. 2017
Yarrowia lipolytica	Covalent	MWCNTs	Feng et al. 2012
Candida antartica	Covalent	MWCNTs	Rastian et al. 2016

Table 4. CNT immobilised lipase for biodiesel production.

Microbe	Type of Nanotube	Substrate	Conversion Efficiency	Reference
Rhizomucor miehei	MWCNT	Waste cooking oil	94%	Fan et al. 2016
Candida antartica	MWCNTs	*J. curcas* L. seed oil	95.2%	Wang et al. 2018
Aspergillus niger	SWCNTs	Biomass	95%	Goh et al. 2012
Candida cylindracea	MWCNTs	–	74.3%	YiQin et al. 2016

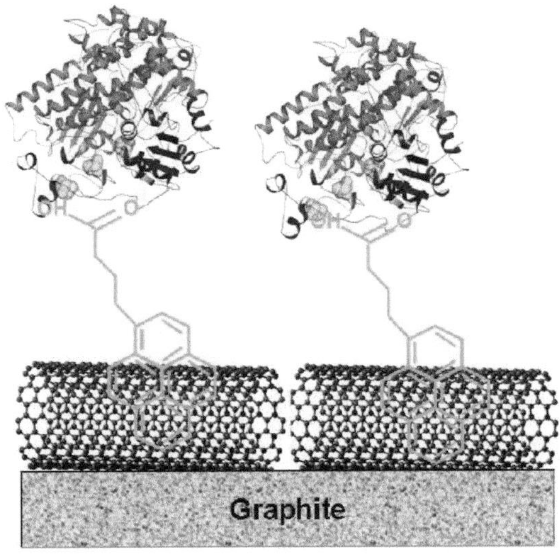

Figure 2. Schematic representation of the build-up of enzyme on SWCNTs via π-π interactions (adapted from Poulpiquet et al. 2013) http://dx.doi.org/10.5772/51782.

tyrosinase) with specific enzymatic potential form the basis of biofuels. The concept was made into reality by Mano et al. (2003) by implanting micro-bioelectrodes to a grape that produced 2.4 μW at 0.54 V. The concept was later extended to insects (Rasmussen et al. 2012), rabbits (Miyake et al. 2011), rats (Sales et al. 2013), etc. However, the efficacy of these enzymatic fuel cells was not up to the level as the power and voltage were far beneath the recommended level, and due to the same reason, they were not able to meet the energy demands of implanted biomedical devices. Since then continuous efforts were made to enhance the performance of biofuel cells.

The initial concept of electrodes with enzymatic activity was extensively modified to develop CNT-based electrodes. The electrodes with compressed CNT were stacked with pyrene-butanoic acid and made active by chemical implantation of biocathode (laccase) and bioanode (glucose dehydrogenase). The modified technique has been validated by implanting enzymatic-CNT based fuel cells into snails (Halámková et al. 2012), rats (Andoralov et al. 2013), clams (Szczupak et al. 2012) and lobsters (Southcott et al. 2013). This validation brings a breakthrough in biofuel cell research as these enzymatic-CNT based fuel cells are sufficient enough to power an implanted biomedical device like a pacemaker. The researchers were successful in surgical implanting of enzymatic-CNT based fuel cells into the retroperitoneal space of rats which were able to produce a voltage of 0.85 V (Zebda et al. 2013).

The obtrusive and stringent protocols followed for the implantation of enzymatic fuel cells particularly in humans should definitely assure benchmarks, such as stable performance, biocompatibility, etc. To overcome this, researchers have developed non-intrusive enzymatic cells that can produce energy utilising metabolic ingredients available on the epidermis or even in subcutaneous level (Valdés-Ramírez et al. 2014, Cosnier et al. 2014).

10. Conclusion

The interesting combination of biology and nanotechnology allows exciting opportunity and active field of interest for enhancing the efficiency of biofuel production. Enzyme immobilisation on CNTs has advanced the field of CNT research. The immobilised enzyme properties depend on the procedure of immobilisation. The non-covalent approach is the most studied approach for the enzyme immobilisation, and this technique preserves the intrinsic electronic structure and properties of CNTs. The functionalisation of CNTs with different biomolecules and polymers are an active area of research. Thus, CNTs attained a growing interest for the immobilisation of catalytic enzymes and development of biofuel cells.

Acknowledgement

Aravind Madhavan acknowledges DST-SERB for providing NPDF. Raveendran Sindhu acknowledges DST for sanctioning a project under DST WOS-B Scheme.

References

Ahmad, R. and S.K. Khare. 2018. Immobilization of *Aspergillus niger* cellulase on multiwall carbon nanotubes for cellulose hydrolysis. Bioresource Technology 252: 72–75.

Andoralov, V., M. Falk, D.B. Suyatin, M. Granmo, J. Sotres, R. Ludwig, P.D. Popov, J. Schouenborg, Z. Blum and S. Shleev. 2013. Biofuel cell based on microscale nanostructured electrodes with inductive coupling to rat brain neurons. Scientific Reports 3: 3270.

Arakaki, A., J. Webb and T. Matsunaga. 2003. A novel protein tightly bound to bacterial magnetic particles in *Magnetospirillum magneticum* strain AMB-1. Journal of Biological Chemistry 278: 8745–8750.

Arepalli, S. 2004. Laser ablation process for single-walled carbon nanotube production. Journal of Nanoscience and Technology 4: 317–325.

Arya, A., K. Gupta, T.S. Chundawat and D. Vaya. 2018. Biogenic synthesis of copper and silver nanoparticles using green alga *Botryococcus braunii* and its antimicrobial activity. Bioinorganic Chemistry and Applications.

Asuri, P., S.S. Karajanagi, H. Yang, T.J. Yim, R.S. Kane and J.S. Dordick. 2006a. Increasing protein stability through control of the nanoscale environment. Langmuir 22: 5833–5836.

Atchudan, R., B.G. Cha, N. Lone, J. Kim and J. Joo. 2019. Synthesis of high-quality carbon nanotubes by using monodisperse spherical mesoporous silica encapsulating iron oxide nanoparticles. Korean Journal of Chemical Engineering 36: 157–165.

Calugay, R.J., H. Miyashita, Y. Okamura and T. Matsunaga. 2003. Siderophore production by the magnetic bacterium Magnetospirillum magneticum AMB-1. FEMS Microbiology Letters 218: 371–375.

Cang-Rong, J. and G. Pastorin. 2009. The influence of carbon nanotubes on enzyme activity and structure: investigation of different immobilization procedures through enzyme kinetics and circular dichroism studies. Nanotechnology 20: 255102.

Çelik, A., A. Dinçer and T. Aydemir. 2016. Characterization of β-glucosidase immobilized on chitosan-multiwalled carbon nanotubes (MWCNTS) and their application on tea extracts for aroma enhancement. International Journal of Biological Macromolecules 89: 406–414.

Chen, R.J., S. Bangsaruntip, K.A. Drouvalakis, N.W.S. Kam, M. Shim, Y. Li, W. Kim, P.J. Utz and H. Dai. 2003. Noncovalent functionalization of carbon nanotubes for highly specific electronic biosensors. Proceedings of the National Academy of Sciences of the United States of America 100: 4984–4989.

Christwardana, M., K.J. Kim and Y. Kwon. 2016. Fabrication of mediatorless/membraneless glucose/oxygen based biofuel cell using biocatalysts including glucose oxidase and laccase enzymes. Scientific Reports 6: 1–10.

Christwardana, M., Y. Chung and Y. Kwon. 2017. A new biocatalyst employing pyrenecarboxaldehyde as an anodic catalyst for enhancing the performance and stability of an enzymatic biofuel cell. NPG Asia Materials 9: e386–9.

Chung, M., T.L. Nguyen, T.Q.N. Tran, H.H. Yoon, I.T. Kim and M.I. Kim. 2018. Ultra rapid sonochemical synthesis of enzyme-incorporated copper nanoflowers and their application to mediator less glucose biofuel cell. Applied Surface Science 429: 203–209.

Cosnier, S., M. Holzinger and A. Le Goff. 2014. Recent advances in carbon nanotube-based enzymatic fuel cells. Frontiers in Bioengineering and Biotechnology 24: 2–45.

Deep, A., A.L. Sharma and P. Kumar. 2015. Lipase immobilized carbon nanotubes for conversion of Jatropha oil to fatty acid methyl esters. Biomass and Bioenergy 81: 83–87.

Durán, N., P.D. Marcato, M. Durán, A. Yadav, A. Gade and M. Rai. 2011. Mechanistic aspects in the biogenic synthesis of extracellular metal nanoparticles by peptides, bacteria, fungi, and plants. Applied Microbiology and Biotechnology 90: 1609–1624.

Fan, Y., G. Wu, F. Su, K. Li, L. Xu, X. Han and Y. Yan. 2016. Lipase oriented-immobilized on dendrimer-coated magnetic multi-walled carbon nanotubes toward catalyzing biodiesel production from waste vegetable oil. Fuel 178: 172–178.

Fan, Y., F. Su, K. Li, C. Ke and Y. Yan. 2017. Carbon nanotube filled with magnetic iron oxide and modified with polyamidoamine dendrimers for immobilizing lipase toward application in biodiesel production. Scientific Reports 7: 45643.

Feng, W. and P. Ji. 2011. Enzymes immobilized on carbon nanotubes. Biotechnology Advances 29(6): 889–895.

Feng, W., X. Sun and P. Ji. 2012. Activation mechanism of *Yarrowia lipolytica* lipase immobilized on carbon nanotubes. Soft Matter 8: 7143.

Fleming, E., F. Du, E. Ou, L. Dai and L. Shi. 2019. Thermal conductivity of carbon nanotubes grown by catalyst-free chemical vapor deposition in nanopores. Carbon 145: 195–200.

Gao, Y. and I. Kyratzis. 2008. Covalent immobilization of proteins on carbon nanotubes using the cross-linker 1-ethyl-3-(3-dimethylaminopropyl) carbodiimide—a critical assessment. Bioconjugate Chemistry 19(10): 1945–1950.

Gomez, J.M., M.D. Romero and T.M. Fernandez. 2005. Immobilization of β-glucosidase on carbon nanotubes. Catalysis Letters 101: 3–4.

Goh, W.J., V.S. Makam, J. Hu, L. Kang, M. Zheng, S.L. Yoong and C.N.B. Udalagama. 2012. Pastoring iron oxide filled magnetic carbon nanotube−enzyme conjugates for recycling of amyloglucosidase: Toward useful applications in biofuel production process. Langmuir 28: 16864−16873.

Guan, Q., Y. Li, Y. Chen, Y. Shi, Gu, B. Li, R. Miao, Q. Chena and P. Ning. 2017. Sulfonated multi-walled carbon nanotubes for biodiesel production through triglycerides transesterification. RSC Advances 7: 7250.

Halámková, L., J. Halámek, V. Bocharova, A. Szczupak, L. Alfonta and E. Katz. 2012. Implanted biofuel cell operating in a living snail. Journal of American Chemical Society 134: 5040–5043.

Herrera-Ramirez, J.M., R. Perez-Bustamante and A. Aguilar-Elguezabal. 2019. An overview of the synthesis, characterization, and applications of carbon nanotubes. pp. 47–75. *In*: Carbon-Based Nanofillers and Their Rubber Nanocomposites. Elsevier.

Holzinger, M., A. Le Goff and S. Cosnier. 2012. Carbon nanotube/enzyme biofuel cells. Electrochimica Acta 82: 179–190.

Iijima, S. and T. Ichihashi. 1993. Single-shell carbon nanotubes of 1-nm diameter. Nature 363: 603–605. Doi: 10.1038/363603a0.

Ingale, A.G. and A. Chaudhari. 2013. Biogenic synthesis of nanoparticles and potential applications: an eco-friendly approach. Journal of Nanomedicine and Nanotechnology 4: 1–7.

Ji, P., H. Tan, X. Xu and W. Feng. 2010. Lipase covalently attached to multi-walled carbon nanotubes as an efficient catalyst in organic solvent. AIChE Journal 56: 3005–3011.

Jiang, K.Y., L.S. Schadler, R.W. Siegel, X.J. Zhang, H.F. Zhang and M. Terrones. 2004. Protein immobilization on carbon nanotubes via a two-step process of diimide-activated amidation. Journal of Materials Chemistry 14: 37–39.

Kang, Z., K. Jiao, J. Cheng, R. Peng, S. Jiao and Z. Hu. 2018. A novel three-dimensional carbonized PANI 1600 @CNTs network for enhanced enzymatic biofuel cell. Biosensors and Bioelectronics 101: 60–65.

Karajanagi, S.S., A.A. Vertegel, R.S. Kane, J.S. Dordick. 2004. Structure and function of enzymes adsorbed onto single-walled carbon nanotubes. Langmuir 20: 11594–9.

Kienle, D.F., R.M. Falatach, J.L. Kaar and D.K. Schwartz. 2018. Correlating structural and functional heterogeneity of immobilized enzymes. ACS Nano 12(8): 8091−8103.

Kim, J.B., J.W. Grate and P. Wang. 2006. Nanostructures for enzyme stabilization. Chemical Engineering Science 61: 1017–1026.

Laurent, N., R. Haddoub and S.L. Flitsch. 2008. Enzyme catalysis on solid surfaces. Trends in Biotechnology 26: 328–337.

Lee, C.A. and Y.C. Tsai. 2009. Preparation of multiwalled carbon nanotube–chitosan–alcoholdehydrogenase nanobiocomposite for amperometric detection of ethanol. Sensors & Actuators, B 138: 518–523.

Lee, D. and J. Seo. 2010. Preparation of carbon nanotubes from graphite powder at room temperature arXiv preprint. arXiv: 10071062.

Lee, H.K., J.K. Lee, M.J. Kim and C.J. Lee. 2010a. Immobilization of lipase on single walled carbon nanotubes in ionic liquid. Bulletin of the Korean Chemical Society 31: 650–652.

Lee, J.H., M. Kim, B. Yoo, N.V. Myung, J. Maeng, T. Lee, A.C. Dohnalkova, J.K. Fredickson, M.J. Sadowsky and H. Hur. 2007. Biogenic formation of photoactive arsenic-sulfide nanotubes by *Shewanella* sp. strain HN-41. Proceedings of the National Academy of Sciences 104: 20410–20415. Doi: 10.1073/pnas.0707595104.

Lee, S.H., T.T.N. Doan, K. Won, S.H. Ha and Y.M. Koo. 2010b. Immobilization of lipase within carbon nanotube–silica composites for non-aqueous reaction systems. Journal of Molecular Catalysis B: Enzymatic 62: 169–172.

Li, S.C., J.H. Chen, H. Cao, D.S. Yao and Liu D.L. 2011. Amperometric biosensor for aflatoxin B1 based on aflatoxin-oxidase immobilized on multiwalled carbon nanotubes. Food Control 22: 43–49.

Lv, P., Q. Feng, Q. Wang, G. Li, D. Li and Q. Wei. 2016. Biosynthesis of bacterial cellulose/carboxylic multi-walled carbon nanotubes for enzymatic biofuel cell application. Materials 9: 183.

MacVittie, K., J. Halámek, L. Halámková, M. Southcott, W.D. Jemison, R. Lobel and E. Katz. 2013. From "Cyborg" lobsters to a pacemaker powered by implantable biofuel cells. Energy and Environmental Science 6: 81–86.

Mano, N., F. Mao and A. Heller. 2003. Characteristics of a miniature compartment-less glucose/O_2 biofuel cell and its operation in a living plant. Journal of American Chemical Society 125: 6588–6594.

Markiton, M., S. Boncel, D. Janas and A. Chrobok. 2017. Highly active nanobiocatalyst from lipase noncovalently immobilized on multiwalled carbon nanotubes for baeyer–villiger synthesis of lactones. ACS Sustainable Chemistry & Engineering 5(2): 1685–1691.

Martin, W., W. Zhu and G. Krilov. 2008. Simulation study of non-covalent hybridization of carbon nanotubes by single-stranded DNA in water. The Journal of Physical Chemistry B 112: 16076–16089.

Matsuura, K., T. Saito, T. Okazaki, S. Ohshima, M. Yumura and S. Iijima. 2006. Selectivity of water soluble proteins in single-walled carbon nanotube dispersions. Chemical Physics Letters 429: 497–502.

Miyake, T., K. Haneda, N. Nagai, Y. Yatagawa, H. Onami, S. Yoshino, T. Abe and M. Nishizawa. 2011. Enzymatic biofuel cells designed for direct power generation from biofluids in living organisms. Energy & Environmental Science 4: 5008–5012.

Moghaddam, M., S. Taylor, M. Gao, S. Huang, L. Dai and M.J. McCall. 2004. Highly efficient binding of DNA on the side walls and tips of carbon nanotubes using photochemistry. Nano Letters 4: 89–93.

Mu, Q., W. Liu, Y. Xing, H. Zhou, Z. Li, Y. Zhang, L. Ji, F. Wang, Z. Si, B. Zhang and B. Yan. 2008. Protein binding by functionalized multiwalled carbon nanotubes is governed by the surface chemistry of both parties and the nanotube diameter. Journal of Physical Chemistry 112: 3300–3307.

Mubarak, N.M., J.R. Wonga, K.W. Tana, J.N. Sahuc, E.C. Abdullah, N.S. Jayakumar and P. Ganesan. 2014. Immobilization of cellulase enzyme on functionalized multiwall carbon nanotubes. Journal of Molecular Catalysis B: Enzymatic 107: 124–131.

Mukherjee, A., B. Debnath and S.K. Ghosh. 2019. Carbon nanotubes as a resourceful product derived from waste plastic—a review. pp. 915–934. *In*: Waste Management and Resource Efficiency. Springer.

Narayanan, K.B. and N. Sakthivel. 2011. Green synthesis of biogenic metal nanoparticles by terrestrial and aquatic phototrophic and heterotrophic eukaryotes and biocompatible agents. Advances in Colloid and Interface Science 169: 59–79.

Nepal, D. and K.E. Geckeler. 2006. pH-sensitive dispersion and debundling of single-walled carbon nanotubes: lysozyme as a tool. Small 2: 406–412.

Nguyen, C.V. 2019. Methods and devices for synthesis of carbon nanotubes. Google Patents.

Oliveira, B.L. and V. Teixeira da Silva. 2014. Sulfonated carbon nanotubes as catalysts for the conversion of levulinic acid into ethyl levulinate. Catalysis Today 234: 257–263.

Ong, C.B. and M.S.M. Annuar. 2018. Immobilization of cross-linked tannase enzyme on multiwalled carbon nanotubes and its catalytic behaviour. Preparative Biochemistry and Biotechnology 48(2): 181–187.

Ozden, S., I.G. Macwan, P.S. Owuor, S. Kosolwattana, P.A.S. Autreto, S. Silwal, R. Vajtai, C.S. Tiwary, A.D. Mohite, P.K. Patra and P.M. Ajayan. 2017. Bacteria as bio-template for 3D carbon nanotube architectures. Scientific Reports 7: 9855.

Pang, H.L., J. Liu, D. Hu, X.H. Zhang and J.H. Chen. 2010. Immobilization of laccase onto 1-aminopyrene functionalized carbon nanotubes and their electrocatalytic activity for oxygen reduction. Electrochimica Acta 55: 6611–6616.

Patolsky, F., Y. Weizmann and I. Willner. 2004. Long-range electrical contacting of redox enzymes by SWNT connectors. Angewandte Chemie 116: 2165–2169.

Paul, S. and S. Samdarshi. 2011. A green precursor for carbon nanotube synthesis. New Carbon Materials 26: 85–88.

Pavlidis, I.V., T. Tsoufis, A. Enotiadis, D. Gournis and H. Stamatis. 2010. Functionalized multi-wall carbon nanotubes for lipase immobilisation. Advanced Engineering Materials 12: B179–B183.

Peigney, A., C. Laurent, E. Flahaut, R.R. Bacsa and A. Rousset. 2001. Specific surface area of carbon nanotubes and bundles of carbon nanotubes. Carbon N.Y. 39: 507–514.

Poulpiquet, A., A. Ciaccafava, S. Benomar, M. Giudici-Orticoni and E. Lojou. 2013. Carbon-nanotube-enzyme biohybrids in a green hydrogen economy. pp. 433–466. *In*: Synthesis and Applications of Carbon Nanotubes and their Composites. Intech Open. http://dx.doi.org/10.5772/51782.

Prasad, K.P., Y. Chen and P. Chen. 2014. Three-dimensional graphene-carbon nanotube hybrid for high-performance enzymatic biofuel cells. ACS Applied Materials & Interfaces 6: 3387–3393.

Prasek, J., J. Drbohlavova, J. Chomoucka, J. Hubalek, O. Jasek, V. Adam and R. Kizek. 2011. Methods for carbon nanotubes synthesis. Journal of Materials Chemistry 21(40): 15872–15884.

Rasmussen, M., R.E. Ritzmann, I. Lee, A.J. Pollack and D. Scherson. 2012. An implantable biofuel cell for a live insect. Journal of American Chemical Society 134: 1458–1460.

Rastian, Z., A.A. Khodadadi, Z. Guo, F. Vahabzadeh and Y. Mortazavi. 2016. Plasma functionalized multiwalled carbon nanotubes for immobilization of *Candida antarctica* lipase B: production of biodiesel from methanolysis of rapeseed oil. Applied Biochemistry and Biotechnology 178(5): 974–89.

Sales, F.C.P.F., R.M. Iost, M.V.A. Martins, M.C. Almeida and F.N. Crespilho. 2013. An intravenous implantable glucose/dioxygen biofuel cell with modified flexible carbon fibre electrodes. Lab on a Chip 13: 468–474.

Shah, S., K. Solanki and M.N. Gupta. 2007. Enhancement of lipase activity in non-aqueous media upon immobilisation on multi-walled carbon nanotubes. BMC Chemistry 1: 30.

Sharma, D., S. Kanchi and K. Bisetty. 2015. Biogenic synthesis of nanoparticles: A review. Arabian Journal of Chemistry Doi: https://doi.org/10.1016/j.arabjc.2015.11.002.

Smart, S.K., A.I. Cassady, G.Q. Lu and D.J. Martin. 2006. The biocompatibility of carbon nanotubes. Carbon N.Y. 44: 1034–1047.

Szczupak, A., J. Halámek, L. Halámková, V. Bocharova, L. Alfontac and E. Katz. 2012. Living battery–biofuel cells operating *in vivo* in clams. Energy and Environmental Science 5: 8891–8895.

Tan, H., W. Feng and P. Ji. 2011. Lipase immobilized on magnetic multi-walled carbon nanotubes. Bioresource Technology 115: 172–176.

Umasankar, Y., B.R. Adhikari and A. Chen. 2017. Effective immobilization of alcohol dehydrogenase on carbon nanoscaffolds for ethanol biofuel cell. Bioelectrochemistry 118: 83–90.

Valdés-Ramírez, G., L. Ya-Chieh, J. Kima, W. Jiaa, A.J. Bandodkara and R. Nuñez-Flores. 2014. Microneedle-based self-powered glucose sensor. Electrochemistry Communications 47: 58–62.

Valentini, L., S.B. Bon, S. Signetti, M. Tripathi, E. Iacob and N.M. Pugno. 2016. Fermentation based carbon nanotube bionic functional composites arXiv preprint. arXiv: 160305407.

Verma, M.L., M. Naebe, C.J. Barrow and M. Puri. 2013. Enzyme immobilisation on amino-functionalised multi-walled carbon nanotubes: structural and biocatalytic characterisation. PLoS ONE 8(9): e73642.

Venkatesan, S., B. Visvalingam, G. Mannathusamy, V. Viswanathan and A.G. Rao. 2019. *In situ* synthesis of multi-walled carbon nanorings by catalytic chemical vapor deposition process. International Nano Letters: 1–8.

Wang, Liu X., Y. Jiang, L. Zhou, L. Ma, Y. He and Y. Gao. 2018. Biocatalytic pickering emulsions stabilized by lipase-immobilized carbon nanotubes for biodiesel production. Catalysts 8(12): 587.

Withey, G.D., J.H. Kim and J. Xu. 2008. DNA-programmable multiplexing for scalable, renewable redox protein bio-nanoelectronics. Bioelectrochemistry 74: 111–117.

Wu, X., B. Zhao, P. Wu, H. Zhang and H. Cai. 2009. Effects of ionic liquids on enzymatic catalysis of the glucose oxidase toward the oxidation of glucose. The Journal of Physical Chemistry B 113: 13365–13373.

Yim, T.J., J. Liu, Y. Lu, R.S. Kane and J.S. Dordick. 2005. Highly active and stable DNAzyme-carbon nanotube hybrids. Journal of American Chemical Society 9: 12200–12201.

YiQin, Z., W. Zheng, T. AiXing, L. QingYun, L. YunRu and L. YouYan. 2016. Immobilization of lipase on functionalized multi-walled carbon nanotubes for biodiesel preparation. Renewable Energy Resources 34(9): 1411–1416.

Yu, C.M., M.J. Yen and L.C. Chen. 2010. A bioanode based on MWCNT/protein-assisted co-immobilization of glucose oxidase and 2, 5-dihydroxybenzaldehyde for glucose fuel cells. Biosensors and Bioelectronics 25: 2515–2521.

Zebda, A., S. Cosnier, J.P. Alcaraz, M. Holzinger, A. Le Goff, C. Gondran, F. Boucher, F. Giroud, K. Gorgy, H. Lamraoui and P. Cinquin. 2013. Single glucose biofuel cells implanted in rats power electronic devices. Scientific Reports 3: 1516.

Zhao, G., Y. Li, J. Wang and H. Zhu. 2011. Reversible immobilization of glucoamylase onto magnetic carbon nanotubes functionalized with dendrimer. Applied Microbiology and Biotechnology 91: 591–601.

Zhao, X., H. Jia, J. Kim and P. Wang. 2009. Kinetic limitations of a bioelectrochemical electrode using carbon nanotube-attached glucose oxidase for biofuel cells. Biotechnology and Bioengineering 104: 1068–1074.

Chapter 4

Role of Chitosan Nanotechnology in Biofuel Production

Meenu Thakur,[1] *Rekha Kushwaha*[2] and *Madan L. Verma*[3,*]

1. Introduction

An increase in population, technological advancement and more urbanization has led to a depletion of energy sources (Noraini et al. 2014). Energy sources are significant for the sustainability of mankind (Huang et al. 2012). This crisis of energy has become a major global issue. Moreover, many aspects of human life depend on fuels as they can produce energy which can be used in transportation (Huang et al. 2012). The main alternative sources of energy are fossil fuels, like oil, coal, petroleum and natural gas. Fossil fuels contribute 80% of the energy requirements of the world. They have been produced over millions of years through activities of various microorganisms. Fossil fuels are non-renewable energy sources. Due to their limited resources, availability and increasing demand, they cannot meet the requirement of our existing population (Demirbas 2009). The increase in oil prices has led to economic recessions and crisises in many countries. Another serious concern is related to the harmful emissions by the combustion of such fuels resulting in environmental pollution and global warming (Agarwal 2007, Zhou et al. 2009, Prabhakar 2009). Thus, the need for renewable environment-friendly alternative sources of energy has received much attention. Development of biological fuels such as biodiesel, biohydrogen and biogas has attained significant attention in previous years. Significant reasons for their popularity are its environment-friendly nature, less toxicity, reduced emissions and overall less cost (Noraini et al. 2014). Most of

[1] Department of Biotechnology, Shoolini Institute of Life Sciences and Business Management, Solan – 173212, Himachal Pradesh, India.

[2] Department of Biochemistry, 242 Christopher S. Bond Life Sciences Center, 1201 Rollins, Street, Columbia, MO 65211, USA.

[3] Department of Biotechnology, School of Basic Sciences, Indian Institute of Information Technology Una – 177220, Himachal Pradesh, India.

* Corresponding author: madanverma@gmail.com; madanverma@iiitu.ac.in

the biofuels can be based on renewable plant sources, viz., algae, soybean, jatropha, corn, coconut, rice bran, linseed, etc. (Ong et al. 2011). All the biofuels can be produced by transesterification reactions. Various enzymes such as lipases, cellulases, etc., enhance the esterification process of hydrolysis of fats/lipids from natural plant-based sources (Lau et al. 2014). Immobilization of catalysts with a suitable matrix produces a qualitative impact on various properties of enzymes, such as reusability, easy recovery and fewer chances of contamination. It significantly improves the operational stability and the performance of the enzyme (Seema and Steven 2002, Cabral et al. 2010, Ines et al. 2011). Chitosan derived from chitin is a natural polymer. It is water-insoluble and provides the most suitable support due to its biocompatibility and high mechanical strength (Lau et al. 2014). As nanobiotechnology is a new and emerging field of biotechnology, novel nano-scaffolds using chitosan nanomaterials can be used for effective immobilization of enzymes (Verma et al. 2013).

Different strategies have been employed for improving the production of biofuels, specifically biodiesel and bioethanol. Biocatalysts such as lipase using magnetic nanoparticles have been reviewed for transesterification processes (Thangaraj et al. 2019). Transesterification has been the preferred method over other existing methods, and it depends upon the biocatalysts. Various fungal cells, such as nematophagous fungal cells *Pochonia chlamydosporia, Metarhizium anisopliae* and *Beauveria bassiana*, are capable of utilizing chitosan as the nutrient. These fungal cells contained alcohol dehydrogenase and pyruvate decarboxylase enzymes that helped in the conversion of simple sugars into ethanol. *Pochonia chlamydosporia* produced maximum ethanol using chitosan. Second-generation biofuel production using lignocellulosic wastes have been a source for bioethanol production using chitosan (Aranda-Martinez et al. 2017). Transesterification reactions using coconut oil have been studied by immobilized lipase on chitosan in a fluidized bed reactor. Immobilization has improved stability by retaining 77.7% of initial activity after 6 months at 5°C. Moreover, the reusability of the immobilized catalyst was increased up to 5 cycles (Silva et al. 2017). In another similar study, synthesis of ethyl esters from coconut oil using lipase from *Pseudomonas fluorescens* immobilized on chitosan with magnetic properties. The unconventional electromagnetic field has been used for easy separation of nanoparticles for enhancing the reuse of biocatalyst (Cubides-Roman et al. 2017). An advanced biocatalyst has been designed for simultaneous hydrolysis, isomerization and fermentation (SHIF) to continuously produce ethanol (Milessi-Esteves et al. 2019).

The chitosan nanotechnology has been applied for immobilization of enzymes and in turn for biofuel production. This has led to the development of novel technology for effective biofuel production. In the present chapter various chitosan nanomaterials and their production technologies as well as their role in biofuel production have been discussed.

2. Chitosan as Nanomaterial

The nanomaterial is a matter with measurements of approximately 1–100 nm which helps in various important applications. Chitosan is a direct starch polymer made of β-1-4 connected D-glucosamine. It was initially revealed by Rouget in 1859. It is polycationic polysaccharide with various properties, such as biocompatibility, biodegradability and low immunogenicity (Abdulhussein and Alsalman 2017). The physical and chemical properties of chitosan can be affected by atomic weight and level of deacetylation (Abdulhussein and Alsalman 2017). However, different factors play an important role in the evaluation of various nanomaterials. Surface area to volume ratio is one of the important factors as nanofibers give more 2/3rd of the surface to volume ratio as compared to spherical particles (Wang 2006, Gupta et al. 2011). Second important factor in the evaluation is enzyme loading. Nanomaterials provide more enzyme loading due to its large surface area. High enzyme loading results in more biocatalytic activity and stability. A third important factor in evaluation is the flow rate. Nanoparticles immobilized on suitable support exhibit Brownian movements (Wang 2006). Other important factors are mass transfer, ease of separation and reactor design (Verma et al. 2013).

2.1 Enzyme Loading

Nanometer-scale materials can give the advantage of enzyme loading due to their large relative surface area. These nano-immobilization techniques are better than conventional methods.

2.2 Flow Rate

The immobilized nanomaterial gives a stable matrix with Brownian motion (Wang 2006). Immobilized enzymes have more diffusibility than free enzymes due to Stokes-Einstein equation. Brownian movements give a difference in activities of free and immobilized enzymes (Jia et al. 2002).

2.3 Mass Transfer

Immobilization also helps to resolve the limitation of mass transfer and thus increases the enzyme activity as compared to free enzymes. Therefore, immobilized enzymes using nanomaterials have low mass transfer resistance which results in high activity and stability (Kim et al. 2006).

2.4 Ease of Separation

Magnetic nanoparticles can be used to immobilize the enzyme for easier separation as magnets can be used for separation of the product (Safarik and Safarikova 2009, Ren et al. 2011). Immobilized enzymes have more reusability and stability as compared to conventional matrices. Moreover, this may provide a leaching reaction (Yiu and Keane 2012). However, the magnetic matrices offer an inexpensive and stable option for immobilization of enzymes.

2.5 Reactor Design

Enzyme efficiency can be enhanced by using appropriate reactors (Sotowa et al. 2008). Enzyme immobilization can be used for developing advanced analysis systems (Song et al. 2012). Higher surface area and flexibility are the main properties exhibited by nanomaterials (Nair et al. 2007). Enzyme immobilized using nanofibers associate added advantages such as less pressure drop and high flow rate as compared to conventional bioreactors. In a similar work, an enzyme immobilized fibrous bioreactor was developed with a continuous study using steady hydrolysis with a constant flow rate using optimum conditions (Huang et al. 2008). Methods for immobilizing the enzymes can use a range of different methods (Figure 2) such as adsorption and cross-linking methods. Various cross-linkers can be used in the immobilization method. The choice of cross-linker will depend upon the application of the enzyme. The best optimized cross-linker for one method may not be the most appropriate for another method. However, covalent immobilization is the best method for immobilization where activation can be achieved using glutaralde (Huang et al. 2003, Verma et al. 2012).

Although there have been several studies on immobilization of enzymes using different matrices, not much work has been done on properties of nanomaterials upon immobilization like composition, morphology and surface chemistry (Asuri et al. 2007). These interactive studies will strengthen the effects of nanomaterial immobilization of enzymes. Several techniques are used to evaluate and characterize enzymes immobilized on a suitable matrix (Ganesan et al. 2009). Conformational changes upon immobilization have been studied by different researchers using various analytical methods such as Brunauer-Emmett-Teller (BET) analysis, atomic force microscopy (AFM), scanning electron microscope (SEM), transmission electron microscopy (TEM), circular dichroism (CD) spectroscopy, x-ray diffraction, x-ray photoelectron spectroscopy and Fourier transform infrared spectroscopy (FTIR). However, the morphology and size of nanomaterial can be observed using TEM and SEM. The binding of lipase can be confirmed by using different analytical techniques. The CD spectroscopy has also been used to analyze the influence of physical/covalent attachments on the secondary structure of enzyme 62% retention of the α-helix content of the native lipase after immobilization has been noticed (Ji et al. 2010). AFM can be used as successful method of analysis (Zhang et al. 2010a). Nanoparticles can enhance the strength of chitosan. Their small size and impact on quantum make them better alternatives for the preparation of nanomaterials. These can offer a variety of surface to volume ratios and ease to scale-up. Mucoadhesive and hydrophilic nature provide additional properties to the chitosan nanoparticles (Abdulhussein and Alsalman 2017). One of the important properties is physical assembling which is based on atomic weight, solid electrostatic attraction, chain adaptability and flexibility to alter the C-2 position in structure (Sinha et al. 2004, Huo et al. 2010). Particles' size immensely affects the performance of nanomaterial as small size will have an impact on surface area (Bhatia 2016). Nanoparticles using chitosan can be subdivided into various classes (see Figure 1).

3. Methods for Immobilization

The major application of nanomaterials is to provide stability and enhanced activity of catalyst along with better reusability. Nanoparticles scaffolds can help in the above-mentioned properties (Verma et al. 2013). Chitosan nanoparticles can be classified in magnetic and nonmagnetic nanoparticles depending on the use of Fe_2O_3

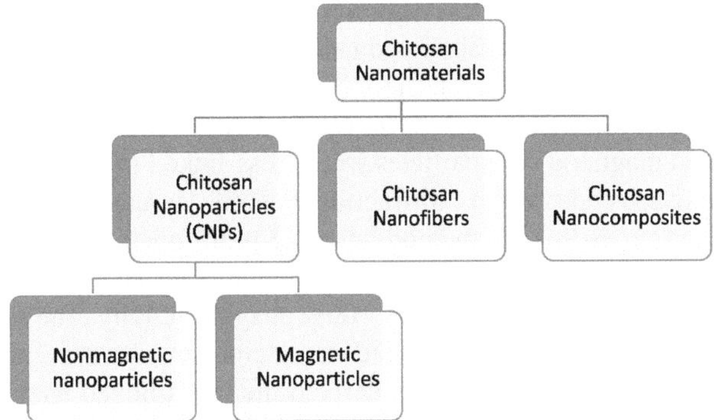

Figure 1. Classification of chitosan nanomaterials.

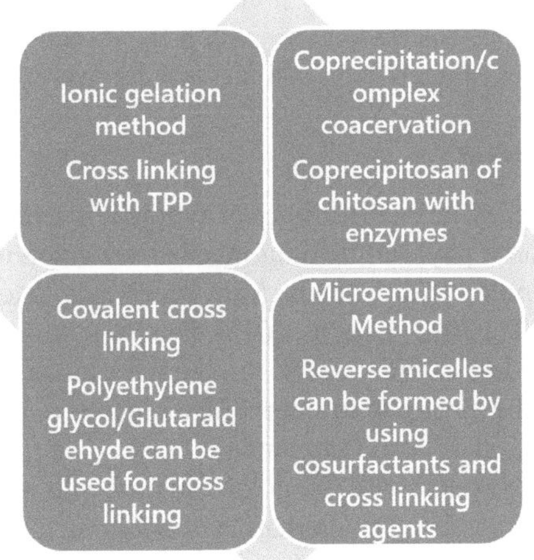

Figure 2. Representation of different methods of preparation of nanoparticles.

nanoparticles. Non-magnetic chitosan nanoparticles need ultra-high centrifugation procedures for separation which limits the use of this type of nanoparticles. Moreover, difficulty in separation increases the cost of operation due to low/less reusability of immobilized enzymes. Magnetic nanoparticles have an advantage like more stability, superparamagnetic property, biocompatibility and low cost. Several strategies have been used to develop different immobilization methods using chitosan.

3.1 Emulsification and Cross-Linking

This method was among first to use for the formation of chitosan nanoparticles. The method involves emulsification of chitosan using emulsifiers such as toluene, span 80 and then cross-linking agents to harden the nanoparticles. The reactive cross-linking agent is used to harden the nanoparticles. Reactive cross-linking was achieved using glutaraldehyde (Ohya et al. 1994). Cross-linking properties and stirring speed influences the final size of particles (Agnihotri et al. 2004, Prabharan and Mano 2005). Chitosan nanoparticles can be fabricated using cross-linking methods (Denkbas and Odobasi 2000). On the other hand, the magnetic nanoparticles consist of magnetic components which have several biotechnological applications (Kefeni et al. 2017). Chitosan-coated magnetic nanoparticles were cross-linked to xylanase and cellulase and were characterized by a x-ray diffraction method (Hernandez et al. 2018).

Most of the cross-linking methods use the multifunctional agents and utilize their stable interactions for cross-linking (Ahmad and Sardar 2015). This method is devoid of support matrix which ensures 100% enzyme activity. The major problems associated are conformational changes and less retainment of enzyme activity (Ahmad and Sardar 2015). Sheldon (2007) has derived the cross-linked enzyme aggregates (CLEAs) method for effective enzyme immobilization. Non-ionic polymers and organic solvents can be used for preparing cross-linked enzyme aggregates which is a simple and effective method. Enzymes such as lipases from *Rhizomucor miehei* and *Thermomyces lanuginosus* can be precipitated using ammonium sulfate, SDS and glutaraldehyde (Lopez-Serrano et al. 2002). By these immobilization methods, the hydrolytic activities of aggregates can increase by two to three-folds.

3.2 Adsorption Method

The adsorption of the enzyme on the surface of a support is a very simple and primitive method. It is based on a physical binding mechanism involving weak interactions, such as a dipole-dipole, hydrophobic, Van der Waals interaction or hydrogen bonding (Hwang and Gu 2013, Es et al. 2015). However, this method involves some disadvantages such as less stability and loss of enzyme molecules. Tang et al. (2004) used a carbon nanotube electrode to immobilize glucose oxidase by the adsorption method. It showed a high response time (~ 5 s), a large determination range (0.1–13.5 mM) and high sensitivity (91 mA/Mcm$_2$) when it was coated with Nefion. The enzyme can be immobilized to alternative matrices such as polyanilin nanofibers to reduce the leaching of enzymes. Kim and co-workers (2011) have compared glucose oxidase (GOD) activity and the stability of enzyme using different adsorption (EA), enzyme adsorption and cross-linking (EAC) and EAPC. The maximum relative activity was reported in EAPC with 100% activity and the highest thermal stability at 50°C.

3.3 Covalent Immobilization

Covalent immobilization involves attaching the enzyme on a suitable nanomatrix by covalent bonding (Ahmad and Sardar 2015). This method is considered as one of the best methods in terms of leaching and also provides thermal stability. However, it involves some of the conformational changes in surface structure which leads to the deactivation of the enzyme by covalent binding to suitable matrics (Hong et al. 2007, Es et al. 2015). In one study based on a comparison between the stability of covalently nanoimmobilized and free enzymes, chymotrypsin immobilized on amine functionalized superparamagnetic nanogel using covalent binding has been studied (Hong et al. 2007). The pH profile was found to be similar in both cases. The immobilized enzyme had more thermal stability as compared to free enzyme.

The free enzyme showed no activity beyond 75°C, whereas the immobilized enzyme retained residual activity (88.7%) until 85°C. Immobilized enzymes displayed high storage stability as compared to free enzyme. In the case of free enzyme, almost full activity was lost after 22 days at 25°C but immobilized enzyme retained 10% of initial activity up to 35 days at 25°C. The Km value of the immobilized enzyme was 1.57 times higher than the native enzyme, and the Vmax value was lower than the free enzyme. This clearly indicated less substrate affinity of the enzyme which might be due to steric hindrance and less diffusibility due to immobilization onto a support. In another study, enzyme aggregates have been developed using chymotrypsin enzyme and immobilizing on nanofibrils (Kim et al. 2005). In the first step, chymotrypsin was attached to the surface of nanofibrils via covalent binding and then remaining free spaces were cross-linked using glutaraldehyde that forms the aggregate coating. The initial activity of the enzyme aggregate was nine times higher using covalent binding as an enzyme coating involves multiple layers of nanofibrils. With respect to enzyme stability, a CT-aggregate-nanofiber was still stable without the loss of its activity during a month. This technology can be applied to various nanomaterials and has potential applications in bioconversion, bioremediation and biosensors (Kim et al. 2005, Liu and Hu 2007).

3.4 Entrapment Immobilization

The entrapment method involves covering the enzyme in a porous gel or fibers (Hwang and Gu 2013). The size of the entrapped magnetite crystallites was found to be approximately 20 nm using TEM analysis. Entrapment can protect the catalyst from external effects due to indirect contact in a confined environment. It minimizes the impact of gas bubbles, mechanical shear and hydrophobic solvents. Entrapment immobilization using nanoparticles were usually based on reverse micelle or sol-gel techniques (Daubresse et al. 1994, Reetz et al. 1998, Yang et al. 2004, Ma et al. 2004, Kim et al. 2008). Entrapment of lipase derived from *Pseudomonas cepacia* on nanostructured magnetite containing hydrophobic sol-gel material has been fabricated. This type of immobilized enzyme had two to three times higher activity than the free enzyme (Reetz et al. 1998).

Yang et al. (2004) reported the simultaneous entrapment of horseradish peroxidase (HPR) and spherical silica-coated nanomagnetite. This technique consisted of two procedures performed in two steps: reverse-micelle and sol-gel processes. Entrapment nano-immobilization using reverse micelle technique produces nanoparticles of uniform size that gives higher and strong dispersion of nanoparticles. Moreover, entrapped biocatalyst using this approach provided high stability at elevated temperature and pH changes. The major problem associated with reverse micelle size is its difficulty in controlling the size and thus rigorous optimization is required (Kim et al. 2008). The sol-gel process involves harsh reaction conditions for the entrapment (Naik et al. 2004). Novel methods involve immobilization using nanomaterial and biomagnetic silica that have given high enzyme stability, loading density and immobilization efficiency (Min and Yoo 2014). A new method was developed involving single enzyme nanoparticle (SEN) (Kim and Grate 2003). Each molecule was surrounded by porous composite with the organic and inorganic networks. SEN fabrication is a two-step procedure in which the first step is surface polymerization using polyvinyl. The next step involves the polymer surface development for entrapment using silanol condensation. This whole entrapment method enhances the catalytic efficiency and its thickness can be controlled. Kinetic parameters involved higher catalytic efficiency and the Km value was similar to free enzyme. This type of entrapment was free of any mass transfer limitation for the substrate. Another method of development of the nanogels using a single enzyme has been studied (Yan et al. 2006). The SENs of HPR showed similar Michaelis-Menten parameters (Km and Kcat), but the thermal stability of the SENs was enhanced up to 65°C. Moreover, the enzyme activity was maintained in the presence of polar organic solvents, such as methanol, tetrahydrofuram and dioxane. The colloidal magnetite-containing lipase was characterized by enzyme activity and was two to three times higher than that of free enzyme. Yan et al. (2006) reported the simultaneous entrapment of horseradish peroxidase (HPR) and spherical silica-coated nanomagnetite. However, this method required a rigorous optimization process because it was difficult to control the reverse micelle size (Kim et al. 2008).

The single enzyme nanoparticles showed similar kinetic parameters with increased thermal stability up to 65°C. Enzyme activity was also maintained in the presence of organic solvents such as methanol, tetrahydrofuran and dioxane.

3.5 Ionic Gelation Method

Chitosan has the property of making hydrogels in the presence of polyanions. Ionic gelation occurs due to inter and intramolecular cross-linkages (Terbojevich and Muzzarelli 2009). Chitosan TPP and vitamin C nanoparticles have been developed by constant magnetic shaking for 1 hour. Different polymer materials such as sodium alginate, carrageenan, dextran sulfate, arabic gum and chondroitin sulfate can be used for ionic gelation (Grenha et al. 2012). The most significant advantage of using the ionic gelation method is the flexibility with which sizes of the nanoparticles can be optimized. In similar studies, chitosan nanoparticles have been fabricated using the ionic gelation method (Agarwal et al. 2018). Furthermore, they were characterized using UV-visible spectroscopy, SEM and FTIR (Agarwal et al. 2018).

3.6 Microemulsion Method

Reverse micelles can be prepared by amphiphilic molecules in the presence of surfactant molecules. The reverse micellar droplets are stabilized by a layer of surfactant molecules and dispersed in oil (Hembram et al. 2016). In most of the cases, glutaraldehyde has been used for the cross-linking agent.

3.6 Other Techniques

Complex coacervation method is among other methods of preparation where coacervates have been formed between cationic chitosan and anionic polyanions. Coprecipitation can be used for precipitating chitosan with a high pH solution for the preparation of monodisperse nanoparticles. Various methods can assess and evaluate nanomaterials on basis of zeta potential, particle size and dimensions, Brownian moments and stretching and bending of atoms in molecules.

4. Role of Enzymes in Biofuels

There are so many advantages for the production of biofuels from lignocellulosic waste/biomass from agriculture and food industries. They are not only cleaner alternatives but also offer several advantages such as:

- Reduces pollution
- Can deal with the shortage of food and feed
- Renewable sources of energy over fossil fuels
- Maintain the economy of countries around the globe

Biomass can be taken as an organic matter grown by photosynthetic conversion of solar energy.

Most of the lignocellulosic biomass comprises cellulose, hemicelluloses and lignin. These lignocellulosic wastes can be converted to simpler monomers by the action of different enzymes (Binod et al. 2019). These biofuels can be classified into first-generation and second-generation based upon the sources. The major drawback of using first-generation biofuels is the limited production of food as well as increasing the prices of food items. The biofuels derived from complex lignocellulosic waste serve as the second generation biofuel. For conversion of lignocellulosic waste into biofuel needs the action of various enzymes. But major technical barrier for adopting enzymes to convert biomass to biofuels is the high cost of the enzymes. An effective strategy is to use some immobilization methods for enzymes which leads to reusability of enzymes and thereby makes them economically viable. There is no single enzyme available which can convert complex lignocellulosic biomass. Various combinations using different enzymes can be used (see Figure 3).

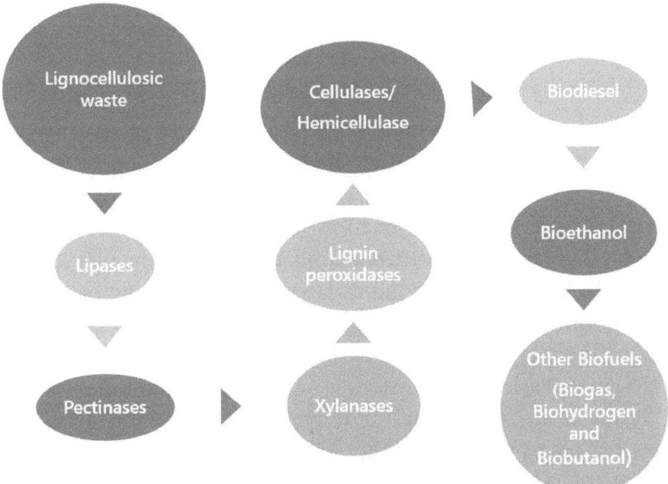

Figure 3. Action of different enzymes on lignocellulosic wastes for biofuel production.

4.1 Cellulases

These are the enzymes responsible for the hydrolysis of cellulose. Conversion of cellulose to simpler glucose is done by three different enzymes:

- Endoglucanases (EC 3.2.1.4) involved in β1-4 glycosidic bonds
- Exocellulases (EC 3.2.1.91) catalyzes the breakage of long chains to short chains
- β-glucosidases (EC 3.2.1.21) hydrolyze the glycosidic bonds

4.2 Xylanases

Xylan is the major component present in hemicelluloses. Xylan degradation in simpler sugars is catalyzed by enzymes known as xylanases. In previous studies, xylanases have been derived from various microorganisms such as *Trichoderma reesei, Humicola insolens* or *Bacillus* (Binod et al. 2019). Different enzymes work singly or via combination with other enzymes for complete hydrolysis of xylan.

- Endo1-4β-xylanases (EC 3.2.1.8) catalyze the glycosidic bonds and results in the production of oligomers.
- β-xylosidase (EC 3.2.1.37) works on oligosaccharides and cellobioses. These enzymes act upon oligosaccharides and cellobioses.
- α-arabinofuranosidase (EC 3.2.1.55) removes the arabinose and glucuronic acids from xylan.
- α-glucuronidase (EC 3.2.1.131) 1,2 glycosidic bonds are hydrolyzed in xylose.
- Esterases (EC 3.1.1.72) catalyzed the xylose units of xylan.

4.3 Lignin Peroxidises (EC 1.11.1.7)

These are lignin-degrading enzymes which cleave lignin.

4.4 Laccases (EC 1.10.3.2)

These are fungal enzymes that can act on lignin and convert them into simpler sugars.

4.5 Lipases (EC 3.1.1.3)

These are enzymes involved in hydrolysis and esterification reactions. These have been studied extensively due to their high stability (Verma et al. 2008, Verma and Kanwar 2008). Lipases can catalyze transesterification reactions in different oils. Transesterification and esterification reactions can be catalyzed by lipases. Transesterification reactions of vegetable oils can convert the biomass into biofuels (Tan et al. 2010). Various recent studies have been carried out on using lipases for the production of biodiesel.

The major limitation of using enzymes for biofuel production is the high cost involved. However, various nanomaterials have been used for the immobilization of enzymes (Verma et al. 2013). It not only provides cost-cutting strategy by reusing the enzymes but also provides stability. Chitosan is more suitable support for the immobilization of enzymes due to low cost, biocompatibility and high strength (Foresti and Ferreira 2007, Huang et al. 2007). In the following section, various immobilization strategies of different enzymes using chitosan nanomaterial for biofuel production have been discussed.

Other problems in the application of enzymes in biodiesel synthesis are enzyme denaturation, high cost and scale-up at reactor level. Enzymes in biodiesel synthesis have inactivation problem due to some substrates/by-products, i.e., methanol and glycerol, respectively (Kumari et al. 2009). Methanol can decrease enzyme activity and biocatalyst efficiency (Ghamguia et al. 2004). Furthermore, a hydrophilic by-product, glycerol can be absorbed on the surface that decreases the enzyme activity. Hence, the selection of an ideal acyl acceptor is an important step in biodiesel production. Several protective measures can be used including stepwise addition of methanol to the reaction mixture, use of methyl or ethyl acetate as acyl acceptors and use of longer alkyl chains alcohols such as t-butanol which can be used to overcome this problem (Royon et al. 2007, Kumari et al. 2009). The high cost involved in enzymatic processes can be reduced using nanomaterials due to high surface area and reusability. Magnetic nanomaterials are materials of choice for efficient separation of products as well as enhanced thermodynamic properties.

5. Chitosan Nanoparticles for Enzyme Immobilization

Nanofibers are solid particles with dimensions ranging from 1–100 nm (Zhao et al. 2011). They are lucrative support matrix for enzyme immobilization. The major advantages are high surface area and more enzyme loading per unit with low diffusion resistance (Wang et al. 2006, Nair et al. 2007). Moreover, the shear stress developed in the batch reactor would disrupt the enzyme carrier. This limits the reusability of immobilized enzymes for a large number of cycles. This limitation can be resolved using a packed bed reactor (PBR) (Wang et al. 2011). Nanomaterials have been successfully employed in biodiesel production. A high conversion (90%) has been achieved in biodiesel production using *T. lanuginosus* lipase (Xie and Ma 2009). Lipase immobilization using GA cross-linker exhibited a lower transesterification rate (40%) as compared to EDC cross-linker (60%). However, maximum biofuel production was achieved using immobilized *Cepacia* lipase (Wang et al. 2009). Butyl biodiesel production has been studied using electrospun polyacrylonitirle nanofiber bound PCL (Sakai et al. 2010). Maximum biodiesel production (94%) has been obtained using immobilized PCL (Li et al. 2011a). During this process, it retained its activity even after 10 cycles.

Thus, the most significant method is immobilizing enzymes using nanocomposites. Magnetic nanocomposite bound lipase gave high conversion of biodiesel (> 90 %) within 30 hours in batch operation and further suggested the importance of novel matrix for immobilization. The immobilized lipase from *Burkholderia* sp. has high methanol tolerance and reusability (Tran et al. 2012). The biofuel production process has been affected by shear stress that has, in turn, created a negative impact on immobilized lipases. A packed bed bioreactor has been developed to minimize shear stress and biodiesel conversion (Wang et al. 2011). Maximum conversion (88%) has been achieved in 192 hours using the packed bed bioreactor. Moreover, this type of reactor showed enhanced stability of nano-immobilized biocatalysts. A nanobiocatalytic system has been developed for biodiesel production using the PBR system for effective and continuous biodiesel production based on soybean oil methanolysis with nano-immobilized lipase using chitosan (Wang et al. 2009). Maximum activity and stability have been reported with new biocomposite in the single-PBR at an optimal flow rate. The optimum conversion rate (75%) was recorded at 12 hours. Wang and coworkers (2011a) developed an effective nanobiocatalytic system for biodiesel production that made highly efficient use of lipase. Multiple PBR gave high conversion than using the single PBR. Thus, a packed bed bioreactor has shown great potential for biodiesel production using nanobiocatalytic systems.

High conversion rate and good stability in the four-packed bed reactor may be due to longer residence time of the reaction mixture in the reactor and elimination in the inhibition of the lipase-nanoparticle by-products. Effective stability and reusability of the immobilized enzyme system lower the cost of biodiesel production. The reactor studies provide the basis of technology for the scale-up of the biodiesel production process.

5.1 Chitosan Nanofibers

Chitosan nanofibers were prepared using an ultrasonic-assisted method with a diameter of 5 nm and length less than 3 μm (Wijesena et al. 2015). These were analyzed by AFM and TEM. These nanofibrils can be used as a precursor for further nanoparticles. Electrospinning technology can be used for spinning synthetic polymers. Electrospinning can be controlled by the suspension of the viscous solution. An electrospun fabric of chitosan was prepared by Ohkawa and coworkers (2004). They discovered that trifluoroacetic acid was the most effective solvent system because of amino groups of chitosan form interactions with triflouroacetic acid. A similar method of nanofibers has been developed by Schifman and Schauer in 2007. In this method, low molecular weight (70 KDa), medium (190–310 KDa) and high molecular weight with 500 KDa chitosans have been used along with trifluoroacetic acid. However, the high molecular weight of fibers had resulted in more length of nanofibers. The length can be increased by cross-linking by using glutaraldehyde. Whereas, in another method, trifluoroacetic acid and dichloromethane have been used for developing chitosan nanofibers (Torres-Giner et al. 2008).

5.2 Chitosan Nanocomposites

Chitosan nanocomposites have been prepared using graphene oxide (Lau et al. 2014). X-ray diffraction studies have confirmed that graphene oxide is well embedded in chitosan which shows its appropriate immobilization. It was further confirmed by thermo-gravimetric analysis. Thus, chitosan lipase immobilized has better esterification with the enzyme activity of 64U. Esterification conversion was improved from 78% to 98% with lipozyme. Carbon nanotubes have been employed for significant drug release using chitosan and nanocomposites (Sharmeen et al. 2018).

5.3 Non-Magnetic Nanoparticles

Nanoparticles with a 1–100 nm nanoscale diameter have been used for biological applications (Verma et al. 2016). Matrices involving various nanoparticles as carrier results in enhanced activity and catalytic efficiencies of enzymes that in turn depend upon the stability, reusability and physical properties (Verma et al. 2013). Adsorption method has been employed to immobilize lipase derived from *Candida antarctica* B (CAL-B). Polystyrene nanoparticles have been used as a carrier and have been synthesized using the nano-precipitation technique. The major interactions involved in adsorption are hydrophobic interactions. However, immobilization efficiency has a greater effect on change in pH. The activity depends upon the ionization state that in turn has a deep impact on the conformation state of the enzyme. The activity of this enzyme was compared with commercial enzyme Novozyme 435 and the activity of immobilized lipase was 1.16-fold higher than commercial preparation and 1.81-fold higher than free enzyme.

5.4 Magnetic Nanoparticles

Reusability is one of the important properties of immobilized enzymes, and non-magnetic particle requires a high speed centrifugation for separation (Chen et al. 2008). However, many workers have preferred using magnetic nanomaterials which can enhance the separation (Dyal et al. 2003, Huang et al. 2003, Lei et al. 2011, Wang et al. 2011, Thangaraj et al. 2016, Marta et al. 2017). Nanoscale magnetic particles possess a unique properties of superparamagnetism (Lu et al. 2007, Vaghari et al. 2016). These do not form aggregates and thus can form solutions (Lu et al. 2007). Most of the iron oxides have been used in the fabrication of magnetic nanoparticles as they have more biocompatibility and less toxicity. Dyal and coworkers (2003) evaluated the activity and stability of *Candida rugosa* lipase (CRL) by using a magnetic nanoparticle that has been immobilized using covalent binding. The main interaction involved in these nanoparticles is when the amine group of lipases reacts with acetyl/amine group of nanoparticles. These types of interactions lead to an improvement in the operational stability of *Candida rugosa* lipase. In another similar study, hydrophobic magnetic nanoparticles have been used to immobilize crude porcine pancreas lipase (Lee et al. 2009). Maximum enzyme activity was observed when lipases were immobilized on the hydrophobic surface of nanoparticles. Sodium dodecyl sulfate can be used as a ligand for hydrophobic magnetic nanoparticles. Immobilized lipases have exhibited more activity with a uniform size that has further enhanced the cthermal stability of the immobilized biocatalyst. Immobilized lipase showed 1.42-fold higher specific activity than the free enzyme. The SDS ligand on the nanoparticles' surface acted as a spacer between the nanoparticles and the enzymes, which resulted in a flexible enzyme structure form. CRL was also immobilized on functionalized superparamagnetic nanoparticles (poly(GMA)-grafted Fe_3O_4/SiOx) by Lei et al. (2011). The diameter of the functionalized magnetic nanoparticles was 100 nm and showed higher saturation magnetization (8.3 kA/m).

Higher thermal stability and pH stability were observed in the case of immobilized lipase that has retained 83% of residual activity after six cycles of reusability. This type of magnetic nanoparticles showed 1.41-fold enhanced activity along with 31-fold high stability than that of free enzyme. The effect of various SiO_2 ratios for coating on Fe_3O_4 and further functionalization of Fe_3O_4/SiO_2 magnetic nanoparticles using organosilane compounds, 3-aminopropyltriethoxysilane (APTES) and 3-mercaptopropyltrimethoxysilane (MPTMS), for lipase immobilization was examined. When the Fe_3O_4:SiO_2 ratio was 1:0.25, the immobilization efficiency was the highest. Nano-immobilized lipases along with silica and magnetic material using APTES exhibited maximum catalytic activity. Moreover, APTES helped in improving surface properties for the characterization of nanoparticles.

5.5 Carbon Nanotubes-Based Lipase Immobilization

Carbon nanotubes are a unique and promising material for enzyme immobilization. Carbon tubes with graphite can be easily rolled into cylinders that can help in easy and stable enzyme immobilization (Yang et al. 2015). Carbon nanotubes can be classified into single-walled carbon nanotubes (SWNTs) and multi-walled carbon nanotubes (MWNTs). Both of these can be used for immobilization (Lee et al. 2010b). Two different solvent systems have been used in the case of pancreatic lipase immobilized onto SWNTs. Both buffer and ionic liquids can be used for immobilization. Ionic liquid enhances the immobilization efficiency with better dispersion of carbon nanotubes than using a buffer solution. Lipases have been immobilized using nanotube-silica composites in multiwalled nanotubes (Lee et al. 2010b) MWNTs were used as additives to prevent lipase inactivation during the sol-gel process. Three different lipase activities from *Candida rugosa*, *Candida antarctica* type B and *Thermomycs lanuginous* immobilized on MWNT increased 10-fold to perform transesterification reaction which may be due to enhanced stability of the catalyst using multiwalled nanotubes. Immobilization involved physical adsorption and displayed high activity retention up to 97%. Transesterification rates increased using both aqueous and organic solvents by 2.2 and 14-fold, respectively (Shah et al. 2015). Mohammad et al. (2017) reported a simple adsorption method to immobilize CRL onto acid-functionalized MWNTs (F-MWNTs). Carboxyl groups as polar groups were induced in MWNT by stirring with an acid mixture containing sulfuric and nitric acids. The charged carboxyl moieties on the MWNTs' surfaces could be connected with other polar moieties (NH_2 and OH) on the CRL. The immobilized CRL on acid F-MWNTs had improved structural integrity and mechanical strength. The activity and thermal stability of the immobilized CRL on MWNTs were twofold enhanced compared to those of the free enzyme.

5.6 Electrospun Nanofibers

The major problem associated with carbon nanotubes for immobilization is mass transfer limitation and difficulty in recycling due to good dispersion (Wang et al. 2009). Electrospun nanofibers have more potential to avoid these above-mentioned problems (Nakane et al. 2007). Lipases were attached to the surface of electrospun nanofibers as the carrier or entrapped in the nanofibers. Nanofiber membranes with carboxyl groups were made from poly(acrylonitrile-co-maleic acid) (PANCMA) via an electrospinning process (Pavlidis et al. 2010). Immobilization has impacted the increase in enzyme activity from 33.9% to 37.6% mg/g with enhanced enzyme loading that reached up to 21.2 mg/g from 2.36 mg/g as compared to the hollow fiber membrane. The efficiency of the biocatalysis increased because the Km value of the immobilized lipase decreased. Huang and coworkers (2011) reported the development of immobilized CRL on an electrospun cellulose nanofiber membrane via covalent binding. The nanofiber was made by electrospun cellulose acetate. These have been stimulated to produce aldehyde group that can enhance the covalent bonding with enzyme molecules. The activity of immobilized CRL was 29.6 U/g under an optimal condition, and the thermal stability was higher than that of free enzyme. Immobilized *Burkholderia cepacia* (BCL) lipase has been immobilized using polycaprolacone nanofibers, and their overall effectiveness was further checked using an aqueous and organic medium (Song et al. 2012). The specific hydrolysis activity was higher than the transesterification activity. The immobilized BCL maintained 50% of its initial activities up to the 10th recycle in a non-aqueous media.

Biofuel production using immobilized lipases onto nanoparticles have been studied (Babaki et al. 2016). Raita et al. (2015) studied biodiesel production from palm oil using immobilized *Thermomyces lanuginosus* lipase (TLL) on magnetic nanoparticles. Optimization of biodiesel production has been performed and a 97.2% increase in yield was obtained with 23.2 mg/g enzyme loading. Moreover, these nano-immobilized catalysts showed stability at 50°C for 24 hours. These enzyme preparations could be reused for five cycles with 80% retained catalytic activity. Waste cooking oil was also converted to biodiesel via transesterification using nano-immobilized lipase (Mehrasbi et al. 2017). Karimi and coworkers (2016) immobilized BCL on superparamagnetic iron-oxide nanoparticles (SIONs) for biodiesel production. Maximum conversion (91%) of waste cooking oil to biodiesel has been observed in 35 hours. Immobilized *Rhizomucor miehei* lipase has been used, and a high diesel yield of 94% from waste cooking oil has been reported. The RML was immobilized onto polyamidoamine (PAMAM) grafted with magnetic MWCNTs (m-MWCN-PAMAM). The immobilized catalysts showed reusability up to 10 cycles along with a 27-fold higher transesterification activity. These results showed great potential for biodiesel production. Tran et al. (2012) developed alkyl-functionalized Fe_3O_4-SiO_2 nanocomposites for lipase immobilization. The immobilized lipase was used for biodiesel production. Maximum biodiesel production (90%) has been observed in batch operation for 30 hours. One-step extraction and transesterification process of biodiesel production from wet microalgae biomass using alkyl-grafted Fe_3O_4-SiO_2-immobilized lipase have been studied. The biodiesel conversion was over 90% under optimal conditions. In another similar study, transesterification of soyabean oil to biofuel has been observed using lipase immobilized on polyacrylonitrile nanofiber. The biodiesel conversion was 90%. The immobilized lipase on the nanofibers maintained 91% of its initial activity for 10 cycles. All these results have strengthened that nanomaterials can be used as best carriers for immobilization of lipase and they have tremendous potential in biodiesel production.

6. Role of Chitosan in Biofuel Production

The biofuel industry has employed the application of nanotechnology and nanomaterial for developing cost-effective and process efficient technology to improve biofuel production. Moreover, it gives eco-friendly and green solutions for all alternative energy resources. Much of the research is focused on designing and fabrication of nanomaterials for efficient conversion of natural resources into biofuels and alternative energy sources (Trindade 2011). Nanomaterials have various properties that enhance the stability of biocatalysts and make them the most promising carriers. They not only enhance the catalytic activity and stability but also increase the loading capacity, durability, reusability and efficient storage that improve the overall biocatalytic activity (Garcia-Martinez 2010). There are numerous successful applications of nanotechnology in different processes of biofuels, including transesterification, anaerobic digestion, pyrolysis, gasification for fatty esters and biogas production. Most of these techniques are possible at a laboratory scale but their transfer to large and commercial scale needs optimization (Zhang et al. 2010). Thus, there is much attention to the optimization of nanotechnology for biofuel production. It includes developing cost-effective and economically viable technology for providing better and improved yields. There are various studies supporting different metal oxides such as titanium oxide (Gardy et al. 2017), calcium oxide (Liu et al. 2008), magnesium oxide (Verziu et al. 2008) and strontium oxide (Liu et al. 2007) as a nanomaterial for biofuel production. Magnetic nanoparticles are better than non-magnetic nanoparticles as the former are easier to separate without involving any equipment, thus they provide an inexpensive technique for biofuel production (Gardy et al. 2017). Different nanomaterials with enhanced structural properties have been successfully used in biofuel production (Yahya et al. 2016). Carbon-based nanotubes (Guan et al. 2017), carbon nanofibers (Stellwagen et al. 2013), grapheme oxide (Mahto et al. 2016) and biochar (Dehkhoda et al. 2010) hold tremendous potential to be used as materials for biofuel production. Another cost-effective technology is to use nanomaterials for immobilization of biocatalysts, such as lipase and cellulases as nanomaterials, which increase the surface area and in turn affect the enzyme loading capacity, stability and reusability of enzymes (Trindade 2011). Another important application of nano-immobilized catalysts is in bioethanol and biodiesel production processes (Kim et al. 2018). Nanomaterials have been used previously in the extraction of oils from algal cells. Moreover, carbon nanotubes, nano clay, metal oxide nanoparticles as well as magnetic nanoparticles can be used for lipid accumulation, extraction and transesterification processes (Zhang et al. 2013).

However, anaerobic digestion uses the nano-iron oxide that gives zero fly ash and less bottom ash with increased biomethane production (Ganzoury and Allam 2015). There are so many potential areas that result in developing cost-effective viable technologies that can be applied to commercial level production. Thus, nanotechnology can be successfully employed for the production of biofuels (Nizami and Rehan 2018). There are various types of biofuels produced using chitosan and nanoparticles that are explained in the following section.

Thus, the development of alternative energy sources like biofuels such as biodiesel, bioethanol and biogas production is the need of the hour. These biofuels are not only renewable sources of energy but also provide cleaner alternatives that are environment-friendly, inexpensive and reduces greenhouse emissions (Noraini et al. 2014). There are other chemical methods of conversion, but enzymatic methods of conversion are safe, less contaminating and are believed to be higher-yielding methods. The main factor in achieving maximum conversion depends upon the effectiveness of the immobilization method and activity of enzymes in biofuel production. Various renewable sources such as algae, soyabean, jatropha, corn palm and lignocellulosic wastes have been used for biofuel production (Ong et al. 2011). The role of chitosan nanoparticles in the immobilization of different enzymes used for converting biomass into bioenergy has been discussed. Different types of biofuels attaining interest are biodiesel, bioethanol, biobutanol and biogas. Lipases derived from microbial sources catalyze the hydrolysis of triglycerides to glycerol and free fatty acids as well as esterification and transesterification reactions (Antczak et al. 2009, Aarthy et al. 2014). Lipases have wide applications and can be used in cosmetics, detergent, food, feed and biodiesel production (Guldhe et al. 2015). Nanomaterials have been used as carriers for enzymes due to their better mass transfer along with high enzyme activity with an increased surface to volume ratio (Hwang et al. 2013).

6.1 Biodiesel Production

Biodiesel has raised interest in recent years due to its non-toxic and renewable nature, lesser emissions, low sulfur content and eco-friendliness (Al-Zuhair 2007, Calero et al. 2015). It produces 78% fewer emissions and is 66% better lubricant than petrodiesel (Noraini et al. 2014). Compared to conventional diesel, biodiesel has a higher combustion efficiency, better flash point and lubrication efficiency (Kim et al. 2018). First-generation biodiesel involves the production of biodiesel using food sources such as vegetable oils which are limited due to their application in food as well as their high cost. Second-generation sources represent a lucrative approach for the production of biodiesel from lignocellulosic wastes as they are abundantly present in nature and are renewable sources. Whereas, third-generation involves microalgae used to convert biomass into bioenergy. Edible and non-edible oils such as jatropha, karanja, mahua and castor oil have been tested (Table 1) for biodiesel production (Kim et al. 2018). Transesterification reactions using oils require short-chain alcohols which produce a high content of free fatty acids (Gui et al. 2008). Conventional commercial methods involve using alkali catalysts such as sodium and potassium hydroxide. Alkali catalysts are neutralized so these cannot be reused (Huang et al. 2012). Moreover, these methods involve using an oil press, osmotic shock or ultrasound for increased extraction (Noraini et al. 2014). Various super-critical conditions have been utilized. All chemicals and physical methods involve a high cost. Lipases can be used for enhancing the biodiesel production and provide better yields than conventional technology. Different types of lipases obtained from various microbial sources can be applied for biodiesel production (Verma et al. 2013).

Table 1. Application of lipase enzyme in production of biodiesel.

Sr. No.	Raw Material	Enzyme	Immobilization Method	Carrier Matrix	Yield/ Conversion	Reference
1.	Jatropha	Lipase from *Candida rugosa*	Covalent binding	Magnetic chitosan	87%	Xie and Wang 2012
2.	Soyabean oil	Lipase from *Rhizopus oryzae*	Adsorption method	Silica and chitosan	98%	Lee et al. 2008
3.	Phenolic acid	Lipase from *Candida rugosa*	Crosslinking	Magnetic chitosan	72.6%	Kumar et al. 2013
4.	Palm oil	Lipase from *Candida rugosa*	covalent	Chitosan	–	Wu et al. 2010
5.	Esters	Lipase from *Burkholderia cepacia*	covalent	chitosan	–	Ghadi et al. 2015
6.	Jatropha oil	Commercial Lipase	Physical adsorption	Chitosan	–	Egwim et al. 2012
7.	Oil from plants	*Candida rugosa*	Cross-linking	Chitosan encapsulated magnetic nanoparticles with genipin and glutaraldehyde	80% up to 5 cycles	Liu et al. 2016
8.	Esters	*Mucor racemosus* NRRL 3631	Adsorption	Ferric oxide chitosan nanoparticles	63% in 5 cycles	Elhadi and Ahmed 2013

Lipases can be immobilized by any of the four methods such as adsorption, covalent binding, entrapment and cross-linking (Tan et al. 2010). It has been reported that lipase from *Candida rugosa* had retained four times better reusability on the magnetic chitosan microsphere for transesterification of oils (Xie and Wang 2012). In a similar study, lipase from *Candida rugosa* and *Rhizopus oryzae* has been mixed in a 3:1 ratio and optimum biodiesel production was achieved at 45°C and 300 rpm. Moreover, 98% conversion was achieved. However, 72.6% of initial activity was retained in lipase immobilized on magnetic Fe_2O_4 chitosan beads for ester synthesis. Reusability up to eight cycles can be achieved using chitosan immobilized lipase (Kumar et al. 2013). Optimization studies on lipase immobilization using chitosan as carriers were performed for biodiesel production (Egwim et al. 2012). Enzyme loading 0.2 g/g chitosan, 40°C temperature, pH 7.0 and immobilization time of 3 hours have been optimized.

Nanomaterials have tremendously revolutionized the immobilization of biocatalysts/enzymes that can convert lignocellulosic wastes to alternative energy sources such as biodiesel and bioethanol. All the current trends related to enzyme immobilization using nanomaterials have been reviewed (Rao et al. 2017). All biofuels such as biodiesel, bioalcohol and biogas involve lipases and cellulases catalyzing the transesterification reactions. Nanomaterials can be synthesized in a number of ways including intercalation, intercalative polymerization and metal-based nanocomposites. However, for biosynthesis of nanomaterials, physical, chemical and biological methods can be used. Biological methods include a variety of methods such as reduction processes using plant extracts like latex of *Jatropha curcas*. Different microorganisms ranging from bacteria like *E. coli, Bacillus sphaericus* to fungi such as *Aspergillus fumigates, Neurospora crassa* to algae like *Chlorella vulgaris* and *Sargassum wightii* can be used for the reduction process. Other methods employ metals such as gold, silver and copper. These methods are based upon reductase enzymes. Nanoparticles can be synthesized via two routes that are extracellular and intracellular. Extracellular production is preferred due to easy downstream processing. These production processes have many advantages of being green synthesis involving no pollution, less time consuming and do not require any stabilizing agent (Pantidos and Horsfall 2014, Shah et al. 2015, Ghanshghaei and Emtiazi 2015, Singh et al. 2016). Recent studies in bioethanol/biodiesel production involve different pre-treatments of lipase and cellulase enzymes. Lipase pre-treatments include ultrasonic irradiation and microwave-assisted treatment and solvent-free immobilization methods. Different nanomaterials were employed for lipase immobilization, out of which chitosan immobilization seems lucrative approach. Lipase from *Pseudomonas fluorescens* LpI has been immobilized using aminosilane modified supermagnetic ferric oxide nanoparticles that have retained 70% enzyme activity (Kanimozhi and Perinbam 2010). In another study, lipases from *Mucor racemosus* NRRL 3631 have been immobilized on ferric oxide chitosan nanoparticles with 63.5% retained activity after eight cycles in biodiesel production. Whereas, 87.73% activity was retained using lipase from *Burkhalderia cepacia* immobilized onto chitosan magnetic ferric oxide core-shell nanoparticles (Ghadi et al. 2015). However, when glutaraldehyde has been used as a cross-linking agent in the encapsulation study of lipase obtained from *Candida rugosa*, it retained 80% of enzyme activity (Liu et al. 2016). In a similar study, using lipase from *Candida rugosa* immobilized on chitosan a grapheme oxide beads. Overall immobilization efficiency increased due to strength provided by grapheme and high enzyme activity (64U). It has enhanced the esterification process significant for biofuel production (Lau et al. 2014). It has provided a thermostable enzyme that provided 98% and 88% conversions. Enhancement in transesterification reactions for biodiesel production from cooking oil using chitosan gel has been studied (Putra et al. 2016). In this, chitosan gel has been prepared using hydrogel and xerogel scheme using soyabean oil and methanol. It has greatly influenced the biodiesel production.

6.2 Bioethanol Production

Bioethanol is another form of biofuel. The conversion is a three-step process involving pretreatment, saccharification and fermentation. Bioethanol has been produced using sugarcane, corn, wheat, etc. (Kim et al. 2018). Most of the substrates are full of complex sugars such as cellulose, sucrose and starch. Pretreatment stage involves the action of the enzyme on complex substrates to easily utilizable forms process known as saccharification. Finally, by fermentation, various microorganisms convert the treated substrate to final product ethanol. Bioethanol is less toxic and

Table 2. Summary of methods of chitosan nanomaterial immobilization.

Sr. No.	Type of Raw Material Used	Composition of Nanomaterial	Method of Preparation	Enzyme/ Organism Used	Applications	Reference
1.	Paddy straw	Magnetic chitosan	Physical adsorption	Cellulase	Bioethanol production	Kumar et al. 2018
2.	Whole cells	Chitosan	Entrapment	*Saccharomyces cerevisiae*	Bioethanol production	Ivanova et al. 2011, Willaert et al. 2006, Jin and Tanaka 2006
3.	Whole cells	Chitosan	Covalent Binding	*Saccharomyces cerevisiae*	Bioethanol production	Ivanova et al. 2011
4.	Whole cells	Magnetite chitosan	Cross-linking method	*Candida utilis, Kluyveromyces lactis*	Bioethanol production	Baldikova et al. 2017
5.	Whole cells	Magnetic chitosan	Cross-linking	*Clostridium beijerenckii*	Biohdyrogen production	Seelert et al. 2015

eco-friendly fuel as compared to petroleum fuel (John et al. 2012). Bioethanol can also be produced from lignocellulosic waste (Ho et al. 2013) which provides long term alternative of fossil fuels as compared to ethanol derived from sucrose and starch (Table 2). As major limitations include less space for agriculture, high water usage has increased the prices of all food items (Guo et al. 2013). Therefore, the next alternative is to use second-generation lignocellulosic mass for conversion to bioethanol. Whereas, limitation associated using lignocellulosic waste is in the presence of more lignin and cellulosic content. Lignocellulosic biomass represents the potential substrate for bioethanol production. It consists of cellulose (40–50%), hemicellulose (25–35%) and lignin (15–20%) (Holzapple 1993). The composition of lignocellulosic varies from plant to plant as well as age of plant. Successful implementation of lignocellulosic bioethanol depends upon economically sustainable and feasible technologies. The major constraints are recalcitrant lignocellulosic waste and high cost of enzymes. Cellulases hydrolyzed β-1-4 glycosidic bonds of cellulose to produce simpler glucose. Cellulases comprise endoglucanases and exoglucanases which became denatured due to unspecific binding to lignin (Chundawat et al. 2011).

Various commercial preparations such as accelerase, cellulase, celic and novozyme 88 are limited due to the high cost involved. Moreover, these enzymes are inhibited by end products. Immobilization of the cellulases to magnetic nanomaterials has been reported to enhance the tolerance and thus reusability of the enzymes (Jordan et al. 2011). Different enzymes have been immobilized on different matrices and checked for bioethanol production (Kumar et al. 2018). Immobilization of enzyme was achieved by the physical adsorption method and covalent binding. Immobilization efficiency for cellulases and xylanase was checked and protein loading was the most significant. Saccharification was studied for paddy straw under optimum conditions. Quantification of reducing sugars was done by HPLC method. Magnetic nanoparticles retained the highest enzyme activities. The highest immobilization efficiency was obtained in the case of cellulase enzyme, i.e., 59.58%. The magnetic nanoparticle was the best in terms of immobilization efficiency. It may help in reducing the cost of overall ethanol production (Kumar et al. 2018).

Earlier studies for bioethanol production involved the whole cell immobilization of *Saccharomyces cerevisiae* in a matrix of alginate and magnetic nanoparticles (Ivanova et al. 2011). Thermostable α-amylase and glucoamylase were immobilized onto chitosan and column reactors used for starch saccharification and production of ethanol thereafter. Ethanol production was 91% and found to depend upon loading rate and feed sugar concentration. The ethanol productivity was 264.0 g/L.h with an optimum loading rate. A significant increase in ethanol production was achieved due to several reasons. Entrapment of yeast cells and enzymes reduce the fermentation time and size of their storage (Willaert et al. 2006). The most important advantage was increased productivity with high dilution rate, high ethanol tolerance and reutilization (Lin and Tanaka 2006). Continuous fermentation systems involving immobilized cells can reduce the overall cost in bioethanol production with maximum volumetric productivity. In this study, yeast cells and enzymes were immobilized using alginate and chitosan nanoparticles further checked for bioethanol production (Ivanova et al. 2011). The covalent bonding method can be employed for immobilization of yeast cells (Table 2). This study was concluded with the fact that immobilized enzymes such as α-amylase from *Aspergillus oryzae* and *Aspergillus niger* as well as glucoamylase from *Aspergillus niger* provide better ethanol production as compared to the conventional process. It makes biofuel production more feasible.

In a similar study, cellulases have been immobilized on chitosan-coated magnetic particles for the conversion of *Agave atrovirens* lignocellulosic waste to bioethanol (Sanchez-Ramirez et al. 2016). The coprecipitation method was used for the immobilization of cellulase on chitosan-coated nanoparticles. Electron microscope with high-resolution technique can be used for the characterization of these nanoparticles. Immobilized enzymes have retained about 80% of activity after 15 cycles. This has provided the basic technology for efficient conversion of biomass to bioenergy. In another study, whole cell biocatalysts were fabricated using magnetic derivatives of chitosan with magnetite and maghemite. Such magnetically responsive biocatalysts can enhance the production of bioethanol (Baldikova et al. 2017). Different magnetic responsive microbial cells such as *Saccharomyces cerevisiae*, *Candida utilis* and *Kluyveromyces lactis* cells composites were employed for checking invertase activity for invert syrup. All the biocatalysts retained 69% initial activity after 8 cycles of use. Maximum activity was noted for *Saccharomyces cerevisiae* followed by *Candida utilis* and *Kluyveromyces lactis*, respectively. The high lignin content of lignocellulosic waste can be hydrolyzed from hemicelluloses into xylose and celluloses into glucose (Canilha et al. 2012). These two fermentable sugars provide a cost-effective and sustainable alternative for biofuel production. Conventionally, *Saccharomyces cerevisiae* is one of the important microorganisms used for the production of ethanol due to the high rate of fermentation, high yield and high tolerance to ethanol and sugar (Zhang et al. 2010, Demeke 2013). However, a wild form of *Saccharomyces cerevisiae* can metabolize xylulose which is an isomer of xylose. This can be used as an alternative process for converting xylose to xylulose using *ex vivo* conditions and followed by subsequent fermentation to ethanol (Silva et al. 2012). An enzyme such as xylose isomerases (EC 5.3.1.5) can be used for this conversion industrially associated with unequal proportions of xylose and xyluose. Simultaneous isomerization and fermentation (SIF) can be an alternative to this unequal proportion as it can lead to a high yield of ethanol due to the complete degradation of xylose (Milessi et al. 2018). The overall cost associated with the process can be reduced by using immobilization of enzyme onto suitable and inexpensive matrix such as chitosan (DiCosimo 2013, Guisan 2013). Production of ethanol using live cells of *Saccharomyces cerevisiae* by coimmobilization with xylose isomerase in calcium alginate gel has been studied (Silva et al. 2012). However, this coimmobilization suffers from contamination from xylose utilizing bacteria. Moreover, the extra steps can lead to an increase in cost (Macrelli 2014). Simultaneous hydrolysis, isomerization and fermentation (SHIF) can a provide better alternatives for the continuous production of second-generation ethanol using wild strains of *Saccharomyces cerevisiae* using cost-effective lignocellulosic material (Milessi et al. 2018). Novel biocatalyst has been designed containing xylose isomerize along with amylase and protease coimmobilized using chitosan matrix for continuous production of ethanol from xylan. In this study, xylanase and xylose isomerize has been covalently immobilized onto chitosan with high enzyme activity (252.5 IU/g) that has provided 93% yield and reusability up to 91% for conversion of xylan into ethanol using fixed-bed bioreactor for SHIF (Milessi-Esteves et al. 2019). Another inexpensive alternative for bioethanol production is chitosan that has huge potential but is associated with one major problem which is the inability of *Saccharomyces cerevisiae* to utilize chitosan (Jaime et al. 2012). This process can be associated with enzymatic hydrolysis of chitosan using chitosanases that can hydrolyze β-(1-4) linkages. Various fungal organisms from the GH75 family contain genes for encoding these chitosanase enzymes (Hoell et al. 2010). These enzymes have a role in fungal nutrition and autolysis (Gruber and Seidl-Seiboth 2012). Different substrates have been evaluated for ethanol production in previous work (Karimi et al. 2006). Two different fungal strains *Pochonia chlamydosporia* and *Metarhizium anisopliae* have been studied for the conversion of chitosan to ethanol (Aranda-Martinez et al. 2016). This bioethanol production process has been influenced by several factors such as temperature, sugar concentration, pH, course of fermentation as well as inoculum size (Azhar et al. 2017, Aranda-Martinez et al. 2017).

6.3 Biogas Production

One renewable bioenergy source is to utilize anaerobic digestion for biogas production. Biogas production can be obtained by the utilization of agricultural as well as animal wastes (Zhang et al. 2013, Bidart et al. 2014). Anaerobic digestion for biogas production involves various steps, viz., hydrolysis, acidogenesis, acetogenesis and methanogenesis. During the hydrolysis step, complex carbohydrates are converted to simple sugars. In the methanogenesis step, intermediate sugars are converted into carbon dioxide and methane by methanogens (Deublein and Steinhouser 2008). An increase in organic matter degrading capacity has been enhanced using nanomaterials and therefore increase in biogas production has been reported (Faisal et al. 2019). In a novel method, biogas production was enhanced by iron nanoparticles. The highest increase in biogas production was obtained. This increase in biogas may be because iron strongly enhances the anaerobic digestion. The optimum conditions have led to an increase in biogas production from 40–200%. Overall nanoparticles can enhance biofuel production by improving the stability of various cellulolytic enzymes.

In a similar study, no significant effect of immobilization of enzymes using chitosan has been reported during methane production. Methane production using three different concentrations of sugar beet wastes was checked and no change in methane and biogas production was obtained (Beiki and Keramati 2019).

6.4 Biohydrogen Production

Out of various biofuels, biohydrogen production is one of the novel biotechnological tools by which organic waste can be converted into biohydrogen. Hydrogen is the cleanest form of energy with 2.75 times higher than conventional fuels. Moreover, biohydrogen involves water vapors as a by-product which does not increase carbon emissions. Nanobiotechnology has helped to increase biohydrogen production (Sivagurunathan et al. 2018). There are various biosynthetic routes for biohydrogen production such as biophotolysis, photofermentation, dark fermentation but practical production is limited due to various problems. Biohydrogen production can be enhanced by implementing nanotechnology (Sivagurunathan et al. 2018).

In another study, *Clostridium beijerinckii* NCIMB8052 immobilized in magnetite particles with chitosan, there was reported an increase in biohydrogen production (Seelert et al. 2015). *Clostridium beijerinckii* produces biohydrogen by anaerobic metabolism. The major problem is with low substrate conversion efficiency. Immobilization has increased stability and can be reused for up to seven batch cycles with 80% retained enzyme activity (Li et al. 2009). Different carbon sources have been optimized for biohydrogen production using *Clostridium beijerinckii* (Ghosh and Hallenbeck 2014). Limited studies have been performed using other matrices. A novel immobilization technique was evaluated using nanoparticle. Immobilization was based on layer by layer method which has resulted in increased accumulation of biohydrogen (Seelert et al. 2015). In this study, ethanol and acetate were major metabolites. It can form the basis for developing technology for commercial production.

6.5 Biobutanol Production

In the 19th century, acetone-butanol-ethanol was a major fermentation industry although butanol ABE fermentations were not feasible due to the high cost involved. However, now interest has been renewed due to problems involved with petrofuels such as pollution problems associated with conventional fuels. *Clostridia* cells can utilize a variety of lignocellulosic waste for biobutanol production (Sukumaran et al. 2011).

7. Improvement in Biofuel Production Using Metabolic Engineering

Biofuels are renewable alternative energy sources with one major problem associated that is undesirable properties due to which these can be blended with other sources in limited amounts. The alternative biofuels have attained much research interest in improving the existing technology, such as straight-chain fuels, higher octane values, better cold flow and lower cloud points that can make them more suitable for existing engines specifically for diesel and jet engines (Bai et al. 2019). Recent progress in microbial engineering has resulted in bioproduction of biofuel molecules like 1-butanol, isobutanol, limonene, hydrogenated farnesene, fatty acid-derived alkanes, alkenes alcohol and esters (Zhang et al. 2008, Connor et al. 2010, Zhang et al. 2011, Wang et al. 2012, Howard et al. 2013, Meadows et al. 2016, Jiang et al., Lin et al. 2017). Based on the chain length or the number of carbon atoms in their molecules, biofuels can be divided into short-chain (C4–C8), medium-chain (C9–C14) and long-chain (C15–C20) fuels. Based on their chain structures, biofuels can be divided into straight-chain or branched-chain biofuels. Compared to straight-chain biofuels, their branched-chain counterparts often have better physical and combustion properties. Branched short-chain alcohols such as isobutanol and 3-methyl-1-butanol have higher octane values than their linear-chain counterparts (Atsumi et al. 2008, Blombach and Eikmanns 2011). Branched long-chain fuels have improved properties such as lower freezing point, better cold flow and lower cloud point compared with their straight-chain counterparts (Knothe and Dunn 2009). These properties have been used for specific purposes such as suitable to be used on low temperature and high altitudes particularly for jet fuels. There have been previous studies for the production of straight-chain esters but metabolic engineering can be used to improve biofuels (Bai et al. 2019).

Most of the existing biofuels involve methyl branches because ethyl esters and higher branched-structures are relatively rare, so they have not been studied extensively. Branched short-chain alcohols and esters like isopentenol, isobutanol and 2-methyl-1-butanol are derived from keto acids. However, long-chain biofuels have been derived from lipid fatty acids. These types of biofuels can be biosynthesized using lipid fatty acid biosynthesis (Youngquist et al. 2013, Mehrer et al. 2018). They can be branched in the middle or terminal position of chain and can be biosynthesized using different routes of lipid metabolic pathways. Recent engineering methods have been used for the overproduction of biofuels. General strategies involve improved titers, yields and productivities (Bai et al. 2019).

Branched short-chain biofuels have been modified either by modification of Ehrlich pathways or by the improvement of precursor pool by optimization of cofactor availability that leads to the overproduction of these biofuels. One of the important producers of biofuels is *Saccharomyces cerevisiae* that can naturally produce fewer titers of biofuels (Dickinson et al. 1998, Hazelwood et al. 2008). Metabolic engineering of the Ehrlich pathway has been performed using α-keto acid decarboxylase (KDC) from *Lactococcus lactis* and alcohol dehydrogenase from *Saccharomyces cerevisiae*. Another enzyme acetolactate synthase Ilv2 that can catalyze valine synthetic pathway has been over-expressed to increase the production of 2-keto-3 methylvalerate. Finally, pyruvate decarboxylase was removed to reduce the flux from pyruvate to ethanol. Thus, the metabolic engineered strain produced isobutanol that is 13-fold higher than that of the wild type strain. In a similar study, isobutanol production was 1.62 g/L with over-expression of malate dehydrognase MAE1 (Matsuda et al. 2013). Similarly, Ehrlich pathways have been modified in some hosts other than *Saccharomyces cerevisiae* such as *E. coli*, *B. subtilis*, *Corynebacterium glutamicum*, *Brevibacterium flavum R. eutropha* and *Synechococcus elongatus* 7942. The major strategies for engineering are modification/reconstruction of the Ehrlich pathway using gene over-expression and optimization. Ehrlich pathway has been reconstructed in *E. coli* using one of the five alpha-keto acid decarboxylases from *Saccharomyces cerevisiae* (Atsumi et al. 2008). Another strategy for the over-production of isobutanol was to delete multiple competing pathways and enzymes such as aldehyde-alcohol dehydrogenase (*adh E*), lactate dehydrogenase (*ldh A*) and fumarate reductase (*frd AB*) and phosphate acetyltransferase (*pta*) to avoid the by-product accumulation in *E. coli.* An increased titer of isobutanol (22 g/L) has been achieved by a combination of Ehrlich and acetolactate synthase genes (Li et al. 2011). Another important strain used was *Corynebacterium glutamicum* that is a major producer for branched amino acids in the fermentation industry and it has a natural capacity to accumulate 2-ketoisovalerate and 2-keto-3-methylvalerate at high intracellular concentrations. An over-expression of α-keto acid decarboxylase from *S. cerevisiae* and alcohol dehydrogenase achieved 2.76 g/L of 3-methyl-1butanol (Vogt et al. 2016). In other similar studies, high titers of isobutanol (140 mg/L) in *R. eutropha* and *Bacillus flavum* have been achieved. However, in *S. elongatus* 7,942 five-step biosynthetic pathway has been engineered with 450 mg/L isobutanol production (Atsumi 2008).

7.1 Use of Chitosan in Biofuel Cells

Biofuel cells are cells that can convert biofuel into electricity. There have been extensive research reports on biofuel cells as an alternative and sustainable energy source. These can be further classified into enzyme biofuel cells (EBC) and microbial fuel cells (MBC) depending upon sources of oxidation energy. The EBCs utilized enzymes for harnessing the redox energy into electricity. Due to the expansion of nanotechnology, advanced engineering material such as chitosan has been used for the fabrication of enzymatic biofuel cells. These biofuel cells are inexpensive and renewable sources of energy. Moreover, these involve no emissions like conventional fuels. These properties of biofuel cells make them the most promising candidates for alternative energy sources. Enzymatic biofuel cells are better than MBCs as the former do not require viable cells (Kumar et al. 2018). As glucose is present in every cell, glucose oxidase has been most frequently used for producing more energy and converting it efficiently into electricity. In one recent study, chitosan modified carbon cloth produced 53% of power density (Duong et al. 2019). A low-cost route using chitosan has been employed for the generation of electricity. However, more studies are required to strengthen these biofuel cells.

8. Conclusion

Biofuel production using lignocellulosic waste is a successful example of converting biomass into bioenergy. This is not only promising green technology for the production of biofuel which gives lesser emissions and hence fewer pollution problems. But it also gives us an efficient conversion process for producing alternative sources of energy using lignocellulosic wastes. Biocatalysts can always fasten these reactions by speeding up the reactions. A variety of different enzymes such as lipases, cellulases, xylanases and lignin peroxidases can be utilized for conversion. However, the major limitation is a high cost involved which can be significantly reduced by using various immobilization techniques using better matrices. In this chapter, the main focus has been on chitosan which is a natural polymer for providing better properties for immobilization of enzymes. Some of the techniques used are still at infancy and need to be explored more. Metabolic engineering and other biotechnological techniques can be used to improve the limitations/resistances faced by enzymes which can further enhance their usage as biocatalysts.

References

Aarthy, M., P. Saravanan, M.K. Gowthaman, C. Rose and N.R. Kamini. 2014. Enzymatic transesterification for production of biodiesel using yeast lipases: An overview. Chemical Engineering Research Design 92: 1591–1601.

Abdulhussien, M.A. and T.M. Alsalman. 2017. Chitosan nanoparticles: Review article. Journal of Interdisciplinary Research 7(3): 233–242.

Agarwal, A.K. 2007. Biofuels (alcohols and biodiesel) applications as fuels for internal combustion engines. Progress in Energy and Combustion Science 33: 233–271.

Agarwal, A.K., R.A. Agarwal, T. Gupta and B.R. Gurjar (eds.). 2017. Biofuels Technology, Challenges and Perspective in Renewable and Green Energy Technology by Springer.

Agrawal, P., G.J. Strijkers and K. Nicolay. 2010. Chitosan-based systems for molecular imaging. Review on Advanced Drug Delivery 62: 42–58.

Agnihotri, S.A., N.N. Mallikarjuna and T.M. Aminabhavi. 2004. Recent advances in chitosan-based micro-and nanoparticles in drug delivery. Journal of Controlled Release 100: 5–28.

Ahmad, R. and M. Sardar. 2015. Enzyme immobilization: An overview on nanoparticles as immobilization matrix. Analytical Biochemistry 4.

Al-Zuhair, S. 2007. Production of biodiesel: Possibilities and challenges. Biofuels Bioproduction and Biorefinery 1: 57–66.

Antczak, M.S., A. Kubiak, T. Antczak and S. Bielecki. 2009. Enzymatic biodiesel synthesis key factors affecting efficiency of the process. Renewable Energy 34: 1185–1194.

Aranda-Martinez, A., N. Lenfant, N. Escudero, E.A. Zavala-Gonzalez, B. Henrissat and L.V. Lopez-Llorca. 2016. CAZyme content of *Pochonia chlamydosporia* reflects that chitin and chitosan modification are involved in nematode parasitism. Environmental Microbiology 18: 4200–4215.

Aranda-Martinez, A., M.A.N. Ortiz, I.S.A. Garcia, E.A. Zavala-Gonzalez and L.V. Lopez-Llorca. 2017. Ethanol production from chitosan by the nematophagous fungus *Pochonia chlamydosporia* and the entomopathogenic fungi Metarhizium anisopliae and Beauveria bassiana. Microbiological Research 204: 30–39.

Asuri, P., S.S. Bale, R.C. Pangule, D.A. Shah, R.S. Kane and J.S. Dordick. 2007. Structure, function, and stability of enzymes covalently attached to single-walled carbon nanotubes. Langmuir 23: 12318–12321.

Atsumi, S., T. Hanai and J.C. Liao. 2008. Non-fermentative pathways for synthesis of branched-chain higher alcohols as biofuels. Nature 451: 86–89.

Babaki, M., M. Yousefi, Z. Habibi, M. Mohammadi, P. Yousefi, J. Mohammadi and J. Brask. 2016. Enzymatic production of biodiesel using lipases immobilized on silica nanoparticles as highly reusable biocatalysts: Effect of water, t-butanol and blue silica gel contents. Renewable Energy 91: 196–206.

Bai, W., W. Geng, S. Wang and F. Zhang. 2019. Biosynthesis, regulation and engineering of microbally produced branched biofuels. Biotechnology for Biofuels 12(84): 1–12.

Baldikova, E., J. Prochazkova, M. Stepanek, J. Hajdnova, K. Paspiskova, M. Safarikova and I. Safarik. 2017. PMAA-stabilized ferrofluid chitosan/yeast composite for bioapplications. Journal of Magnetism and Magnetic Materials 427: 29–33.

Basha, J.S. and R.B. Anand. 2011. Role of nanoadditive blended biodiesel emulsion fuel on the working characteristics of a diesel engine. Journal of Renewable and Sustainable Energy 3(2): 023106.

Beiki, H. and M. Keramati. 2019. Improvement of methane production from sugar beet wastes using TiO_2 and Fe_3O_4 nanoparticles and chitosan micropowder additives. Applied Biochemistry and Biotechnology. Doi: 10.1007/s12010-019-02987-2.

Bhatia, S. 2016. Nanoparticles types, classification, characterization, fabrication methods and drug delivery applications. Natural Polymer Drug Delivery Systems 3: 33–92.

Bidart, C., M. Fröhling and F. Schultmann. 2014. Livestock manure and crop residue for energy generation: Macro-assessment at a national scale. Renewable and Sustainable Energy Reviews 38: 537–550.

Binod, P., E. Gnansounou, R. Sindhu and A. Pandey. 2019. Enzymes for 2nd generation biofuels, recent developments and future perspectives. Bioresource Technology Reports 5: 317–325.

Blombach, B. and B.J. Eikmanns. 2011. Current knowledge on isobutanol production with *Escherichia coli*, *Bacillus subtilis* and *Corynebacterium glutamicum*. Bioengineered Bugs 2: 346–350.

Cabral, P.P., M.M.R. Fonseca and S.F. Dias. 2010. Esterification activity and operational stability of *Candida rugosa* lipase immobilized in polyurethane foams in the production of ethyl butyrate. Biochemical Engineering Journal 48: 246–252.

Canilha, L., A.K. Chandel, S.D.S. Milessi, F.A.F. Antunes, W. Luiz da Costa Freitas, das Graças Almeida Felipe and S.S. da Silva. 2012. Bioconversion of sugarcane biomass into ethanol: An overview about composition, pretreatment methods, detoxification of hydrolysates, enzymatic saccharification, and ethanol fermentation. Journal of Biomedicine and Biotechnology 1–15.

Calero, J., D. Luna, E.D. Sancho, C. Luna, F.M. Bautista, A.A. Romero, A. Posadillo, J. Berbel and C. Verdugo-Escamillla. 2015. An overview on glycerol-free processes for the production of renewable liquid biofuels, applicable in diesel engines. Renewable and Sustainable Energy Reviews 42: 1437–1452.

Chen, Y.Z., C.T. Yang, C.B. Ching and R. Xu. 2008. Immobilization of lipases on hydrophobilized zirconia nanoparticles: Highly enantioselective and reusable biocatalysts. Langmuir 24: 8877–8884.

Chundawat, S.P.S., G.T. Beckham, M.E. Himmel and B.E. Dale. 2011. Deconstruction of lignocellulosic biomass to fuels and chemicals. Annual Review of Chemical and Biomolecular Engineering 2: 121–145.

Connor, M.R., A.F. Cann and J.C. Liao. 2010. 3-Methyl-1-butanol production in *Escherichia coli*: random mutagenesis and two-phase fermentation. Applied Microbiology and Biotechnology 86: 1155–64.

Cubides-Roman, D.C., V.H. Perez, H.F.D. Castro, C.E. Orrego, O.H. Gerraldo, E.G. Silveira and G.F. David. 2017. Ethyl esters (biodiesel) production by *Pseudomonas fluorescens* lipase immobilized on chitosan with magnetic properties in a bioreactor assisted by lectromagnetic field. Fuel 96: 481–487.

Daubresse, C., C. Grandfils, R. Jerome and P. Teyssie. 1994. Enzyme immobilization in nanoparticles produced by inverse microemulsion polymerization. Journal of Colloid and Interface Science 168: 222–229.

Dehkhoda, A.M., A.H. West and N. Ellis. 2010. Biochar based solid acid catalyst for biodiesel production. Applied Catalysis A: General 382(2): 197–204.

Demirbas, A. 2009. Biofuels securing the planet's future energy needs. Energy Conversion and Management 50: 2239–2249.

Demeke, M.M., H. Dietz, Y. Li, M.R. Foulquié-Moreno, S. Mutturi, S. Deprez, T. Den Abt, B.M. Bonini, G. Liden and F. Dumortier. 2013. Development of a D-xylose fermenting and inhibitor tolerant industrial *Saccharomyces cerevisiae* strain with high performance in lignocellulose hydrolysates using metabolic and evolutionary engineering. Biotechnology for Biofuels 6–89.

Denkbas, E.B. and M. Odabasi. 2000. Chitosan microspheres and sponges: preparation and characterization. Journal Applied Polymer Science 76(11): 1637–1642.

Deublein, D. and A. Steinhauser. 2008. Biogas from Waste and Renewable Resources. Wiley Online Library: Weinheim, Germany. pp. 13–27.

Dickinson, J.R., S.J. Harrison and M.J. Hewlins. 1998. An investigation of the metabolism of valine to isobutyl alcohol in *Saccharomyces cerevisiae*. Journal of Biological Chemistry 273: 25751–25756.

DiCosimo, R., J. McAuliffe, A.J. Poulose and G. Bohlmann. 2013. Industrial use of immobilized enzymes. Chemical Society Reviews 42: 6437–6474.

Duong, N.B., C.L. Wang, L.Z. Huang, W.T. Fang and H. Yang. 2019. Development of a facile and low-cost chitosan modified carbon cloth for efficient self-pumping enzymatic biofuel cells. Journal of Power Sources 111–119.

Dyal, A., K. Loos, M. Noto, S.W. Chang, C. Spagnoli, K.V. Shafi, A. Ulman, M. Cowman and R.A. Gross. 2003. Activity of *Candida rugosa* lipase immobilized on -Fe_2O_3 magnetic nanoparticles. Journal of the American Chemical Society 125(7): 1684–1685.

Es, I., J.D. Vieira and A.C. Amaral. 2015. Principles, techniques, and applications of biocatalyst immobilization for industrial application. Applied Microbiology and Biotechnology 99: 2065–2082.

Egwim, E.C., A. Austin, O.A. Oyewole and I.N. Okoliegbe. 2012. Optimization of lipase immobilization on chitosan beads for biodiesel production. Global Research Journal of Microbiology 2(2): 103–112.

El-Hadi, A. and H. Ahmed. 2013. Application of Fe_3O_4 chitosan nanoparticles for *Mucor racemosus* NRRL 3631 lipase immobilization. Egyptian Pharmaceutical Journal 12(2): 155–160.

Faisal, S., Y. Hafeez, Y. Zafar, S. Majeed, X. Leng, S. Zhao, I. Saif, K. Malik and X. Li. 2019. A review on nanoparticles as biogas producer nano fuels and biosensing monitoring. Applied Sciences 9(59): 1–19.

Foresti, M.L. and M.L. Ferreira. 2007. Chitosan-immobilized lipases for the catalysis of fatty acid esterifications. Enzyme and Microbial Technology 40: 769–777.

Ganesan, A., B.D. Moore, S.M. Kelly, N.C. Price, O.J. Rolinski, D.J.S. Birch, I.R. Dunkin and P.J. Halling. 2009. Optical spectroscopic methods for probing the conformational stability of immobilised enzymes. ChemPhysChem 10: 1492–1499.

Ganzoury, M.A. and N.K. Allam. 2015. Impact of nanotechnology on biogas production: A mini-review. Renewable & Sustainable Energy Reviews 50: 1392–1404.

García-Martínez, J.E. 2010. Nanotechnology for the Energy Challenge. John Wiley & Sons.

Gardy, J., A. Hassanpour, X. Lai, M.H. Ahmed and M. Rehan. 2017. Biodiesel production from used cooking oil using a novel surface functionalised TiO_2 nano-catalyst. Applied Catalysis B: Environmental 207: 297–310.

Gardy, J., A. Osatiashtiani, O. Céspedes, A. Hassanpour, X. Lai, A.F. Lee, K. Wilson and M. Rehan. 2018. A magnetically separable SO_4/Fe-Al-TiO_2 solid acid catalyst for biodiesel production from waste cooking oil. Applied Catalysis B: Environmental 234: 268–278.

Ghadi, A., F. Tabadeh, S. Mahjoub, A. Mohsenifar, F.T. Roshan and R.S. Alavije. 2015. Fabrication and characterization of core-shell magnetic chitosan nanoparticles as a novel carrier for immobilization of *Burkholderia cepacia* lipase. Journal of Oleo Science 64(4): 423–430.

Ghamguia, H., M. Karra-Chaabouni and Y. Gargouri. 2004. 1-butyl oleate synthesis by immobilised lipase from *Rhizopus oryzae*: A comparative study between n-hexane and solvent-free system. Enzyme and Microbial Technology 35: 355–363.

Ghashghaei, S. and G. Emtiazi. 2015. The methods of nanoparticles synthesis using bacteria as biological nanofactories their mechanisms and major applications. Current Bionanotechnology 1(1): 3–17.

Ghosh, D. and P.C. Hallenbeck. 2014. Metabolic engineering: key for improving biological hydrogen production. pp. 1–46. *In*: Lu, X. (ed.). Biofuels from Microbes to Molecules. 1st edn. Horizon Scientific Press, United Kingdom.

Grenha, A. 2012. Chitosan nanoparticles: A survey of preparation methods. Journal of Drug Targeting 1: 1–38.

Gruber, S.G. and V. Seidl-Seiboth. 2012. Self versus non-self: fungal cell wall degradation in Trichoderma. Microbiology 158: 26–34.

Guan, Q., Y. Li, Y. Chen, Y. Shi, J. Gu, B. Li, R. Miao, Q. Chen and P. Ning. 2017. Sulfonated multi-walled carbon nanotubes for biodiesel production through triglycerides transesterification. RSC Advances 7(12): 7250–7258.

Gui, M.M., K.T. Lee and S. Bhatia. 2008. Feasibility of edible oil vs non edible oil vs waste edible oil as biodiesel feedstock. Energy 33: 1646–1653.

Guisan, J.M. 2013. New opportunities for immobilization of enzymes. pp. 1–13. *In*: Guisan, J.M. (ed.). Methods in Molecular Biology, Springer: Clifton, NJ, USA.

Guldhe, A., B. Singh, T. Mutanda, K. Permaul and F. Bux. 2015. Advances in synthesis of biodiesel via enzyme catalysis: novel and sustainable approaches. Renewable & Sustainable Energy Reviews 41: 1447–1464.

Guo, H., M. Daroch, L. Liu, G. Qiu, S. Geng and G. Wang. 2013. Biochemical features and bioethanol production of microalgae from coastal waters of Pearl River Delta. Bioresource Technology 127: 422–428.

Gupta, M.N., M. Katoli, M. Kapoor and K. Solanki. 2011. Nanomaterials as matrices for enzyme immobilization. Artificial Cells Blood Substitutes and Biotechnology 39: 98–109.

Hazelwood, L.A., J.M. Daran, A.J.A. van Maris, J.T. Pronk and J.R. Dickinson. 2008. The ehrlich pathway for fusel alcohol production: a century of research on *Saccharomyces cerevisiae* metabolism. Applied and Environmental Microbiology 74: 2259–2266.

Hernandez, A.D., G. Gracida, B.E. Garcia-Almendarez, C. Regalado, R. Nune and A.A. Reyes. 2018. Characterization of magnetic nanoparticles coated with chitosan: A potential approach for enzyme immobilization. Hindawi. Journal of Nanomaterials 1–11.

Ho, S.H., S.W. Huang, C.Y. Chen, T. Hasunuma, A. Kondo and J.S. Chang. 2013. Bioethanol production using carbohydrate-rich microalgae biomass as feedstock. Bioresource Technology 135: 191–198.

Hoell, I.A., G. Vaaje-Kolstad and V.G.H. Eijsink. 2010. Structure and function of enzymes acting on chitin and chitosan. Biotechnology and Genetic Engineering Reviews 27: 331–366.

Holzapple, M. 1993. Cellulose. Encylopedia of Food Science, Food Technology and Nutrition 2: 2731–2738.

Hong, J., P. Gong, D. Xu, L. Dong and S. Yao. 2007. Stabilization of chymotrypsin by covalent immobilization on amine-functionalized superparamagnetic nanogel. Journal of Biotechnology 128: 597–605.

Howard, T.P., S. Middelhaufe, K. Moore, C. Edner, D.M. Kolak, G.N. Taylor, D.A. Parker, R. Lee, N. Smirnoff, S.J. Aves and J. Love. 2013. Synthesis of customized petroleum-replica fuel molecules by targeted modification of free fatty acid pools in *Escherichia coli*. Proceedings of the National Academy of Sciences of the United States of America 110: 7636–7641.

Huang, D., H. Zhou and L. Lin. 2012. Biodiesel; an alternative to conventional fuel. 2012 International conference on future energy, environment and materials. Energy Procedia 16: 1874–1885.

Huang, S.H., M.H. Liao and D.H. Chen. 2003. Direct binding and characterization of lipase onto magnetic nanoparticles. Biotechnology Progress 19: 1095–1100.

Huang, X.J., G. Dan and Z.K. Xu. 2007. Preparation and characterization of stable chitosan nanofibers membrane for lipase immobilization. European Polymer Journal 43: 3710–3718.

Huang, X.J., A.G. Yu and Z.K. Xu. 2008. Covalent immobilisation of lipase from *Candida rugosa* onto poly(acrylonitrile-co-2-hydroxyethyl methacrylate) electrospun fibrous membranes for potential bioreactor application. Bioresource Technology 99: 5459–5465.

Huang, X.J., P.C. Chen, F. Huang, Y. Ou, M.R. Chen and Z.K. Xu. 2011. Immobilization of *Candida rugosa* lipase on electrospun cellulose nanofiber membrane. Journal of Molecular Catalysis B: Enzymatic 70: 95–100.

Huo, M., Y. Zhang and J. Zhou. 2010. Synthesis and characterization of low-toxic amphiphilic chitosan derivatives and their application as micelle carrier for antitumor drug. International Journal of Pharmaceutics 394(1-2): 162–173.

Hwang, E.T. and M.B. Gu. 2013. Enzyme stabilization by nano/microsized hybrid materials. Engineering in Life Sciences 13: 49–61.

Ines Belhaj, B.R., B.R. Zamen, G. Ali and B. Hafedh. 2011. Esterification activity and stability of Talaromyces thermophilus lipase immobilized onto chitosan. Journal of Molecular Catalysis B: Enzymatic 68: 230–239.

Ivanova, V., P. Petrova and J. Hristov. 2011. Application in ethanol fermentation of immobilized yeast cells in matrix of agniate/magnetic nanoparticles on chitosan-magnetite microparticles and cellulose coated magnetic nanoparticles. International Journal of Chemical Engineering 3(2): 121–145.

Jaime, M.D.L.A., L.V. Lopez-Llorca, A. Conesa, A.Y. Lee, M. Proctor, L.E. Heisler, M. Gebbia, G. Giaever, J.T. Westwood and C. Nislow. 2012. Identification of yeast genes that confer resistance to chitosan oligosaccharide (COS) using chemogenomics. BMC Genomics 13: 267–275.

Ji, P., H.S. Tan, X. Xu and W. Feng. 2010. Lipase covalently attached to multiwalled carbon nanotubes as an efficient catalyst in organic solvent. AIChE Journal 56: 3005–3011.

Jia, H., G. Zhu, B. Vugrinovich, W. Kataphinan, D.H. Reneker and P. Wang. 2002. Enzyme-carrying polymeric nanofibers prepared via electrospinning for use as unique biocatalysts. Biotechnology Progress 18: 1027–1032.

Jiang, W., J.B. Qiao, G.J. Bentley, D. Liu and F. Zhang. 2017. Modular pathway engineering for the microbial production of branched-chain fatty alcohols. Biotechnology for Biofuels 10: 244–251.

John, R.P., G.S. Anisha, K.M. Nampoorthi and A. Pandey. 2012. Micro and macroalgal biomass: A renewable source for bioethanol. Bioresource Technology 102(1): 186–193.

Jordan, J., C. Kumar and C. Theegala. 2011. Preparation and characterization of cellulose bound magnetite nanoparticle. Molecular Catalysis B-Enzymatic 68(2): 139–146.

Kanimozhi, S. and K. Perinbam. 2010. Biodiesel production from *Pseudomonas fluorescens* Lp I lipase immobilized on amino-silane modified super paramagnetic Fe_3O_4 nanoparticles. Journal of Physics: Conference Series 1(1): 43–51.

Karimi, K., G. Emtiazi and M.J. Taherzadeh. 2006. Ethanol production from dilute-acid pretreated rice straw by simultaneous saccharification and fermentation with Mucor indicus *Rhizopus oryzae*, and *Saccharomyces cerevisiae*. Enzyme and Microbial Technology 40: 138–144.

Kefeni, K.K., T.A.M. Msagati and B.B. Mamba. 2017. Ferrite nanoparticle synthesis, characterization and application in electronic device. Materials Science and Engineering: B 215: 37–55.

Kim, B.C., S. Nair, J. Kim, J.H. Kwak, J.W. Grate, S.H. Kim and M.B. Gu. 2005. Preparation of biocatalytic nanofibres with high activity and stability via enzyme aggregate coating on polymer nanofibres. Nanotechnology 16: S382–S388.

Kim, H., I. Lee, Y. Kwon, B.C. Kim, S. Ha, J.H. Lee and J. Kim. 2011. Immobilization of glucose oxidase into polyaniline nanofiber matrix for biofuel cell applications. Biosensors and Bioelectronics 26: 3908–3913.

Kim, J. and J.W. Grate. 2003. Single enzyme nanoparticles armored by a nanometer scale organic/inorganic network. Nanotechnology Letters 3: 1219–1222.

Kim, J., J.W. Grate and P. Wang. 2006. Nanostructures for enzyme stabilisation. Chemical Engineering Science 61: 1017–1026.

Kim, J., J.W. Grate and P. Wang. 2008. Nanobiocatalysis and its potential applications. Trends in Biotechnology 26: 639–646.

Kim, K.H., O.K. Lee and E.Y. Lee. 2018. Nano-immobilized biocatalysts for biodiesel production from renewable and sustainable resources. Catalysts 8(68): 1–13.

Knothe, G. and R.O. Dunn. 2009. A comprehensive evaluation of the melting points of fatty acids and esters determined by differential scanning calorimetry. Journal of the American Oil Chemists' Society 86: 843–856.

Kondo, T., H. Tezuka, J. Ishii, F. Matsuda, C. Ogino and A. Kondo. 2012. Genetic engineering to enhance the Ehrlich pathway and alter carbon flux for increased isobutanol production from glucose by *Saccharomyces cerevisiae*. Journal of Biotechnology 159: 32–37.

Kumar, V., F. Jahan, S. Raghuwanshi, R. Mahajan and R.K. Saxena. 2013. Immobilization of *Rhizopus oryzae* lipase on magnetic Fe_3O_4 chitosan beads and its potential in phenolic acids ester synthesis. Biotechnology and Bioprocess Engineering 18: 787–795.

Kumar, A., S. Sharma, L.M. Pandey and P. Chandra. 2018. Advance engineered materials in fabrication of biosensing electrodes of enzymatic biofuel cells. Materials Science for Energy Technologies (18): 1–44.

Kumari, A., P. Mahapatra, V.K. Garlapati and R. Banerjee. 2009. Enzymatic transesterification of Jatropha oils. Biotechnology for Biofuels 2: 1.

Lau, S.C., H.N. Lim, M. Basri, H.R.G. Masoumi, A.A. Tajudin, N.M. Huang, A. Pandikumar, C.H. Chia and Y. Andou. 2014. Enhanced biocataytic esterification with lipase immobilized chitosan/graphene oxide beads. Plos One 9(8): 1–10.

Lee, D.G., K.M. Ponvel, M. Kim, S. Hwang, I.S. Ahn and C.H. Lee. 2009. Immobilization of lipase on hydrophobic nano-sized magnetite particles. Journal of Molecular Catalysis B: Enzymatic 57: 62–66.

Lee, H.K., J.K. Lee, M.J. Kim and C.J. Lee. 2010a. Immobilization of lipase on single walled carbon nanotubes in ionic liquid. Bulletin of the Korean Chemical Society 31: 650–652.

Lee, J.H., D.H. Lee, J.S. Lim, B.H. Um, C. Park, S.W. Kang and S.W. Kim. 2008. Optimization of the process for biodiesel production using a mixture of immobilized *Rhizopus oryzae* and *Candida rugosa* lipases. Journal of Microbiology and Biotechnology 18(12): 1927–1931.

Lee, S.H., T.T.N. Doan, K. Won, S.H. Ha and Y.M. Koo. 2010b. Immobilization of lipase within carbon nanotube–silica composites for non-aqueous reaction systems. Journal of Molecular Catalysis B: Enzymatic 62: 169–172.

Lei, L., X. Liu, Y. Li, Y. Cui, Y. Yang and G. Qin. 2011a. Study on synthesis of poly (GMA)-grafted Fe_3O_4/SiOx magnetic nanoparticles using atom transfer radical polymerization and their application for lipase immobilization. Materials Chemistry and Physics 125(3): 866–871.

Li, S.F., Y.H. Fan, R.F. Hu and W.T. Wu. 2011b. *Pseudomonas cepacia* lipase immobilised onto the electrospun PAN nanofibrous membranes for biodiesel production from soybean oil. Journal of Molecular Catalysis B: Enzymatic 72: 40–45.

Li, Y.G., H.S. Gao, W.L. Li, J.M. Xing and H.Z. Liu. 2009. *In situ* magnetic separation and immobilization of dibenzothiophene-desulfurizing bacteria. Bioresource Technology 100(21): 5092–5096.

Li, S.S., J.P. Wen and X.Q. Jia. 2011. Engineering *Bacillus subtilis* for isobutanol production by heterologous Ehrlich pathway construction and the biosynthetic 2-ketoisovalerate precursor pathway overexpression. Applied Microbiology and Biotechnology 91: 577–589.

Lin, P.C., R. Saha, F. Zhang and H.B. Pakrasi. 2017. Metabolic engineering of the pentose phosphate pathway for enhanced limonene production in the *Cyanobacterium Synechocysti s* sp. PCC 6803. Scientific Reports 7: 17503–17513.

Lin, Y. and S. Tanaka. 2006. Ethanol fermentation from biomass resources: current state and prospects. Applied Microbiology and Biotechnology 69: 627–642.

Liu, H. and N. Hu. 2007. Study on direct electrochemistry of glucose oxidase stabilized by cross-linking and immobilized in silica nanoparticle films. Electroanalysis 19: 884–892.

Liu, X., H. He, Y. Wang and S. Zhu. 2007. Transesterification of soybean oil to biodiesel using SrO as a solid base catalyst. Catalysis Communications 8(7): 1107–1111.

Liu, X., H. He, Y. Wang, S. Zhu and X. Piao. 2008. Transesterification of soybean oil to biodiesel using CaO as a solid base catalyst. Fuel 87(2): 216–221.

Liu, Y., H. Zhou and L. Wang. 2016. Stability and catalytic properties of lipase immobilized on chitosan encapsulated magnetic nanoparticles cross-linked with genipin and glutaraldehyde. Journal of Chemical Technology & Biotechnology 91(5): 159–1367.

Lopez-Serrano, P., L. Cao, F. Van Rantwijk and R.A. Sheldon. 2002. Cross-linked enzyme aggregates with enhanced activity: Application to lipases. Biotechnology Letters 24: 1379–1383.

Lu, A.H., E.E. Salabas and F. Schüth. 2007. Magnetic nanoparticles: Synthesis, protection, functionalization, and application. Angewandte Chemie International Edition 46: 1222–1244.

Ma, D.I.N.G., M. Li, A.J. Patil and S. Mann. 2004. Fabrication of protein/silica core–shell nanoparticles bymicroemulsion-based molecular wrapping. Advanced Materials 16: 1838–1841.

Macrelli, S., M. Galbe and O. Wallberg. 2014. Effects of production and market factors on ethanol profitability for an integrated first- and second-generation ethanol plant using the whole sugarcane as feedstock. Biotechnology for Biofuels 7–26.

Mahto, T.K., R. Jain, S. Chandra, D. Roy, V. Mahto and S.K. Sahu. 2016. Single step synthesis of sulfonic group bearing graphene oxide: A promising carbo-nano material for biodiesel production. Journal of Environmental Chemical Engineering 4(3): 2933–2940.

Marta, Z., C. Dorota, S. Tomasz, S. Adam, W. Katarzyna, S. Joanna, K. Halina and M.P. Marsza. 2017. Chitosan–collagen coated magnetic nanoparticles for lipase immobilization-new type of enzyme friendly polymer shell crosslinking with squaric acid. Catalysts 7: 26–31.

Matsuda, F., I. Jun, K. Takashi, I. Kengo, H. Tezuka and A. Kondo. 2013. Increased isobutanol production in *Saccharomyces cerevisiae* by eliminating competing pathways and resolving cofactor imbalance. Microbial Cell Factories 12(119): 1–11.

Meadows, A.L., K.M. Hawkins, Y. Tsegaye, E. Antipov, Y. Kim, L. Raetz, R.H. Dahl, A. Tai, T. Mahatdejkul-Meadows and L. Xu. 2016. Rewriting yeast central carbon metabolism for industrial isoprenoid production. Nature 537: 694–697.

Mehrasbi, M.R., J. Mohammadi, M. Peyda and M. Mohammadi. 2017. Covalent immobilization of Candida Antarctica lipase on core-shell magnetic nanoparticles for production of biodiesel from waste cooking oil. Renewable Energy 101: 593–602.

Mehrer, C.R., M.R. Incha, M.C. Politz and B.F. Pfleger. 2018. Anaerobic production of medium chain fatty alcohol via a beta reduction pathway. Metabolic Engineering 48: 63–71.

Mileti, C.N., V. Abetz, K. Ebert and K. Loos. 2010. Immobilization of *Candida antarctica* lipase B on polystyrene nanoparticles. Macromolecular Rapid Communications 31: 71–74.

Milessi, T.S., P.M. Aquino, C.R. Silva, G.S. Moraes, T.C. Zangirolami, R.C. Giordano and R.L.C. Giordano. 2018. Influence of key variables on the simultaneous isomerization and fermentation (SIF) of xylose by a native *Saccharomyces cerevisiae* strain co-encapsulated with xylose isomerase for 2G ethanol production. Biomass & Bioenergy 119: 277–283.

Milessi, T.S., F.A.S. Corradini, W. Kopp, T.C. Zangirolami, P.W. Tardoli, R.C. Giordano and R.L.C. Giordano. 2019. An innovative biocatalyst for continuous 2G Ethanol production from Xylo-oligomers by *Saccharomyces cerevisiae* through simultaneous hydrolysis, isomerization and fermentation (SHIF). Catalysts 9: 225–231.

Min, K. and Y.J. Yoo. 2014. Recent progress in nanobiocatalysis for enzyme immobilization and its application. Biotechnology and Bioprocess Engineering 19: 553–567.

Mohd Azhar, S.H., R. Abdulla, S.A. Jambo, H. Marbawi, J.A. Gansau, A.A. Mohd Faik and K.F. Rodrigues. 2017. Yeasts in sustainable bioethanol production: a review. Biochemistry and Biophysics Reports 10: 52–61.

Naik, R.R., M.M. Tomczak, H.R. Luckarift, J.C. Spain and M.O. Stone. 2004.Entrapment of enzymes and nanoparticles using biomimetically synthesized silica. Chemical Communications 15: 1684–1685.

Nair, S., J. Kim, B. Crawford and S.H. Kim. 2007. Improving biocatalytic activity of enzyme-loaded nanofibers by dispersing entangled nanofiber structure. Biomacromolecules 8: 1266–1270.

Nakane, K., T. Hotta, T. Ogihara, N. Ogata and S.J. Yamaguchi. 2007. Synthesis of (Z)-3-Hexen-1-yl acetate by lipase immobilised in polyvinylalcohol nanofibers. Journal of Applied Polymer Science 106: 863–867.

Nizami, A.S. and M. Rehan. 2018. Towards nanotechnology based biofuel industry. Biofuel Research Journal 18: 798–799.

Noraini, M.Y., H.C. Ong, M.J. Badrul and W.T. Chong. 2014. A review on potential enzymatic reaction for biofuel production from algae. Renewable and Sustainable Energy Reviews 39: 24–34.

Ohkawa, K., D.I. Cha, H. Kim, A. Nishida and H. Yamamoto. 2004. Electrospinning of chitosan. Macromolecular Rapid Communications 25: 1600–1605.

Ohya, Y., M. Shiratani, H. Kobayashi and T. Ouchi. 1994. Release behaviour of 5-fluorouracil from chitosan-gel nanospheres immobilizing 5-fluorouracil coated with polysaccharides and their cell specific cytotoxicity. Pure and Applied Chemistry A31: 629–642.

Ong, H.C., T.M.L. Mahila, H.H. Masjuki and R.S. Norhasyima. 2011. Comparison of palm oil, jatropha curcas and Calophyllum inophyllum for biodiesel: a review. Renewable and Sustainable Energy Reviews 15(8): 3501–15.

Pantidos, N. and L.E. Horsfall. 2014. Biological synthesis of metallic nanoparticles by bacteria, fungi and plants. Journal of Nanomedicine & Nanotechnology 5(5): 1–10.

Pavlidis, I.V., T. Tsoufis, A. Enotiadis, D. Gournis and H. Stamatis. 2010. Functionalized multi-wall carbon nanotubes for lipase immobilisation. Advanced Engineering Materials 12: B179–B183.

Prabaharan, M. and J.F. Mano. 2005. Chitosan-based particles as controlled drug delivery systems. Drug Delivery 12: 41–57.

Prabhakar, S.V.R.K. and M. Elder. 2009. Biofuels and resource use efficiency in developing Asia: Back to basics. Applied Energy 86: 30–36.

Putra, R.S., P. Kharis, Y. Antonoa, M. Idris, J. Ruaa and H. Ramadhani. 2016. Enhanced electrocatalytic biodiesel production with chitosan gel (hydrogel and xerogel). Procedia Engineering 148: 609–614.

Raita, M., J. Arnthong, V. Champreda and N. Laosiripojana. 2015. Modification of magnetic nanoparticle lipase designs for biodiesel production from palm oil. Fuel Processing Technology 134: 189–197.

Rao, A. A. Sathiavelu and S. Mythili. 2017. Mini review on nanoimmobilization of lipase and cellulase for biofuel production. Biofuels 1–10.

Reetz, M.T., A. Zonta, V. Vijayakrishnan and K. Schimossek. 1998. Entrapment of lipases in hydrophobic magnetite-containing sol-gel materials: Magnetic separation of heterogeneous biocatalysts. Journal of Molecular Catalysis A: Chemical 134: 251–258.

Ren, Y., J.G. Rivera, L. He, H. Kulkarni, D.K. Lee and P.B. Messersmith. 2011. Facile, high efficiency immobilisation of lipase enzyme on magnetic iron oxide nanoparticle via a biomimetic coating. BMC Biotechnology 11: 63.

Royon, D., M. Daz, G. Ellenrieder and S. Locatelli. 2007. Enzymatic production of biodiesel from cotton seed oil using t-butanol as a solvent. Bioresource Technology 98: 648–653.

Safarik, I. and M. Safarikova. 2009. Magnetic nano and microparticles in biotechnology. Chemical Papers 63: 497–505.

Sakai, S., Y. Liu, T. Yamaguchi, R. Watanabe, M. Kawabe and K. Kawakami. 2010. Immobilisation of Pseudomonas cepacia lipase onto electrospun polyacrylonitrile fibers through physical adsorption and application to transesterification in nonaqueous solvent. Biotechnology Letters 32: 1059–1062.

Sanchez-Ramirez, J., J.L. Martinez Hernandez, P. Segura-cericeros, G. Lope, H. Saade, M.A. Medina-Morales, R. Ramroz-Gonzales, C.N. Anguilar and A. Ilyina. 2016. Cellulases immobilization of chitosan coated magnetic nanoparticles: application for *Agave atovirens* lignocellulosic biomass hydrolysis. Bioprocess and Biosystems Engineering 1–14.

Schiffman, J.D. and C.L. Schauer. 2007. One step electrospinning of crosslinked chitosan fibers. Biomacromolecules 8: 2665–2667.

Seelert, T., D. Ghosh and V. Yargeau. 2015. Improving biohydrogen production using *Clostridium beijerinckii* immobilized with magnetite nanoparticles. Applied Microbiology and Biotechnology 99(9): 4107–16.

Seema, S.B. and S. Steven. 2002. Immobilization of lipase using hydrophilic polymers in the form of hydrogel beads. Biomaterials 23: 3627–3636.

Sharmeen, S., A.F.M. Mustafizur Rahman, M.M. Lubna, K.S. Salem, R. Islam and M.A. Khan. 2018. Polyethylene glycol functionalized carbon nanotubes/gelatine chitosan nanocomposite: An approach for significant drug release. Bioactive Materials 3: 236–244.

Shah, M., D. Fawcett and S. Sharma. 2015. Green synthesis of metallic nanoparticles via biological entities. Materials 8(11): 7278–7308.

Sheldon, R.A. 2007. Cross-linked enzyme aggregates (CLEAs): Stable and recyclable biocatalysts. Biochemical Society Transactions 35: 1583–1587.

Silva, C.R., T.C. Zangirolami, J.P. Rodrigues, K. Matugi, R.C. Giordano and R.L.C. Giordano. 2012. An innovative biocatalyst for production of ethanol from xylose in a continuous bioreactor. Enzyme and Microbial Technology 50: 35–42.

Sinha, V.R., A.K. Singla and S. Wadhawan. 2004. Chitosan microspheres as a potential carrier for drugs. International Journal of Pharmaceutics 274(1-2): 1–33.

Singh, P., Y.J. Kim and D. Zhang. 2016. Biological synthesis of nanoparticles from plants and microorganisms. Trends in Biotechnology 34(7): 588–599.

Sivagurunathan, P., A. Kadier, A. Mudhoo, G. Kumar, K. Chandrasekhar, T. Kobayashi and K. Xu. 2018. Nanomaterials for biohydrogen production. biomedical, environmental, and engineering applications. Ed Suvardhan Kanchi. Wiley Sci. 217–237.

Song, J., D. Kahveci, M.L. Chen, Z. Guo, E.Q. Xie, X.B. Xu, F. Besenbacher and M.D. Dong. 2012. Enhanced catalytic activity of lipase encapsulated in PCL nanofibers. Langmuir 28: 6157–6162.

Sotowa, K.I., K. Takagi and S. Sugiyama. 2008. Fluid flow behavior and the rate of an enzyme reaction in deep microchannel reactor under high-throughput condition. Chemical Engineering Journal 135S: S30–S36.

Stellwagen, D.R., F. van der Klis, D.S. van Es, K.P. de Jong and J.H. Bitter 2013. Functionalized carbon nanofibers as solid-acid catalysts for transesterification. ChemSusChem 6(9): 1668–1672.

Sukumaran, R.K., L.D. Gottumkkala, K. Rajasrer, D. Alex and A. Pandey. 2011. Butanol fuel from biomass: revisiting ABE fermentations. Biofuels. Alternative Feedstock and Conversion Processes 571–586.

Tan, T., J. Lu, K. Nie, L. Deng and F. Wang. 2010. Biodiesel production with immobilized lipase: A review. Biotechnology Advances 28(5): 628–634.

Tang, H., J. Chen, S. Yao, L. Nie, G. Deng and Y. Kuang. 2004. Amperometric glucose biosensor based on adsorption of glucose oxidase at platinum nanoparticle-modified carbon nanotube electrode. Analytical Biochemistry 331: 89–97.

Terbojevich, M. and R.A.A. Muzzarelli. 2009. Chitosan. pp. 367–378. *In*: Phillips, G.O. and P. Williams (eds.). Handbook of Hydrocolloids. Cambridge: Woodhead Publishing Ltd.

Thangaraj, B., Z. Jia, L. Dai, D. Liu and W. Du. 2016. Effect of silica coating on Fe_3O_4 magnetic nanoparticles for lipase immobilization and their application for biodiesel production. Arabian Journal of Chemistry 12(8): 4694–4706.

Thangaraj, B., P.R. Solomon, B. Muniyandi, S. Ranganathan and L. Lin. 2019. Catalysis in biodiesel production: a review. Clean Energy 3(1): 2–23.

Torres-Giner, S., E. Gimenez and J.M. Lagaoron. 2008. Characterization of morphology and thermal properties of prolamine nanostructures. Food Hydrocolloides 22601–14.

Tran, D.T., C.L. Chen and J.S. Chang. 2012. Immobilization of Burkholderia sp. lipase on a ferric silica nanocomposite for biodiesel production. Journal of Biotechnology 158: 112–119.

Trindade, S.C. 2011. Nanotech biofuels and fuel additives. pp. 103–114. *In*: Dos Santos Bernardes, M.A. (ed.). Biofuel's Engineering Process Technology. Intech Open.

Vaghari, H., H. Jafarizadeh-Malmiri, M. Mohammadlou, A. Berenjian, N. Anarjan, N. Jafari and S. Nasiri. 2016. Application of magnetic nanoparticles in smart enzyme immobilization. Biotechnology Letters 38: 223–233.

Verma, M.L. and S.S. Kanwar. 2008. Properties and application of poly (MAc-co-DMA-cl- MBAm) hydrogel immobilised *Bacillus cereus* MTCC 8372 lipase for synthesis of geranyl acetate. Journal of Applied Polymer Science 110: 837–846.

Verma, M.L., W. Azmi and S.S. Kanwar. 2008. Microbial Lipases: at the interface of aqueous and non-aqueous media: a review. Acta Microbiologica et Immunologica Hungarica 55: 265–293.

Verma, M.L., C.J. Barrow, J.F. Kennedy and M. Puri. 2012. Immobilisation of β-D-galactosidase from Kluyveromyces lactis on functionalized silicon dioxide nanoparticles: characterization and lactose hydrolysis. International Journal of Biological Macromolecules 50: 432–437.

Verma, M.L., C.J. Barrow and M. Puri. 2013. Nanobiotechnology as a novel paradigm for enzyme immobilisation and stabilisation with potential applications in biodiesel production. Applied Microbiology and Biotechnology 97: 23–39.

Verma, M.L., M. Puri and C.J. Barrow. 2016. Recent trends in nanomaterials immobilised enzymes for biofuel production. Critical Reviews in Biotechnology 36: 108–119.

Verziu, M., B. Cojocaru, J. Hu, R. Richards, C. Ciuculescu and P. Filip Parvulescu. 2008. Sunflower and rapeseed oil transesterification to biodiesel over different nanocrystalline MgO catalysts. Green Chemistry 10(4): 373–381.

Vogt, M., C. Brusseler, J. van Ooyen, M. Bott and J. Marienhagen. 2016. Production of 2-methyl-1-butanol and 3-methyl-1-butanol in engineered *Corynebacterium glutamicum*. Metabolic Engineering 38: 436–445.

Wang, B., J. Wang, W. Zhang and D.R. Meldrum. 2012. Application of synthetic biology in cyanobacteria and algae. Frontiers in Microbiology 3: 344–349.

Wang, P. 2006. Nanoscale biocatalyst systems. Current Opinion in Biotechnology 17: 574–579.

Wang, X., P. Dou, P. Zhao, C. Zhao, Y. Ding and P. Xu. 2009a. Immobilisation of lipases onto magnetic Fe_3O_4 nanoparticles for application in biodiesel production. ChemSusChem 2: 947–950.

Wang, X., X. Liu, C. Zhao, Y. Ding and P. Xu. 2011a. Biodiesel production in packed-bed reactors using lipase–nanoparticle biocomposite. Bioresource Technology 102: 6352–6355.

Wang, X., X. Liu, X. Yan, P. Zhao, Y. Ding and P. Xu. 2011b. Enzymenanoporous gold biocomposite: excellent biocatalyst with improved biocatalytic performance and stability. PLoS One 6(9): e24207.

Wang, Z.G., J.Q. Wang and Z.K. Xu. 2006. Immobilization of lipases from *Candida rugosa* on electrospun polysulfone nanofibrous membranes by adsorption. Journal of Molecular Catalysis B: Enzymatic 42: 45–51.

Wang, Z.G., L.S. Wan, Z.M. Liu, X.J. Huang and Z.K. Xu. 2009b. Enzyme immobilization on electrospun polymer nanofibers: An overview. Journal of Molecular Catalysis B: Enzymatic 56: 189–195.

Wijesena, R.N., N. Tissera, Y.Y. Kannangara, Y. Lin, G.A.J. Amaratunga and D.S. Nalin. 2015. A method for top down preparation of chitosan nanoparticles and nanofibers. Carbohydrate Polymers 117: 731–738.

Willaert, R. and V. Nedovic. 2006. Primary beer fermentation by immobilized yeast-a review on flavour formation and control strategies. Journal of Chemical Technology & Biotechnology 81: 1353–1367.

Xie, W. and N. Ma. 2009. Immobilised lipase on Fe_3O_4 nanoparticles as biocatalyst for biodiesel production. Energy & Fuel 23: 1347–1353.

Xie, W. and J. Wang. 2012. Immobilized lipase on magnetic chitosan microspheres for transesterification of soybean oil. Biomass & Bioenergy 36: 373–380.

Yahya, N.Y., N. Ngadi, M. Jusoh and N.A.A. Halim. 2016. Characterization and parametric study of mesoporous calcium titanate catalyst for transesterification of waste cooking oil into biodiesel. Energy Conversion and Management 129: 275–283.

Yan, M., J. Ge, Z. Liu and P. Ouyang. 2006. Encapsulation of single enzyme in nanogel with enhanced biocatalytic activity and stability. Journal of the American Chemical Society 128: 11008–11009.

Yang, H.H., S.Q. Zhang, X.L. Chen, Z.X. Zhuang, J.G. Xu and X.R. Wang. 2004. Magnetite-containing spherical silica nanoparticles for biocatalysis and bioseparations. Analytical Chemistry 76: 1316–1321.

Yang, N., X. Chen, T. Ren, P. Zhang and D. Yang. 2015. Carbon nanotube based biosensors. Sensors and Actuators B: Chemical 207: 690–715.

Yiu, H.H.P. and M.A. Keane. 2012. Enzyme-magnetic nanoparticle hybrids: new effective catalysts for the production of high value chemicals. Journal of Chemical Technology & Biotechnology 87: 583–594.

Youngquist, J.T., M.H. Schumacher, J.P. Rose, T.C. Raines, M.C. Politz, M.F. Copeland and B.F. Pfleger. 2013. Production of medium chain length fatty alcohols from glucose in *Escherichia coli*. Metabolic Engineering 20: 177–86.

Zhang, F., S. Rodriguez and J.D. Keasling. 2011. Metabolic engineering of microbial pathways for advanced biofuels production. Current Opinion in Biotechnology 22: 775–83.

Zhang, K., M.R. Sawaya, D.S. Eisenberg and J.C. Liao. 2008. Expanding metabolism for biosynthesis of nonnatural alcohols. Proceedings of the National Academy of Sciences of the United States of America 105: 20653–20658.

Zhang, X., S. Yan, R. Tyagi and R. Surampalli. 2013. Biodiesel production from heterotrophic microalgae through transesterification and nanotechnology application in the production. Renewable and Sustainable Energy Reviews 26: 216–223.

Zhang, X.L., S. Yan, R.D. Tyagi, R.Y. Surampalli and T.C. Zhang. 2010a. Application of nanotechnology and nanomaterials for bioenergy and biofuel production. pp. 478–496. *In*: Khanal, S.K., R.Y. Surampalli, T.C. Zhang, B.P. Lamsal, R.D. Tyagi and C.M. Kao (eds.). Bioenergy and Biofuel from Biowastes and Biomass. American Society of Civil Engineers (ASCE).

Zhang, T., Z. Chi, C.H. Zhao, Z.M. Chi and F. Gong. 2010. Bioethanol production from hydrolysates of inulin and the tuber meal of Jerusalem artichoke by Saccharomyces sp. W0. Bioresource Technology 101: 8166–8170.

Zhang, J., F. Zhang, H. Yang, X. Huang, H. Liu, J. Zhang and S. Guo. 2010b. Graphene oxide as a matrix for enzyme immobilisation. Langmuir 26: 6083–6085.

Zhao, L.M., L.E. Shi, Z.L. Zhang, J.M. Chen, D.D. Shi, J. Yang and Z.X. Tang. 2011. Preparation and application of chitosan nanoparticles and nanofibers. Brazilian Journal of Chemical Engineering 28: 353–362.

Zhou, A. and E. Thomson. 2009. The development of biofuels in Asia. Applied Energy 86: 11–20.

Chapter 5

Advancements in Nanobiocatalysis for Bioenergy and Biofuel Production

G. Velvizhi,[1] ***J. Ranjitha,***[1] ***S. Vijayalakshmi***[1] **and** ***G.N. Nikhil***[2,*]

1. Introduction

By nature, enzymes are biocatalysts and they have gained prominence over inorganic catalysts such as high specificity, high reaction rate even under mild reaction conditions of pH and temperature, low energy consumption, water-solubility, biodegradability, few side reactions and non-toxicity (de Jesús Rostro-Alanis et al. 2016). Hence, enzymes are regarded as ideal catalysts for mercantile purposes and biocatalysis is considered as an attractive tool of biotechnology, enfolding an intense commercial and societal niche in areas, viz., energy, food, health, etc. (de Jesús Rostro-Alanis et al. 2016, Illanes et al. 2012). Yet, these applications are frequently restricted due to enzyme instability all through the outfitted practice and complexity in harvesting and recycling the enzyme. Sustaining their structural confirmation during any biochemical reaction is exceptionally exigent. These hindrances can be surmounted by stabilizing the enzyme while immobilizing it over solid matrix which facilitates easy recovery and is a rational strategy for accomplishing this objective (Sheldon and van Pelt 2013, Sheldon 2007). Consequently, the marketable use of biocatalysts relies on the development of effectual means of immobilization and innovative matrix materials. The major bottleneck is to construct catalysts that hold worthy enzymatic properties and are tough enough to endure harsh conditions (Cantone et al. 2013, Eş et al. 2015).

The current progress in biocatalysis is noted as an enzymatic course of action performed in organic solvents and aqueous media (Clouthier and Pelletier 2012).

[1] Carbon Dioxide and Green Technologies Centre (CGTC), Vellore Institute of Technology, Vellore (Tamil Nadu), India - 632014.

[2] Department of Biotechnology, Dr. B.R. Ambedkar National Institute of Technology, Jalandhar (Punjab), India - 144 011.

* Corresponding author: nikhilgn@nitj.ac.in

Enzymes play a key role in the recent trends in 'white biotechnology' including green chemistry and sustainable energy (Patel et al. 2006). The advent of nanotechnology has offered a range of miscellaneous means such as nano-scaffolds as supports for enzyme immobilization (Saallah and Lenggoro 2018, Mohamad et al. 2015). Enzyme immobilization is valuable for marketable applications due to expediency in management, ease of enzyme purification and re-use, economic product cost and resistant toward heat and pH fluctuations (Sheldon and van Pelt 2013, Sheldon 2007). The advancement of nanotechnology took place by the swift escalation of nanobiotechnology research for diverse applications. In the beginning, nanotechnology and biotechnology were merely merged lacking any mutual benefits (Fei et al. 2009). Later, symbiotic exchanges between the dual technologies have resulted in additional groundbreaking advances; 'Nanobiocatalysis' is one such characteristic paradigm. The core lead of nanobiocatalysts is the elevated ratio of surface/volume that augments the accessibility of the biocatalyst (de Jesús Rostro-Alanis et al. 2016). In this way, the architecture and relevance of nanostructured materials signify an imperative area of budding interest. Nano-biocatalysis shows the potential and stimulating field for perceiving and creating new biotransformations by adjoining to composite molecular communications at the sub-atomic level amid the surroundings and all reaction components (de Jesús Rostro-Alanis et al. 2016, Illanes et al. 2012, Misson et al. 2015). This chapter discusses the recent trends in nano-biocatalysis aiding to improve biofuels/bioenergy production.

2. Nano-Immobilized Biocatalysis

In the recent advances, nano-biocatalysis has experienced an escalation in the field of nanotechnology since the biocatalytic process is limited due to the unsteadiness of the enzymes and their short lifetime (Bommarius and Riebel 2004, Verma et al. 2013). Consequently, immobilization is a key factor in the thriving of engineering processes based on the type of enzymes (DiCosimo et al. 2013) (Figure 1). To boost the steadiness, enzymes are immobilized on the functionalized nanostructured materials to give higher stability to nano-biocatalysts. The concentration of the enzyme immobilized by the nanoparticle was significantly high when compared to

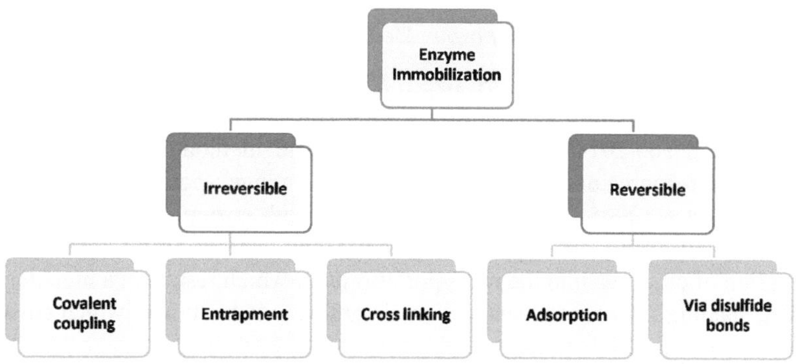

Figure 1. Typical flow chart of different enzyme immobilization methods.

that provided by investigational procedures based on immobilization on planar two-dimensional surfaces due to greater surface/volume ratio (Ansari and Husain 2012). Besides, when compared with native enzymes, the enzyme immobilized on nanoparticle has good thermal stability and has a wider operational pH and temperature conditions (Cipolatti et al. 2016). The hypothesis of using nanoscale configurations for immobilization is to diminish dispersion limitations and exploit the exposed surface to enhance enzyme load (Qu et al. 2013).

2.1 Enzyme Stabilization Methods

2.1.1 Nano-Entrapment

In the majority of techniques for acquiring nanoparticles, that enclose enzymes, are based on the 'nano-entrapment' method, which using the water-in-oil microemulsion scheme (also called reverse micelles) directs isolated nanoparticles through polymerization in the water phase or water-oil interface (Krishnamoorthi et al. 2015). One of the bottlenecks of this method is the intricacy in planning the magnitude of reverse micelles as well as the number of enzyme molecules inside each reverse micelle which will directly influence the finishing properties of enzyme-entrapped nanoparticles. Therefore, the triumph in this method necessitates the meticulous optimization process that addresses these concerns. This method of enzyme nano-entrapment was used to integrate magnetic nanoparticles jointly with enzymes into sphere-shaped silica nanoparticles (Kim et al. 2008b, Yang et al. 2004).

An innovative artificial method was detailed as 'single enzyme nanoparticles' (SENs) in which an organic-inorganic fusion polymer arrangement of dimensions, less than a few nanometers, was constructed upon the enzyme surface (Bhinder and Dadra 2009). Another customized method using a fundamental route was reported as 'single enzyme nano-gels' that consist of two steps: enzyme acryloylation followed by aqueous polymerization and cross-linking to capture a solo enzyme within each nano-gel formed (Ge et al. 2008). Another appealing method for constructing a nanostructure around an enzyme molecule is by self-grouping of organoclay-enfolded enzymes. In this method, the enzyme is enfolded with a thin layer of cationic organoclay oligomers by fusing the enzyme and the organoclay (Kim et al. 2008b).

2.1.2 Immobilization into Nanoporous Media

Nano-porous media with aperture size 2–50 nm have captured a huge interest as it hosts for enzyme immobilization due to their controlled porosity and high surface area (Min and Yoo 2014). Newly, a 'ship-in-a-bottle' method proved to be a simple but effectual resource of enzyme immobilization into nano-porous media (Kim et al. 2008b, Lee et al. 2005). This method enhanced both enzyme loading and enzyme activity by efficiently foiling the enzyme leakage. The foremost step entails the adsorption of enzymes into the nano-porous matter which results in a high degree of enzyme loading. The subsequent step is glutaraldehyde treatment, which covalently links the enzyme molecules, resulting in cross-linked enzyme aggregates at the nanometer-scale within the nanopores.

2.1.3 Coatings on Nanomaterials

Enzyme varnish comprises of the covalent enzyme as an accessory on diverse nano-materials with enzyme cross-linking, resulting in an augmentation in enzyme loading, overall enzyme activity and enzyme steadiness (Kim et al. 2006a). Enzyme-coating systems could be frequently recycled and have been productively tested in continuous reactors over extensive periods of operation as in biofuel cells (Kim et al. 2006b, Fischback et al. 2006). The study reported alleviates the activity of glucose oxidase coating on carbon nano-tubes which enabled the continuous operation of a biofuel cell for over 16 hours. This high operational constancy is predictable to overlay the means for the augmented exploitation of biofuel cells but the functioning of which is presently classified due to their short life span.

2.2 Allied Biocatalysis

The majority of industrial enzymes are usually short of interfacial assembling, which is preferred for *in vitro* bio-transformations among immiscible chemicals in biphasic reaction media. Nanoscale production of enzymes can initiate preferred similarity in communications to force the congregation of enzyme molecules at the oil-water boundary. For instance, enzymes suitably conjugated with hydrophobic polymer groups were able to self-assemble at oil-water interfaces where they could catalyze interfacial reactions involving chemicals dissolved in the oil and water phases (Zhu and Wang 2004, Zhu and Wang 2005). The catalytic competence of the enzyme is appreciably improved for the reason that the interface-assembled biocatalysts have an instantaneous way into chemicals suspended in both phases. In recent times, carbon nano-tubes and numerous additional categories of nanoparticles were also found to be able to assemble at the interface when they were conjugated with enzymes (Asuri et al. 2006). Interfacial enzyme assemblies can be mono-layered or multi-layered; however, higher enzyme activities were appreciated when thinner assemblies were formed (Zhu and Wang 2004, Zhu and Wang 2005). It was deduced that thick assemblies might hinder mass-transfer across the interface, consequently averting the reactants to arrive at the active-site of the enzymes.

Enzymes that have been assembled at the interface showed improved stability when compared to native enzymes simply placed in the same oil-water biphasic systems. This has been attributed to a reduction in interfacial stress caused by the conjugated enzymes acting as macromolecular surfactants (Zhu and Wang 2004). The stability of the interface-assembled enzymes could be further improved by adding stabilizers such as polyols and sugars which create a more favorable environment for enzyme assemblies (Narayanan et al. 2007). Assemblies of enzyme-carrying nanoparticles on solid surfaces have been further explored for the development of bioactive materials (Qhobosheane et al. 2004, Shipway et al. 2000). In one example, a 2D assembly of silica nanoparticles with covalently attached enzymes was achieved in a pattern-controlled manner on the surface of silica plates (Qhobosheane et al. 2004). It is expected that such assemblies might find applications in the development of biosensors and biochips through the combination of the highly selective activities of the enzymes and the optical and/or electrochemical properties of the supporting materials. Moreover, multiple layers of nanoparticles could also be formed, thus introducing unique optical or electrical properties to the surface in addition to their enzymatic bioactivity (Shipway et al. 2000).

3. Nano-Biocatalysis for Biofuel and Bioenergy Generation

The world is in the exploration of eco-friendly energy resources to alleviate approaching environmental changes and the deficiency of readily accessible fuel to offer the energy essential for current and upcoming anthropogenic activities (Hallenbeck et al. 2012). The major barricade is the restrictions of the usual metabolic process that is accessible in the fermentative production processes. For instance, in the case of biohydrogen production about one-third of the substrate can be used for hydrogen production while the remaining two-thirds forming other fermentation products, viz., acetate, butyrate, butanol, acetone, etc. (Hallenbeck 2009). Besides, low-energy translation, the competence of the substrates and disappointing constancy limit the realistic applicability of the process. To address these challenges, in the recent years, nanoparticles for biological applications are targeted to develop the conversion efficiency of the substrate to fuel to augment the bioactivity of microorganisms and to build a steady biofuel production above an extended period. It was noticed that few microbes take the edge of nanoparticles principally in anaerobic situations by efficiently transferring more electrons to acceptors (Beckers et al. 2013). Some nanoparticles have revealed that they can escalate the kinetics of cell metabolism through the competence to react swiftly with electron donors and boost the activity of microbes as biocatalysts (Xu et al. 2012). Following are a few renewable biofuels/ bioenergy options influenced by nanotechnology by application of nanoparticles at different outlooks.

3.1 Biohydrogen

Synthesis of biocatalyzed hydrogen (H_2) through fermentative route is measured as one of the pledging substitutes for sustainable and renewable energy as it demonstrates ecological virtues (Nikhil et al. 2014). The route is feasible from a realistic viewpoint and can function under typical conditions. An increasing rate of hydrogen production and hydrogen yield are the decisive factors to uphold cost-effective production (Rittmann and Herwig 2012). Industrial-scale operations cause few methodical hindrances, viz., feed-inhibition of un-dissociated volatile fatty acids (VFA) (Diamantis et al. 2013, Mohan et al. 2012). Besides, hydrogen storage in a secure and for reversible usage poses a limitation to research. Typically, it can be stored with solid absorbent materials such as palladium under pressure, which further adds unnecessary load to automobiles. Utilizing steady organic substances like formic acid containing high hydrogen by composition may surmount the storage difficulty owing to the trouble-free discharge of hydrogen by a single step dehydrogenation reaction (Sur 2012). Fascinatingly, bacteria there in nature have enzymes that are capable to catalyze the oxidation of formic acid to hydrogen with high turnover frequencies. This is much efficient than the presently identified industrial hydrogenation/dehydrogenation chemical catalysts (Johnson et al. 2010). Formate is an important metabolite for energy generation for various bacteria. A shortest single step decomposition of formate is catalyzed by formate dehydrogenase (FDH) which yields hydrogen. For the reason that redox potential is low (–420 mV), it can serve as an electron donor for the anaerobic reduction of fumarate and nitrate or nitrite (Leonhartsberger et al. 2002). Recent studies have shown the feasibility of hydrogen production from formate and obtained high hydrogen production rates using pure cultures (Yoshida et al. 2005, Shin et al. 2010). On the other hand, using mixed consortia as biocatalyst for H_2 production is a potential and promising option for scale-up, especially when wastewater is used as a substrate (Mohan 2008).

In a recent study, the production of formic acid by using formate dehydrogenase/alcohol dehydrogenase enzyme complex through enzymatic CO_2 transformation technology is being reported as a novel technique (Marpani et al. 2017). A gas substrate was used for biological reaction and the entire rate of reaction was affected by slow mass transfer rate (liquid to gas) and minimum solubility of the gas. In a biological reaction, the slow mass transfer in the liquid-gas will result in poor efficiency and less cost-effectiveness. For a bioprocess, the transfer rate of gas-liquid mass is pretentious by the operating conditions, dispenser and geometry of the reactor and examined for engineering attitudes to enhance the bioprocess productivity. Some other studies with nanoparticles for the improvement of mass transfer in the gas-liquid interface in CO, syngas and O_2 have been analyzed to overcome the limitation in engineering methods (Mondal et al. 2012). By absorbing the interface in gas-liquid, the nanoparticles can enhance the coefficient in gas-liquid mass transfer. The effect of nano-particles on the mass transfer coefficient of CO_2 in gas-liquid aqueous form with or without functional groups was also reported (Ashrafmansouri and Esfahany 2014). A novel effort has been made to improve the rate of FDH (formaldehyde dehydrogenase) reaction for the production of formic acid with the addition of nano-particles from carbon dioxide (Kim et al. 2016). The lower productivity in the enzyme reaction with gas as a substrate and the mass transfer rate in gas-liquid restrict the complete rate of reaction. Magnetic nano-particles with cobalt ferrite silica (MSNs) and cobalt ferrite silica-methyl-functionalized (methyl-MSNs) were formed and studied in gas-liquid mass transfer in the CO_2-water system. The production of formic acid has been enhanced by 12% by adding methyl-MSNs in the reaction with formate dehydrogenase with less reaction time starting with 1 hour to 1.5 hours. Hence, it can be concluded that by adding methyl-MSNs the overall productivity has been increased in the enzyme reaction (Kim et al. 2016).

In a different study, the addition of 5-nm gold nanoparticles enhanced the competence of hydrogen fermentation significantly, and the maximum hydrogen yield reached 36.3% in contrast with the blank test (Zhang and Shen 2007). Alike studies were carried out using silver nanoparticles on fermentative hydrogen production using mixed bacterial culture (Zhao et al. 2013). The experiments with silver nanoparticles revealed higher hydrogen yields than the blank. The existence of silver nanoparticles in the fermentation medium reduced the yield of ethanol but increased the yield of acetic acid. The addition of silver nanoparticles resulted in a higher cell biomass production rate. Besides, the study inferred that silver nanoparticles could not only increase the hydrogen yield but reduce the lag phase for hydrogen production concurrently (Zhao et al. 2013). An extensive review of the application of nanotechnology (nanoparticles) in dark fermentative hydrogen production gave abundant information on utilizing inorganic/organic nanoparticles in the bioreactors and reported remarkable developments in the dark fermentative hydrogen productivities (Pugazhendhi et al. 2018).

The activity of photosynthetic hydrogen-producing microalgae was also improved using nanoparticles (Sekoai et al. 2019). The importance of nanoparticles in algal biotechnology showed augmentation in biomass density and physiological processes such as photosynthetic activity, nitrogen metabolism and protein level (Eroglu et al. 2013). These nanoparticles serve up as catalysts and persuade metabolic pathways which endorses the synthesis of photosynthetic pigments (chlorophyll a, chlorophyll b, carotenoids, anthocyanin, etc.), lipid production and nitrogen metabolism (Pádrová et al. 2015, Raliya et al. 2015). Nanoparticles also support the production of carbohydrates which in turn promotes the growth of algal cells (Ahmad et al. 2018, Mishra et al. 2014b). Besides, nanoparticles improve the activity of the main enzymes such as glutamate dehydrogenase, glutamate-pyruvate transaminase, glutamine synthase and nitrate reductase, and these enzymes are essential for the metabolism of microalgal species (Yang et al. 2006, Mishra et al. 2014a).

3.2 Biogas

It is well known that typical anaerobic digestion involves a cascade of metabolic reactions catalyzed by specific enzymes that translate different organic resources into biogas and its components (alcohols and VFAs) (Zhao and Chen 2011, Romero-Güiza et al. 2016). The metallo-enzymes are influenced when the respective trace metals are supplemented as co-factors, thus augmenting the biogas production during the digestion process (Hoelzle et al. 2014). Trace metals work as an electron donor in the anaerobic fermentation process and they augment the overall methanogen population and their enzyme activity (Myszograj et al. 2018). They can also optimize the microbial population, change the hydrolysis fermentation types and stimulate the acetic acid content (Weiland 2010). Comparison of the effects of micronutrients of $NiCl_2$, Fe_2O_3, $CoCl_2$ and $(NH_4)_6Mo_7O_{24}$ along with their nanoparticles on biogas production was reported (Juntupally et al. 2017). These reports inferred that trace metals as nanoparticles have an augmenting effect on biogas as compared to their micronutrients. The influence of trace elements such as nickel (Ni) and cobalt (Co) nanoparticles on methane production has been studied in a specifically constructed batch anaerobic digester (Abdelsalam et al. 2016, Abdelsalam et al. 2017). It was observed that the biogas production was enhanced by 1.74 times. In another study, iron (II) oxide nanoparticles were used in anaerobic granular sludge bioreactor for beet-sugar wastewater remediation. It was reported that due to the utilization of iron (II) oxide nanoparticles as conduits for electron transfer in methanogens, thus it improved biogas production (Ambuchi et al. 2017). Addition of nanoparticles has revealed promising outcomes in anaerobic processes, particularly concerning electron donors/acceptors and cofactor of key enzymes such as [Fe]- and [Ni-Fe]-hydrogenase (Sekoai et al. 2019) (Table 1). The use of nanoparticles increases the hydrolysis of organic matter. The enhancement in substrate hydrolysis is because the nanoparticles provide a large surface area to volume ratio for microbes to bind at active sites of enzymes that excites their biochemical processes, viz., the activity of hydrogenase and ferredoxins (Myszograj et al. 2018).

Table 1. Effect of nano-additives on the rate of biogas production.

NPs Type	Feedstock Type	Temperature (°C)	Concentration of NPs	Incubation Time (day)	Observations	References
CuO	AGS Cattle manure	30 36	1.4 mg/l	83 14	15% Decrease in methane production	Otero-González et al. 2014
ZnO	Cattle manure	36	240 mg/l	14	72% reduction seen in the production of biogas	Luna-delRisco et al. 2011
TiO_2	Waste activated sludge	35	6, 30, 150 mg/g total suspended solids	Different fermentation time	No effect	Mu et al. 2011
Al_2O_3	Waste activated sludge	35	6, 30, 150 mg/g total suspended solids	Different fermentation time	No effect	Mu et al. 2011
SiO_2	Waste activated sludge	35	6, 30, 150 mg/g total suspended solids	Different fermentation time	No effect	Mu et al. 2011
Fe_3O	Waste water sludge	37	100 ppm	60	Enhancement of biogas production up to 180% and improvement of methane production by 234%	Casals et al. 2014
Nano zerovalent iron	Waste activated sludge	37	0.1 wt%	17	Increase in the production biogas by 30.4% and production of methane by 40.4%	Su et al. 2013
Fe/SiO_2	–	55	10^{-5}mol/l	–	Improvement of methane production by 7%	Al-Ahmad et al. 2014
Au	Waste water treatment sludge	37	100 mg/l	50	No effect	García et al. 2012
Pt/SiO_2	–	55	10^{-5}mol/l	–	7% Enhancement in the production of methane	Al-Ahmad et al. 2014
Co/SiO_2	–	55	10^{-5}mol/l	–	8% Enhancement in the production of methane	Al-Ahmad et al. 2014
Ni/SiO_2	–	55	10^{-5}mol/l	–	70% Increase in methane production	Al-Ahmad et al. 2014
Micro/nano fly ash	Municipal Solid waste	35	3 g/g volatile solids	90	Biogas production was improved by 2.9 times	Lo et al. 2012

3.3 Bioethanol

Ethanol is a potential substitute for conventional petroleum-derived fuels in the automobile segment owing to its cost-effective and ecological benefits (Puppan 2002). During biological metabolism, macromolecules are converted into organic solvents like alcohol, acetones and acids by solventogenic microbes like *C. butylica, C. acetobutylicum, C. beijericum, C. pasteurianum, C. felsenium, C. propylbutylicum, C. barkeri, C. glycolycum, C. thermosccharolyticum, C. saccharoperbutylacetonicum, Zygomonasmobilis*, etc. In a typical anaerobic fermentation, the microbial metabolism proceeds from an early acidogenic phase, producing predominantly acetate and butyrate, then to a solventogenic phase were acetate and butyrate are reabsorbed and finally converted to acetone, butanol and ethanol. A sequential bioethanol and biogas production were investigated to improve the yield of utilized biomass (Liu et al. 2015). A recent technological development involved in solventogenesis process is the usage of different types of nanoparticles for the enhancement of biofuel production (Sekoai et al. 2019). In another study, the syngas fermentation process was supplemented with nanoparticles to enhance the bioethanol production using *Clostridium ljungadahlii* (Kim and Lee 2016). Six different types of nanoparticles, viz., palladium on carbon, palladium on alumina, silica, hydroxyl-functionalized single-walled carbon nanotubes, alumina and iron (III) oxide were used in this study. Among these, silica nanoparticles enhanced the bioethanol production in the gas-fermentation process using *Clostridium ljungadahlii.* The methyl-functionalized silica nanoparticles increased the quantity of biomass, ethanol and acetic acid production.

3.4 Biofuel Cells

Biofuel cell is an appliance that translates chemical energy to electrical energy using a biological constituent such as bacteria or enzymes (Shukla et al. 2004) (Figure 2). Enzyme-based biofuel cells comprise of the isolated enzyme as the catalyst either at the anode/cathode compartment or even both the compartments (Minteer et al. 2007). Oxido-reductase enzymes such as glucose oxidase, glucose dehydrogenase, alcohol dehydrogenase, cellobiose dehydrogenase and fructose dehydrogenase were used as biocatalysts at the anode compartment. Laccases and other oxidases were used as biocatalysts in the cathode compartment. Enzyme immobilized biofuel cells have the prospective to be a power source for low-wattage sensors, medical implants and communication devices. Yet, their realistic applications are restricted by a few bottlenecks such as their short shelf life, improper constancy of enzyme activity, etc. Nanostructured materials offer a huge surface area for enzyme immobilization, thus attaining a high charge density on the electrode resulting in speedy electron transfer (Kim et al. 2006b). Conductive nanomaterials such as carbon-based (carbon nanotubes and graphite) and metallic nanoparticles (gold and silver) contribute to augmentation in power density. In a study, electro-deposited graphite oxide/cobalt hydroxide/chitosan composite-enzyme electrode was used for developing a biofuel cell (Yin et al. 2016). A high-power density (up to 517 $\mu W/cm^2$ at 0.46 V) was obtained when laccase was used as the cathodic enzyme and glucose oxidase was used as the anodic enzyme. A new generation of biofuel cell, i.e., H_2/O_2 cell-based

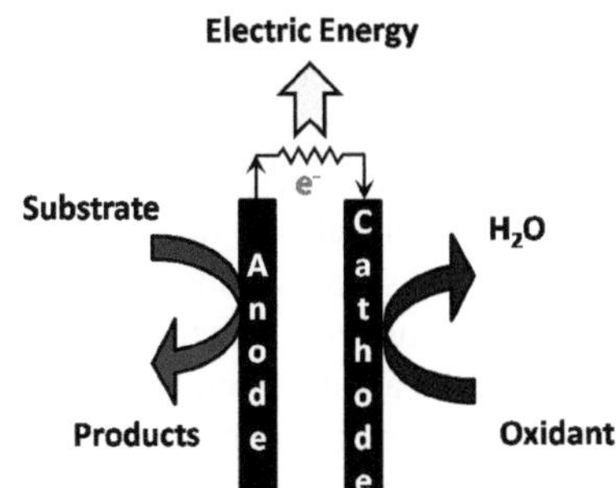

Figure 2. Typical schematic diagram of a biofuel cell.

on thermophilic enzymes were designed electrodes by immobilizing biocatalysts on a fishbone carbon nanofiber (De Poulpiquet et al. 2014). Hydrogenase from the hyperthermophilic bacterium *Aquifexaeolicus* (MbH1) was used as the anode biocatalyst, while thermostable bilirubin oxidase from *Bacillus pumilus* was used as the cathodic biocatalyst. A high power density of up to 1.5 mW/cm^2 at 60°C was achieved without any mediator.

In another study, the life span of biofuel cells was delayed; dehydrogenase enzymes were immobilized within the well-controlled aperture arrangement of modified Nafion membranes to avert enzyme denaturation and enhance enzyme constancy (Kim et al. 2008b, Moore et al. 2004). The advancement of enzyme varnish established to be successful in alleviating the glucose oxidase activity and extending the operational lifetime of biofuel cells. It is anticipated that augmented enzyme loading which might be attained through diverse nano-immobilization methods could lead to increased power densities. The application of conductive nanomaterials, such as carbon nanotubes and gold nanoparticles, also contributes to an augmented power density of biofuel cells by mediating the charge transfer amid the enzyme and the electrode. A new efficient method described the reconstitution of apoenzymes on cofactor functionalized nanomaterials that are conductive and associated with electrodes (Trifonov et al. 2013). This would allow the enzyme molecules to be arranged on the conductive surface with optimal electrical contact with electrodes. Such advancement would efficiently accelerate the electron transport from the enzyme molecules to the electrodes, thus appreciably escalating the power density of biofuel cells.

Other recent bioelectrochemical technology, i.e., microbial fuel cells (MFC) have put on immense attention as an unconventional energy translation method for producing bioelectricity (Mohan et al. 2014b, Du et al. 2007, Logan et al. 2006). Besides power generation, MFCs are also applicable for wastewater remediation and nutrients recovery (Mohanakrishna et al. 2010, Zhou et al. 2013, Velvizhi and Venkata Mohan 2017). Other models of bioelectrochemical systems such as microbial electrolysis cell, microbial electrochemical cells, etc., are also considered as a substitute for bioenergy generation with simultaneous pollutant removal (Gude 2016, Velvizhi and Venkata Mohan 2015). Therefore, bioelectrochemical technology possesses the capability to treat wastewater with electricity generation and referred to be a potential solution for water and energy crisis (Nikhil et al. 2018, Mohan et al. 2014a). In general, MFCs demonstrated momentous benefits; for instance, (i) substrate is directly converted into energy by utilizing biotic electro-catalyst, (ii) functions at a range of temperatures, pH and assorted biomass conditions, (iii) produces little amounts of activated sludge from wastewater compared to routine treatment technologies and (iv) external energy for aeration is not required (He 2012). Moreover, organics contained in the surplus sludge resulted from different treatment strategies as well were used for producing electricity in MFC (Du et al. 2007, Ge et al. 2013, Slate et al. 2019). At the anode, microbes degrade the organic matter and thus releasing electrons and protons which transfer through an external circuit and a membrane to the cathode, ensuing water formation with oxygen as terminal electron acceptor (Singh et al. 2019, Mohan et al. 2019).

The efficiency and financial viability of MFC are chiefly linked with the nature and composition of electrode materials of anode and cathode compartments (Arunasri and Mohan 2019). The functioning of MFC depends on the elementary and structural properties of anodes that affect their outcome on the biofilms formation (i.e., bacterial adhesion) and the efficacy of the interfacial electron transport from bacteria to electrodes and substrate oxidation (Busalmen et al. 2008). These attributes are linked with extracellular electron transfer (EET) competence by an augmented exterior area that is precise for biofilms development and by the communications of the microbe-electrode-electrolyte system (Schröder 2007, Logan 2010). Furthermore, the metabolic rates of anaerobic bacteria have also been improved by anode material that provides anaerobic terminal electron acceptors for oxidizing organic matter in wastes (Franks and Nevin 2010). Therefore, the choice of suitable electrode materials and their composites are necessary to boost MFC power production (Zhou et al. 2011). Amendments or implementation of nanostructured materials into or on conventional carbon structured electrode materials, such as 3D carbon black, carbon nanotube (CNT), porous carbon and graphene and polyaniline (PANI) have been recognized as prospective materials (Ghasemi et al. 2013a, Minteer et al. 2012). Diminished ohmic loss and improved bacterial cell adhesion are influenced by the variations of anodes with metals or their oxide-based nanocomposites (Ghasemi et al. 2012). Attachment of iron (III) oxide to CNTs produces a multilayered network that improves bacterial growth and electron transfer (Ghasemi et al. 2013b). Hematite has been coated onto carbon cloth which considerably has improved the EET competence and has provided an augmented current density in comparison with a pristine carbon cloth electrode in MFC (Ren et al. 2016).

3.5 Biodiesel

Microalgae-based research has received an incredible consideration because the prime cellular components, like carbohydrates, proteins, lipids and pigments, have been efficiently utilized for mercantile applications (Kim et al. 2013, Wijffels and Barbosa 2010, Lee et al. 2015). Amongst the microalgal components, lipid production for biodiesel has been the key focus. It involves the process that comprises of cultivation, harvesting, oil extraction and conversion to fuel (Figure 3). Research studies have reported that the cost for harvesting microalgae is approximated to be soaring as 20–30% of the total biodiesel making cost, despite the fact that it varies

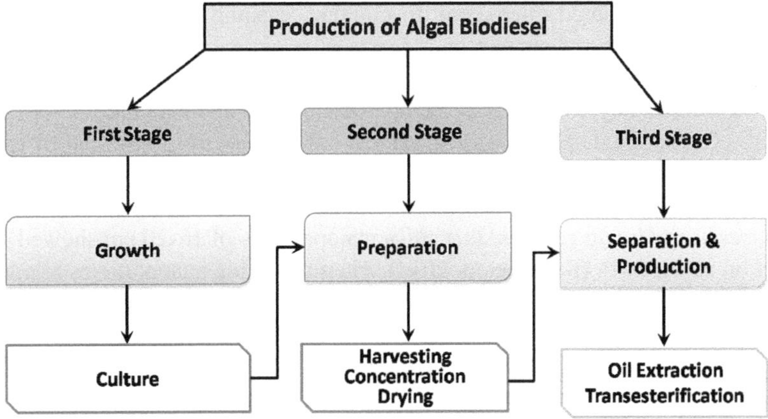

Figure 3. Typical process flow chart for microalgal biodiesel production.

according to the kind of harvesting method, the bulkiness of microalgal culture and the nature of microalgal species (Mata et al. 2010, Rashid et al. 2014). In recent times, multifaceted approaches that surmount the techno-economic restrictions of the existing microalgal biodiesel production process. For illustration, instantaneous harvesting/cell-disruption using nanoparticles have been reported (Lee et al. 2014a). Umpteen studies have been carried out to improve biodiesel product yield and ensuing fuel quality via a variety of heterogeneous nanocatalysts (Pattarkine and Pattarkine 2012, Pugh et al. 2011, Sani et al. 2013, Zhang et al. 2013). Nanoparticle manufacturing methods involved for microalgal cultivation and harvesting processes such as light back-scattering (Eroglu et al. 2013), micronutrient supplementation (Kadar et al. 2012), lipid synthesis inducement (Kang et al. 2014), magnetic separation (Lee et al. 2014a) and natural and artificial nanoparticle flocculation have been reported (Farid et al. 2013, Lee et al. 2013).

3.6 Microalgal Cultivation Using Nanoparticles

Research studies were carried out using nanoparticles as nutrient and lipid inducer. Nanoparticles are added to media as nutritional minerals (e.g., iron nanoparticles and magnesium nanoparticles) are essential for photosynthetic microalgae. Moreover, iron has been accountable to enhance lipid amassing (Liu et al. 2008). A comparative study was performed between usual soluble iron chelated with ethylenediaminetetraacetic acid (EDTA) and synthetic nanoscale zero-valent iron (nZVI) nanoparticles for the cultivation of marine microalgae and noticed usual growth within ZVI and improved growth with Fe-EDTA (Kadar et al. 2012). Iron nanoparticles such as nZVI can generate various reactive oxygen species (ROS, including singlet oxygen, superoxide, hydrogen peroxide and hydroxyl radicals) via Fenton-type reactions that could cause oxidative stress to microorganisms and thus induce stress conditions for lipid build-up in algae (Kadar et al. 2012, Lee et al. 2013). In a different study, magnesium sulfate nanoparticles were supplemented as a constituent required for chlorophyll synthesis in mixotrophic cultivation of *C. vulgaris* using glycerol as an organic carbon source (Sarma et al. 2014). It was pragmatic that the addition of $MgSO_4$

nanoparticles stimulated microalgal flocculation, ensuing enhanced photosynthesis and increased glycerol consumption. Hence, oil yield enhanced comparative to the use of predictable $MgSO_4$ salt.

In another study, the effect of nanoparticle dosage and nanoparticle size on the growth of microalgal species was investigated. In one of the studies on *C. vulgaris* reported that silica nanoparticles of 38–190 nm size enhanced cell growth compared to its control, i.e., without nanoparticles (San et al. 2014). While in another study on *Scenedesmus obliquus* reported that silica nanoparticles of 10–20 nm showed a toxic effect on cell growth (Sarma et al. 2014). Titanium oxide nanoparticles at optimum concentrations can induce oxidative stress under an ultra-violet environment which in turn improves lipid accumulation but high concentrations can pessimistically influence the feasibility of algal cells (Kang et al. 2014). Aminoclay nanoparticles are composed of metal cations such as Ca^{2+}, Mg^{2+} and Fe^{3+} at the center, sandwiched by or paired with amino functional groups linked via covalent bonding (Lee et al. 2014b). Aminoclays have copious primary amine clusters that persuade solubilization of protonated amine in aqueous solution and ensuing adsorption onto microalgal cells. They are noted to not restrain light infiltration due to their apparent and water-soluble properties in aqueous solution (Lee et al. 2011) and to give higher CO_2 accessibility at sites of primary amine groups by converting CO_2 into carbonate ions (Song et al. 2014).

3.6.1 Microalgae Harvesting Using Nanoparticles

Small size and low concentrations of microalgae in addition to the susceptibility of microalgal species on harvesting competence depend on culture medium and pH circumstances (Yoo et al. 2014, Farooq et al. 2013). In recent times, nanoparticles with their exclusive attributes perhaps are considered as choice flocculants for microalgae because usual inorganic flocculants such as $FeCl_3$ and alum have shown both incompetent microalgae harvesting effectiveness and a huge reliance on nature of microalgal species (Abdelaziz et al. 2013). Several studies have reported that nanoparticles as flocculants have been engaged in either the solitary or cross-breed form adorned with various cationic polymers and chemicals.

Magnetic separation is recognized as one of the most versatile separation processes due to their exclusive benefits such as the processing is speedy and mildly handled and there are trouble-free scale-up and mechanization (Borlido et al. 2013). Usually, metal oxides such as iron (III) oxide, as covered by hydroxyl species, have a pH-sensitive surface charge that turns out to be positive below the isoelectric point or negative exceeding it by gaining or losing protons, respectively (Lee et al. 2014a, Xu et al. 2011). This permits flocculation of negative-zeta-potential microalgae with positive-zeta-potential metal oxides by electrostatic attraction followed by magnetic separation (Xu et al. 2011, Hu et al. 2013). Yet, good dispersibility of magnetic nanoparticles is an essential aspect to attain a high harvesting capacity. A comparative study was carried out with and without poly(sodium 4-styrene sulfonate) coating to understand the dispersibility and magnetic separation efficiency of magnetic iron oxides. It was noticed that the dispersibility was significantly improved with the hybrid magnetic particles compared with the bare ones from their suspension.

This was attributed to the stability of colloid which is inversely comparative to the competence of magnetophoretic separation (Yeap et al. 2012).

3.6.2 Microalgal Cell-Disruption Assisted by Nanoparticles

Most oleaginous microalgae own a substantial cell wall made-up of composite carbohydrates and glycoproteins and possess chemical resistance and soaring mechanical strength (Kim et al. 2013). As the cell wall is sturdy, lipid mining has remained a costly and energy-intensive procedure in biodiesel production from microalgae. Exploiting nanoparticles for cell-disruption facilitated ease of lipid removal from wet algal biomass. One of the research studies showed the presence of aminoclay improved the oil extraction efficiency of oleaginous *Chlorella* sp. KR-1 due to the destabilization effect of algal cells resulting in increased microalgae harvesting efficiency (Lee et al. 2013).

3.6.3 Harvesting Lipase Enzyme Using Nanoparticles

Lipase enzyme is classified into three types depending on the enzyme specificity to region or position, i.e., specificity toward fatty acid types and specificity for a particular type of acylglycerols (mono-, di-, or triglycerides) (Teo et al. 2014). It was stated that the manufacturing price of biodiesel could be minimized and simultaneously the biodiesel yield can be increased using immobilized lipases like Novozyme 435 and Lipozyme TL-IM (Bajaj et al. 2010). The biocatalyst can be immobilized in both magnetic and non-magnetic nanoparticles but magnetic nanoparticles have rewards over non-magnetic nanoparticles as discussed earlier. Magnetic iron (III) oxide has gained more attention toward biofuels for the reason that its biocompatibility, stability, large surface area, superparamagnetic property and low cost of synthesis (Goh et al. 2012). The magnetic nanoparticles provide a high surface-to-volume ratio which increases the binding capacity and catalytic specificity of the conjugated enzyme and also as the catalyst is separated by using magnets the final product will be free of contamination (Ansari et al. 2009). Similarly, lipase enzyme was immobilized on magnetic chitosan microspheres, and the catalyst was used in the transesterification of soybean oil which ensured 87% of yield was obtained. The catalyst could be used without a significant decrease in the activity four times (Xie and Wang 2012). The lipid extracted from the *Scenedesmus* sp. was transesterified into biodiesel using nano-biocatalyst. The percentage yield of produced biodiesel by lipase enzyme showed a high yield of 95.1% compared with another catalyst such as sodium hydroxide and potassium hydroxide (Tripathi et al. 2014). It was reported that immobilized lipase can be used four times without a significant decrease in enzyme activity (Tran et al. 2012). In another study, the biocatalytic activity of vegetable oil by immobilizing *Thermomyces lanuginosus* lipase on iron (III) oxide magnetic nanoparticle using carrier modified by 3-aminopropyl triethyoxysilane and covalently linked by 1-ethyl-3-(3-dimethylaminopropyl) carbodiimide and N-hydroxysuccinimide achieved 97.2% yield (Raita et al. 2015). The catalyst was separated by the magnet and it was reused for five cycles without any change in stability and activity. The enzymes immobilized on nanoparticles obtained high enzyme loading capacity and significant mass transfer efficiency when compared with enzyme immobilized on micrometric support (Ansari and Husain 2012). Consequently, the encapsulation of lipase enzyme were done using an organic-inorganic solid with spherical morphology and the nanospheres was composed of liposomes internally. Porous inorganic silica shell was covering the organic phase which has the responsibility of stabilizing liposomal phase and also protecting the lipase enzyme. Using this enzyme transesterification was carried out to produce biodiesel with 86% yield. This encapsulated enzyme was reused and its activity remained even after five cycles (Macario et al. 2013).

3.7 *Perspectives, Possibilities and Challenges*

The relevance of nanotechnology in diverse segments which includes biofuel/bioenergy production has fascinated an immense pact of consideration (Verma et al. 2013). Research studies to date have recommended that magnetic nanobiocatalysts are promising. It is thought that the application of magnetic nanomaterials for purposes such as the functionalization and immobilization of enzymes would be a path-breaking substitute to usual methods since such nanobiocatalysts can be used frequently for the above one hydrolysis reaction (Min and Yoo 2014). The frequent utilization of nanobiocatalysts with immobilized enzymes would assist to construct the procedure commercial as the same enzyme can be recycled. Nanomaterials can improve the chemistry of microbial metabolism at a molecular level. Moreover, it would also assist in better metabolic activities due to the high surface-to-volume ratio. It has also been demonstrated that the alteration or functionalization of nanomaterials with diverse molecules makes the procedure additionally vigorous (Verma et al. 2016).

3.7.1 Nanoparticles in Fermentative Production of Biofuels

Biofuels are a potential solution to the pressing problems to mankind due to a decrease in oil resources and an increase in energy demand (Sekoai et al. 2019). Biohydrogen production is carried out by a diverse group of anaerobic bacteria which uses various metabolic routes to generate molecular hydrogen (Das and Veziroglu 2008). The process is highly dependent on operational conditions such as substrate concentration, pH, temperature and hydraulic retention time which must be optimally controlled to maximize its yield as highlighted (Mohan 2009, Ghimire et al. 2015). These parameters are important because they stimulate the activity of biohydrogen-producing bacteria. It has also been shown that the activity of microorganisms can be enhanced by using nanoparticles in anaerobic conditions because they increase the transfer of electrons in metabolic processes (Mullai et al. 2013, Zhang and Shen 2007). They also improve the kinetics of biohydrogen-producing processes due to their ability to react faster with electron donors. Nanoparticles which include metal and metal oxides, like copper, gold, palladium, silver, iron-iron oxide, nickel-nickel oxide, silica, titanium oxide and carbon nanoparticles (CNTs), and composites can altogether impact the microbial metabolic movement for hydrogen generation by redox exchange of electrons (Kumar et al. 2019). They also enhance the microbial processes, i.e., the activity of biohydrogen-producing enzymes such as [Fe-Fe] and [Ni-Fe]-hydrogenases and ferredoxins (proteins that mediate the transfer of electrons in biohydrogen-producers) is improved as a result of these nano-based additives

(Faisal et al. 2019). Nanoparticles may empower the biohydrogen production by their surface and quantum measure impact (Kumar et al. 2019). The degree of the quantum measure has been specifically correlated to the rate of electron exchange amongst nanoparticles and catalyst particles, like hydrogenase, which catalyzes the reduction of molecular hydrogen to proton and vice versa, resulting in hydrogen oxidation. There is a certain challenge to be addressed for successful commercial utilization and futuristic scope (Rathore et al. 2019, Show et al. 2019).

3.7.2 Biofuel Cells for Bioenergy

Microbial fuel cell (MFC) is a new bio-catalyzed eco-friendly remediation strategy at the nexus of water and energy crisis (Nikhil and Mohan 2019). Yet, it is obligatory to look ahead of the sustainable energy yields earlier to the commercialization of MFC in a wastewater treatment plant. The perception of energy issues during bio-electrochemical treatment of wastewater is essential for recognizing appropriate application fortes and for additional progression of technology (Mohan et al. 2014a). The most remarkable obstacle to MFC operation lies in that the process is innate susceptibility to electron losses, which turn out to be most important during scaling-up. A serious concern is a spatial design of electrodes and electron losses, viz., ohmic, activation and mass transfer (Oliveira et al. 2013). The functioning of an MFC system relies on multifaceted connections between highly coupled electro-active biofilms and the electrodes (Logan 2009). Bacteria may act alone or in syntrophic cooperation that has the fortuitous ability to oxidize many different substrates by making use of their adaptable metabolic pathways. Recently, comprehensive research activities on MFC about the advancement in reactor configurations, electrode materials, enrichment of electrochemically active bacteria and optimization of process confines calation in power densities in quite a few orders of scale (Kim et al. 2008a).

Commercializing MFC necessitates the external energy required for starting the circuit for MFC energy extraction which is another major limitation. A variety of strategies may be investigated for better MFC functioning by inventing new kind of electrodes, viz., (i) composite nitrogen-doped carbon materials with high surface area (Palanisamy et al. 2019), (ii) nitrogen-doped 3D porous carbon with ultra-fine metal nanoparticles (Fogel and Limson 2016), (iii) binary or ternary metal composition over the large surface area of carbon structures (Min et al. 2008), (iv) nano-sized transition metal additives with carbon additives (Fraiwan et al. 2014) and (v) incorporation of metal nanoparticles into the porous structures of metal oxides (Zhou et al. 2011). The invention of innovative electrodes might offer competent electrical conductivity, assist in the development of quicker biofilm formation and augment the constancy of MFCs for enduring operation while wastewater is being treated and simultaneous producing electricity (Qiao et al. 2010, Wei et al. 2011).

3.7.3 Microalgae for Biodiesel

The consequences of the nanoparticle development are reliant mostly on the nature of microalgal species, operating conditions and nature of nanoparticle (Miazek et al. 2015). Despite the momentous significance of systematic research and engineering fields, the expertise is still naive. Besides, logical investigation on distinctive nanoparticles for better growth and/or most favorable oil-accumulation conditions in cells is predictable in the outlook (Lee et al. 2015). Sustaining a steady culture in terms of a microalgal community and co-existing bacteria (Praveenkumar et al. 2012) or averting growth against foreign bacterial and fungal contaminations (Wijffels and Barbosa 2010) is a demanding concern for microalgal mass-cultivation. The anti-bacterial/anti-fungal agents are required for protection from other microorganisms' invasion into the microalgal culture system (Kim et al. 2013). For instance, aminoclay-nanoparticle-coated microalgal cells, which mimic silica-coated cells (Drachuk et al. 2013), can be used to protect from harsh conditions including the stressful conditions formed by nutrient-uptake competitors. Importantly, the re-culture of microalgae in the presence of aminoclay using supernatant media after harvesting of microalgal biomass shows that aminoclay has almost no toxicity or any negative effects on microalgal culture (Farooq et al. 2013). Such re-culture is also significant for minimization of the ecotoxicity of engineered aminoclay nanoparticles to be released into bodies of water (Choi et al. 2014).

Several risky concerns stay behind to be resolved before realistic applications of nanoparticles to the microalgae harvesting process can be realized. Magnetic metal-oxide-based flocculants are to be economically developed with steady nanoparticles. The efficient revival and reclaim of nanoparticles by energy competent and economical methods are major themes of investigation. The utmost use of magnetic nanoparticles would be their recycling without discard. Yet, the current schemes necessitate significant dosages of acid and alkali for pH control (Lee et al. 2014a) as well as strong acids (Xu et al. 2011). For realistic comprehensive use, an innovative recycling process ought to be developed and their prices could be minimized by using inexpensive, bulk industrial chemicals and/or CO_2-rich chimney gas from industry (Lee et al. 2014a). The competence of nanoparticles is mainly reliant on the species used and the culture-depended conditions that are maintained. Supplementary research on comprehensive microalgal species, real culture media equipped with fertilizers and/or wastewater nutrients and up-gradation toward industry scale are seriously encouraged to assist the advancement of nanoparticle engineering moving forward.

Although nanoparticle flocculants have exposed high-quality harvesting performances, their research expenses with reference to possible commercialization possibly high to date. Therefore, supplementary innovation of resourceful and economical nanoparticles is essentially a prelude to the success of any reasonable harvesting process. A blend with diverse harvesting techniques such as bio-flocculation can create synergistic effects over price and competence. Improvements in harvesting, lipid extraction and/or conversion processes would signify breakthroughs in the efforts to reduce price and increase efficiencies within the overall biodiesel production. Technical combination advancement for concurrent harvesting/cell-wall disruption (Lee et al. 2014a) and solvent-free nanoparticle-based oil extraction methods (Zhang et al. 2013) are hot topics for further investigation.

4. Conclusion

The current development of nanobiocatalysis has apparent benefits of nanobiocatalytic methods and their vivid prospect as a combination of biocatalysis and nanotechnology. Nanobiocatalytic strategies of enzyme immobilization have already established remarkable enzyme stabilization—which will offer novel prospects for both conventional and innovative promising areas of enzyme applications—as well as other attractive attributes of nanomaterials such as magnetic separation and electron conductivity. For the concluding achievements of these methods, connections between enzymes and materials at the nanoscale require to be additionally elucidated. The exhaustive understanding of these nano-scale communications will assist in the improvement of innovative approaches that could escort to many realistic applications of enzyme technology.

The exploitation of nanotechnology in these diverse areas is chiefly credited to the new properties of nanoparticles which include their nanoscale size, structure/morphology and high reactivity. The exceptionally diminished size of nanoparticles allows: (i) for a large surface-area-to-volume-ratio and causes an increased number of active sites which are essential for producing different reactions and processes, (ii) the ability of nanoparticles to exhibit different and (iii) nanoscale materials react at a faster rate with other molecules compared to large particles. Besides, nanoparticles demonstrate other encouraging features, for instance, a high degree of crystallinity, catalytic activity, chemical constancy and high adsorption competence. They can either be metallic, semiconductor or polymeric and are synthesized using a top-down or bottom-up approach. These exclusive attributes have made nanoparticles to be striking materials for superior biofuel production. They are mainly used as catalytic agents and play a significant job in the transport of electrons, diminish inhibitory compounds and develop the activity of anaerobic consortia. Metallic nanoparticles are gaining escalating eminence in the technical fraternity due to the reality that these materials can be synthesized using a mixture of functional groups which augments their constancy and facilitates them to be conjugated with ligands, antibodies and drugs, hence expanding their relevance in themes such as diagnostic imaging, biotechnology, drug delivery and magnetic separation.

The future outlook is 'Biorefineries' which are amenities that process biomass into fuels, power and value-added chemicals and with the rising demographics and exhausting petroleum raw materials they are promptly becoming more significant to humanity. The skill required to process a broad diversity of biomass types can be extremely multifaceted owing to their unknown, changing or complexity to breakdown chemical structures within them. One of the potential routes to a triumphant biorefinery that can treat an extensive range of biomass and generate products with good selectivity is the use of nanoparticles as assorted catalysts. The possibility of nanoparticles to catalyze and amend chemical processes, thus influencing both the nature of the products and their distribution is seen as highly promising.

Acknowledgments

Authors thank the Editor for the opportunity and kind assistance.

References

Abdelaziz, A.E., G.B. Leite and P.C. Hallenbeck. 2013. Addressing the challenges for sustainable production of algal biofuels: II. Harvesting and conversion to biofuels. Environmental Technology 34: 1807–1836.

Abdelsalam, E., M. Samer, Y. Attia, M. Abdel-Hadi, H. Hassan and Y. Badr. 2016. Comparison of nanoparticles effects on biogas and methane production from anaerobic digestion of cattle dung slurry. Renewable Energy 87: 592–598.

Abdelsalam, E., M. Samer, Y. Attia, M. Abdel-Hadi, H. Hassan and Y. Badr. 2017. Effects of Co and Ni nanoparticles on biogas and methane production from anaerobic digestion of slurry. Energy Conversion and Management 141: 108–119.

Ahmad, B., A. Shabbir, H. Jaleel, M.M.A. Khan and Y. Sadiq. 2018. Efficacy of titanium dioxide nanoparticles in modulating photosynthesis, peltate glandular trichomes and essential oil production and quality in Mentha piperita L. Current Plant Biology 13: 6–15.

Ahmad, R. and M. Sardar. 2015. Enzyme immobilization: an overview on nanoparticles as immobilization matrix. Biochemistry and Analytical Biochemistry 4: 1.

Al-Ahmad, A.E., S. Hiligsmann, S. Lambert, B. Heinrichs, W. Wannoussa, L. Tasseroul, F. Weekers and P. Thonart. 2014. Effect of encapsulated nanoparticles on thermophillic anaerobic digestion. http://hdl.handle.net/2268/169020.

Ambuchi, J.J., Z. Zhang, L. Shan, D. Liang, P. Zhang and Y. Feng. 2017. Response of anaerobic granular sludge to iron oxide nanoparticles and multi-wall carbon nanotubes during beet sugar industrial wastewater treatment. Water Research 117: 87–94.

Ansari, F., P. Grigoriev, S. Libor, I.E. Tothill and J.J. Ramsden. 2009. DBT degradation enhancement by decorating Rhodococcus erythropolis IGST8 with magnetic Fe_3O_4 nanoparticles. Biotechnology and Bioengineering 102: 1505–1512.

Ansari, S.A. and Q. Husain. 2012. Potential applications of enzymes immobilized on/in nano materials: A review. Biotechnology Advances 30: 512–523.

Arunasri, K. and S.V. Mohan. 2019. Biofilms: Microbial life on the electrode surface. Microbial Electrochemical Technology. Elsevier, 295–313.

Ashrafmansouri, S.-S. and M.N. Esfahany. 2014. Mass transfer in nanofluids: A review. International Journal of Thermal Sciences 82: 84–99.

Asuri, P., S.S. Karajanagi, J.S. Dordick and R.S. Kane. 2006. Directed assembly of carbon nanotubes at liquid−liquid interfaces: Nanoscale conveyors for interfacial biocatalysis. Journal of the American Chemical Society 128: 1046–1047.

Bajaj, A., P. Lohan, P.N. Jha and R. Mehrotra. 2010. Biodiesel production through lipase catalyzed transesterification: an overview. Journal of Molecular Catalysis B: Enzymatic 62: 9–14.

Beckers, L., S. Hiligsmann, S.D. Lambert, B. Heinrichs and P. Thonart. 2013. Improving effect of metal and oxide nanoparticles encapsulated in porous silica on fermentative biohydrogen production by Clostridium butyricum. Bioresource Technology 133: 109–117.

Bhinder, S.S. and P. Dadra. 2009. Application of nanostructures and new nano particles as advanced biomaterials. Asian J. Chem. 21: S167.

Bommarius, A.S. and B.R. Riebel. 2004. Biocatalysis: Fundamentals and Applications. John Wiley & Sons.

Borlido, L., A. Azevedo, A. Roque and M. Aires-Barros. 2013. Magnetic separations in biotechnology. Biotechnology Advances 31: 1374–1385.

Busalmen, J.P., A. Esteve-Núñez, A. Berná and J.M. Feliu. 2008. C-type cytochromes wire electricity-producing bacteria to electrodes. Angewandte Chemie International Edition 47: 4874–4877.

Cantone, S., V. Ferrario, L. Corici, C. Ebert, D. Fattor, P. Spizzo and L. Gardossi. 2013. Efficient immobilisation of industrial biocatalysts: criteria and constraints for the selection of organic polymeric carriers and immobilisation methods. Chemical Society Reviews 42: 6262–6276.

Casals, E., R. Barrena, A. García, E. González, L. Delgado, M. Busquets-Fité, X. Font, J. Arbiol, P. Glatzel and K. Kvashnina. 2014. Programmed iron oxide nanoparticles disintegration in anaerobic digesters boosts biogas production. Small 10: 2801–2808.

Choi, M.-H., Y. Hwang, H.U. Lee, B. Kim, G.-W. Lee, Y.-K. Oh, H.R. Andersen, Y.-C. Lee and Y.S. Huh. 2014. Aquatic ecotoxicity effect of engineered aminoclay nanoparticles. Ecotoxicology and Environmental Safety 102: 34–41.

Cipolatti, E.P., A. Valerio, R.O. Henriques, D.E. Moritz, J.L. Ninow, D.M. Freire, E.A. Manoel, R. Fernandez-Lafuente and D. de Oliveira. 2016. Nanomaterials for biocatalyst immobilization–state of the art and future trends. RSC Advances 6: 104675–104692.

Clouthier, C.M. and J.N. Pelletier. 2012. Expanding the organic toolbox: a guide to integrating biocatalysis in synthesis. Chemical Society Reviews 41: 1585–1605.

Das, D. and T.N. Veziroglu. 2008. Advances in biological hydrogen production processes. International Journal of Hydrogen Energy 33: 6046–6057.

de Jesús Rostro-Alanis, M., E.I. Mancera-Andrade, M.B.G. Patiño, D. Arrieta-Baez, B. Cardenas, S.O. Martinez-Chapa and R.P. Saldívar. 2016. Nanobiocatalysis: nanostructured materials–a minireview. Biocatalysis 2: 1–24.

De Poulpiquet, A., A. Ciaccafava and E. Lojou. 2014. New trends in enzyme immobilization at nanostructured interfaces for efficient electrocatalysis in biofuel cells. Electrochimica Acta 126: 104–114.

Diamantis, V., A. Khan, S. Ntougias, K. Stamatelatou, A.G. Kapagiannidis and A. Aivasidis. 2013. Continuous biohydrogen production from fruit wastewater at low pH conditions. Bioprocess and Biosystems Engineering 36: 965–974.

DiCosimo, R., J. McAuliffe, A.J. Poulose and G. Bohlmann. 2013. Industrial use of immobilized enzymes. Chemical Society Reviews 42: 6437–6474.

Drachuk, I., M.K. Gupta and V.V. Tsukruk. 2013. Biomimetic coatings to control cellular function through cell surface engineering. Advanced Functional Materials 23: 4437–4453.

Du, Z., H. Li and T. Gu. 2007. A state of the art review on microbial fuel cells: a promising technology for wastewater treatment and bioenergy. Biotechnology Advances 25: 464–482.

Eroglu, E., P.K. Eggers, M. Winslade, S.M. Smith and C.L. Raston. 2013. Enhanced accumulation of microalgal pigments using metal nanoparticle solutions as light filtering devices. Green Chemistry 15: 3155–3159.

Eş, I., J.D.G. Vieira and A.C. Amaral. 2015. Principles, techniques, and applications of biocatalyst immobilization for industrial application. Applied Microbiology and Biotechnology 99: 2065–2082.

Faisal, S., F. Yusuf Hafeez, Y. Zafar, S. Majeed, X. Leng, S. Zhao, I. Saif, K. Malik and X. Li. 2019. A review on nanoparticles as boon for biogas producers—nano fuels and biosensing monitoring. Applied Sciences 9: 59.

Farid, M.S., A. Shariati, A. Badakhshan and B. Anvaripour. 2013. Using nano-chitosan for harvesting microalga Nannochloropsis sp. Bioresource Technology 131: 555–559.

Farooq, W., Y.-C. Lee, J.-I. Han, C.H. Darpito, M. Choi and J.-W. Yang. 2013. Efficient microalgae harvesting by organo-building blocks of nanoclays. Green Chemistry 15: 749–755.

Fei, G., G.H. Ma, P. Wang and Z.G. Su. 2009. Enzyme immobilization, biocatalyst featured with nanoscale structure. Encyclopedia of Industrial Biotechnology: Bioprocess, Bioseparation, and Cell Technology, 1–26.

Fischback, M.B., J.K. Youn, X. Zhao, P. Wang, H.G. Park, H.N. Chang, J. Kim and S. Ha. 2006. Miniature biofuel cells with improved stability under continuous operation. Electroanalysis: An International Journal Devoted to Fundamental and Practical Aspects of Electroanalysis 18: 2016–2022.

Fogel, R. and J. Limson. 2016. Applications of nanomaterials in microbial fuel cells. Nanomaterials for Fuel Cell Catalysis. Springer, 551–575.

Fraiwan, A., S. Adusumilli, D. Han, A. Steckl, D. Call, C. Westgate and S. Choi. 2014. Microbial power-generating capabilities on micro-/nano-structured anodes in micro-sized microbial fuel cells. Fuel Cells 14: 801–809.

Franks, A.E. and K.P. Nevin. 2010. Microbial fuel cells, a current review. Energies 3: 899–919.

García, A., L. Delgado, J.A. Torà, E. Casals, E. González, V. Puntes, X. Font, J. Carrera and A. Sánchez. 2012. Effect of cerium dioxide, titanium dioxide, silver, and gold nanoparticles on the activity of microbial communities intended in wastewater treatment. Journal of Hazardous Materials 199: 64–72.

Ge, J., D. Lu, J. Wang, M. Yan, Y. Lu and Z. Liu. 2008. Molecular fundamentals of enzyme nanogels. The Journal of Physical Chemistry B 112: 14319–14324.

Ge, Z., F. Zhang, J. Grimaud, J. Hurst and Z. He. 2013. Long-term investigation of microbial fuel cells treating primary sludge or digested sludge. Bioresource Technology 136: 509–514.

Ghasemi, M., S. Shahgaldi, M. Ismail, Z. Yaakob and W.R.W. Daud. 2012. New generation of carbon nanocomposite proton exchange membranes in microbial fuel cell systems. Chemical Engineering Journal 184: 82–89.

Ghasemi, M., W.R.W. Daud, S.H. Hassan, S.-E. Oh, M. Ismail, M. Rahimnejad and J.M. Jahim. 2013a. Nano-structured carbon as electrode material in microbial fuel cells: a comprehensive review. Journal of Alloys and Compounds 580: 245–255.

Ghasemi, M., M. Ismail, S.K. Kamarudin, K. Saeedfar, W.R.W. Daud, S.H. Hassan, L.Y. Heng, J. Alam and S.-E. Oh. 2013b. Carbon nanotube as an alternative cathode support and catalyst for microbial fuel cells. Applied Energy 102: 1050–1056.

Ghimire, A., L. Frunzo, F. Pirozzi, E. Trably, R. Escudie, P.N. Lens and G. Esposito. 2015. A review on dark fermentative biohydrogen production from organic biomass: process parameters and use of by-products. Applied Energy 144: 73–95.

Goh, W.J., V.S. Makam, J. Hu, L. Kang, M. Zheng, S.L. Yoong, C.N. Udalagama and G. Pastorin. 2012. Iron oxide filled magnetic carbon nanotube–enzyme conjugates for recycling of amyloglucosidase: Toward useful applications in biofuel production process. Langmuir 28: 16864–16873.

Gude, V.G. 2016. Wastewater treatment in microbial fuel cells–an overview. Journal of Cleaner Production 122: 287–307.

Hallenbeck, P.C. 2009. Fermentative hydrogen production: principles, progress, and prognosis. International Journal of Hydrogen Energy 34: 7379–7389.

Hallenbeck, P.C., M. Abo-Hashesh and D. Ghosh. 2012. Strategies for improving biological hydrogen production. Bioresource Technology 110: 1–9.

He, Z. 2012. Microbial fuel cells: now let us talk about energy. ACS Publications.

Hoelzle, R.D., B. Virdis and D.J. Batstone. 2014. Regulation mechanisms in mixed and pure culture microbial fermentation. Biotechnology and Bioengineering 111: 2139–2154.

Hu, Y.-R., F. Wang, S.-K. Wang, C.-Z. Liu and C. Guo. 2013. Efficient harvesting of marine microalgae Nannochloropsis maritima using magnetic nanoparticles. Bioresource Technology 138: 387–390.

Illanes, A., A. Cauerhff, L. Wilson and G.R. Castro. 2012. Recent trends in biocatalysis engineering. Bioresource Technology 115: 48–57.

Johnson, T.C., D.J. Morris and M. Wills. 2010. Hydrogen generation from formic acid and alcohols using homogeneous catalysts. Chemical Society Reviews 39: 81–88.

Juntupally, S., S. Begum, S.K. Allu, S. Nakkasunchi, M. Madugula and G.R. Anupoju. 2017. Relative evaluation of micronutrients (MN) and its respective nanoparticles (NPs) as additives for the enhanced methane generation. Bioresource Technology 238: 290–295.

Kadar, E., P. Rooks, C. Lakey and D.A. White. 2012. The effect of engineered iron nanoparticles on growth and metabolic status of marine microalgae cultures. Science of the Total Environment 439: 8–17.

Kang, N.K., B. Lee, G.-G. Choi, M. Moon, M.S. Park, J. Lim and J.W. Yang. 2014. Enhancing lipid productivity of Chlorella vulgaris using oxidative stress by TiO_2 nanoparticles. Korean Journal of Chemical Engineering 31: 861–867.

Kim, I.S., K.-J. Chae, M.-J. Choi and W. Verstraete. 2008a. Microbial fuel cells: recent advances, bacterial communities and application beyond electricity generation. Environmental Engineering Research 13: 51–65.

Kim, J., J.W. Grate and P. Wang. 2006a. Nanostructures for enzyme stabilization. Chemical Engineering Science 61: 1017–1026.

Kim, J., H. Jia and P. Wang. 2006b. Challenges in biocatalysis for enzyme-based biofuel cells. Biotechnology Advances 24: 296–308.

Kim, J., J.W. Grate and P. Wang. 2008b. Nanobiocatalysis and its potential applications. Trends in Biotechnology 26: 639–646.

Kim, J., G. Yoo, H. Lee, J. Lim, K. Kim, C.W. Kim, M.S. Park and J.-W. Yang. 2013. Methods of downstream processing for the production of biodiesel from microalgae. Biotechnology Advances 31: 862–876.

Kim, Y.-K. and H. Lee. 2016. Use of magnetic nanoparticles to enhance bioethanol production in syngas fermentation. Bioresource Technology 204: 139–144.

Kim, Y.-K., S.-Y. Lee and B.-K. Oh. 2016. Enhancement of formic acid production from CO_2 in formate dehydrogenase reaction using nanoparticles. RSC Advances 6: 109978–109982.

Krishnamoorthi, S., A. Banerjee and A. Roychoudhury. 2015. Immobilized enzyme technology: potentiality and prospects. Journal of Enzymology and Metabolism 1: 010–104.

Kumar, G., T. Mathimani, E.R. Rene and A. Pugazhendhi. 2019. Application of nanotechnology in dark fermentation for enhanced biohydrogen production using inorganic nanoparticles. International Journal of Hydrogen Energy 44: 13106–13113.

Lee, J., J. Kim, J. Kim, H. Jia, M.I. Kim, J.H. Kwak, S. Jin, A. Dohnalkova, H.G. Park and H.N. Chang. 2005. Simple synthesis of hierarchically ordered mesocellular mesoporous silica materials hosting crosslinked enzyme aggregates. Small 1: 744–753.

Lee, K., S.Y. Lee, R. Praveenkumar, B. Kim, J.Y. Seo, S.G. Jeon, J.-G. Na, J.-Y. Park, D.-M. Kim and Y.-K. Oh. 2014a. Repeated use of stable magnetic flocculant for efficient harvest of oleaginous Chlorella sp. Bioresource Technology 167: 284–290.

Lee, Y.-C., W.-K. Park and J.-W. Yang. 2011. Removal of anionic metals by amino-organoclay for water treatment. Journal of Hazardous Materials 190: 652–658.

Lee, Y.-C., B. Kim, W. Farooq, J. Chung, J.-I. Han, H.-J. Shin, S.H. Jeong, J.-Y. Park, J.-S. Lee and Y.-K. Oh. 2013. Harvesting of oleaginous Chlorella sp. by organoclays. Bioresource Technology 132: 440–445.

Lee, Y.-C., S.Y. Oh, H.U. Lee, B. Kim, S.Y. Lee, M.-H. Choi, G.-W. Lee, J.-Y. Park, Y.-K. Oh and T. Ryu. 2014b. Aminoclay-induced humic acid flocculation for efficient harvesting of oleaginous Chlorella sp. Bioresource Technology 153: 365–369.

Lee, Y.-C., K. Lee and Y.-K. Oh. 2015. Recent nanoparticle engineering advances in microalgal cultivation and harvesting processes of biodiesel production: a review. Bioresource Technology 184: 63–72.

Leonhartsberger, S., I. Korsa and A. Bock. 2002. The molecular biology of formate metabolism in enterobacteria. Journal of Molecular Microbiology and Biotechnology 4: 269–276.

Liu, Y., J. Xu, Y. Zhang, Z. Yuan, M. He, C. Liang, X. Zhuang and J. Xie. 2015. Sequential bioethanol and biogas production from sugarcane bagasse based on high solids fed-batch SSF. Energy 90: 1199–1205.

Liu, Z.-Y., G.-C. Wang and B.-C. Zhou. 2008. Effect of iron on growth and lipid accumulation in Chlorella vulgaris. Bioresource Technology 99: 4717–4722.

Lo, H., H. Chiu, S. Lo and F. Lo. 2012. Effects of micro-nano and non micro-nano MSWI ashes addition on MSW anaerobic digestion. Bioresource Technology 114: 90–94.

Logan, B.E., B. Hamelers, R. Rozendal, U. Schröder, J. Keller, S. Freguia, P. Aelterman, W. Verstraete and K. Rabaey. 2006. Microbial fuel cells: methodology and technology. Environmental Science & Technology 40: 5181–5192.

Logan, B.E. 2009. Exoelectrogenic bacteria that power microbial fuel cells. Nature Reviews Microbiology 7: 375.

Logan, B.E. 2010. Scaling up microbial fuel cells and other bioelectrochemical systems. Applied Microbiology and Biotechnology 85: 1665–1671.

Luna-delRisco, M., K. Orupõld and H.-C. Dubourguier. 2011. Particle-size effect of CuO and ZnO on biogas and methane production during anaerobic digestion. Journal of Hazardous Materials 189: 603–608.

Macario, A., F. Verri, U. Diaz, A. Corma and G. Giordano. 2013. Pure silica nanoparticles for liposome/lipase system encapsulation: Application in biodiesel production. Catalysis Today 204: 148–155.

Marpani, F., M. Pinelo and A.S. Meyer. 2017. Enzymatic conversion of CO_2 to CH_3OH via reverse dehydrogenase cascade biocatalysis: Quantitative comparison of efficiencies of immobilized enzyme systems. Biochemical Engineering Journal 127: 217–228.

Mata, T.M., A.A. Martins and N.S. Caetano. 2010. Microalgae for biodiesel production and other applications: a review. Renewable and Sustainable Energy Reviews 14: 217–232.

Miazek, K., W. Iwanek, C. Remacle, A. Richel and D. Goffin. 2015. Effect of metals, metalloids and metallic nanoparticles on microalgae growth and industrial product biosynthesis: a review. International Journal of Molecular Sciences 16: 23929–23969.

Min, B., Ó.B. Román and I. Angelidaki. 2008. Importance of temperature and anodic medium composition on microbial fuel cell (MFC) performance. Biotechnology Letters 30: 1213–1218.

Min, K. and Y.J. Yoo. 2014. Recent progress in nanobiocatalysis for enzyme immobilization and its application. Biotechnology and Bioprocess Engineering 19: 553–567.

Minteer, S.D., B.Y. Liaw and M.J. Cooney. 2007. Enzyme-based biofuel cells. Current Opinion in Biotechnology 18: 228–234.

Minteer, S.D., P. Atanassov, H.R. Luckarift and G.R. Johnson. 2012. New materials for biological fuel cells. Materials Today 15: 166–173.

Mishra, A., M. Kumari, S. Pandey, V. Chaudhry, K. Gupta and C. Nautiyal. 2014a. Biocatalytic and antimicrobial activities of gold nanoparticles synthesized by Trichoderma sp. Bioresource Technology 166: 235–242.

Mishra, V., R.K. Mishra, A. Dikshit and A.C. Pandey. 2014b. Interactions of nanoparticles with plants: an emerging prospective in the agriculture industry. Emerging Technologies and Management of Crop Stress Tolerance. Elsevier, 159–180.

Misson, M., H. Zhang and B. Jin. 2015. Nanobiocatalyst advancements and bioprocessing applications. Journal of the Royal Society Interface 12: 20140891.

Mohamad, N.R., N.H.C. Marzuki, N.A. Buang, F. Huyop and R.A. Wahab. 2015. An overview of technologies for immobilization of enzymes and surface analysis techniques for immobilized enzymes. Biotechnology & Biotechnological Equipment 29: 205–220.

Mohan, S.V. 2008. Fermentative hydrogen production with simultaneous wastewater treatment: influence of pretreatment and system operating conditions. Journal of Scientific and Industrial Research 67: 950–961.

Mohan, S.V. 2009. Harnessing of biohydrogen from wastewater treatment using mixed fermentative consortia: process evaluation towards optimization. International Journal of Hydrogen Energy 34: 7460–7474.

Mohan, S.V., P. Chiranjeevi and G. Mohanakrishna. 2012. A rapid and simple protocol for evaluating biohydrogen production potential (BHP) of wastewater with simultaneous process optimization. International Journal of Hydrogen Energy 37: 3130–3141.

Mohan, S.V., J.S. Sravan, S.K. Butti, K.V. Krishna, J.A. Modestra, G. Velvizhi, A.N. Kumar, S. Varjani and A. Pandey. 2019. Microbial electrochemical technology: emerging and sustainable platform. Microbial Electrochemical Technology. Elsevier, 3–18.

Mohan, S.V., G. Velvizhi, K.V. Krishna and M.L. Babu. 2014a. Microbial catalyzed electrochemical systems: a bio-factory with multi-facet applications. Bioresource Technology 165: 355–364.

Mohan, S.V., G. Velvizhi, J.A. Modestra and S. Srikanth. 2014b. Microbial fuel cell: critical factors regulating bio-catalyzed electrochemical process and recent advancements. Renewable and Sustainable Energy Reviews 40: 779–797.

Mohanakrishna, G., S.V. Mohan and P. Sarma. 2010. Bio-electrochemical treatment of distillery wastewater in microbial fuel cell facilitating decolorization and desalination along with power generation. Journal of Hazardous Materials 177: 487–494.

Mondal, M.K., H.K. Balsora and P. Varshney. 2012. Progress and trends in CO_2 capture/separation technologies: a review. Energy 46: 431–441.

Moore, C.M., N.L. Akers, A.D. Hill, Z.C. Johnson and S.D. Minteer. 2004. Improving the environment for immobilized dehydrogenase enzymes by modifying Nafion with tetraalkylammonium bromides. Biomacromolecules 5: 1241–1247.

Mu, H., Y. Chen and N. Xiao. 2011. Effects of metal oxide nanoparticles (TiO_2, Al_2O_3, SiO_2 and ZnO) on waste activated sludge anaerobic digestion. Bioresource Technology 102: 10305–10311.

Mullai, P., M. Yogeswari and K. Sridevi. 2013. Optimisation and enhancement of biohydrogen production using nickel nanoparticles–A novel approach. Bioresource Technology 141: 212–219.

Myszograj, S., A. Stadnik and E. Płuciennik-Koropczuk. 2018. The influence of trace elements on anaerobic digestion process. Civil and Environmental Engineering Reports 28: 105–115.

Narayanan, R., G. Zhu and P. Wang. 2007. Stabilization of interface-binding chloroperoxidase for interfacial biotransformation. Journal of Biotechnology 128: 86–92.

Nikhil, G., S.V. Mohan and Y. Swamy. 2014. Behavior of acidogenesis during biohydrogen production with formate and glucose as carbon source: substrate associated dehydrogenase expression. International Journal of Hydrogen Energy 39: 7486–7495.

Nikhil, G., D.K. Chaitanya, S. Srikanth, Y. Swamy and S.V. Mohan. 2018. Applied resistance for power generation and energy distribution in microbial fuel cells with rationale for maximum power point. Chemical Engineering Journal 335: 267–274.

Nikhil, G. and S.V. Mohan. 2019. Bioelectrochemical energy transitions persuade systemic performance. microbial electrochemical technology. Elsevier, 437–449.

Oliveira, V., M. Simões, L. Melo and A. Pinto. 2013. Overview on the developments of microbial fuel cells. Biochemical Engineering Journal 73: 53–64.

Otero-González, L., J.A. Field and R. Sierra-Alvarez. 2014. Inhibition of anaerobic wastewater treatment after long-term exposure to low levels of CuO nanoparticles. Water Research 58: 160–168.

Pádrová, K., J. Lukavský, L. Nedbalová, A. Čejková, T. Cajthaml, K. Sigler, M. Vítová and T. Řezanka. 2015. Trace concentrations of iron nanoparticles cause overproduction of biomass and lipids during cultivation of cyanobacteria and microalgae. Journal of Applied Phycology 27: 1443–1451.

Palanisamy, G., H.-Y. Jung, T. Sadhasivam, M.D. Kurkuri, S.C. Kim and S.-H. Roh. 2019. A comprehensive review on microbial fuel cell technologies: Processes, utilization, and advanced developments in electrodes and membranes. Journal of Cleaner Production 221: 598–621.

Patel, M., M. Crank, V. Dornberg, B. Hermann, L. Roes, B. Huesing, L. van Overbeek, F. Terragni and E. Recchia. 2006. Medium and long-term opportunities and risk of the biotechnological production of bulk chemicals from renewable resources—the potential of white biotechnology. Utrecht University, Department of Science, Technology and Society (STS)/Copernicus Institute. https://dspace.library.uu.nl/.

Pattarkine, M.V. and V.M. Pattarkine. 2012. Nanotechnology for algal biofuels. The Science of Algal Fuels. Springer, 147–163.

Praveenkumar, R., K. Shameera, G. Mahalakshmi, M.A. Akbarsha and N. Thajuddin. 2012. Influence of nutrient deprivations on lipid accumulation in a dominant indigenous microalga Chlorella sp., BUM11008: Evaluation for biodiesel production. Biomass and Bioenergy 37: 60–66.

Pugazhendhi, A., S. Shobana, D.D. Nguyen, J.R. Banu, P. Sivagurunathan, S.W. Chang, V.K. Ponnusamy and G. Kumar. 2018. Application of nanotechnology (nanoparticles) in dark fermentative hydrogen production. International Journal of Hydrogen Energy 44: 1431–1440.

Pugh, S., R. McKenna, R. Moolick and D.R. Nielsen. 2011. Advances and opportunities at the interface between microbial bioenergy and nanotechnology. The Canadian Journal of Chemical Engineering 89: 2–12.

Puppan, D. 2002. Environmental evaluation of biofuels. Periodica Polytechnica Social and Management Sciences 10: 95–116.

Qhobosheane, M., P. Zhang and W. Tan. 2004. Assembly of silica nanoparticles for two-dimensional nanomaterials. Journal of Nanoscience and Nanotechnology 4: 635–640.

Qiao, Y., S.-J. Bao and C.M. Li. 2010. Electrocatalysis in microbial fuel cells—from electrode material to direct electrochemistry. Energy & Environmental Science 3: 544–553.

Qu, X., P.J. Alvarez and Q. Li. 2013. Applications of nanotechnology in water and wastewater treatment. Water Research 47: 3931–3946.

Raita, M., J. Arnthong, V. Champreda and N. Laosiripojana. 2015. Modification of magnetic nanoparticle lipase designs for biodiesel production from palm oil. Fuel Processing Technology 134: 189–197.

Raliya, R., P. Biswas and J. Tarafdar. 2015. TiO_2 nanoparticle biosynthesis and its physiological effect on mung bean (Vigna radiata L.). Biotechnology Reports 5: 22–26.

Rashid, N., M.S.U. Rehman, M. Sadiq, T. Mahmood and J.-I. Han. 2014. Current status, issues and developments in microalgae derived biodiesel production. Renewable and Sustainable Energy Reviews 40: 760–778.

Rathore, D., A. Singh, D. Dahiya and P.S.-N. Nigam. 2019. Sustainability of biohydrogen as fuel: Present scenario and future perspective. AIMS Energy 7: 1–19.

Ren, G., H. Ding, Y. Li and A. Lu. 2016. Natural hematite as a low-cost and earth-abundant cathode material for performance improvement of microbial fuel cells. Catalysts 6: 157.

Rittmann, S. and C. Herwig. 2012. A comprehensive and quantitative review of dark fermentative biohydrogen production. Microbial Cell Factories 11: 115.

Romero-Güiza, M., J. Vila, J. Mata-Alvarez, J. Chimenos and S. Astals. 2016. The role of additives on anaerobic digestion: a review. Renewable and Sustainable Energy Reviews 58: 1486–1499.

Saallah, S. and I.W. Lenggoro. 2018. Nanoparticles carrying biological molecules: Recent advances and applications. KONA Powder and Particle Journal 35: 89–111.

San, N.O., C. Kurşungöz, Y. Tümtaş, Ö. Yaşa, B. Ortac and T. Tekinay. 2014. Novel one-step synthesis of silica nanoparticles from sugarbeet bagasse by laser ablation and their effects on the growth of freshwater algae culture. Particuology 17: 29–35.

Sani, Y.M., W.M.A.W. Daud and A.A. Aziz. 2013. Solid acid-catalyzed biodiesel production from microalgal oil—The dual advantage. Journal of Environmental Chemical Engineering 1: 113–121.

Sarma, S.J., R.K. Das, S.K. Brar, Y. Le Bihan, G. Buelna, M. Verma and C.R. Soccol. 2014. Application of magnesium sulfate and its nanoparticles for enhanced lipid production by mixotrophic cultivation of algae using biodiesel waste. Energy 78: 16–22.

Schröder, U. 2007. Anodic electron transfer mechanisms in microbial fuel cells and their energy efficiency. Physical Chemistry Chemical Physics 9: 2619–2629.

Sekoai, P.T., C.N.M. Ouma, S.P. Du Preez, P. Modisha, N. Engelbrecht, D.G. Bessarabov and A. Ghimire. 2019. Application of nanoparticles in biofuels: an overview. Fuel 237: 380–397.

Sheldon, R.A. 2007. Enzyme immobilization: the quest for optimum performance. Advanced Synthesis & Catalysis 349: 1289–1307.

Sheldon, R.A. and S. van Pelt. 2013. Enzyme immobilisation in biocatalysis: why, what and how. Chemical Society Reviews 42: 6223–6235.

Shin, J.-H., J.H. Yoon, S.H. Lee and T.H. Park. 2010. Hydrogen production from formic acid in pH-stat fed-batch operation for direct supply to fuel cell. Bioresource Technology 101: S53–S58.

Shipway, A.N., E. Katz and I. Willner. 2000. Nanoparticle arrays on surfaces for electronic, optical, and sensor applications. ChemPhysChem. 1: 18–52.

Show, K., Y. Yan and D.-J. Lee. 2019. Biohydrogen production: status and perspectives. Biohydrogen. Elsevier, 391–411.

Shukla, A., P. Suresh, S. Berchmans and A. Rajendran. 2004. Biological fuel cells and their applications. Curr. Sci. 87: 455–468.

Singh, H.M., A.K. Pathak, K. Chopra, V. Tyagi, S. Anand and R. Kothari. 2019. Microbial fuel cells: a sustainable solution for bioelectricity generation and wastewater treatment. Biofuels 10: 11–31.

Slate, A.J., K.A. Whitehead, D.A. Brownson and C.E. Banks. 2019. Microbial fuel cells: An overview of current technology. Renewable and Sustainable Energy Reviews 101: 60–81.

Song, D., Y.-C. Lee, S.B. Park and J.-I. Han. 2014. Gaseous carbon dioxide conversion and calcium carbonate preparation by magnesium phyllosilicate. RSC Advances 4: 4037–4040.

Su, L., X. Shi, G. Guo, A. Zhao and Y. Zhao. 2013. Stabilization of sewage sludge in the presence of nanoscale zero-valent iron (nZVI): abatement of odor and improvement of biogas production. Journal of Material Cycles and Waste Management 15: 461–468.

Sur, U.K. 2012. Efficient storage of hydrogen fuel in formic acid using an active iron-based catalytic system. Current Science 102: 384.

Teo, C.L., H. Jamaluddin, N.A.M. Zain and A. Idris. 2014. Biodiesel production via lipase catalysed transesterification of microalgae lipids from Tetraselmis sp. Renewable Energy 68: 1–5.

Tran, D.-T., K.-L. Yeh, C.-L. Chen and J.-S. Chang. 2012. Enzymatic transesterification of microalgal oil from Chlorella vulgaris ESP-31 for biodiesel synthesis using immobilized Burkholderia lipase. Bioresource Technology 108: 119–127.

Trifonov, A., K. Herkendell, R. Tel-Vered, O. Yehezkeli, M. Woerner and I. Willner. 2013. Enzyme-capped relay-functionalized mesoporous carbon nanoparticles: effective bioelectrocatalytic matrices for sensing and biofuel cell applications. ACS Nano 7: 11358–11368.

Tripathi, R., J. Singh, R. kumar Bharti and I.S. Thakur. 2014. Isolation, purification and characterization of lipase from Microbacterium sp. and its application in biodiesel production. Energy Procedia 54: 518–529.

Velvizhi, G. and S. Venkata Mohan. 2015. Bioelectrogenic role of anoxic microbial anode in the treatment of chemical wastewater: Microbial dynamics with bioelectro-characterization. Water Research 70: 52–63.

Velvizhi, G. and S. Venkata Mohan. 2017. Influence of multi-electrode bioelectrochemical system (ME-BET) for the treatment of chemical based high saline wastewater. Bioresource Technology 242: 77–86.

Verma, M.L., C.J. Barrow and M. Puri. 2013. Nanobiotechnology as a novel paradigm for enzyme immobilisation and stabilisation with potential applications in biodiesel production. Applied Microbiology and Biotechnology 97: 23–39.

Verma, M.L., M. Puri and C.J. Barrow. 2016. Recent trends in nanomaterials immobilised enzymes for biofuel production. Critical Reviews in Biotechnology 36: 108–119.

Wei, J., P. Liang and X. Huang. 2011. Recent progress in electrodes for microbial fuel cells. Bioresource Technology 102: 9335–9344.

Weiland, P. 2010. Biogas production: current state and perspectives. Applied Microbiology and Biotechnology 85: 849–860.

Wijffels, R.H. and M.J. Barbosa. 2010. An outlook on microalgal biofuels. Science 329: 796–799.

Xie, W. and J. Wang. 2012. Immobilized lipase on magnetic chitosan microspheres for transesterification of soybean oil. Biomass and Bioenergy 36: 373–380.

Xu, L., C. Guo, F. Wang, S. Zheng and C.-Z. Liu. 2011. A simple and rapid harvesting method for microalgae by *in situ* magnetic separation. Bioresource Technology 102: 10047–10051.

Xu, S., H. Liu, Y. Fan, R. Schaller, J. Jiao and F. Chaplen. 2012. Enhanced performance and mechanism study of microbial electrolysis cells using Fe nanoparticle-decorated anodes. Applied Microbiology and Biotechnology 93: 871–880.

Yang, F., F. Hong, W. You, C. Liu, F. Gao, C. Wu and P. Yang. 2006. Influence of nano-anatase TiO_2 on the nitrogen metabolism of growing spinach. Biological Trace Element Research 110: 179–190.

Yang, H.-H., S.-Q. Zhang, X.-L. Chen, Z.-X. Zhuang, J.-G. Xu and X.-R. Wang. 2004. Magnetite-containing spherical silica nanoparticles for biocatalysis and bioseparations. Analytical Chemistry 76: 1316–1321.

Yeap, S.P., A.L. Ahmad, B.S. Ooi and J. Lim. 2012. Electrosteric stabilization and its role in cooperative magnetophoresis of colloidal magnetic nanoparticles. Langmuir 28: 14878–14891.

Yin, J., H.U. Lee and J.Y. Park. 2016. An electrodeposited graphite oxide/cobalt hydroxide/chitosan ternary composite on nickel foam as a cathode material for hybrid supercapacitors. RSC Advances 6: 34801–34808.

Yoo, G., Y. Yoo, J.-H. Kwon, C. Darpito, S.K. Mishra, K. Pak, M.S. Park, S.G. Im and J.-W. Yang. 2014. An effective, cost-efficient extraction method of biomass from wet microalgae with a functional polymeric membrane. Green Chemistry 16: 312–319.

Yoshida, A., T. Nishimura, H. Kawaguchi, M. Inui and H. Yukawa. 2005. Enhanced hydrogen production from formic acid by formate hydrogen lyase-overexpressing *Escherichia coli* strains. Appl. Environ. Microbiol. 71: 6762–6768.

Zhang, X., S. Yan, R.D. Tyagi and R.Y. Surampalli. 2013. Biodiesel production from heterotrophic microalgae through transesterification and nanotechnology application in the production. Renewable and Sustainable Energy Reviews 26: 216–223.

Zhang, Y. and J. Shen. 2007. Enhancement effect of gold nanoparticles on biohydrogen production from artificial wastewater. International Journal of Hydrogen Energy 32: 17–23.

Zhao, W., Y. Zhang, B. Du, D. Wei, Q. Wei and Y. Zhao. 2013. Enhancement effect of silver nanoparticles on fermentative biohydrogen production using mixed bacteria. Bioresource Technology 142: 240–245.

Zhao, Y. and Y. Chen. 2011. Nano-TiO_2 enhanced photofermentative hydrogen produced from the dark fermentation liquid of waste activated sludge. Environmental Science & Technology 45: 8589–8595.

Zhou, M., M. Chi, J. Luo, H. He and T. Jin. 2011. An overview of electrode materials in microbial fuel cells. Journal of Power Sources 196: 4427–4435.

Zhou, M., H. Wang, D.J. Hassett and T. Gu. 2013. Recent advances in microbial fuel cells (MFCs) and microbial electrolysis cells (MECs) for wastewater treatment, bioenergy and bioproducts. Journal of Chemical Technology & Biotechnology 88: 508–518.

Zhu, G. and P. Wang. 2004. Polymer-enzyme conjugates can self-assemble at oil/water interfaces and effect interfacial biotransformations. Journal of the American Chemical Society 126: 11132–11133.

Zhu, G. and P. Wang. 2005. Novel interface-binding chloroperoxidase for interfacial epoxidation of styrene. Journal of Biotechnology 117: 195–202.

Chapter 6

Recent Trends in Nanobiocatalysis for Biofuel Production

Nisha Singh,[1] *Shweta Singh*[1] and *Motilal Mathesh*[2,*]

1. Introduction

The fossil energy sources (such as coal, natural gas, oil and petroleum) are non-renewable and their excessive use has led to depleting fuel reserves, global warming and rising fuel prices (Raman et al. 2015, Joshi et al. 2017). With these limitations of current excessively used fuels, there is a quest for alternative fuel sources, such as biofuels bioethanol, biobutanol, biodiesel, biohydrogen and biogas, that can be derived from renewable resources (Rodionova et al. 2017, Ho et al. 2014) (Figure 1). Regarding this pressing issue, both developed and developing countries have made biofuel development mandatory (Saravanan et al. 2018).

Biofuels can be produced from a variety of feedstocks based on which they are categorized into three groups, namely first, second, and third-generation biofuels (Saladini et al. 2016, Naik et al. 2010, Leong et al. 2018, Alaswad et al. 2015) (Figure 2). First-generation biofuels, derived from feed-based resources such as edible crops, are superior and easier to ferment compared to second and third-generation biofuels. However, first-generation biofuel production has drawbacks in terms of food versus fuel conflict, long-term availability, need for productive land and other social challenges (Ahmed and Sarkar 2018).

Lignocellulosic biomass (e.g., agricultural residues, energy crops, forest residues, cellulosic materials in the form of food waste and municipal solid waste) represents an abundantly available resource that are low-cost renewable feedstock used to produce second-generation biofuels (Bansal et al. 2016, Kang et al. 2014). With respect to sustainability and environmental impact, second-generation biofuels

[1] DBT-IOC Centre for Advanced Bioenergy Research, R&D Indian oil Corporation LTD., Sector 13 Faridabad, 121002, India.

[2] Institute of Molecules and Materials, Radboud University, Heyendaalseweg 135, 6525 AJ, Nijmegen, The Netherlands.

* Corresponding author: motilalmathesh@gmail.com

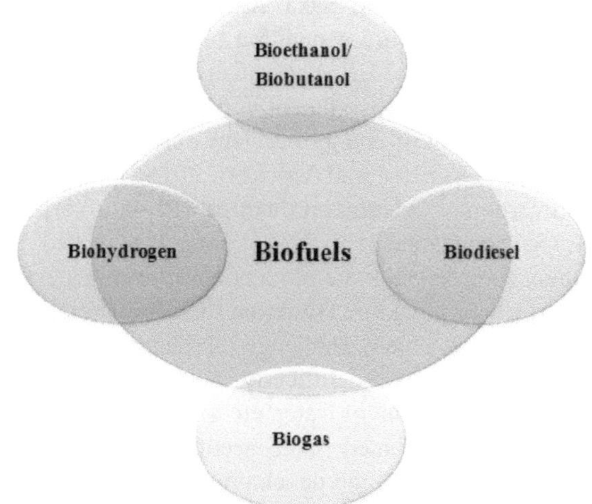

Figure 1. Common types of biofuels.

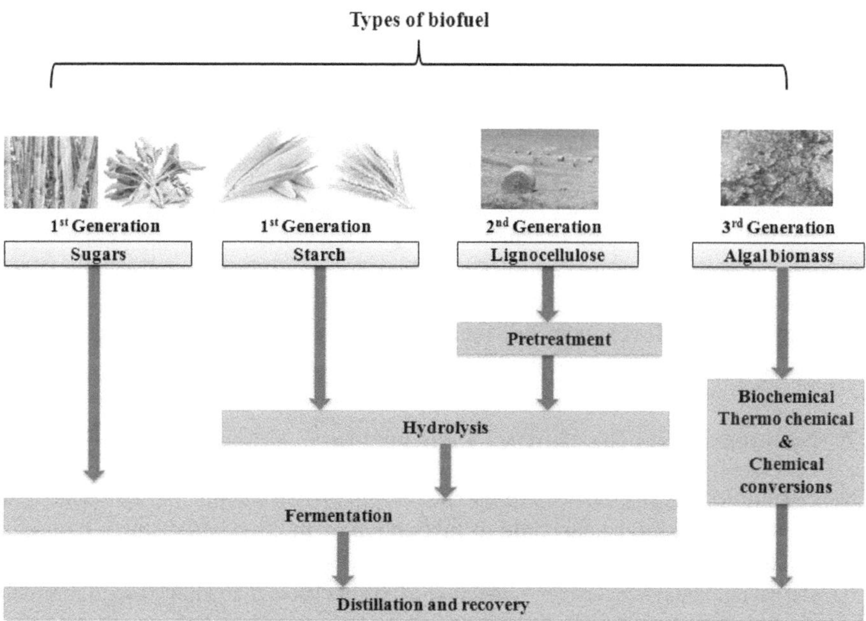

Figure 2. Classification of biofuels based on feedstock types.

are highly advantageous (Ahmed and Sarkar 2018). Despite the above-mentioned benefits, economic feasibility of lignocellulosic biofuels is a huge challenge. The commercial acceptance of lignocellulose-based biofuels is still hindered by various technical barriers that must be overcome before they can compete with fossil fuels (Balan 2014, Oh et al. 2018, Kim 2018). One such barrier is the natural recalcitrance of biomass due to which harnessing its energy content at industrial level is quite challenging (Rabemanolontsoa and Saka 2016, Sindhu et al. 2016). The above factors led to development of third-generation biofuels that are algae-based (Jones and Mayfield 2012, Leong et al. 2018, Alaswad et al. 2015). In recent years, research on algae has increased in both government and private sectors due to their ability to produce biofuels in the most energy efficient manner without causing negative impact on environment (Vassilev and Vassileva 2016). Furthermore, algae play a significant role in carbon alleviation and can be fed on a variety of substrates to produce biofuels (Satyanarayana et al. 2011, Leong et al. 2018). Due to above said benefits, achieving a commercial scale production of third-generation biofuels have gained worldwide research attention. However, the commercial production of third-generation biofuels still faces many challenges and uncertainties, including weather dependence, high infrastructure cost, difficulty in algae harvesting, improper nutrient ratio, requirement of large amount of water, etc. (Alam et al. 2015).

All the limitations associated with second and third-generation biofuel production processes necessitates the development of novel biotechnological tools that can pave the way for a viable biofuel industry. In recent years, nanotechnology has been in the limelight for development of next generation technology providing advancements in biofuel production (Rai and Da Silva 2017, Tripathi et al. 2018, Akia et al. 2014, Sekoai et al. 2018a, Ziolkowska 2018). Nanotechnology could assist in (a) mitigating existing limitations related to enzyme stability and (b) recycling and improving biomass pretreatment and (c) improving the efficacy of participating microorganisms that will be discussed in more details in Section 3 of this chapter. This chapter will also discuss the application of nanotechnology tools for second and third-generation biofuel production and throw light on ways how nanotechnology can impact their production and processing. Development of cutting-edge technologies dealing with the application of nanotechnology tools for enzyme improvement, algal cultivation and harvesting, transesterification and microbial immobilization have been reviewed. Although role of nanotechnology is advancing very fast in different areas of biofuel, this chapter will focus on the application of nanotechnology for bioethanol, biohydrogen, biodiesel and biogas production.

2. Nanocatalysts for Biofuel Production

Nanotechnology is an advanced research area, which helps in manipulating materials at the atomic and molecular scale to fabricate new molecular structures termed as nanomaterials (1–100 nm) (Sekoai et al. 2018a). Nanomaterial-enabled enzymes, termed as nanocatalysts, are renowned to be highly efficient and have found immense application in the field of biofuel production. The nanomaterials are mostly used as a catalytic agent and could be metallic, semiconductor or polymeric. Currently, for industrial applications, various nanomaterials such as nanoparticles, nanosheets, nanotubes, nanopores, nanocomposites, nanofibers, etc., are being used (Leo and Singh 2018) as nanocatalysts (Figure 3). They offer several favorable attributes such as high adsorption capacity, large surface-area-to-volume ratio, higher catalytic activity, chemical stability and quantum confinement (Rai and Da Silva 2017). Different types of nanomaterials that are gaining increased interest in biofuel production are discussed below.

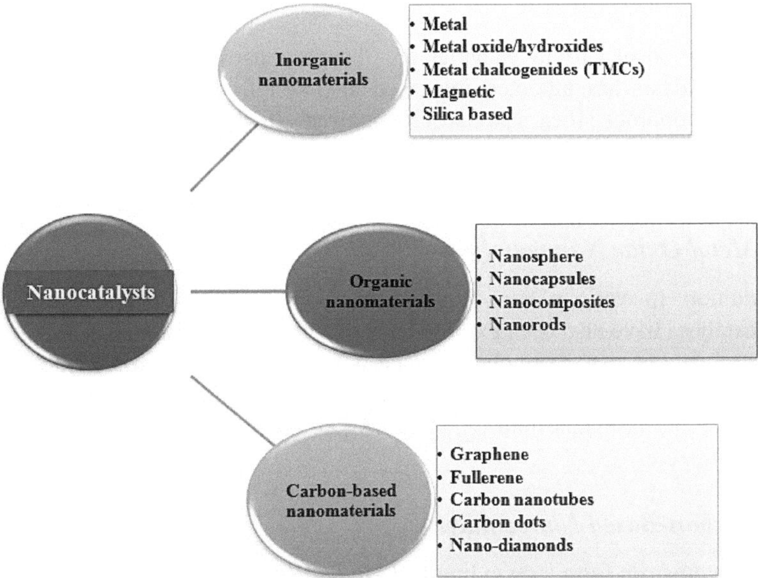

Figure 3. Different types of nanocatalysts.

2.1 *Magnetic Nanoparticles*

Magnetic nanoparticles have attracted significant attention for biofuel production compared to other nanomaterial-based catalysts. The use of magnetic nanoparticles is advantageous, mainly due to their ability to easily separate from reaction media by applying an external magnetic field and subsequent reusability leading to more economical biofuel production. In addition to that, the capability of magnetic particles to conjugate with enzymes and biological system make them an ideal candidate for industrial application (Gardy et al. 2018).

2.2 *Metallic Nanoparticles*

The use of metallic nanoparticles has been increasing significantly for applications such as biofuel production due to their strong ability to adsorb electrons and exceptional electrical, optical and catalytic properties. The positive effect of various nanoparticles, including copper (Cu), gold (Au), silver (Ag), iron (Fe), nickel (Ni), palladium (Pd), etc., were observed on various biofuel production (Patel et al. 2018).

2.3 *Acid-Functionalized Nanoparticles*

Similar to other nanoparticles, acid-functionalized nanoparticles do have an immense application in biofuel production due to their fascinating characteristics, such as high catalytic activity in gas and liquid phase reaction (Tripathi et al. 2018, Sekoai et al. 2018a). Acid-functionalized nanoparticles can be promising catalysts for pretreatment and hydrolysis of lignocellulosic feedstocks.

2.4 Silica Nanoparticles

Immobilization of various enzymes on silica nanoparticles and their use as silica nanobiocatalysts has attracted increased attention from researchers all over the world. Mesoporous silica nanoparticles showed a better affinity for the adsorption of enzymes like cellulase during lignocellulose degradation and thus have improved bioethanol production (Zhou et al. 2018b, Misson et al. 2015).

2.5 Metal Oxide Nanocatalysts

In addition to various nanocatalysts mentioned above, different metal-oxide nanocatalysts have also been explored to produce different biofuels like bioethanol, biodiesel, and biogas. Examples of metal-oxide nanocatalysts employed for biofuel production include titanium oxide (Gardy et al. 2017), calcium oxide (Pattarkine and Pattarkine 2012), magnesium oxide (Chang et al. 2014) and strontium oxide (Liu et al. 2007), etc.

2.6 Carbon-Based Nanocatalysts

Carbon atoms can form long cylindrical tubular structures that can be single-walled or multi-walled with a diameter of around 1 nanometre (nm). These carbon nanotubes possess unique properties to serve as an ideal nanocatalyst, like large surface area, high stability, low toxicity, high mechanical strength, etc. Few studies have shown the effective role of multi-walled carbon nanotubes in biodiesel production (El-Seesy et al. 2016, Guan et al. 2017).

Other carbon-based nanocatalysts such as carbon nanofibers (Lee et al. 2009) and graphene oxide (Zhao et al. 2014, Mathesh et al. 2016) also hold great potential for biofuel production, especially biodiesel. The application of carbon-based catalysts for biofuel production is promising due to their renewable nature, ease of synthesis and cost-effectiveness.

2.7 Graphene-Based Nano-Biocatalysts

Graphene (and its derivatives) is a 2D nanomaterial that has gained tremendous interest in the field of catalysis due to its unique features, like high surface area, high thermal stability and exceptional physical and chemical properties (Wang et al. 2014). They also serve as a good template for binding enzymes and contains various functional group anchoring on them that could be used for functionalization (Zhu et al. 2015). Apart from this, the hydrophobicity of these nanomaterials could be tailored at the molecular level (Ramakrishna et al. 2018) that could act beneficial for enzyme immobilization, such as lipases. This change in hydrophobicity could help in tuning the enzyme activity making them more active than the native form (Mathesh et al. 2016). Because of all the above properties, graphene has been utilized by various researchers to develop novel materials for producing biofuels to catch up with the increasing fuel demand throughout the world. In this section, we will discuss the recent advancements in the field of graphene-related biofuel production.

The cleavage of cellulose into glucose acts as an entry point for bio-refineries that can be transformed into biofuels later on (Rinaldi and Schüth 2009). The hydrolysis of cellulose has been demonstrated with the use of graphene oxides (GO). GO adsorbs the cellulose by forming hydrogen bonds and causes esterification of the hydroxyl and carbonyl groups present on them. The sulfated group generated was able to carry out hydrolysis to produce glucose. This also resulted in aggregation of GO due to the removal of hydroxyl groups and thus could be easily removed from the reaction mixture (Zhao et al. 2014). A 50% glucose yield was obtained at 150°C for 24 hours with 58.6% conversion of cellulose. The superior hydrolytic property was attributed to the carboxylic/phenolic groups present on them together with its soft and layered structure. There have also been studies conducted to produce furfural that acts as an important intermediate to synthesize important biofuels like 2-methylfuran, 2-methyltetrahydrofuran, etc. (Dutta et al. 2012). One such example is the one-pot synthesis of 5-ethoxymethylfurfural (EMF) from carbohydrates that gave yields of 71%, 34% and 66% in conversion of fructose, sucrose and inulin in one pot, respectively (Wang et al. 2013a). The excellent catalytic performance is due to the sulfonic and oxygen functional groups present on GO that have a synergistic effect in the production of biofuels.

Biodiesel is yet another renewable biofuel that has been produced with the help of enzyme immobilization on GO. One such study involved immobilization of *Candida rugose* lipase on GO using covalent binding on the functional groups present on GO with Fe_3O_4 nanocomposites tethered on them (Figure 4). The immobilized enzyme could catalyze transesterification of soyabean oil to produce biodiesel with a 92.8% yield without any side reactions (Xie and Huang 2018). The catalyst could also be recycled with the help of an external magnetic field and can be reused five times without any significant loss of activity. Similarly, biodiesel was also produced by using graphene-like nanoporous carbon functionalized with strong ionic liquids and sulfonic groups. The catalytic activity was attributed to the control over-acidity, unique layered structure that enhanced the fast diffusion of reactants and products and good accessibility to active sites. The biodiesel yield was up to 78.8% after

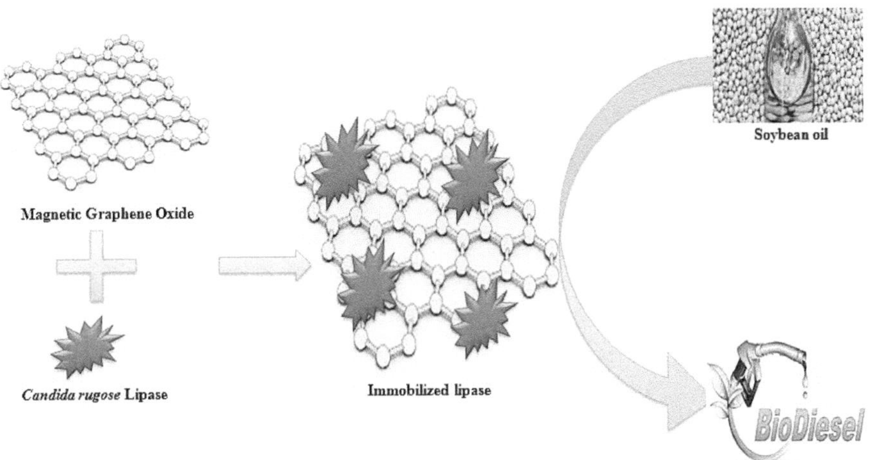

Figure 4. Schematic illustration of the preparation of lipase immobilization on graphene oxide Fe_3O_4 nanocomposites for transesterification of soybean oil [Figure adapted from (Xie and Huang 2018)].

10 hours of reaction that is much more than the commercially available catalysts with no apparent change in activity even after five cycles.

Recently, biodiesel production has also been achieved from wet microalgae through transesterification of its high lipid content. The current problems associated with using wet microalgae for biodiesel production is the use of sulfuric acid as acid catalyst, i.e., hard to separate from the reaction mixture, its lack of reusability and high production cost due to high energy consumption and non-environment friendly nature (Vicente et al. 2007). This problem was overcome with the help of GO that has SO_3H groups which can act as a solid acid catalyst. The presence of SO_3H groups released H^+ ions that helped in transesterification of lipids into glycerol and fatty acid methyl esters with upto 96% conversion efficiency (Cheng et al. 2016).

Hydrogen, another fuel, that is a promising and clean renewable source of energy (Nielsen et al. 2013) is known to be produced from biomass processing such as formic acid (FA) through dehydrogenation reactions in the presence of catalysts (Loges et al. 2010). For this purpose, reduced GO decorated with gold (Au)@palladium (Pd) nanoclusters has been studied as a catalyst. This work showed high catalytic efficiency with 240 ml of gas produced in 56 minutes with turnover frequency (TOF) of 89.1 mol H_2mol^{-1} catalyst h^{-1} with good stability even after three cycles (Wang et al. 2013c). The enhanced activity was due to the charge transfer between Au@Pd that is effective for dehydrogenation of FA together with the presence of GO that provides abundant catalytic sites due to its large surface area decorated with Au@Pd nanoclusters. The above reaction can also proceed through undesired dehydrogenation into water and carbon monoxide that inhibits the H_2 production. In order to prevent this, a study was carried out by growing silver (Ag)@Pd nanoparticles on reduced GO (Figure 5). The presence of functional groups helped in capturing the metal ions and to control the size of the nanoparticles. The catalyst produced 303 ml of gas within 120 minutes and showed high selectivity for FA dehydrogenation with TOF of 105.2 mol H_2mol^{-1} catalyst h^{-1} and prevented the side reaction pathway (Ping et al. 2013). The plausible reason was the prevention of nanoparticle aggregation and surface densities of functional groups resulting in reduced GO microstructure which is favorable for dehydrogenation of FA.

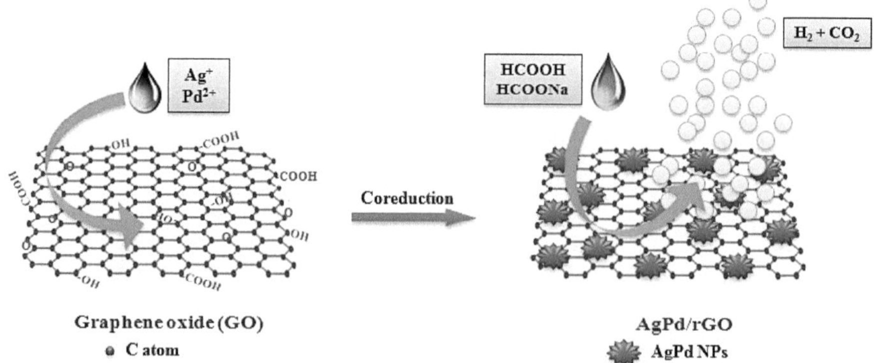

Figure 5. Schematic representation for preparation of AgPd/rGO that act as catalyst to produce Hydrogen [Figure adapted from (Ping et al. 2013)].

The production of biogas such as biomethane has been carried out previously with the help of microorganisms through anaerobic digestion (AD) of organic materials. This is particularly accomplished by acidogenic and methanogenic bacteria but with lower efficiency due to limited electron transfer during AD. This problem was overcome with the help of using graphene that can increase the direct interspecies electron transfer, thereby contributing to increased efficiency. In one such study, graphene nanomaterial enhanced biomethane production by 28% with a highest production rate of 27.9 ml/g/day when glycine (Figure 6) was used as an organic material source (Lin et al. 2018). In comparison, when ethanol was used as an organic material source, biomethane production increased by 25% (Lin et al. 2017) due to the larger specific surface area that allows better interaction with microbes.

All the nanocatalysts mentioned above can be synthesized by various methods, mainly divided into green and non-green approaches. Non-green approaches include physical and chemical means of nanocatalysts synthesis, such as electrochemical reduction, co-precipitation, heat evaporation, photochemical reduction, microemulsion, thermal decomposition, hydrothermal synthesis, etc. The greener approaches involve the synthesis of nanocatalysts using plant materials and biological organisms like fungi and algae. All these approaches have their advantages and disadvantages. Nevertheless, greener approaches for nanocatalysts production are highly recommended due to their minimal inhibitory effect on biocatalysts involved in biofuel production, non-toxicity and environmentally friendly nature.

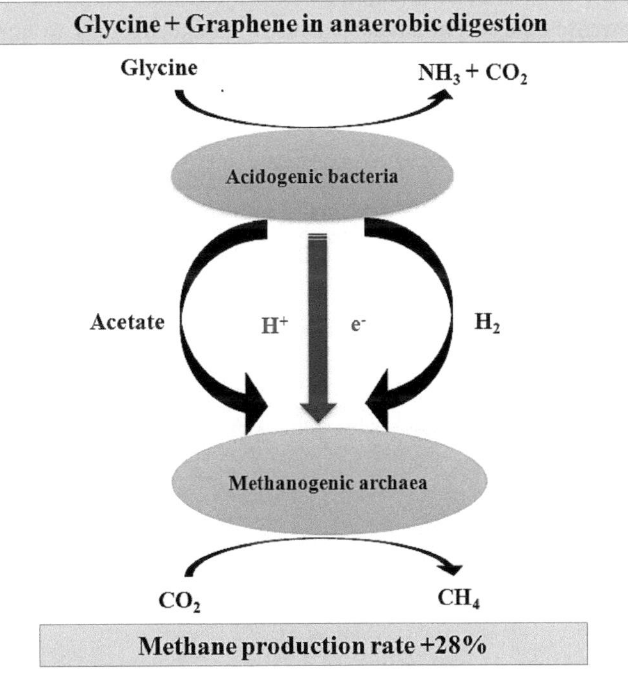

Figure 6. Schematic representation of biomethane production pathway using graphene and glycine as source of organic material [Figure adapted from (Lin et al. 2018)].

3. Role of Nanotechnology for Enhanced Biofuel Production

In the past, the potential application of different kinds of nanocatalysts has been found in many disciplines to solve complex environmental issues. The emerging field of nanotechnology could offer potential solutions to various challenges involved in conventional biofuel processes (Figure 7). There have been numerous reports for the combination of nanocatalysts and biomass to improve the production of different biofuels (Leo and Singh 2018, Patel et al. 2018, Antonio Fernandes Antunes et al. 2017). It is possible to produce eco-friendly third-generation biofuels with algae by applying a nanocatalyzed reaction (Farooq et al. 2013, Hu et al. 2013, Sekoai et al. 2018a). More specifically, for algae-based biofuel production, nanomaterials like silica, single-walled carbon nanotubes, metal-oxide nanoparticles and nano-clay have been successfully used during various stages of lipid accumulation, extraction and transesterification (Zhang et al. 2013). Similarly, to break methane molecules to hydrogen and carbon dioxide for biogas production, nanomaterials can be employed (Ganzoury and Allam 2015). In recent studies, several engineered nanomaterials have been successfully exploited to produce glycerol and biodiesel from animal fats and vegetable oils (Tripathi et al. 2018, Hama et al. 2018, Khan et al. 2014). Another promising application of nanotechnology in the biofuel industry is enzyme immobilization particularly during lipase-catalyzed biodiesel production and cellulase-catalyzed bioethanol production. These nano-immobilized enzymes are reported to have a larger surface area for high enzyme loading, enhanced stability and reusability even under extreme conditions (Verma et al. 2013). The following section showcases the recent advancements in nanotechnology to improve second and third-generation biofuel production, highlighting the advancement in enzyme improvement and pretreatment methodology.

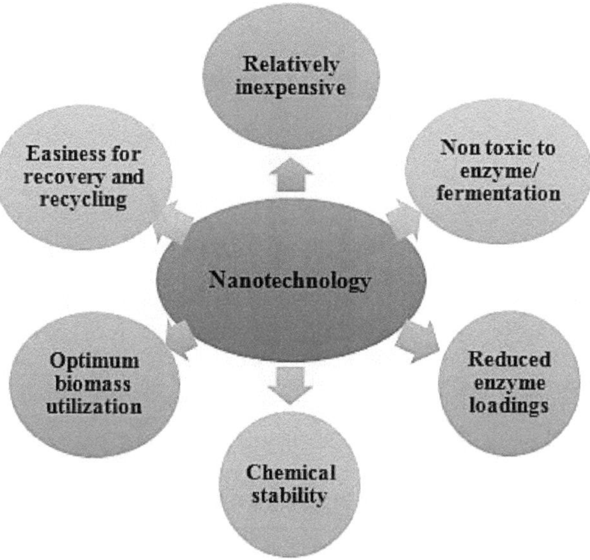

Figure 7. Benefits of nanocatalysts in biofuel industry.

3.1 Role of Nanobiocatalysts in Bioethanol Production

Among different biofuels, bioethanol is considered an eco-friendly substitute for petroleum due to its several advantages (Aditiya et al. 2016). Bioethanol is mainly produced via enzyme-based microbial fermentation of lignocellulosic biomass or energy crops planted, specifically for bioethanol production. Due to the large scale availability of this feedstock, bioethanol derived from lignocellulosic biomass has enormous potential to meet the global energy demand (Gaurav et al. 2017, Ho et al. 2014).

The principal constituents of lignocellulosic biomass are cellulose-hemicellulose (sugar part) and lignin (non-sugar part) that are arranged in a highly compact structure. Due to this, physicochemical pretreatment and enzymatic saccharification steps are essential before fermentation in order to extract the monomeric sugar molecules before they can be fermented to bioethanol (Kim 2018, Aditiya et al. 2016, Kang et al. 2014) making these additional steps cost-intensive and energy-consuming. For example, various pretreatment approaches based on acid-alkali treatment, steam explosion, ammonia fiber explosion, organosolv treatment, ionic liquid treatment, hydrodynamic cavitation, wet oxidation, etc., have been proposed to deconstruct the biochemical structure of lignocellulose (Kumar and Sharma 2017, Terán Hilares et al. 2017). However, these pretreatment methods not only impart additional process costs but also lead to the formation of undesirable fermentation inhibitors (Jönsson and Martín 2016, Xu and Huang 2014). Fermentation inhibitors such as furfural, hydroxymethyl furfural, phenolic and aliphatic compounds and various inorganic ions are formed during such pretreatment that affects the subsequent steps in bioethanol production by inhibiting cellulolytic enzymes and fermenting microorganisms, thus reducing the final bioethanol yields.

Likewise, during enzymatic saccharification, costly cellulase enzymes are required in making them cost-inefficient (Klein-Marcuschamer et al. 2012, Binod et al. 2018). Irrespective of its simpler structure, the disintegration of crystalline cellulose to its monomeric form mainly involves in the co-operative action of three enzymes, namely endoglucanases, cellobiohydrolases and β-glucosidases. Apart from these in biorefinery, a plethora of glycoside hydrolases and auxiliary enzymes in surplus quantity are also needed to achieve efficient saccharification of cellulose. At present, the high production cost of cellulases, their lower catalytic efficiency (k_{cat}) on real lignocellulosic substrates compared to model cellulosic substrates and longer reaction time are the major obstacles hindering the commercialization of second-generation bioethanol (Klein-Marcuschamer et al. 2012). Cellulase engineering by rational and directed evolution to decrease enzyme dosage has been the major emphasis of the research program for biofuel production over the past few decades. However, so far no dramatic activity enhancement required for the industrial demand has been reported using these approaches.

Current research indicates that thermal stability, high efficiency in terms of decreased enzyme dosage and reusability are the prerequisite attributes for cellulase production technology to make it commercial. Due to these limitations, the development of nanotechnology-based tools has attracted worldwide research attention. In this section, the influential role of nanobiocatalysts in cellulase enzyme improvement and biomass pretreatment to achieve enhanced bioethanol production is briefly discussed.

3.1.1 Nanobiocatalysts for Cellulase Improvement

The cost-effectiveness of an enzyme-based bioethanol production process can be improved by cellulase immobilization (Puri et al. 2013). Enzyme immobilization enables the separation of enzymes from the product, thus diminishing the effect of product inhibition and promotes enzyme recovery for further use. In addition, enzyme immobilization often results in enhancing operational conditions, improving thermal stability and increased shell life of enzymes. Several support carriers like cyanogen, bromide-activated agarose, nylon, liposome, etc., have been used for the immobilization of enzymes; however, they usually result in leaking of enzymes (Mohamad et al. 2015).

Recently, cellulase immobilization using a support matrix like nanomaterials have attracted increased attention (Zdarta et al. 2018, Datta et al. 2013, Molina et al. 2018) as nano-immobilization can improve enzyme activity, reusability and stability, thus reducing the overall production cost. The large surface-area-to-volume ratio of nanomaterials increases the anchoring area of enzymes, providing robust cross-linking, better separation from the reaction mixture and finally enhanced catalytic reaction (Kim et al. 2018). Among various nanomaterials, magnetic nanoparticles, metal nanoparticles, acid-functionalized nanoparticles, etc., have been successfully exploited directly or indirectly for the production of different biofuels mentioned above. Among these, magnetic nanoparticles (MNPs) loaded with cellulase are the foremost choice of researchers because the catalyst can be easily recovered using an external magnetic field (Zhou et al. 2018a, Schröfel et al. 2014). In the past, various attempts have been made for immobilization of cellulase on MNPs to improve enzyme stability and catalytic activity.

One such study demonstrated the immobilization of β-glucosidase on iron oxide nanoparticles via covalent bonding to improve bioethanol production. The immobilized enzyme had an excellent binding efficiency of 93%, improved thermostability and higher efficacy toward cellobiose hydrolysis compared to free enzymes (Verma et al. 2013). In another study, the immobilization of cellulase on Fe_3O_4-chitosan nanoparticles was demonstrated having high magnetic sensitivity using glutaraldehyde as the coupling agent (Zang et al. 2014). The immobilized cellulase showed good recyclability and retained 50% of the initial activity for 10 cycles. Later, cellulase from *Aspergillus niger* was immobilized onto β-cyclodextrin-conjugated MNPs that retained 85% of its activity during the hydrolysis of rice straw biomass. The immobilized enzyme was recycled by the external magnetic field for a total of 16 cycles of hydrolysis (Huang et al. 2015). Similarly, immobilized cellulase produced by *Aspergillus fumigatus* onto MnO_2 nanoparticles was utilized for bioethanol production from agricultural waste (Cherian et al. 2015). The immobilized cellulase had improved thermal stability at temperatures between 30–80°C, wide pH range 4–8 and thus higher bioethanol yield of 21.96 g/L.

Furthermore, the immobilized enzyme was stable up to five cycles and retained 60% of catalytic activity. Recently, simultaneous immobilization of β-glucosidase A (BglA) and cellobiohydrolases D (CelD) on superparamagnetic nanoparticles were carried out (Song et al. 2016). Both the immobilized enzymes exhibited higher enzyme activity and better thermostability than the non-immobilized counterpart. Moreover, the immobilized enzyme retained 85% of the total enzyme activity after being recycled for three times. More recently, cellulase immobilized on zinc ferrite nanoparticles significantly improved enzymatic saccharification by ultrasound-assisted alkaline pretreated biomass compared to the free enzyme while retaining its activity up to three cycles (Manasa et al. 2017). In another study, *Trichoderma reesei* cellulase was immobilized via covalent binding on chitosan-coated magnetic nanoparticles and glutaraldehyde as the coupling agent (Sánchez-Ramírez et al. 2017). The immobilized cellulase not only had improved thermal and storage ability but also retained 80% of initial activity after being recycled 15 times during the hydrolysis of carboxymethyl cellulase.

Apart from MNPs, other metallic nanoparticles have also been used for enzyme immobilization. In this context, there have been few reports on the use of mesoporous silica nanoparticles (Chang et al. 2011), zinc oxide nanoparticles (Srivastava et al. 2016), functionalized magnetic nanosphere (Zhang et al. 2015b), gold and silver nanoparticles (Mishra and Sardar 2015) and magnetic nanofibers (Lee et al. 2009) for the immobilization of cellulase. In all these studies, immobilization protected the enzymes from the inhibitory effects of the final product and the intermediate metabolites, thus improving ethanol yield.

Apart from enzyme immobilization, there are reports on improved bioethanol production using microorganisms immobilized on nanoparticles. *Saccharomyces cerevisiae* and *Zymomonas mobilis* are the two most common ethanologenic microorganisms that have huge potential in industrial bioethanol production. Few reports on nano-immobilization of these organisms are available reflecting the benefits of nanotechnology for bioethanol production. For example, immobilization of cells from *Saccharomyces cerevisiae* and hydrolytic enzymes in a matrix of alginate/magnetic nanoparticles, chitosan-magnetite nanoparticles and cellulose-coated magnetic nanoparticles (Ivanova et al. 2011) have been studied that showed high bioethanol production from corn starch hydrolysate for more than 42 days. Recently, immobilized β-galactosidase from dairy yeast in silicon dioxide nanoparticles was used to produce bioethanol using co-immobilized cultures of *Kluyveromyces marxianus* and *Saccharomyces cerevisiae* (Beniwal et al. 2018). The immobilized enzyme was reused up to 15 times without any significant loss in catalytic activity with a high bioethanol yield of 63.9 g/L from concentrated cheese whey (150 g/L). All these studies suggest the vast scope of nanotechnology for bioethanol production.

3.1.2 Nanobiocatalysts for Efficient Biomass Pretreatment

Biomass pretreatment is the foremost step in conventional biomass processing for biofuels and its effectiveness defines the extent of saccharification of biomass in subsequent steps. Generally, acidic pretreatment in combination with high temperature is the most commonly employed biomass pretreatment method. However, acidic pretreatment is disadvantageous due to loss of hemicellulose, which requires substantial capital investment and the need for neutralization before biomass saccharification (Jönsson and Martín 2016). In this context, engineered acid-functionalized nanoparticles could be used as a catalyst for pretreatment and hydrolysis (Leo and Singh 2018, Molina et al. 2018, Srivastava et al. 2017). These acid-functionalized nanoparticles can be recovered from the reaction media by physical separation. In previous reports, the role of various acid-functionalized nanoparticles such as silica-coated MNPs functionalized with alkylsulfonic acid (AS-SiMNPs) and perfluoroalkyl-sulfonic acid (PS-SiMNPs) (Gill et al. 2007), acid-functionalized perfluoroalkylsulfonic acid, alkylsulfonic acid, butylcarboxylic acid (Pena et al. 2012), etc., have been described for effective conversion of biomass into fermentable sugars which further could be converted to different biofuels with ease.

Apart from these, the nano-shear hybrid alkaline technique (NSHA) has also been used in improving the efficiency of pretreatment. The technique involves the application of high-speed shear-specific nanomixer combined with chemical reagents and mild temperature for the separation of lignin entities from lignocellulosic biomass, thereby promoting cellulose disruption (Wang et al. 2013b).

3.2 Role of Nanobiocatalysts in Biodiesel Production

Biodiesel (mainly composed of ethyl esters and fatty acid methyl esters) is another cleaner form of bioenergy whose global production has expanded rapidly in past years. Current technology for biodiesel production involves transesterification of oils and animal fats through the addition of methanol in the presence of homogeneous or heterogeneous catalysts, which generates glycerol as a by-product, that must be disposed of carefully or used for suitable applications (Abbaszaadeh et al. 2012). Furthermore, the catalyst plays a major role in determining the rate of transesterification reaction and final biodiesel yields. Traditional catalysts used for transesterification involve materials like metal oxide, zeolite, hydrotalcite and alkali metal ions. However, the inefficiency of these catalysts in terms of high cost, weak strength, small surface area, low catalytic performance, little or no resistance to atmospheric CO_2 and requirement of high temperature and pressure are the prime reasons hindering the large-scale production of biodiesel (Chang et al. 2014).

In recent years, nanotechnology evolved as one of the promising and extremely influential technology for the conversion of biomass to biodiesel. Various investigations have been carried out using different nanomaterials including magnetic nanoparticles, acid/alkali functionalized nanocatalysts, silica functionalized nanoparticles, gold nanoparticles, bimetallic nanoparticles, nanorods, nano-immobilized lipases, nanocrystalline metal oxides, etc., to improve biodiesel production.

A recent study revealed the effectiveness of sulfamic and sulfonic silica-coated crystalline Fe/Fe_3O_4 core/shell MNPs to produce biodiesel from low-grade feedstock, such as waste cooking oil (Wang et al. 2015a). During the transesterification reaction, the acid-functionalized magnetic nanoparticles presented high activity (> 95% conversion) throughout five continuous runs. In another study, nanoparticle was synthesized with a mixture of magnetic iron/cadmium and iron/tin oxide by co-precipitation method for the hydrolysis and transesterification of soybean oil and their fatty acids (Alves et al. 2014). The synthesized nanocatalyst was superparamagnetic that led to easy recovery and 90% of FAME conversion was achieved in 3 hours of reaction with no significant loss in activity even after four cycles. In another study, the effect of novel magnetic nanocatalyst Ca/Fe_3O_4@SiO_2, synthesized via a combination of sol-gel and incipient wetness impregnation method, for production of biodiesel from sunflower oil (Mostafa and Norouzi 2016). The nanocatalyst improved biodiesel yield and was reused several times without any significant loss in catalytic activity. In another study, a highly active calcium oxide CaO nanocatalyst (nano-CaO) was synthesized through a calcination-hydration-dehydration technique (Anr et al. 2016). The CaO nanocatalyst was employed for transesterification of Jatropha oil into biodiesel with a maximum yield of 98.5%. In another study, calcium oxide nanoparticles (CaONPs) and magnesium oxide nanoparticles (MgONPs) prepared by sol-gel and sol-gel self-combustion methods, respectively, were used for the production of biodiesel and glycerine via transesterification reaction (Tahvildari et al. 2015). Recently, a solid base catalyst, nanoKF/Al_2O_3, was synthesized using λ-Al_2O_3 nanoparticles as support for glycerol free production of biodiesel (Tang et al. 2017). The synthesized nanocatalyst was used for the transesterification of methanol, canola oil and dimethyl carbonate (DMC). The maximum yield of biodiesel reported was 98.8% that was suggested to be enhanced due to the large surface area of λ-Al_2O_3-supported nanoparticles. Overall the process involving biodiesel production using novel nanotechnology tools is highly advantageous over the existing/traditional processes.

In the context of biodiesel production, the lipase-catalyzed process is involved in the transesterification of non-edible oil and waste cooking oil having high free fatty acid content. The advantages of the lipase-catalyzed processes over conventional catalytic process include broad feedstock specificity, lower energy requirement and lower pretreatment cost (Amini et al. 2017, Sekoai et al. 2018a); however, high cost of lipases is a major drawback. To address this issue, the application of nano-immobilized lipases for enhancing the economic viability of enzymatic transesterification has been attempted. The use of nano-immobilized lipases for biodiesel production primarily offers improved thermal and functional stability of the enzyme, apart from other benefits like reduced production cost and improved substrate selectivity (Tacias-Pascacio et al. 2017, Ahmed Ansari and Husain 2011).

Another study reported immobilization of lipases on amino-functionalized magnetic nanoparticles (Fe_3O_4NPs) by using glutaraldehyde as the coupling agent with 84% of binding efficacy during transesterification of soybean oil by the immobilized lipase (Xie and Ma 2009). The immobilized enzyme could be used four times without losing catalytic activity. Later, immobilization of lipase from *Candida rugosa* on Fe_3O_4/Poly (styrene-methacrylic acid) magnetic polymer-coated microspheres using EDAC [1-ethayl-3-(3-(dimethylamino)propyl)carbodiimide hydrochloride] as a coupling reagent was demonstrated (Xie and Wang 2014). The immobilized lipase presented high activities to soybean oil transesterification and was stable up to four cycles without any significant loss in initial activity. Later, biodiesel production from palm oil using immobilized lipase from *Thermomyces lanuginosus* on MNPs was demonstrated (Raita et al. 2015) with a biodiesel production yield of 97.2% under optimized conditions while the immobilized lipase was able to retain 80% of its activity up to five cycles. In another study, lipase from *Burkholderia cepacia* was immobilized on superparamagnetic iron oxide nanoparticles (SPION) to achieve efficient enzymatic transesterification of waste cooking oil (WCO) (Karimi 2016). The immobilized lipase was recovered easily and WCO to biodiesel conversion reached 91% within 35 hours.

Apart from MNPs and nanocomposites, magnetic multi-walled carbon nanotubes (CNTs), metal oxide nanoparticles have been used for lipase immobilization to enhance biodiesel production. CNTs are cost-effective and offer essential traits for ideal catalysis such as renewability, chemical stability, large surface area, low toxicity, etc. (Srivastava et al. 2017). Although very few studies have been reported that utilizes CNTs for biodiesel production. One such report demonstrated a biodiesel yield of 95% from waste cooking oil using *Rhizomucor miehei* lipase immobilized onto polyamidoamine grafted with magnetic multi-walled carbon nanotubes (MWCNTs) (Fan et al. 2016). The immobilized lipase presented a 27-fold higher activity than free lipase and retained 100% activity up to 10 recycles. Similarly, high-performing sulfonated multi-walled carbon nanotube solid acid catalyst (S-MWCNTs) via facile synthesis for biodiesel production was developed (Guan et al. 2017). The S-MWCNTs showed significant catalytic activity for transesterification of triglycerides and achieved an overall conversion of 97.8% in 1 hour.

Lipase immobilization on polyacrylonitrile nanofibers (PAN) was also carried out for the transesterification of soybean oil and biodiesel conversion (Li et al. 2011) showing promising results. Another study demonstrated the application of lipase immobilized alkyl-functionalized Fe_3O_4-SiO_2 nanocomposites for biodiesel production in which about 90% biodiesel conversion was achieved in 30 hours (Tran et al. 2012). Later, they also evaluated biodiesel production from wet microalgae biomass using alkyl-grafted Fe_3O_4-SiO_2 immobilized lipase and achieved over 90% biodiesel conversion (Tran et al. 2013). All these reports suggested that nanomaterials can be used as a highly effective support matrix for lipase immobilization, showing the potential of nano-immobilization technology in the biodiesel industry.

3.3 Role of Nanobiocatalysts in Biohydrogen Production

Biohydrogen is another renewable source of energy that is considered as the cleanest and most energy-efficient biofuel. Microbial biohydrogen production is usually carried out by a diverse group of anaerobic bacteria and the process is termed as dark fermentation (Rai and Singh 2016). The dark fermentation reaction of molecular hydrogen production is catalyzed by the enzyme hydrogenase that mediates the transfer of electrons in biohydrogen-producers and is categorized as [NiFe], [FeFe] and [Fe]-hydrogenases (Sekoai et al. 2018b). Dark fermentative biohydrogen production provides a cost-effective and eco-friendly means of bioprocessing as it can be produced using a variety of renewable feedstocks under mild fermentation conditions (Łukajtis et al. 2018). However, incomplete substrate utilization and lower yields by microbes during dark fermentation are the major limitations impeding the scalability of biological hydrogen production (Patel et al. 2018). The operational parameters such as pH, temperature, initial substrate concentration and hydraulic retention time critically regulate the final biohydrogen yield during dark fermentation that requires to be optimized. In addition to this, genetic engineering of anaerobic microorganisms involved in dark-fermentative hydrogen production is also underway to overcome the limitation of lower biohydrogen yield (Hallenbeck and Ghosh 2012).

In some recent studies, enhancement in the biological hydrogen production via supplementing the media with nanoparticles has been carried out and is being recognized as a promising approach. These nanoparticles have a stimulatory effects on biohydrogen-production processes because they (i) assist hydrogenase enzymes to overcome the energy barrier making them more reactive, (ii) anti-microbial properties of these nanoparticles selectively enrich only hydrogen producers, (iii) increases the transfer of electrons and (iv) provides a large surface-area-to-volume-ratio for effective binding of bacteria to the substrate (Patel et al. 2018). Thus, a positive impact of various nanoparticles such as MNPs (e.g., gold, silver, copper, iron, nickel, titanium, palladium, etc.), silica (SiO_2), carbon nanotubes, activated carbon and graphene nanocomposite were observed on biological hydrogen production. Among the various synthetic approaches applied, the synthesis of nanoparticles via biological methods has been suggested to be most suitable for biohydrogen production (Mohanraj et al. 2016). However, they must be used at optimum concentrations to prevent microbial growth inhibitions. In this section, we have described the potential of various nanoparticles synthesized for biohydrogen production using model substrates (pure sugars) and biological waste as feedstock.

The positive impact of silver nanoparticles (AgNPs) on acidogenesis and dark fermentative biohydrogen production by glucose-fed mixed bacteria was investigated (Zhao et al. 2013). Herein, silver nanoparticles stimulated higher cell biomass production and altered the production of other intermediates like ethanol, acetate, butyrate, acetate, valerate, and propionate. It was observed that the addition of silver nanoparticles (20 nmol/L) led to a maximum biohydrogen yield of 2.48 mol H_2/mol glucose. In addition to that, these nanoparticles were beneficial to maintain the acidogenic phase throughout the process. Later, the enhancement in substrate utilization by bacteria and increased biohydrogen yield by incorporation of gold nanoparticles (AuNPs) during dark fermentation of artificial wastewater (Zhang and Shen 2007). At 5 mM concentration of AuNPs, the anaerobic culture resulted in 2.48 mol H_2/mol hexose production which was 62.3% higher than the control. Another study reported the effect of phytogenic palladium nanoparticles (prepared using *Cortandrum sattvum* leaf extract) on dark fermentative biohydrogen production by *Enterobacter cloacae* using glucose as the substrate with a maximum hydrogen yield of 2.48 mol H_2/mol glucose due to reduced lag phase time (Mohanraj et al. 2014).

Many bacterial cultures have been evaluated for biohydrogen production in the presence of iron nanoparticles using both sugars and bio-waste materials as substrates (Malik et al. 2014, Zhang et al. 2015a, Taherdanak et al. 2015). The linear and interactive effect of initial pH, glucose concentration and nickel nanoparticle concentration on the dark fermentation process using microflora in batch tests were studied (Mullai et al. 2013). Consequently, a maximum cumulative biohydrogen production of 4,400 mL and a yield of 2.54 mol H_2/mol glucose was obtained at a glucose concentration of 14.01 g/L. Another study demonstrated the additive effect of biologically synthesized iron nanoparticles (FeNPs) on enhanced biohydrogen yield and glucose utilization using *Enterobacter cloacae* (Nath et al. 2015) with maximum biohydrogen yield of 1.9 mol H_2/mol glucose at 100 mg/L concentration of FeNPs. In another study, the stimulatory effect of TiO_2 and magnetic hematite nanoparticles on the enzyme activity and gene expression of *Clostridium pasteurianum* during dark fermentative biohydrogen production (Hsieh et al. 2016). The synthesized nanoparticles significantly improved the enzyme activity by assisting efficient electron transfer with a maximum hydrogen production rate of 45.2 mL/h. Likewise, the final biohydrogen yield from glucose was maximized by supplementing media with ferric oxide nanoparticles (FONPs) during dark fermentation involving pure cultures of *Enterobacter aerogenes* (ATCC13408) (Lin et al. 2016). SEM and TEM images revealed cellular internalization of FONPs, resulting in the release of iron element and enhanced hydrogenase synthesis and activity. Here, FONPs helped in enhancing electron transfer and allowed to shift the metabolic pathway for more hydrogen production. Recently, the effects of Ni nanoparticles and Ni-graphene nanocomposite on anaerobic digestion of industrial wastewater containing mono-ethylene glycol were investigated (Elreedy et al. 2017). Here, a maximum biohydrogen yields of 24.73 ± 1.12 mL $H_2/gCOD_{initial}$ and 41.28 ± 1.69 mL H_2/g $COD_{initial}$ were achieved using Ni and Ni-graphene nanoparticles at a dosage of 60 mg/L, respectively. Similarly, optimization of operational conditions, substrate concentration and iron nanoparticle concentration for dark fermentative biohydrogen production from sweet potato starch was investigated (Vi et al. 2017) that resulted in a cumulative biohydrogen yield of 3,501 mg/L under optimized conditions. This study highlighted the positive effect of iron nanoparticles as a co-factor in active sites of hydrogenase enzymes to maximize the activity of biohydrogen producing bacteria. In a recent study, the mechanism and microbial community dynamics of biohydrogen production from grass using zero-valent iron nanoparticles were studied (Yang and Wang 2018). It showed that the supplementation of zero-valent iron nanoparticles favored the metabolic pathway of predominant bacteria toward higher hydrogen production with a maximum biohydrogen yield and production rate of 64.7 mL/g dry grass and 12.1 mL/h, respectively.

Apart from this, nano-based additives have also been used to enhance dark fermentative biohydrogen production. For example, the enhancement effect of three metallic (Pb, Ag and Cu) and metal oxide (FeO) nanoparticles on biohydrogen production from glucose were studied using *Clostridium butyricum* strain by encapsulating them in porous silica (SiO_2) (Beckers et al. 2013). An increased biohydrogen production rate by 58% and biohydrogen yield by 38% were observed for a strain that was encapsulated in FeO nanoparticles as compared to the cultures without FeO nanoparticles. The improvement in hydrogenase activity and electron transfer was suggested as the possible reason for higher yield.

In past years, a few studies have been reported on the application of mesoporous silica nanoparticles for enhancement of biohydrogen-producing biochemical pathways (Zhou et al. 2018b). In one such study, the application of SBA-15 silica nanoparticles for improving fermentative biohydrogen production by a mixed consortium using chemical wastewater as substrate (Mohan et al. 2008) was studied. The immobilization of consortia on SBA-15 material showed remarkable improvement in the production of biohydrogen by nine times (7.29 mol/kg COD_R-day) in comparison with control experiments at a high organic loading rate of 2.55 kg COD/m^3-day. Santa Barbara Amorphous (SBA-15), mesoporous silica, is another attractive material that provides large surface-area-to-volume-ratio, uniform pore size, thermal stability and non-toxicity, which could be used for the enhancement of dark fermentative biohydrogen production (Huirache-Acuña et al. 2013, Patel et al. 2018).

However, some dark fermentation studies also demonstrated the negative effects of nanoparticles on biohydrogen producing pathways of some bacterial strains (Mohanraj et al. 2016, Beckers et al. 2013). In this context, the influence of copper and $CuSO_4$ nanoparticles on biohydrogen production from glucose using *Clostridium acetobutylicum* and *Enterobacter cloacae* was compared (Mohanraj et al. 2016). It was observed that both Cu and $CuSO_4$ nanoparticles supplementation had a negative effect on acetate and butyrate fermentation pathways. It can, therefore, be concluded from the above studies that combining dark fermentation with nanoparticles or nano-based additives can play an important role in the advancement of biohydrogen production technologies.

3.4 Role of Nanobiocatalysts in Biogas Production

Biogas is commonly referred to as a mixture of different gases produced by the anaerobic digestion of organic matter using anaerobic organisms like methanogenic bacteria (Ganzoury and Allam 2015). The main feedstock utilized to produce biogas includes agricultural waste, food waste, municipal solid waste, sewage, etc. Different metal ions are believed to play an important role in the metabolism of methanogenic bacteria during the process of anaerobic digestion (Feng et al. 2010). Thus, their respective nanoparticles form could be employed for biogas production suggesting another important application of nanobiotechnology in the field of bioenergy production. For example, studies like the engineered iron oxide nanoparticles to improve biogas production during anaerobic digestion process (Casals et al. 2014) and the efficacy of varying concentrations of cobalt and nickel nanoparticles toward enhanced methane gas production (Abdelsalam et al. 2015). Furthermore, the study of the later example showed that nickel and cobalt nanoparticles stimulated the activity of methanogenic bacteria, thereby reducing the time of biogas production (Abdelsalam et al. 2017).

3.5 Role of Nanobiocatalysts in Algae-Based Biofuel Production

Microalgae have emerged as a more desirable feedstock for biodiesel production due to their high lipid content and ability to grow even in saline or wastewater (Satyanarayana et al. 2011). However, like other biofuels, algae-based biofuel production faces some key barriers preventing their commercialization. These include mainly the high production cost, associated with lower biomass yield and harvesting difficulties. The development of advanced harvesting technologies, effective lipid extraction methods and efficient transesterification process at low temperature are the proposed possible way out for a cost-effective algal biofuel production process (Pattarkine and Pattarkine 2012, Mandotra et al. 2018). In recent years, the application of novel nanotechnology tools to improve the overall process of algae-based biofuel production is gaining increased research interest. The research advancement dealing with the application of nanotechnology for algae biofuel production include enhanced CO_2 accumulation and biomass production, improved illumination, easier harvesting due to supplementation of magnetic nanoparticles to algal culture, modification in algal cell wall for maximum lipid recovery, stable enzymes production for transesterification, etc. In this context, green microalga *Chlorella vulgaris* culture was supplemented with $MgSO_4$ nanoparticles and organic carbon sources (Sarma et al. 2014). The incorporation of $MgSO_4$ nanoparticles resulted in improved photosynthetic efficiency, easier flocculation and enhanced lipid yield. In another study, metal nanoparticles together with localized surface plasmon resonances (LSPR) were used to amplify light at a specific wavelength outside closed photobioreactor (PBR) that improved the light uptake efficiency of algae during growth and harvest (Torkamani et al. 2010). In another study, the application of spheroidal silver nanoparticles and gold nanorods improved the illumination of *Chlorella vulgaris* (Eroglu et al. 2013) which was due to an increased accumulation of chlorophyll and carotenoid pigments by nanoparticle supplementation. Recently, an increase in the oil content of *Chlorella vulgaris* from 8.44 to 17.68% with the increasing concentration of silver nanoparticles (AgNPs) from 50 to 150 μg/g was also demonstrated (Abdul Razack et al. 2016).

Algal biomass harvesting from wet biomass is a difficult and highly expensive procedure at the industrial scale (Wang et al. 2015b, Mandotra et al. 2018). Although flocculation is considered as the most cost-effective means of harvesting algae, they render inefficient for smaller algal cells (Vandamme et al. 2013). Recently, the application of nanoparticles was suggested to achieve quick, inexpensive and energy-efficient flocculation of biomass. In this context, microalga *Nannochloropsis maritime* culture was supplemented with Fe_3O_4 nanoparticles and as a result microalgae harvesting was improved to 95% at a dose of 120 mg/L (Hu et al. 2013). Later, supplementation of CTAB-decorated Fe_3O_4 nanoparticles to *Chlorella* sp. culture was described that resulted in 96.6% of microalgae harvesting (Seo et al. 2016).

Apart from this, nanotechnology tools have improved lipid extraction as well which is essential for effective transesterification (Lee et al. 2015). At present, a combination of various solvents such as chloroform, methanol, hexane, etc., are used to achieve lipid extraction from algal biomass. Moreover, the solvent combination and extraction conditions have a huge impact on final lipid yield and thus need optimization (Zhang et al. 2013). These studies suggested that combining microalgae-based biodiesel production and nanotechnology tools could reduce the cost of downstream processing and improve the overall process economics.

4. Conclusions and Future Perspectives

This chapter gives a comprehensive understanding of the role of nanobiocatalysts to improve the production of bioethanol, biodiesel, biohydrogen, biogas and algae-based biofuels. The supplementation of various nanomaterials provides solid support and shows their high potential and significant role in the field of biofuel. The immobilization of cellulase and lipase enzymes on metal nanoparticles allows easy recovery and repeated use and seems very promising for large scale applications. Overall, nanotechnological tools are highly efficient, convenient and environmentally friendly. However, most of the studies are at the laboratory level and more thorough research is required to understand their suitability at large-scale in terms of production cost, eco-friendly nature and non-toxicity.

Conflict of Interest

The authors declare no financial or commercial conflict of interest.

References

Abbaszaadeh, A., B. Ghobadian, M.R. Omidkhah and G. Najafi. 2012. Current biodiesel production technologies: A comparative review. Energy Conversion and Management 63: 138–148.

Abdelsalam, E., M. Samer, M.E. Abdel-Hadi, H. Hassan and Y. Badr. 2015. Effects of $CoCl_2$, $NiCl_2$ and $FeCl_3$ additives on biogas and methane production. Misr Journal of Agricultural Engineering 32(2): 843–862.

Abdelsalam, E., M. Samer, Y.A. Attia, M.A. Abdel-Hadi, H.E. Hassan and Y. Badr. 2017. Influence of zero valent iron nanoparticles and magnetic iron oxide nanoparticles on biogas and methane production from anaerobic digestion of manure. Energy 120: 842–853.

Abdul Razack, S., S. Duraiarasan and V. Mani. 2016. Biosynthesis of silver nanoparticle and its application in cell wall disruption to release carbohydrate and lipid from *C. vulgaris* for biofuel production. Biotechnology Reports 11: 70–76.

Aditiya, H.B., T.M.I. Mahlia, W.T. Chong, H. Nur and A.H. Sebayang. 2016. Second generation bioethanol production: a critical review. Renewable and Sustainable Energy Reviews 66: 631–653.

Ahmed Ansari, S. and Q. Husain. 2011. Potential applications of enzymes immobilized on/in nano materials: A review. Biotechnology Advances 30: 512–523.

Ahmed, W. and B. Sarkar. 2018. Impact of carbon emissions in a sustainable supply chain management for a second generation biofuel. Journal of Cleaner Production 186: 807–820.

Akia, M., F. Yazdani, E. Motaee, D. Han and H. Arandiyan. 2014. A review on conversion of biomass to biofuel by nanocatalysts. Biofuel Research Journal 1: 16–25.

Alam, F., S. Mobin and H. Chowdhury. 2015. Third generation biofuel from algae. Procedia Engineering 105: 763–768.

Alaswad, A., M. Dassisti, T. Prescott and A.G. Olabi. 2015. Technologies and developments of third generation biofuel production. Renewable and Sustainable Energy Reviews 51: 1446–1460.

Alves, M.B., F.C.M. Medeiros, M.H. Sousa, J.C. Rubim and P.A.Z. Suarez. 2014. Cadmium and tin magnetic nanocatalysts useful for biodiesel production. Journal of the Brazilian Chemical Society 25: 2304–2313.

Amini, Z., Z. Ilham, H.C. Ong, H. Mazaheri and W.H. Chen. 2017. State of the art and prospective of lipase-catalyzed transesterification reaction for biodiesel production. Energy Conversion and Management 141: 339–353.

Anr, R., A.A. Saleh, M.S. Islam, S. Hamdan and M.A. Maleque. 2016. Biodiesel production from crude Jatropha oil using a highly active heterogeneous nanocatalyst by optimizing transesterification reaction parameters. Energy and Fuels 30: 334–343.

Antonio Fernandes Antunes, F., S. Gaikwad, A. Ingle, R. Pandit, J. Santos, M. Rai and S. Da Silva. 2017. Bioenergy and biofuels: nanotechnological solutions for sustainable production. pp. 3–18. *In*: Rai, M. and S. da Silva (eds.). Nanotechnology for Bioenergy and Biofuel Production. Green Chemistry and Sustainable Technology. Springer, Cham.

Balan, V. 2014. Current challenges in commercially producing biofuels from lignocellulosic biomass. International Scholarly Research Notices Biotechnology 2014: 31.

Bansal, A., P. Illukpitiya, F. Tegegne and S.P. Singh. 2016. Energy efficiency of ethanol production from cellulosic feedstock. Renewable and Sustainable Energy Reviews 58: 141–146.

Beckers, L., S. Hiligsmann, S.D. Lambert, B. Heinrichs and P. Thonart. 2013. Improving effect of metal and oxide nanoparticles encapsulated in porous silica on fermentative biohydrogen production by *Clostridium butyricum*. Bioresource Technology 133: 109–117.

Beniwal, A., P. Saini, A. Kokkiligadda and S. Vij. 2018. Use of silicon dioxide nanoparticles for β-galactosidase immobilization and modulated ethanol production by co-immobilized *K. marxianus* and *S. cerevisiae* in deproteinized cheese whey. LWT 87: 553–561.

Binod, P., E. Gnansounou, R. Sindhu and A. Pandey. 2018. Enzymes for second generation biofuels: recent developments and future perspectives. Bioresource Technology Reports (in press).

Casals, E., R. Barrena, A. García, E. González, L. Delgado, M. Busquets-Fité, X. Font, J. Arbiol, P. Glatzel, K. Kvashnina, A. Sánchez and V. Puntes. 2014. Programmed iron oxide nanoparticles disintegration in anaerobic digesters boosts biogas production. Small 10: 2801–2808.

Chang, F., Q. Zhou, H. Pan, X.-F. Liu, H. Zhang, W. Xue and S. Yang. 2014. Solid mixed-metal-oxide catalysts for biodiesel production: a review. Energy Technology 2: 865–873.

Chang, R.H.-Y., J. Jang and K.C.W. Wu. 2011. Cellulase immobilized mesoporous silica nanocatalysts for efficient cellulose-to-glucose conversion. Green Chemistry 13: 2844–2850.

Cheng, J., Y. Qiu, R. Huang, W. Yang, J. Zhou and K. Cen. 2016. Biodiesel production from wet microalgae by using graphene oxide as solid acid catalyst. Bioresource Technology 221: 344–349.

Cherian, E., M. Dharmendirakumar and G. Baskar. 2015. Immobilization of cellulase onto MnO_2 nanoparticles for bioethanol production by enhanced hydrolysis of agricultural waste. Chinese Journal of Catalysis 36: 1223–1229.

Datta, S., L.R. Christena and Y.R.S. Rajaram. 2013. Enzyme immobilization: an overview on techniques and support materials. 3 Biotech. 3: 1–9.

Dutta, S., S. De, B. Saha and M.I. Alam. 2012. Advances in conversion of hemicellulosic biomass to furfural and upgrading to biofuels. Catalysis Science and Technology 2: 2025–2036.

El-Seesy, A.I., A.K. Abdel-Rahman, M. Bady and S. Ookawara. 2016. The influence of multi-walled carbon nanotubes additives into non-edible biodiesel-diesel fuel blend on diesel engine performance and emissions. Energy Procedia 100: 166–172.

Elreedy, A., E. Ibrahim, N. Hassan, A. El-Dissouky, M. Fujii, C. Yoshimura and A. Tawfik. 2017. Nickel-graphene nanocomposite as a novel supplement for enhancement of biohydrogen production from industrial wastewater containing mono-ethylene glycol. Energy Conversion and Management 140: 133–144.

Eroglu, E., P. Eggers, M. Winslade, S. Smith and C. Raston. 2013. Enhanced accumulation of microalgal pigments using metal nanoparticle solutions as light filtering devices. Green Chemistry 15: 3155–3159.

Fan, Y., G. Wu, F. Su, K. Li, L. Xu, X. Han and Y. Yan. 2016. Lipase oriented-immobilized on dendrimer-coated magnetic multi-walled carbon nanotubes toward catalyzing biodiesel production from waste vegetable oil. Fuel 178: 172–178.

Farooq, W., Y.-C. Lee, J.-I. Han, C.H. Darpito, M. Choi and J.W. Yang. 2013. Efficient microalgae harvesting by organo-building blocks of nanoclays. Green Chemistry 15: 749–755.

Feng, X.M., A. Karlsson, B.H. Svensson and S. Bertilsson. 2010. Impact of trace element addition on biogas production from food industrial waste-linking process to microbial communities. FEMS Microbiology Ecology 74: 226–240.

Ganzoury, M.A. and N.K. Allam. 2015. Impact of nanotechnology on biogas production: a mini-review. Renewable and Sustainable Energy Reviews 50: 1392–1404.

Gardy, J., A. Hassanpour, X. Lai, M.H. Ahmed and M. Rehan. 2017. Biodiesel production from used cooking oil using a novel surface functionalised TiO_2 nano-catalyst. Applied Catalysis B 207: 297–310.

Gardy, J., A. Osatiashtiani, O. Céspedes, A. Hassanpour, X. Lai, A.F. Lee, K. Wilson and M. Rehan. 2018. A magnetically separable SO_4/Fe-Al-TiO_2 solid acid catalyst for biodiesel production from waste cooking oil. Applied Catalysis B 234: 268–278.

Gaurav, N., S. Sivasankari, G.S. Kiran, A. Ninawe and J. Selvin. 2017. Utilization of bioresources for sustainable biofuels: a review. Renewable and Sustainable Energy Reviews 53: 205–214.

Gill, C.S., B.A. Price and C.W. Jones. 2007. Sulfonic acid-functionalized silica-coated magnetic nanoparticle catalysts. Journal of Catalysis 251: 145–152.

Guan, Q., Y. Li, Y. Chen, Y. Shi, J. Gu, B. Li, R. Miao, Q. Chen and P. Ning. 2017. Sulfonated multi-walled carbon nanotubes for biodiesel production through triglycerides transesterification. RSC Advances 7: 7250–7258.

Hallenbeck, P.C. and D. Ghosh. 2012. Improvements in fermentative biological hydrogen production through metabolic engineering. Journal of Environmental Management 95: S360–S364.

Hama, S., H. Noda and A. Kondo. 2018. How lipase technology contributes to evolution of biodiesel production using multiple feedstocks. Current Opinion in Biotechnology 50: 57–64.

Ho, D.P., H.H. Ngo and W. Guo. 2014. A mini review on renewable sources for biofuel. Bioresource Technology 169: 742–749.

Hsieh, P.-H., Y.-C. Lai, K.-Y. Chen and C.-H. Hung. 2016. Explore the possible effect of TiO_2 and magnetic hematite nanoparticle addition on biohydrogen production by *Clostridium pasteurianum* based on gene expression measurements. International Journal of Hydrogen Energy 41: 21685–21691.

Hu, Y.-R., F. Wang, S.-K. Wang, C.-Z. Liu and C. Guo. 2013. Efficient harvesting of marine microalgae Nannochloropsis maritima using magnetic nanoparticles. Bioresource Technology 138: 387–390.

Huang, P.-J., K.-L. Chang, J.-F. Hsieh and S.-T. Chen. 2015. Catalysis of rice straw hydrolysis by the combination of immobilized cellulase from *Aspergillus niger* β-cyclodextrin-Fe_3O_4 nanoparticles and ionic liquid. BioMed Research International 2015: 9.

Huirache-Acuña, R., R. Nava, C. Peza-Ledesma, J. Lara-Romero, G. Alonso-Núez, B. Pawelec and E. Rivera-Muñoz. 2013. SBA-15 mesoporous silica as catalytic support for hydrodesulfurization catalysts—review. Materials (Basel) 6: 4139–4167.

Ivanova, V., P. Gencheva and J. Hristov. 2011. Application in the ethanol fermentation of immobilized yeast cells in matrix of alginate/magnetic nanoparticles, on chitosan-magnetite microparticles and cellulose-coated magnetic nanoparticles. International Review of Chemical Engineering 3: 289–299.

Jones, C.S. and S.P. Mayfield. 2012. Algae biofuels: versatility for the future of bioenergy. Current Opinion in Biotechnology 23: 346–351.

Jönsson, L.J. and C. Martín. 2016. Pretreatment of lignocellulose: formation of inhibitory by-products and strategies for minimizing their effects. Bioresource Technology 199: 103–112.

Joshi, G., J.K. Pandey, S. Rana and D.S. Rawat. 2017. Challenges and opportunities for the application of biofuel. Renewable and Sustainable Energy Reviews 79: 850–866.

Kang, Q., L. Appels, T. Tan and R. Dewil. 2014. Bioethanol from lignocellulosic biomass: current findings determine research priorities. The Scientific World Journal 2014: 298153–298153.

Karimi, M. 2016. Immobilization of lipase onto mesoporous magnetic nanoparticles for enzymatic synthesis of biodiesel. Biocatalysis and Agricultural Biotechnology 8: 182–188.

Khan, T.M.Y., A.E. Atabani, I.A. Badruddin, A. Badarudin, M.S. Khayoon and S. Triwahyono. 2014. Recent scenario and technologies to utilize non-edible oils for biodiesel production. Renewable and Sustainable Energy Reviews 37: 840–851.

Kim, D. 2018. Physico-chemical conversion of lignocellulose: inhibitor effects and detoxification strategies: a mini review. Molecules 23: 309.

Kim, K., O. Lee and E. Lee. 2018. Nano-immobilized biocatalysts for biodiesel production from renewable and sustainable resources. Catalysts 8: 68.

Klein-Marcuschamer, D., P. Oleskowicz-Popiel, B.A. Simmons and H.W. Blanch. 2012. The challenge of enzyme cost in the production of lignocellulosic biofuels. Biotechnology and Bioengineering 109: 1083–1087.

Kumar, A.K. and S. Sharma. 2017. Recent updates on different methods of pretreatment of lignocellulosic feedstocks: a review. Bioresources and Bioprocessing 4: 7.

Lee, S.-M., L.H. Jin, J.H. Kim, S.O. Han, H.B. Na, T. Hyeon, Y.-M. Koo, J. Kim and J.-H. Lee. 2009. β-Glucosidase coating on polymer nanofibers for improved cellulosic ethanol production. Bioprocess and Biosystems Engineering 33: 141–147.

Lee, Y.-C., K. Lee and Y.-K. Oh. 2015. Recent nanoparticle engineering advances in microalgal cultivation and harvesting processes of biodiesel production: A review. Bioresource Technology 184: 63–72.

Leo, V.V. and B.P. Singh. 2018. Prospectus of nanotechnology in bioethanol productions. pp. 129–139. *In*: Srivastava, N., M. Srivastava, H. Pandey, P.K. Mishra and P.W. Ramteke (eds.). Green Nanotechnology for Biofuel Production. Cham: Springer International Publishing.

Leong, W.-H., J.-W. Lim, M.-K. Lam, Y. Uemura and Y.-C. Ho. 2018. Third generation biofuels: A nutritional perspective in enhancing microbial lipid production. Renewable and Sustainable Energy Reviews 91: 950–961.

Li, S.-F., Y.-H. Fan, R.-F. Hu and W.-T. Wu. 2011. *Pseudomonas cepacia* lipase immobilized onto the electrospun PAN nanofibrous membranes for biodiesel production from soybean oil. Journal of Molecular Catalysis B: Enzymatic 72: 1–2.

Lin, R., J. Cheng, L. Ding, W. Song, M. Liu, J. Zhou and K. Cen. 2016. Enhanced dark hydrogen fermentation by addition of ferric oxide nanoparticles using *Enterobacter aerogenes*. Bioresource Technology 207: 213–219.

Lin, R., J. Cheng, J. Zhang, J. Zhou, K. Cen and J.D. Murphy. 2017. Boosting biomethane yield and production rate with graphene: The potential of direct interspecies electron transfer in anaerobic digestion. Bioresource Technology 239: 345–352.

Lin, R., C. Deng, J. Cheng, A. Xia, P.N.L. Lens, S.A. Jackson, A.D.W. Dobson and J.D. Murphy. 2018. Graphene facilitates biomethane production from protein-derived glycine in anaerobic digestion. iScience 10: 158–170.

Liu, X., H. He, Y. Wang and S. Zhu. 2007. Transesterification of soybean oil to biodiesel using SrO as a solid base catalyst. Catalysis Communications 8: 1107–1111.

Loges, B., A. Boddien, F. Gärtner, H. Junge and M. Beller. 2010. Catalytic generation of hydrogen from formic acid and its derivatives: useful hydrogen storage materials. Topics in Catalysis 53: 902–914.

Łukajtis, R., I. Hołowacz, K. Kucharska, M. Glinka, P. Rybarczyk, A. Przyjazny and M. Kamiński. 2018. Hydrogen production from biomass using dark fermentation. Renewable and Sustainable Energy Reviews 91: 665–694.

Malik, S.N., V. Pugalenthi, A.N. Vaidya, P.C. Ghosh and S.N. Mudliar. 2014. Kinetics of nano-catalysed dark fermentative hydrogen production from distillery wastewater. Energy Procedia 54: 417–430.

Manasa, P., P. Saroj and N. Korrapati. 2017. Immobilization of cellulase enzyme on zinc ferrite nanoparticles in increasing enzymatic hydrolysis on ultrasound-assisted alkaline pretreated *Crotalaria juncea* biomass. Indian Journal of Science and Technology 10.

Mandotra, S., R. Kumar, S. Upadhyay and P. Ramteke. 2018. Nanotechnology: a new tool for biofuel production. pp. 17–28. *In*: Srivastava, N., M. Srivastava, H. Pandey, P. Mishra and P. Ramteke (eds.). Green Nanotechnology for Biofuel Production. Biofuel and Biorefinery Technologies. Springer, Cham.

Mathesh, M., B. Luan, T.O. Akanbi, J.K. Weber, J. Liu, C.J. Barrow, R. Zhou and W. Yang. 2016. Opening lids: modulation of lipase immobilization by graphene oxides. ACS Catalysis 6: 4760–4768.

Mishra, A. and M. Sardar. 2015. Cellulase assisted synthesis of nano-silver and gold: application as immobilization matrix for biocatalysis. International Journal of Biological Macromolecules 77: 105–113.

Misson, M., H. Zhang and B. Jin. 2015. Nanobiocatalyst advancements and bioprocessing applications. Journal of the Royal Society Interface 12: 20140891–20140891.

Mohamad, N.R., N.H.C. Marzuki, N.A. Buang, F. Huyop and R.A. Wahab. 2015. An overview of technologies for immobilization of enzymes and surface analysis techniques for immobilized enzymes. Biotechnology Biotechnological Equipment 29: 205–220.

Mohan, S.V., G. Mohanakrishna, S.S. Reddy, B.D. Raju, K.R. Rao and P.N. Sarma. 2008. Self-immobilization of acidogenic mixed consortia on mesoporous material (SBA-15) and activated carbon to enhance fermentative hydrogen production. International Journal of Hydrogen Energy 33: 6133–6142.

Mohanraj, S., K. Anbalagan, S. Kodhaiyolii and V. Pugalenthi. 2014. Comparative evaluation of fermentative hydrogen production using *Enterobacter cloacae* and mixed culture: Effect of Pd (II) ion and phytogenic palladium nanoparticles. Journal of Biotechnology 192: 87–95.

Mohanraj, S., K. Anbalagan, P. Rajaguru and V. Pugalenthi. 2016. Effects of phytogenic copper nanoparticles on fermentative hydrogen production by *Enterobacter cloacae* and *Clostridium acetobutylicum*. International Journal of Hydrogen Energy 41: 10639–10645.

Molina, G., G. Pagotto Borin, F.M. Pelissari and F.J. Contesini. 2018. Nanotechnology applied for cellulase improvements. pp. 93–114. *In*: Srivastava, N., M. Srivastava, H. Pandey, P.K. Mishra and P.W. Ramteke (eds.). Green Nanotechnology for Biofuel Production. Cham: Springer International Publishing.

Mostafa, F. and L. Norouzi. 2016. Preparation and kinetic study of magnetic Ca/Fe_3O_4@SiO_2 nanocatalysts for biodiesel production. Renewable Energy 94: 579–586.

Mullai, P., M.K. Yogeswari and K. Sridevi. 2013. Optimisation and enhancement of biohydrogen production using nickel nanoparticles—a novel approach. Bioresource Technology 141: 212–219.

Naik, S.N., V.V. Goud, P.K. Rout and A.K. Dalai. 2010. Production of first and second generation biofuels: a comprehensive review. Renewable and Sustainable Energy Reviews 14: 578–597.

Nath, D., A.K. Manhar, K. Gupta, D. Saikia, S.K. Das and M. Mandal. 2015. Phytosynthesized iron nanoparticles: effects on fermentative hydrogen production by *Enterobacter cloacae* DH-89. Bulletin of Materials Science 38: 1533–1538.

Nielsen, M., E. Alberico, W. Baumann, H.-J. Drexler, H. Junge, S. Gladiali and M. Beller. 2013. Low-temperature aqueous-phase methanol dehydrogenation to hydrogen and carbon dioxide. Nature 495: 85.

Oh, Y.-K., K.-R. Hwang, C. Kim, J.R. Kim and J.-S. Lee. 2018. Recent developments and key barriers to advanced biofuels: a short review. Bioresource Technology 257: 320–333.

Patel, S.K.S., J.-K. Lee and V.C. Kalia. 2018. Nanoparticles in biological hydrogen production: an overview. Indian Journal of Microbiology 58: 8–18.

Pattarkine, M.V. and V.M. Pattarkine. 2012. Nanotechnology for algal biofuels. pp. 147–163. *In*: Gordon, R. and J. Seckbach (eds.). The Science of Algal Fuels: Phycology, Geology, Biophotonics, Genomics and Nanotechnology. Dordrecht: Springer Netherlands.

Pena, L., M. Ikenberry, K. Hohn and D. Wang. 2012. Acid-functionalized nanoparticles for pretreatment of wheat straw. Journal of Biomaterials and Nanobiotechnology 3: 342–352.

Ping, Y., J.-M. Yan, Z.-L. Wang, H.-L. Wang and Q. Jiang. 2013. Ag0.1-Pd0.9/rGO: an efficient catalyst for hydrogen generation from formic acid/sodium formate. Journal of Materials Chemistry A 1: 12188–12191.

Puri, M.J., C. Barrow and M. Verma. 2013. Enzyme immobilization on nanomaterials for biofuel production. Trends in Biotechnology 31: 215–6.

Rabemanolontsoa, H. and S. Saka. 2016. Various pretreatments of lignocellulosics. Bioresource Technology 199: 83–91.

Rai, M. and S.S. Da Silva. 2017. Nanotechnology for Bioenergy and Biofuel Production, Springer.

Rai, P.K. and S.P. Singh. 2016. Integrated dark- and photo-fermentation: Recent advances and provisions for improvement. International Journal of Hydrogen Energy 41: 19957–19971.

Raita, M., J. Arnthong, V. Champreda and N. Laosiripojana. 2015. Modification of magnetic nanoparticle lipase designs for biodiesel production from palm oil. Fuel Processing Technology 134: 189–197.

Ramakrishna, T.R.B., D.P. Killeen, T.D. Nalder, S.N. Marshall, W. Yang and C.J. Barrow. 2018. Quantifying graphene oxide reduction using spectroscopic techniques: a chemometric analysis. Applied Spectroscopy 72: 1764–1773.

Raman, S., A. Mohr, R. Helliwell, B. Ribeiro, O. Shortall, R. Smith and K. Millar. 2015. Integrating social and value dimensions into sustainability assessment of lignocellulosic biofuels. Biomass and Bioenergy 82: 49–62.

Rinaldi, R. and F. Schüth. 2009. Acid hydrolysis of cellulose as the entry point into biorefinery schemes. ChemSusChem. 2: 1096–1107.

Rodionova, M.V., R.S. Poudyal, I. Tiwari, R.A. Voloshin, S.K. Zharmukhamedov, H.G. Nam, B.K. Zayadan, B.D. Bruce, H.J.M. Hou and S.I. Allakhverdiev. 2017. Biofuel production: challenges and opportunities. International Journal of Hydrogen Energy 42: 8450–8461.

Saladini, F., N. Patrizi, F.M. Pulselli, N. Marchettini and S. Bastianoni. 2016. Guidelines for energy evaluation of first, second and third generation biofuels. Renewable and Sustainable Energy Reviews 66: 221–227.

Sánchez-Ramírez, J., J.L. Martínez-Hernández, P. Segura-Ceniceros, G. López, H. Saade, M.A. Medina-Morales, R. Ramos-González, C.N. Aguilar and A. Ilyina. 2017. Cellulases immobilization on chitosan-coated magnetic nanoparticles: application for Agave Atrovirens lignocellulosic biomass hydrolysis. Bioprocess and Biosystems Engineering 40: 9–22.

Saravanan, A.P., T. Mathimani, G. Deviram, K. Rajendran and A. Pugazhendhi. 2018. Biofuel policy in India: a review of policy barriers in sustainable marketing of biofuel. Journal of Cleaner Production 193: 734–747.

Sarma, S.J., R.K. Das, S.K. Brar, Y. Le Bihan, G. Buelna, M. Verma and C.R. Soccol. 2014. Application of magnesium sulfate and its nanoparticles for enhanced lipid production by mixotrophic cultivation of algae using biodiesel waste. Energy 78: 16–22.

Satyanarayana, K.G., A.B. Mariano and J.V.C. Vargas. 2011. A review on microalgae, a versatile source for sustainable energy and materials. International Journal of Energy Research 35: 291–311.

Schröfel, A., G. Kratošová, I. Šafařík, M. Šafaříková, I. Raška and L.M. Shor. 2014. Applications of biosynthesized metallic nanoparticles—a review. Acta Biomaterialia 10: 4023–4042.

Sekoai, P., N. Moro, S. Du Preez, P. Modisha, N. Engelbrecht, D. Bessarabov and A. Ghimire. 2018a. Application of nanoparticles in biofuels: an overview. Fuel 237: 380–397.

Sekoai, P., K.O. Yoro, M. Bodunrin, A.O. Ayeni and M.O. Daramola. 2018b. Integrated system approach to dark fermentative biohydrogen production for enhanced yield, energy efficiency and substrate recovery. Reviews in Environmental Science and Biotechnology 17: 501.

Seo, J.Y., R. Praveenkumar, B. Kim, J.-C. Seo, J.-Y. Park, J.-G. Na, S.G. Jeon, S.B. Park, K. Lee and Y.-K. Oh. 2016. Downstream integration of microalgae harvesting and cell disruption by means of cationic surfactant-decorated Fe_3O_4 nanoparticles. Green Chemistry 18: 3981–3989.

Sindhu, R., P. Binod and A. Pandey. 2016. Biological pretreatment of lignocellulosic biomass–an overview. Bioresource Technology 199: 76–82.

Song, Q., Y. Mao, M. Wilkins, F. Segato and R. Prade. 2016. Cellulase immobilization on superparamagnetic nanoparticles for reuse in cellulosic biomass conversion. AIMS Bioengineering 3: 264–276.

Srivastava, N., M. Srivastava, P.K. Mishra and P.W. Ramteke. 2016. Application of ZnO nanoparticles for improving the thermal and ph stability of crude cellulase obtained from *Aspergillus fumigatus* AA001. Frontiers in Microbiology 7: 514–514.

Srivastava, N., M. Srivastava, A. Manikanta, P. Singh, P.W. Ramteke and P.K. Mishra. 2017. Nanomaterials for biofuel production using lignocellulosic waste. Environmental Chemistry Letters 15: 179–184.

Tacias-Pascacio, V.G., J.J. Virgen-Ortíz, M. Jiménez-Pérez, M. Yates, B. Torrestiana-Sanchez, A. Rosales-Quintero and R. Fernandez-Lafuente. 2017. Evaluation of different lipase biocatalysts in the production of biodiesel from used cooking oil: critical role of the immobilization support. Fuel 200: 1–10.

Taherdanak, M., H. Zilouei and K. Karimi. 2015. Investigating the effects of iron and nickel nanoparticles on dark hydrogen fermentation from starch using central composite design. International Journal of Hydrogen Energy 40: 12956–12963.

Tahvildari, K., Y.N. Anaraki, R. Fazaeli, S. Mirpanji and E. Delrish. 2015. The study of CaO and MgO heterogenic nano-catalyst coupling on transesterification reaction efficacy in the production of biodiesel from recycled cooking oil. Journal of Environmental Health Science and Engineering 13: 73–73.

Tang, Y., H. Ren, F. Chang, X. Gu and J. Zhang. 2017. Nano KF/Al_2O_3 particles as an efficient catalyst for no-glycerol biodiesel production by coupling transesterification. RSC Advances 7: 5694–5700.

Terán Hilares, R., G.F. De Almeida, M.A. Ahmed, F.A.F. Antunes, S.S. Da Silva, J.-I. Han and J.C.D. Santos. 2017. Hydrodynamic cavitation as an efficient pretreatment method for lignocellulosic biomass: A parametric study. Bioresource Technology 235: 301–308.

Torkamani, S.N., S. Wani, Y.J. Tang and R. Sureshkumar. 2010. Plasmon-enhanced microalgal growth in miniphotobioreactors. Applied Physics Letters 97: 043703.

Tran, D.-T., C.-L. Chen and J.-S. Chang. 2012. Immobilization of Burkholderia sp. lipase on a ferric silica nanocomposite for biodiesel production. Journal of Biotechnology 158: 112–119.

Tran, D.-T., C.-L. Chen and J.-S. Chang. 2013. Effect of solvents and oil content on direct transesterification of wet oil-bearing microalgal biomass of *Chlorella vulgaris* ESP-31 for biodiesel synthesis using immobilized lipase as the biocatalyst. Bioresource Technology 135: 213–221.

Tripathi, S.K., R. Kumar, S.K. Shukla, A. Qidwai and A. Dikshit. 2018. Exploring application of nanoparticles in production of biodiesel. pp. 141–153. *In*: Srivastava, N., M. Srivastava, H. Pandey, P.K. Mishra and P.W. Ramteke (eds.). Green Nanotechnology for Biofuel Production. Cham: Springer International Publishing.

Vandamme, D., I. Foubert and K. Muylaert. 2013. Flocculation as a low-cost method for harvesting microalgae for bulk biomass production. Trends in Biotechnology 31: 233–239.

Vassilev, S.V. and C.G. Vassileva. 2016. Composition, properties and challenges of algae biomass for biofuel application: an overview. Fuel 181: 1–33.

Verma, M.L., R. Chaudhary, T. Tsuzuki, C.J. Barrow and M. Puri. 2013. Immobilization of β-glucosidase on a magnetic nanoparticle improves thermostability: Application in cellobiose hydrolysis. Bioresource Technology 135: 2–6.

Vi, L.V.T., A. Salakkam and A. Reungsang. 2017. Optimization of key factors affecting bio-hydrogen production from sweet potato starch. Energy Procedia 138: 973–978.

Vicente, G., M. Martínez and J. Aracil. 2007. Optimisation of integrated biodiesel production. Part I. A study of the biodiesel purity and yield. Bioresource Technology 98: 1724–1733.

Wang, H., T. Deng, Y. Wang, X. Cui, Y. Qi, X. Mu, X. Hou and Y. Zhu. 2013a. Graphene oxide as a facile acid catalyst for the one-pot conversion of carbohydrates into 5-ethoxymethylfurfural. Green Chemistry 15: 2379–2383.

Wang, H., J. Covarrubias, H. Prock, X. Wu, D. Wang and S.H. Bossmann. 2015a. Acid-functionalized magnetic nanoparticle as heterogeneous catalysts for biodiesel synthesis. The Journal of Physical Chemistry C 119: 26020–26028.

Wang, S.-K., A.R. Stiles, C. Guo and C.-Z. Liu. 2015b. Harvesting microalgae by magnetic separation: a review. Algal Research 9: 178–185.

Wang, W., S. Ji and I. Lee. 2013b. Fast and efficient nanoshear hybrid alkaline pretreatment of corn stover for biofuel and materials production. Biomass and Bioenergy 51: 35–42.

Wang, X., G. Sun, P. Routh, D.-H. Kim, W. Huang and P. Chen. 2014. Heteroatom-doped graphene materials: syntheses, properties and applications. Chemical Society Reviews 43: 7067–7098.

Wang, Z.-L., J.-M. Yan, H.-L. Wang, Y. Ping and Q. Jiang. 2013c. Au@Pd core–shell nanoclusters growing on nitrogen-doped mildly reduced graphene oxide with enhanced catalytic performance for hydrogen generation from formic acid. Journal of Materials Chemistry A 1: 12721–12725.

Xie, W. and N. Ma. 2009. Immobilized Lipase on Fe_3O_4 Nanoparticles as biocatalyst for biodiesel production. Energy and Fuels 23: 1347–1353.

Xie, W. and J. Wang. 2014. Enzymatic production of biodiesel from soybean oil by using immobilized lipase on Fe_3O_4/Poly(styrene-methacrylic acid) magnetic microsphere as a biocatalyst. Energy and Fuels 28: 2624–2631.

Xie, W. and M. Huang. 2018. Immobilization of *Candida rugosa* lipase onto graphene oxide Fe_3O_4 nanocomposite: characterization and application for biodiesel production. Energy Conversion and Management 159: 42–53.

Xu, Z. and F. Huang. 2014. Pretreatment methods for bioethanol production. Applied Biochemistry and Biotechnology 174: 43–62.

Yang, G. and J. Wang. 2018. Improving mechanisms of biohydrogen production from grass using zero-valent iron nanoparticles. Bioresource Technology 266: 413–420.

Zang, L., J. Qiu, X. Wu, W. Zhang, E. Sakai and Y. Wei. 2014. Preparation of magnetic chitosan nanoparticles as support for cellulase immobilization. Industrial and Engineering Chemistry Research 53: 3448–3454.

Zdarta, J., A. Meyer, T. Jesionowski and M. Pinelo. 2018. A general overview of support materials for enzyme immobilization: characteristics, properties, practical utility. Catalysts 8: 92.

Zhang, L., L. Zhang and D. Li. 2015a. Enhanced dark fermentative hydrogen production by zero-valent iron activated carbon micro-electrolysis. International Journal of Hydrogen Energy 40: 12201–12208.

Zhang, W., J. Qiu, H. Feng, L. Zang and E. Sakai. 2015b. Increase in stability of cellulase immobilized on functionalized magnetic nanospheres. Journal of Magnetism and Magnetic Material 375: 117–123.

Zhang, X.L., S. Yan, R.D. Tyagi and R.Y. Surampalli. 2013. Biodiesel production from heterotrophic microalgae through transesterification and nanotechnology application in the production. Renewable and Sustainable Energy Reviews 26: 216–223.

Zhang, Y. and J. Shen. 2007. Enhancement effect of gold nanoparticles on biohydrogen production from artificial wastewater. International Journal of Hydrogen Energy 32: 17–23.

Zhao, W., Y. Zhang, B. Du, D. Wei, Q. Wei and Y. Zhao. 2013. Enhancement effect of silver nanoparticles on fermentative biohydrogen production using mixed bacteria. Bioresource Technology 142: 240–245.

Zhao, X., J. Wang, C. Chen, Y. Huang, A. Wang and T. Zhang. 2014. Graphene oxide for cellulose hydrolysis: how it works as a highly active catalyst? Chemical Communications 50: 3439–3442.

Zhou, Y., C. Jin, Y. Li and W. Shen. 2018a. Dynamic behavior of metal nanoparticles for catalysis. Nano Today 20: 101–120.

Zhou, Y., G. Quan, Q. Wu, X. Zhang, B. Niu, B. Wu, Y. Huang, X. Pan and C. Wu. 2018b. Mesoporous silica nanoparticles for drug and gene delivery. Acta Pharmaceutica Sinica B 8: 165–177.

Zhu, S., J. Wang and W. Fan. 2015. Graphene-based catalysis for biomass conversion. Catalysis Science and Technology 5: 3845–3858.

Ziolkowska, J.R. 2018. Introduction to biofuels and potentials of nanotechnology. *In*: Srivastava, N., M. Srivastava, H. Pandey, P.K. Mishra and P.W. Ramteke (eds.). Green Nanotechnology for Biofuel Production. Cham: Springer International Publishing.

Chapter 7

Nanobiotechnology of Ligninolytic and Cellulolytic Enzymes for Enhanced Bioethanol Production

***Pardeep Kaur*[1,*] and *Gurvinder Singh Kocher*[2]**

1. Introduction

Biofuels, generally defined as any energy-enriched chemical derived from biomass, combine various unique characteristics such as renewable energy sources, biodegradability, low toxicity, diversity and easy and local availability. The biofuels not only represent an alternative to the depletion of fossil fuel resources, but their combustion is considered to be carbon neutral unlike the combustion processes of fossil fuels which produce a majority of CO_2 emissions in the Earth's atmosphere (Zuliani et al. 2018). In recent years, almost all bioethanol produced in the world has a starch or sugar-based plant origin which comes under the category of first-generation biofuels for which edible plants are used as feedstock, such as starch from corn, wheat, rice, etc., and sucrose from sugarcane and sugar beet (Naik et al. 2010). Even though these primary sources of biomass are the most exploited, inherent competition between foods versus fuels is highly debatable (Leo et al. 2016). The second-generation biofuel, on the other side, does not compete against food supplies as they are based on the non-food raw material. The second-generation biofuel production is typically from lignocellulosic biomasses like wood wastes, perennial grasses, forest litters, agricultural residues and others (Robak and Balcerek 2018) that are majorly dominated by cellulosic components followed by hemicelluloses and less amount of lignin. As these polymeric carbohydrates are relatively difficult to hydrolyze to simpler sugar forms and subsequently, their conversion into ethanol is not only challenging but also slightly time-consuming and costly (Wongwatanapaiboon et al.

[1] Department of Biotechnology, Sri Guru Granth Sahib World University, Fatehgarh Sahib-140407, India.
[2] Department of Microbiology, Punjab Agricultural University, Ludhiana-141004, India.
* Corresponding author: pardeepkaur2108@gmail.com

2012). Thus, given these facts, there is an expansive demand to develop proficient technologies capable of resolving the issues that have raised in the field of bioethanol production. Nanotechnology is one of the most growing areas of research in biofuels and bioenergy fields, especially the subarea of nano-cellulosic materials which have attracted considerable attention from researchers in the past decade. Nanotechnology has different applications, such as modification in feedstocks, development of more efficient catalysts, development of nanomaterials and others that hold great potential to resolve the major bottlenecks of biofuels and bioenergy field (Rai et al. 2016). The nanomaterials possess exceptional characteristics such as high surface areas, a high degree of crystallinity, catalytic activity, stability, adsorption capacity, durability and efficient storage which could collectively help to optimize the overall system and can widely be exploited in biofuel systems. They also have a high potential for recovery, reusability and recycling. Another promising application of nanotechnology in the biofuel industry is enzyme (biocatalysts) immobilization during lignocellulosic ethanol production processes. The advantages nanostructures offer in this area include a large surface area for high enzyme loading, higher enzymatic stability and possibility of enzyme reusability which could reduce the operational cost of large-scale biofuel production plants (Kim et al. 2018, Nizami and Rehan 2018). In this context, the chapter aims to focus on the application of nanotechnology to biofuels production, especially highlight the immobilization of ligninolytic and cellulolytic enzymes followed by a discussion about safety issues concerning this technology.

2. Bioethanol

The tremendous increase in the world's population and growing world economy are the key drivers behind the expansion in global energy demand. According to the U.S. Energy Information Administration (EIA), energy consumption is projected to increase by an average of 1.4% per year. Energy consumption rates for China, India and Africa—the three most populous places on the globe—are projected to grow faster than the rest of the world by 2040 and pose significant implications for global energy markets (US EIA 2018). In India, oil takes the lead, after coal, as the largest energy source sharing about 30.5% of primary energy consumption (BP Energy outlook 2030, 2013). The increase in urbanization, infrastructure development and rising per capita income have caused a manifold increase in vehicle density that has consequently increased the demand for gasoline. The transport sector's stake in the total primary energy consumption will increase from 8.1%, as noted in 2010, to 11.3% by 2030 (UNEP 2015). Currently, an estimated 46% of transportation fuel demand is fulfilled by diesel alone followed by gasoline at 24%, the demand of which is estimated to increase from 134 billion liters in 2015 to 225 billion liters by 2026 (Biofuels Annual 2016). To date, the energy needs of the global economy are met solely by fossil fuels, which is estimated to contribute 60% of the growth in energy and 80% of total energy supply in 2035 (BP Energy outlook 2017). The burning of fossil fuels, especially gasoline, releases carbon monoxide, nitrogen oxides, particulate matters and unburned hydrocarbons which are responsible for environmental pollution. Besides, carbon dioxide, a greenhouse gas linked to global warming and ammonia are also emitted, which in excess can enter into water bodies and cause problems of algal blooms and oxygen-deprived aquatic zones (US EPA 2016).

Ethanol, an oxygenated fuel with high octane value, is an attractive substitute to not only petroleum but also to the common toxic supplement in gasoline, i.e., methyl tertiary butyl ether (MTBE) (Wheals et al. 1999). Petroleum-based automobile fuels blended with ethanol can be used to run combustion engines at higher compression, thus providing superior performance. The bioethanol blending strategy will not only help to limit the dependency on petroleum but also reduces greenhouse gas (GHG) emission to an extent of 85% (da Silva and Alvarez 2004). Bioethanol is considered as a potential alternative to fossil fuels as it is a clean, safe and renewable resource (Rezania et al. 2015). Its production from different renewable sources has already been introduced on a large scale in Brazil, Europe, the USA and China. It is expected that around 33% of the energy needs of Europe and the USA for different transportation purposes will be satisfied by converting biomass to biofuels by 2030 (Goncalves et al. 2015). The Government of India (GoI) in January 2003 launched an Ethanol Blended Petrol Program (EBPP) in nine states and four union territories (UTs) to promote the blending of bioethanol and non-edible oils with gasoline and diesel, respectively. During the year 2004–05, the blending mandate was made discretionary due to its shortage but reinstituted in the second stage of EBPP (October 2006) in twenty states and seven UTs. These ad-hoc policy changes continued until December 2009 when GoI introduced an all-inclusive 'National Policy on Biofuels' frame worked by the Ministry of New and Renewable Energy (MNRE). While diesel biofuel blending is nearly zero, the petrol blending stands at an overall of about 3.0% in the form of first-generation (1G) or molasses-based ethanol. The annual requirement of 1G-ethanol stands at about 500 crore liters, but the current total installed capacity is about 265 crore liters. In such a scenario, the target of 20% blending by 2020 looks remote unless agricultural waste-based bioethanol, i.e., second-generation (2G) bioethanol production technologies are successfully demonstrated (IESS 2047 2016).

3. Lignocellulosic Biomass: Substrate for Bioethanol Production

Lignocellulose is the main and most abundant component of the renewable biomass produced by photosynthesis; it is estimated that 200 billion tons are produced annually in the world. It consists of three main biopolymers that form the cell wall of plants: cellulose, hemicellulose and lignin (Valdez-Vazquez and Sanchez 2017). Cellulose is a long-chain homopolymer of anhydrous glucose molecules joined together by ß-1,4 linkages. Hemicellulose is a heterogeneous short-chain polymer of different sugars (xylans, mannans and galactans) and sugar acids. It is amorphous in its natural form. Lignin, on the other hand, is a highly branched polymer of phenyl-propane units joined by carbon-carbon and ether bonds. The composition of lignin (ratio of different phenyl-propane monomers) varies among different types of biomass and woods. In general, lignin from softwood has higher guaiacyl units compared to syringyl units while hardwood lignin contains guaiacyl and syringyl

Table 1. Types of lignocelluloses and their chemical composition.

Lignocellulosic Biomass	Examples	Cellulose (%)	Hemicellulose (%)	Lignin (%)	References
Hard wood	Poplar	50.8–53.3	26.2–28.7	15.5–16.3	Malherbe and Cloete 2002
	Oak	40.4	35.9	24.1	
	Eucalyptus	54.1	18.4	21.5	
	Hardwood bark	22–40	20–38	30–35	
Soft wood	Pine	42.0–50.0	24.0–27.0	20.0	Malherbe and Cloete 2002
	Douglas fir	44.0	11.0	27.0	
	Spruce	45.5	22.9	27.9	
	Softwood bark	18–38	15–33	30–60	
Agricultural waste	Wheat Straw	35.0–39.0	23.0–30.0	12.0–16.0	McKendry (2002)
	Barley Straw	36.0–43.0	24.0–33.0	6.3–9.8	Saini et al. 2015
	Rice Straw	29.2–34.7	23.0–25.9	17.0–19.0	Prasad et al. 2007
	Rice Husks	28.7–35.6	12.0–29.3	15.4–20.0	Saini et al. 2015
	Corn Cobs	33.7–41.2	31.9–36.0	6.1–15.9	Saini et al. 2015
	Corn Stalk	35.0–39.6	16.8–35.0	7.0–18.4	Prasad et al. 2007
	Sugarcane	25.0–45.0	28.0–32.0	15.0–25.0	Saini et al. 2015
	Bagasse	32.0–35.0	24.0–27.0	15.0–21.0	
	Nut Shells	25–30	25–30	5–30	Kim and Day 2011
	Cotton seed hairs	34–35	18–28	14–22	Howard et al. 2003
	Sweet Sorghum Bagaase	36–43	0.15–0.25	41–45	Saini et al. 2015
	Coir				Saini et al. 2015
Grasses	Grasses	25.0–40.0	25.0–50.0	10.0–30.0	Lee et al. 2014,
	Switchgrass	35.0–40.0	25.0–30.0	15.0–20.0	Malherbe and Cloete 2002
Others	Primary wastewater solids	8–15	NA	24–29	Lee et al. 2014
	Newspaper	40–55	25–40	18–30	Howard et al. 2003
	Solid cattle manure	1.6–4.7	1.4–3.3	2.7–5.7	Sun and Cheng 2002
	Sorted refuse	60	20	20	Sun and Cheng 2002
	Swine waste	6.0	28	NA	
	Waste paper from chemical pulps	60–70	10–20	5–10	Saini et al .2015

units in comparable amounts (Brebu and Vasile 2010). The lignin does not consist of fermentable sugars and plays an important role in providing a recalcitrant structure that is difficult to disrupt. These structural properties of lignocelluloses make the pretreatment step essential to improve its digestibility and increase the release of fermentable sugars (Chen et al. 2017).

3.1 Cellulose

The major component of lignocellulosic biomass is cellulose. Unlike glucose in other glucan polymers, the repeating unit of the cellulose chain is the disaccharide cellobiose. Its structure consists of extensive intramolecular and intermolecular hydrogen bonding networks, which tightly binds the glucose units. Since about half of the organic carbon in the biosphere is present in the form of cellulose, the conversion of cellulose into fuels and valuable chemicals has paramount importance.

3.2 Hemicellulose

Hemicellulose is the second most abundant polymer. Unlike cellulose, hemicellulose has a random and amorphous structure, which is composed of several heteropolymers including xylan, galactomannan, glucuronoxylan, arabinoxylan, glucomannan and xyloglucan. They differ in composition too: hardwood hemicelluloses contain mostly xylans, whereas softwood hemicelluloses contain mostly glucomannans. The heteropolymers of hemicellulose are composed of different 5- and 6-carbon monosaccharide units: pentoses (xylose and arabinose), hexoses (mannose, glucose and galactose) and acetylated sugars. Hemicelluloses are embedded in the plant cell walls to form a complex network of bonds that provide structural strength by linking cellulose fibers into microfibrils and cross-linking with lignin.

3.3 Lignin

The lignin is a three-dimensional polymer of phenylpropanoid units. It functions as the cellular glue which provides compressive strength to the plant tissue and the individual fibers, stiffness to the cell wall and resistance against insects and pathogens. The oxidative coupling of three different phenylpropane building blocks; monolignols, *p*-coumaryl alcohol, coniferyl alcohol and sinapyl alcohol, forms the structure of lignin. The corresponding phenylpropanoid monomeric units in the lignin polymer are identified as *p*-hydroxyphenyl (H), guaiacyl (G) and syringyl (S) units, respectively (Kocher et al. 2017).

4. Bioethanol Production

The production process of bioethanol from lignocellulosic biomass consists of four major operations: (1) pretreatment, (2) hydrolysis of carbohydrates to sugar monomers, (3) fermentation of sugars (4) and ethanol recovery. Cellulose is embedded in a complex matrix of hemicellulose and lignin necessitating a pretreatment (physical, chemical, phyiscochemical or biological) process to reduce the biomass recalcitrance. Cellulose is difficult to degrade due to its complex and highly crystalline structure. During pretreatment, the biomass matrix is opened, improving the enzyme accessibility to cellulose. During hydrolysis (also known as 'saccharification') of cellulose, long chains of glucose molecules are broken down into glucose monomers which are further fermented to ethanol. The hydrolysis of cellulose and sugar fermentation can be combined into a single step process known as simultaneous saccharification and fermentation (SSF). A large fraction of hemicellulose is usually hydrolyzed during pretreatment processes performed at $pH < 7.0$ which can be fermented separately or in combination with the SSF process

(SSCoF process). Hydrolysis of cellulose is a crucial step in the overall bioethanol production pathway due to its large cost contribution, mainly because of its much higher amount of enzyme usage (40 to 100 times than that of starch hydrolysis). The two main approaches for the conversion of lignocellulose to ethanol are: 'acid-based' and 'enzyme-based' (Galbe and Zacchi 2002, Licht 2006, Hahn-Hägerdal et al. 2007). Biomass hydrolysis, i.e., the depolymerization of the biomass polysaccharides to fermentable sugars must be performed via environmentally friendly and economically feasible technologies (Lynd et al. 2005). The enzyme-based ethanol production has an advantage over chemical procedure because of its higher conversion efficiency, the absence of substrate loss due to chemical modifications and the use of more moderate and non-corrosive physical-chemical operating conditions (Acharaya and Chaudhary 2012).

4.1 *Biological Delignification*

Several pretreatment technologies have been developed to overcome the lignin barrier. These pretreatments produce other effects apart from increasing the digestibility of lignocellulose, such as hemicellulose solubilization and/or degradation by wet oxidation, acid pretreatment and steam explosion and cellulose decrystallization by ammonia fiber explosion (Alvira et al. 2010). Biological delignification is a promising technology due to the low environmental impact, higher product yield, mild reaction conditions, few side reactions, less energy demand and reduced reactor requirements to resist pressure and corrosion (Kocher et al. 2017). Moreover, bio-delignification also avoids the formation of degradation compounds that inhibit the subsequent steps. A good number of microorganisms have been reported for their delignification potential using a wide range of biomass (Tuomela et al. 2000). Table 2 presents different ligninolytic microorganisms involved in the delignification process. Some of the most promising microorganisms for biological delignification are white-rot fungi which can mineralize lignin into CO_2 and water in pure culture (Lundquist et al. 1977, Hatakka 1983, Martínez et al. 2005, Isroi et al. 2011). These fungi have developed an extracellular and unspecific oxidative enzymatic system for lignin degradation which involves the action of manganese peroxidise, laccase, versatile peroxidise and lignin peroxidase.

4.2 *Ligninolytic Enzymes*

4.2.1 Laccase

Yoshida, by the end of the nineteenth century, reported extraction of laccases from exudates of Japanese tree, *Rhusvernicifera*. Laccases (E.C. 1.10.3.2), classified as benzenediol: oxygen oxidoreductase or p-diphenol oxidases, are blue copper-containing oxidases that bring about the single-electron oxidation of aromatic amine and polyphenolic sub-structures of lignin. Laccase is glycoproteins with a molecular weight of 60–70 Kda and consist of three cupredoxin like domains. Their primary structure consists of 500 amino acid residues, organized in three successive domains

Table 2. Ligninolytic microorganisms.

Organisms	Subdivision	Lignin Degradation	Environment	Genera	References
Brown rot fungi	Basidiomycotina	Lignin modification	Softwoods and coniferous ecosystems	*Gloeophyllum Trabeum, Postia placenta, Serpula Lacrymans, Coniophora Puteana* and *Coniophoraputeana*	Dashtban et al. 2010, Daniel 2016, Hammel 1996, Kantharaj et al. 2017
Soft rot fungi	Ascomycotina or deuteromycotina	Limited lignin degradation	Aquatic environment, plant litter	*Chaetomium Globosum, Trichoderma reesei, Paecilomyces* and *Fusarium*	Daniel 2016, Kantharaj et al. 2017
White rot fungi	Basidiomycotina (Ascomycotina)	Lignin mineralization, selective or nonselective delignification	Hardwood of angiosperms	*Ceriporiopsissubvermispora, Coriolus versicolor, Cyathusstercoreus, Phanerochaete chrysosporium, PhlebiasubserialisAuricularia auricula-judae* and *Pleurotusostreatus*	Daniel 2016, Dashtban et al. 2010, Hammel 1996, Placido and Capareda 2015, Pozdnyakova 2012
Bacteria	Actinomycetes, myxobacteria	Low lignin degradation	Sapwood, water-saturated wood, wood at late stage of decomposition, plant litter	*Streptomyces, Nocardia, Pseudomonas, Cupriavidus Basilensis* B-8	Dashtban et al. 2010, de Gonzalo et al. 2016

with β barrel topology. They have four copper (Cu) molecules that participate in oxygen reduction and water production. Laccases catalyze the single-electron oxidation of phenols to phenoxy radicals by transferring four electrons to O_2 in the process. The results of the electron oxidation of lignin-related phenols include Cα-oxidation, limited demethoxylation and aryl-Cα cleavage. Laccase oxidation can also lead to Cα-Cβ cleavage in certain phenolic structures. However, coupling/polymerization is a major consequence of laccase oxidation of lignin-associated phenols and isolated lignins (Kocher et al. 2017).

4.2.2 Lignin Peroxidase

LiPs (E.C. 1.11.1.14) were originally discovered in nitrogen and carbon-limited cultures of *Phanerochaete chrysosporium*. LiP possess high redox potential (700 to 1,400 mV), low optimum pH 3 to 4.5 and the ability to catalyze the degradation of a wide number of aromatic structures such veratryl alcohol (3,4-dimethoxybenzyl) and methoxybenzenes (Piontek et al. 1993, Angel 2002, Dias et al. 2007). They oxidize aromatic rings moderately activated by electron-donating substitutes; in contrast, common peroxidase participates in the catalysis of highly activated aromatic substrates (ammine, hydroxyl, etc.). An explanation for this type of catalysis is the production of veratrylalcohol radicals. These radicals have higher redox potential than LiP's compounds I and II and can participate in the degradation of compounds with high redox potential (Khindaria et al. 1996). LiPs are monomeric glycosylated enzymes of 40 kDa, with 343 amino acid residues, 370 water molecules, heme group, four carbohydrates and two calcium ions. Their secondary structure is principally helicoidal. It contains eight major helixes, eight minor helixes and two anti-parallel beta-sheets. LiPs contain two domains at both sides of the heminic group. This group is inlaid in the protein but accessible to solvents via two small channels (Choinowski et al. 1999). The heminic cavity includes 40 residues, and it bonds to the protein via hydrogen bridges. Additionally, the heminic iron (Fe) coordinates with a His and a molecule of water. This His is associated with the high redox potential of LiP. The enzyme's redox potential rises when the His has a reduced imidazol character. In addition to that, a greater distance between the His and the heminic group increases the redox potential of the enzyme. This increment in the redox potential is a response to the electronic deficiency in the Fe of the porphyrin ring. In fact, this distance causes most differences among enzymes with similar porphyrin cores (Choinowski et al. 1999, Hammel and Cullen 2008). Another characteristic related to LiPs' high redox potential is the invariant presence of a tryptophan residue (Trp171) in the enzymes' surface. The Trp171 seems to facilitate electronic transference to the enzyme from substrates that cannot access into the heminic oxidative group (Hammel and Cullen 2008). Additionally, the Trp171 participates with the catalysis of veratryl alcohol, a metabolite produced by some ligninolytic fungi. Veratryl alcohol participates in the oxidation of different aromatic molecules. Some researchers conceptualized that this alcohol protects the enzyme from the action of H_2O_2 and participates as a redox mediator between the enzyme and substrates which cannot get inside the heminic center (Choinowski et al. 1999, Angel 2002).

4.2.3 Manganese Peroxidase

Kuwahara in 1984 found the first MnP (E.C. 1.11.1.13) in batch cultures of *Phanerochaete chrysosporium* (Dias et al. 2007). They are glycoprotein with a molecular weight between 38 and 62.5 kD, approximately 350 amino acid residues and a 43% of identity with LiP sequence. MnP structure has two domains with the heminic group in the middle, ten major helixes, a minor helix and five disulfide bridges. One of these bridges participates in the manganese (Mn) bonding site. This site is a characteristic feature that distinguishes MnP from other peroxidases (Sundaramoorthy et al. 1994). The enzyme's catalytic cycle starts with the transference of two electrons from the heminic group to H_2O_2; it produces compound I and water. After that, compound I catalyzes the oxidation of one substrate molecule with the production of a free radical and compound II. Compound II oxidizes Mn^{2+} to produce Mn^{3+}, the cation responsible to oxidize aromatic compounds. The compound II demands Mn^{2+} presence for its reaction; in contrast, compound I can oxidize Mn^{2+} or the substrate. After Mn^{3+} is stabilized by organic acids, it reacts non-specifically with organic molecules by removing an electron and a proton from the substrates (Martin 2002). Mn^{3+} is a small-sized compound with high redox potential which diffuses easily in the lignified cell wall. Therefore, Mn^{3+} starts the attack inside the plant cell wall which facilitates the penetration and action of the other enzymes.

4.2.4 Versatile Peroxidases

VP (EC 1.11.1.16) is a glycoprotein that is capable of oxidizing substrates of other basidomycetes' peroxidases, viz., Mn(II) and VA for MnP and LiP, respectively. Due to its dual oxidative ability, VP forms an attractive ligninolytic enzyme group (Ruiz-Duenas et al. 2009). It also oxidizes high redox-potential compounds such as the dye, Reactive Black 5 (RB5) and a wide variety of phenols including hydroquinones (Heinfling et al. 1998). The fact that makes VP superior to both LiP and MnP is that it can oxidize substrates with a wide range of redox potential. This is a consequence of its hybrid molecular structure that provides multiple binding sites for the substrates. The Mn(III) is released from VP that acts as a diffusible oxidizer of phenolic lignin as well as free phenol substrates. The heme of the enzyme is placed in the interior which is made available to the outside medium via two channels. The function of the first channel is akin to the one mentioned for LiP. The second channel participates in the oxidation of Mn(II) to Mn(III). Typical examples include the species of *Pleurotus* such as *P. eryngii* (Ruiz-Duenas et al. 2009), *P. ostreatus* (Cohen et al. 2011), *Bjerkandera adusta* (Mester and Field 1998) and *Bjerkandera fumosa.*

4.2.5 Cellulases

The cellulase is the key enzyme for the conversion of cellulose into simple sugars (Chinedu et al. 2005). Cellulase is a family of enzymes hyrolysing β-1,4-glycosidic bonds of intact cellulose and other related cello-oligosaccharide derivatives. They are found significant potentials for industrial applications, especially in sectors of foods, chemicals, detergents, cosmetics, pulp, paper, etc. (Sette et al. 2008). Synergistic action of three principal types of the enzymes, viz., endoglucanase (EC 3.2.1.4), exoglucanase (EC 3.2.1.91) and β-glucosidase (EC 3.2.1.21) are required to accomplish the degradation of intact hydrogen-bond-ordered cellulose.

4.3 Classification of Cellulases

Cellulase is a complex enzyme system comprising of endo-1,4-β-Dglucanase (endoglucanase, EC 3.2.1.4), exo-1,4-β-D-glucanase (exoglucanase, EC 3.2.1.91) and β-D-glucosidase (β-D-glucoside glucanhydrolase, EC 3.2.1.21) (Joshi and Pandey 1999).

4.3.1 Endoglucanase

Endoglucanase (endo-β-1,4-D-glucanase, endo-β-1,4-D-glucan-4-glucano-hydrolase), often called CMCase which hydrolyzes carboxymethyl cellulose (CMC) or swollen cellulose randomly. Accordingly, the length of the polymer decreases, resulting in an increase of reducing sugar concentration (Robson 1989, Begum et al. 2009, Szijártó et al. 2004). Endoglucanase also acts on cellodextrins, the intermediate product of cellulose hydrolysis and converts them to cellobiose (disaccharide) and glucose. These enzymes are inactive against crystalline celluloses such as cotton or avicel.

4.3.2 Exoglucanase

Exoglucanase (exo-β-1,4-D glucanase, cellobiohydrolase) degrades cellulose by splitting-off the cellobiose units from the non-reducing end of the chain. It is also active against swollen, partially degraded amorphous substrates and cellodextrins but does not hydrolyze soluble derivatives of cellulose like carboxymethyl cellulose and hydroxyethyl cellulose. Some cellulase systems also contain glucohydrolase (exo-1,4-D-glucan-4-glucohydrolase) as a minor component (Joshi and Pandey 1999).

4.3.3 β-glucosidase

β-glucosidase completes the process of hydrolysis of cellulose by cleaving cellobiose and removing glucose from the non-reducing end (i.e., with a free hydroxyl group at C-4) of oligosaccharides. The enzyme also hydrolyzes alkyl and aryl β-glucosides (Kubicek et al. 1993). Apart from these, there are several other enzymes, viz., glucuronide, acetylesterase, xylanase, β-xylosidase, galactomannase and glucomannase that disrupts hemicelluloses. The combination of these enzymes can be used for the hydrolysis of both cellulose and hemicelluloses (Verardi et al. 2012).

4.4 Cellulolytic Microorganisms

Cellulase can be produced via biological route using bacterial or fungal fermentation. There is a wide range of microorganisms, capable of producing cellulases such as aerobic and anaerobic bacteria, soft rot fungi, white rot fungi (WRF) and brown rot fungi (BRF). Bacteria belonging to genera of *Clostridium, Cellulomonas, Bacillus, Thermomonospora, Ruminococcus, Bacteriodes, Erwinia, Acetovibrio, Microbispora* and *Streptomyces* are known to produce cellulose (Bisaria 1998). Anaerobic bacterial species such as *Clostridium phytofermentans, Clostridium thermocellum, Clostridium hungatei* and *Clostridium papyrosolvens* produces cellulases with high specific activity (Duff and Murray 1996, Bisaria 1998). Most of the fungi can produce a complete cellulase system as compared to bacteria. The commercial cellulase is most commonly produced from two strains of soft rot fungi (SRF), namely *Trichoderma reesei* and *Aspergillus niger* (Kaur et al. 2007). Fungi known to produce cellulases include *Sclerotium rolfsii, Phanerochaete chrysosporium* and various species of *Trichoderma, Aspergillus, Schizophyllum* and *Penicillium* (Fan et al. 1987, Duff and Murray 1996).

5. Nanobiotechnology: Application in Bioethanol Production

Nanobiotechnology is the developing branch of science, which is applied for the assessment of new technological replacements. It is the most significant study in modern science, which allows chemists, engineers, biotechnologists and physicians to work at molecular and cellular levels. Nanoparticles have been used extensively in many applications such as biomedicine, targeted drug delivery biosensors, water purification, protein immobilization and environmental remediation. Current researches have indicated that nanotechnology applied in nanomaterials can exhibit advanced properties that are exceptional in science (Engelmann et al. 2013). Even though the new generation of biofuels have advantages such as the high availability of agricultural and forest residues around the world, its production technologies also present many challenges such as (i) the cost for the production is high compared to the first-generation biofuels; (ii) the existing infrastructure is not sufficient for the production process; (iii) the production has many technological barriers and therefore, more efforts are needed to develop the processes related to enzymes, pretreatment and fermentation in order to make these more cost-effective and energy-efficient. Considering the need for research efforts for the evaluation of new technological alternatives, nanotechnology could offer meaningful solutions by changing the characteristics of feed materials and by the intervention of biocatalysts. Different nanomaterials such as carbon nanotubes, magnetic and metal oxide nanoparticles and others are advantageous to become an essential part of sustainable bioenergy production (Rai et al. 2016).

5.1 Nanomaterials and their Advantages

For industrial biotechnological applications, various nanomaterials such as nanofibers, nanoparticles, nanosheets, nanotubes, nanopores and nanocomposites are being used. Their high surface area can be used for a higher amount of enzyme to be immobilized; as a result, better biocatalytic activity and stability could be achieved. An enzyme immobilized on nanoparticle offers low mass transfer resistance in comparison to macro-scale matrices which result in enhanced stability and activity of an enzyme. Moreover, magnetic nanomaterials allow quick and efficient removal by the use of a magnet that facilitates the quick separation of the enzyme from the product. Unlike other conventional methods (centrifugation and filtration) that lead to enzyme instability due to mechanical shearing, the use of magnetic nanoparticles provides an efficient and economical option for the process (Ren et al. 2011, Yiu and Keane 2012).

5.2 Enzyme Immobilization Techniques

The nanoparticle-based immobilized enzymes are collectively known as 'nanobiocatalyst'. The development of nanobiocatalyst is an example of emerging and innovative breakthroughs of nanotechnology and biotechnology (Rai et al. 2018). Many metal oxides are currently used for the immobilization of cellulose (Mei et al. 2009). The following are the techniques adopted for immobilization (Kim et al. 2018).

5.2.1 Cross-Linking Immobilization

The cross-linking immobilization process attaches the enzymes using a multifunctional reagent (Ahmad and Sardar 2015). This method does not require a support matrix and the resulting enzyme maintains 100 percent activity (Sheldon 2007). However, a loss of enzyme activity via conformational change can occur during immobilization. The control of the cross-linking reaction is difficult; thus, it is not easy to obtain an enzyme with high activity retention.

5.2.2 Adsorption Immobilization

The adsorption of the enzyme on the surface of a support is old technology and a simple method. It is based on a physical binding mechanism, such as a dipole-dipole, hydrophobic, or Van der Waals interaction or hydrogen bonding (Hwang and Gu 2013, Eş et al. 2015). Physical binding can be performed in relatively ambient conditions and show a high enzyme loading (Ahmad and Sardar 2015, Eş et al. 2015). Adsorption immobilization does not provide high stability and might cause a loss of enzyme molecules during operation and washing because of weak binding between the enzyme and the supports (Tang et al. 2004).

5.2.3 Covalent Immobilization

The covalent immobilization of the biocatalyst is the attachment of enzyme to the nanomatrix by covalent bonding between the enzyme and the supports (Ahmad and

Table 3. Type of enzymes and respective immobilized nanoparticles for bioethanol production.

Enzyme	**Immobilization of Nanoparticle**	**Reference**
Cellulase	Fe_3O_4 TiO_2	Ahmad and Sardar 2014, Gao et al. 2008, Khoshnevisan et al. 2017
Xylanase	SiO_2	Dhiman et al. 2012
Laccase	Magnetic chitosan NPs Silica NPs Titania NPs	Fang et al. 2008, Ji et al. 2017 Wang et al. 2009, Zhang et al. 2017
Hemicellulase	Trimethoxysilyl	Hegedus 2011
ß-D-glucosidase	Fe_3O_4 Superparamagnetic magnetite	Chen et al. 2014, Vaenzuela et al. 2014

Sardar 2015). The strong binding of the enzyme to the support matrix via the covalent bond prevents enzyme leaching from the surface and improves the thermal stability in some cases (Ahmad and Sardar 2015, Eş et al. 2015). This technique, however, often provokes the deactivation of the enzyme because of the conformational restriction of the enzyme by covalent binding (Eş et al. 2015, Hong et al. 2007).

5.2.4 Entrapment Immobilization

The entrapment technology entraps the enzyme in a porous gel or fibers (Hwang and Gu 2013, Ahmad and Sardar 2015). The entrapment process can protect enzyme activity because of the indirect contact with the confined environment, which minimizes the effects of gas bubbles, mechanical sheer and hydrophobic solvents (Hwang and Gu 2013). Entrapment immobilizations using nanoparticles are generally based on the reverse-micelle or sol-gel technique (Daubresse et al. 1994, Reetz et al. 1998, Ma et al. 2004, Yang et al. 2004, Kim et al. 2008).

5.3 Potential Nanoparticles for Bioethanol Production

Due to unique size and physicochemical properties, nanoparticles have advantageous applications in bioethanol production. In current nanotechnology, the development of trustworthy protocols for the synthesis of different nanomaterials with small size and high monodispersity are interesting issues (Mandal et al. 2005). Many nanoparticles like Fe_3O_4, TiO_2, ZnO, SnO_2, carbon, fullerene and graphene have been used in sugar and alcohol production.

5.3.1 Magnetic Nanoparticles

Magnetic nanoparticles also have extensive uses in biotechnology, biomedical, material science, engineering and environmental areas. Magnetic nanoparticles retain exceptional properties including their high surface-to-volume ratio, quantum properties and ability to carry other molecules due to their small size. A most important advantage of magnetic nanoparticles over other metal nanoparticles is that these can be easily removed or recovered by applying an appropriate magnetic field, which reduces the probabilities of nanotoxicity (Ahmed and Douek 2013). Studies that were carried out by using magnetic nanoparticles were reported for the immobilization of enzymes involved in bioethanol production. Generally, enzyme immobilization on nanoparticles is achieved by covalent binding or physical adsorption. However, the covalent binding method is more suitable because it reduces protein desorption due to the formation of covalent bonds between enzyme and nanoparticles (Abraham et al. 2014). For stable immobilization of the enzyme on nanomaterials, these compounds need to be modified or coated with a chemically active polymer to provide the functional group for linkage of the enzyme. The effect of MNPs on the catalytic behavior and stability of enzyme strongly depend on the nanomaterials' characteristics, such as size, structure and functionalization (Schwaminger et al. 2017). The surface chemistry of these nanomaterials can influence their interactions

Table 4. Different approaches for cellulase immobilization.

Immobilization Approach	Characteristics	References
Silica-based surface functionalization	Improved chemical stability and biocompatibility, preventing the aggregation of nanoparticles.	Li et al. 2014, Lee et al. 2014, Tao et al. 2016
Graphene-based surface functionalization	Unique physical and electronic properties, providing large surface area for biomolecules to anchor, high immobilization yield and efficiency.	Li et al. 2015, Gokhale et al. 2015, Gao et al. 2018
Silanization-based surface functionalization	Generation a modified surface exposing amino groups as adsorbent or as coupling sites.	Huang et al. 2015
Chitosan-based surface functionalization	Providing suitable surface for biomolecules to anchor.	Sánchez-Ramírez et al. 2017, Zhang et al. 2014
Co-immobilization	Simultaneous co-immobilization of various biomolecules.	Cho et al. 2012, Honda et al. 2015, Sojitra et al. 2016
Carrier-free immobilization	Novel strategies to improve the robust immobilized biocatalysts.	Jafari Khorshidi et al. 2016
Physical adsorption and covalent coupling on Iron oxide	Better immobilization efficiency, thermo-tolerance and catalytic efficiency.	Kumar et al. 2018
Electrostatic assembly of cellulase and low-priced silica-coated magnetic nanoparticles	Non-leaching of biocatalyst and high recovery yields.	Roth et al. 2016
Chitosan-coated magnetic nanoparticles modified with alpha-ketoglutaric acid (alpha-KA-CCMNPs)	A broader pH range of high activity and thermostability.	Zhou et al. 2009

with protein molecules and thus affect the adsorption as well as the conformation and biological function of conjugated enzymes (Patila et al. 2016). For instance, the immobilization of cellulase on superparamagnetic nanoparticles (SPMNPs) via ionic bound was achieved to enhance enzyme stability (Khoshnevisan et al. 2011). In another study, activated magnetic support using zinc doped was applied cellulase immobilization to increase the saturation and the magnetization of nanoparticles (Abraham et al. 2014). A series of porous terpolymers with cross-linking via suspension and polymerization was synthesized for increased loading of the enzyme (Qi et al. 2015).

Lee et al. (2009) demonstrated the immobilization of the β-glucosidase enzyme on polymer magnetic nanofibers by entrapment method for cellulosic ethanol production. β-glucosidase is the enzyme responsible for the conversion of cellobiose into glucose which can be metabolized by microorganisms to produce bioethanol. In fact, the entrapment of β-glucosidase on magnetic nanofibers provides stability to the enzyme and also the possibility of repeated use, separating them by applying a magnetic field. Verma et al. (2013) evaluated β-glucosidase (isolated from fungus) immobilization on magnetic nanoparticles used as nanobiocatalyst for bioethanol production. The authors verified that 93% of enzyme-binding efficiency was recorded, showing about 50% of its initial activity at the 16th cycle. In another study, Goh et al. (2012) demonstrated that enzyme involved in bioethanol production was immobilized in single-walled carbon nanotubes which were already incorporated by magnetic iron oxide nanoparticles to give magnetic properties. In this study, the performance of the immobilized enzymes could be controlled by altering the concentration of iron oxide nanoparticles in nanotubes. Thus, the immobilized enzyme can be stored in acetate buffer at 4°C for its longer storage. Moreover, enzyme immobilized on TiO_2 nanoparticles by adsorption methods was also successfully used for the hydrolysis of lignocellulosic materials, aiming for the use of bioethanol production (Ahmad and Sardar 2014). In another study, Cherian et al. (2015) reported the immobilization of cellulase recovered from *Aspergillus fumigatus* on manganese dioxide nanoparticles by covalent binding. The authors verified that the immobilized enzyme showed potential to enhance its thermostability property compared to free enzymes, presenting stability up to 70°C. Immobilized cellulase mediated hydrolysis followed by the use of yeast resulted into the production of 21.96 g/L of bioethanol from agricultural waste. After repeated use for about five cycles, the immobilized enzyme showed 60% of its activity. Sanchez-Ramirez et al. (2017) covalently immobilized cellulase obtained from *Trichoderma reesei* on chitosan-coated magnetic nanoparticles using glutaraldehyde as a coupling molecule. The immobilized enzyme was able to retain about 80% of its activity even after 15 cycles of repeated use in the hydrolysis of carboxymethyl cellulose. Kumar et al. (2017) reported immobilization of holocellulase recovered from *Aspergillus niger*on magnetic iron oxide (Fe_2O_3) nanoparticles for the efficient hydrolysis of pretreated paddy straw. The immobilization was achieved by both physical adsorption and covalent coupling. The comparative study on saccharification using immobilized and free cellulose showed that the immobilized enzyme showed 52%, whereas free enzyme showed 47% saccharification efficiency.

5.3.2 Carbon, Silica, Gold and other Nanoparticles

Apart from magnetic nanoparticles, other nanomaterials can be used in the nanotechnology processes, such as silica and TiO_2, polymeric nanoparticles and carbon materials such as fullerene, graphene, carbon nanotubes and others. These materials have been successfully reported for the immobilization of different enzymes regarding the processes of bioethanol production (Huang et al. 2011, Cho et al. 2012, Pavlidis et al. 2012). For instance, a metal such as nickel can be used in the degradation of lignocellulosic material into hemicelluloses (Chandel et al. 2011). Srivastava et al. (2014) reported the synthesis of different concentrations of nickel-cobaltite nanoparticles and evaluated their effect on the production of cellulases. The study demonstrated that a 1 mM concentration of nanoparticles showed maximum production of cellulase with a decrease in the production of cellulase at higher concentrations. Furthermore, at 1mM concentration, nanoparticles showed the stability of enzyme at 80°C for 8 hours whereas the stability of the enzyme was less in the absence of nanoparticles. Lupoi and Smith (2011) studied the immobilization of cellulase on silica nanoparticles, demonstrating the efficacy of immobilized and free enzymes in the hydrolysis of cellulose into glucose. The authors observed that immobilized cellulase enzyme showed an increased yield of glucose when compared to the free enzyme, verifying that immobilized enzymes can be used in simultaneous saccharification and fermentation. Chiang et al. (2011) demonstrated the immobilization of two mesoporous silica nanoparticles having a different particle size, pore size and surface area by physical adsorption and chemical binding. It was reported that cellulase immobilized on mesoporous silica nanoparticles having large pore size by chemical binding showed effective cellulose-to-glucose conversion exceeding 80% yield and excellent stability. Carbon nanotubes (CNTs) owing to their attractive properties such as high mechanical strength and thermostability have received great attention for enzyme immobilization (Ahmad et al. 2018, Mubarak et al. 2017). Pan et al. (2007) demonstrated the use of CNTs entrapped with Rh particles to enhanced catalytic activity for the production of ethanol. The free cores available on the carbon nanotubes are reported as a way to facilitate the incorporation of materials of different interests. Mubarak et al. (2017) demonstrated the immobilization of cellulase on functionalized multi-walled carbon nanotubes (MWCNTs) with a maximum of 97% binding efficacy at the enzyme concentration of 4 mg/mL. Moreover, the catalytic activity of the immobilized enzyme remained about 52% even after six cycles of hydrolysis. Ahmad et al. (2018) reported immobilization of cellulase recovered from *Aspergillus niger* on functionalized MWCNTs through carbodiimide coupling. The immobilization of cellulase via covalent linking was found to be effective in the hydrolysis of cellulose with 85% binding efficacy. Furthermore, the studies on reusability of immobilized enzyme suggested that it can be reused multiple times without much loss in enzyme activity because immobilized enzyme retained 75% of its original activity after the 6th cycle.

5.4 Conversion of Lignocelluloses to Nanocelluloses

The global focus in the current times is directed not only toward the conversion of lignocellulose to biofuel production but also toward the synthesis of nanocellulose. Nanocellulose is a single crystalline cellulose which is a rigid, needle-shaped structure with 1–100 nm in diameter and is categorized into nanocrystalline cellulose (NCC) and nanofibrillated cellulose (NFC). Both types of nanocellulose are chemically similar to dissimilar physical characteristics. The nanocellulose exhibit novel attractive properties such as high specific surface area, high aspect ratio, nontoxicity, low thermal expansion; apart from these, they have low cost come from natural and abundant sources and have even the ability to replace synthetic fibers (Lee et al. 2014, Siro et al. 2010). The relative degree of crystallinity and the geometrical aspect ratio (length to diameter; L/d) are very important parameters controlling the properties of nanocellulose. Different strategies for top-down destructuring of cellulosic fibers like mechanical reaction (cryocrushing, grinding and high-pressure homogenization), biological reaction (enzymatic treatment) and chemical reaction (oxidation and acid hydrolysis) have widely been used. Among the cellulose depolymerization treatments, oxidation pretreatment is one of the common techniques used to disintegrate cellulose into nanocellulose by applying 2,2,6,6-tetramethylpiperidinyl-1-oxyl (TEMPO) radicals. TEMPO mediated oxidation method generates sinter-fibrillar repulsive forces between fibrils by modifying the surface of native cellulose. This led to the conversion of primary hydroxyls in cellulose into carboxylate groups which later became negatively charged and resulted in repulsion of the nanofiber and thus contributing to easy and fast fibrillation (Saito et al. 2007). Biological-based hydrolyzing agent, cellulase which is composed of a multicomponent enzyme system, allows restrictive and selective hydrolysis of the specified components in the cellulosic fibers. The multi-step enzymatic process involves the synergistic action of endoglucanases and exoglucanases/cellobiohydrolases for the initial disorderedness of solid crystal of cellulose at the solid-liquid interface. Although, the nanocellulose synthesis by enzymatic route offers many advantages such as higher yields, higher selectivity, lower energy costs and milder operating conditions than chemical processes, there are many economical (high cost of cellulase enzyme) and technical (rate-limiting step of cellulose degradation with long processing period) obstacles in the success path of this technology (Lee et al. 2014).

5.5 Considerations About Human and Environmental Safety

The main causes of the toxicity of nanoparticles in humans are regarded as their unique properties which also make them useful. For instance, nanoparticle surface, which may act as a nanocatalyst can also cause oxidative reactions and as they are small enough to enter human cells, these reactions can generate many cytotoxic effects on various human tissues. Moreover, most nanoparticles are composed of metals, which are already known to be extremely toxic to some organs and easily bio-accumulated by the human body (Fischer et al. 2007). The main exposure of humans to nanoparticles is through inhalation, ingestion or dermal contact (Gupta et al. 2012). After entering the bloodstream through any of the aforesaid routes, nanoparticles can reach any organ in the human body causing toxic effects that are highly dependent on their concentration, size and also activity level of the respective organ. Owing to the high metabolic rate of the kidney, lungs and liver, these organs are generally at a higher risk to interact with any material entering into the body including nanoparticles. Additionally, the worker manufacturing along with the processing of nanoparticles, has the chances of getting exposed through their dermal contact or breathing. Platinum nanoparticles have also been investigated for their exposure effects on the early stage of development. Literature reported that depending on their concentration, they lower the heart rate, delay the hatching process and also affect the touch response and axis curvature (Asharani et al. 2011). Therefore, it is imperative to study the hazardous effect of the nanoparticles on such organs in animal models. In the present scenario, various approaches are being made for assessing the toxicity of nanoparticles. Most of the studies involve *in vitro* evaluation of nanotoxicity. However, extensive studies are needed to focus on *in vivo* interaction of nanoparticles particularly used for biofuel and bioenergy production.

6. Conclusion

In the present scenario where the scarcity of fossil fuels looms large, exploring energy alternatives in the form of biofuels, namely, bioethanol is of utmost importance. Hence, biofuel is considered the future fuel, and nanotechnology plays a pivotal role to enable the production of next-generation biofuels. Various nanocatalysts like magnetic nanoparticles, carbon nanotubes, metal oxide nanoparticles, engineered nanomaterials, etc., are promising to become an integral part of sustainable bioenergy production. The application of enzymes immobilized in nanoparticles is pioneering research in the field of biofuel production, allowing reusability of not only enzymes but whole-cell systems. The rampant use of nanoparticles raises many concerns as it could have deleterious impacts on human health and environment, notwithstanding their advantages. These concerns have prompted studies to determine the safety measures required in order to limit the exposure to nanoparticles or to overcome problems related to humans and the environment.

References

Abraham, R.E., M.L. Verma, C.J. Barrow and M. Puri. 2014. Suitability of magnetic nanoparticle immobilised cellulases in enhancing enzymatic saccharification of pretreated hemp biomass. Biotechnology for Biofuels 7: 90. DOI: 10.1186/1754-6834-7-90.

Acharaya, S. and A. Chaudhary. 2012. Bioprospecting thermophiles for cellulase production: A review. Brazilian Journal of Microbiology 43(3): 844–856.

Ahmed, M. and M. Douek. 2013. The role of magnetic nanoparticles in the localization and treatment of breast cancer. BioMed Research International 11 pages. DOI: 10.1155/2013/281230.

Ahmad, R. and M. Sardar. 2014. Immobilization of cellulase on TiO_2 nanoparticles by physical and covalent methods: A comparative study. Indian Journal of Biochemistry and Biophysics 51: 314–320.

Ahmad, R. and M. Sardar. 2015. Enzyme immobilization: An overview on nanoparticles as immobilization matrix. Biochemistry and Analytical Biochemistry 4: 178. DOI: 10.4172/2161-1009.1000178.

Ahmad, R. and S.K. Khare. 2018. Immobilization of *Aspergillus niger* cellulase on multiwall carbon nanotubes for cellulose hydrolysis. Bioresource Technology 252: 72–75.

Alvira, P., E. Tomàs-Péjo, M. Ballesteros and M.J. Negro. 2010. Pretreatment technologies for an efficient bioethanol production based on enzymatic hydrolysis: A review. Bioresource Technology 101: 4851–4861.

Angel, T.M. 2002 Molecular biology and structure-function of lignin-degrading heme peroxidases. Enzyme and Microbial Technology 30: 425–444.

Asharani, P.V., Y. Lianwu, Z. Gong and S. Valiyaveettil. 2011. Comparison of the toxicity of silver, gold and platinum nanoparticles in developing zebrafish embryos. Nanotoxicology 5(1): 43–54.

Begum, F., N. Absar and M.S. Alam. 2009. Purification and characterization of extracellular cellulase from *Aspergillus oryzae* ITCC-4857.01. Journal of Advanced and Scientific Research 5: 1645–1651.

Biofuel Annual. 2016. Avaliable at http://www.agrochart.com/en/news/5717/india-biofuels-annual. html.

Bisaria, V.S. 1998. Bioprocessing of agro-residues to value added products. pp. 197–246. *In*: Martin, A.M. (ed.). Bioconversion of Waste Materials to Industrial Products, 2nd edn. Chapman and Hall, UK.

BP Energy Outlook. 2017. Available at https://www.bp.com/content/dam/bp/pdf/energy-economics/energy-outlook-2017/bp-energy-outlook-2017.pdf.

BP Energy Outlook. 2030. 2013. Available at https://www.bp.com/content/ dam/bp/pdf/energy-economics/energy-outlook-2015/bp-energy-outlook-booklet_2013.pdf.

Brebu, M. and C. Vasile. 2010. Thermal degradation of lignin—Review. Cellulose Chemistry and Technology 44: 353–363.

Chandel, A., S.S. Da Silva and O.V. Singh. 2011. Detoxification of lignocellulosic hydrolysates for improved bioethanol production. pp. 225–246. *In*: Dos Santos Bernardes, M.A. (ed.). Biofuel Production-recent Developments and Prospects. In tech: Croatia.

Chen, H., J. Liu, X. Chang, D. Chen, Y. Xue, P. Liu, H. Lin and S. Han. 2017. A review on the pretreatment of lignocellulose for high-value chemicals. 3 Biotech. 160: 196–206.

Chen, T., W. Yang, Y. Guo, R. Yuan, L. Xu et al. 2014. Enhancing catalytic performance of β-glucosidase via immobilization on metal ions chelated magnetic nanoparticles. Enzyme and Microbial Technology 63: 50–57.

Cherian, E., M.K. Dharmendirak and G. Baskar. 2015. Immobilization of cellulase onto MnO_2 nanoparticles for bioethanol production by enhanced hydrolysis of agricultural waste. Chinese Journal of Catalysis 36(8): 1223–1229.

Chiang, D.Y., H.Y. Lian, S.Y. Leo, S.G. Wang, Y. Yamauchi and K.C.W. Wu. 2011. Controlling particle size and structural properties of mesoporous nanoparticles using Taguchi method. Journal of Physical Chemistry 115(27): 131558–131565.

Chinedu, S.N., V. Okochi, H. Smith and O. Omidiji. 2005. Isolation of cellulolytic microfungi involved in wood-waste decomposition: Prospects for enzymatic hydrolysis of cellulosic wastes. International Journal of Biomedical and Health Sciences 1(2): 1–6.

Cho, E.J., S. Jung, H.J. Kim, Y.G. Lee, K.C. Nam, H.J. Lee and H.J. Bae. 2012. Co-immobilization of three cellulases on Au-doped magnetic silica nanoparticles for the degradation of cellulose. Chemical Communications 48: 886–888.

Choinowski, T., W. Blodig, K.H. Winterhalter and K. Piontek. 1999. The crystalstructure of lignin peroxidase at 1.70 Å resolution reveals a hydroxy groupon the C β of tryptophan 171: A novel radical site formed during the redoxcycle. Journal of Molecular Biology 286: 809–827.

Cohen, R., Y. Hadar and O. Yarden. 2011. Transcript and activity levels of different *Pleurotusostreatus* peroxidases are differentially affected by Mn^{2+}. Environmental Microbiology 3: 312–322.

Daniel, G. 2016. Fungal degradation of wood cell wall. Secondary Xylem Biology 131–167.

Daubresse, C., C. Grandfils, R. Jerome and P. Teyssie. 1994. Enzyme immobilization in nanoparticles produced by inverse microemulsion polymerization. Journal of Colloid and Interface Science 168: 222–229.

da Silva, M.L.B. and P.J.J. Alvarez. 2004. Enhanced anaerobic biodegradation of benzene-toluene-ethylbenzene-xylene-ethanol mixtures in bioaugmented aquifer columns. Applied and Environmental Microbiology 70: 4720–26.

Dashtban, M., H. Schraft, T.A. Syed and W. Qin. 2010. Fungal biodegradation and modification of lignin. International Journal of Biochemistry and Molecular Biology 1(1): 36–50.

de Gonzalo, D., I.D. Colpa, M.H.M. Habib and M.W. Fraaije. 2016. Bacterial enzymes involved in lignin degradation. Journal of Biotechnology 236: 110–119.

Dhiman, S.S., S.S. Jagtap, M. Jeya, J.R. Haw, Y.C. Kang and J.K. Lee. 2012. Immobilization of *Pholiotaadiposa* xylanase onto SiO_2 nanoparticles and its application for production of xylooligosaccharides. Biotechnology Letters 34: 1307–1313.

Dias, A., A. Sampaio and R. Bezerra. 2007. Environmental applications of fungal and plant systems: decolourisation of textile wastewater and related dyestuffs. pp. 445–463. *In*: Singh, S. and R. Tripathi (eds.). Environmental Bioremediation Technologies, Springer Berlin Heidelberg.

Duff, S.J.B. and W.D. Murray. 1996. Bioconversion of forest products industry waste cellulosics to fuel ethanol: A review. Bioresource Technology 55: 1–33.

Engelmann, W., A. Aldrovandi, A. Guilherme and B. Filho. 2013. Prospects for the regulation of nanotechnology applied to food and biofuels. Vigilancia Sanitaria em Debate 1(4): 110–121.

Eş, I., J.D.G. Vieira and A.C. Amaral. 2015. Principles, techniques, and applications of biocatalyst immobilization for industrial application. Applied Microbiology and Biotechnology 99: 2065–2082.

Fan, L.T., M.M. Gharpuray and Y.H. Lee. 1987. Cellulose Hydrolysis Biotechnology Monographs. Springer, Berlin, p. 57.

Fang, H., J. Huang, L. Ding, M. Li and Z. Chen. 2009. Preparation of magnetic chitosan nanoparticles and immobilization of laccase. Journal of Wuhan University of Technology Material Science Edition 24: 42–47.

Fischer, H.C. and W.C. Chan. 2007. Nanotoxicity: The growing need for *in vivo* study. Current Opinions in Biotechnology 18: 565–571.

Galbe, M. and G. Zacchi. 2002. A review of the production of ethanol from softwood. Applied Microbiology and Biotechnology 59: 618–628.

Gao, Y. and I. Kyratzis. 2008. Covalent immobilization of proteins on carbon nanotubes using the cross-linker 1-ethyl-3-(3-dimethylaminopropyl) carbodiimide—a critical assessment. Bioconjugate Chemistry 19: 1945–1950.

Gao, J., C.L. Lu, Y. Wang, S.S. Wang, J.J. Shen, J.X. Zhang and Y.W. Zhang. 2018. Rapid immobilization of cellulase onto graphene oxide with a hydrophobic spacer. Catalysts 8: 180. DOI: 10.3390/catal8050180.

Goh, W.J., V.S. Makam, J. Hu, L. Kang, M. Zheng, S.L. Yoong, C.N. Udalagama and G. Pastorin. 2012. Iron oxide filled magnetic carbon nanotube-enzyme conjugates for recycling of amyloglucosidase: Toward useful applications in biofuel production process. Langmuir 28(49): 16864–16873.

Gokhale, A.A., J. Lu and I.J. Lee. 2013. Immobolization of cellulose on magnetoresposive grapheme nanosupports. Journal of Molecular Catalysis B: Enzymology 90: 76–86.

Gonçalves, F.A., E.S. Santos and G.R. Maced. 2015. Use of cultivars of low cost, agroindustrial and urban waste in the production of cellulosic ethanol in Brazil: A proposal to utilization of microdistillery. Renewable and Sustainable Energy Reviews 50: 1287–1303.

Gupta, I., N. Duran and M. Rai. 2011. Nano-silver toxicity: emerging concerns and consequences in human health. *In*: Rai, M. and N. Cioffi (eds.). Nanoantimicrobials: Progress and Prospects.

Hahn-Hägerdal, B., M. Galbe, M.F. Gorwa-Grauslund, G. Lidén and G. Zacchi. 2007. Bio-ethanol-the fuel of tomorrow from the residues of today. Trends in Biotechnology 24(12): 549–556.

Hammel, K.E. 1996. Extracellular free radical biochemistry of ligninolytic fungi. New Journal of Chemistry 20: 195–198.

Hammel, K.E. and D. Cullen. 2008. Role of fungal peroxidases in biological ligninolysis. Current Opinions in Plant Biology 11: 349–355.

Hatakka, A.I. 1983. Pretreatment of wheat straw by white-rot fungi for enzymatic saccharification of cellulose. European Journal of Applied Microbiology and Biotechnology 18: 350–357.

Hegedus, I., E. Nagy, J. Kukolya, T. Barna and A.C.S. Fekete. 2011. Cellulase and hemicellulose enzymes as single molecular nanobiocomposites. Hungarian Journal of Industrial Chemistry 39: 341–349.

Heinfling, A., F.J. Ruiz-Duenas, M.J. Martinez, M. Bergbauer, U. Szewzyk and A.T. Martinez. 1998. A study on reducing substrates of manganese oxidizing peroxidases from *Pleurotuseryngii* and *Bjerkandera adusta*. FEBS Letters 428: 141–146.

Honda, T., T. Tanaka and T. Yoshino. 2015. Stoichiometrically controlled immobilization of multiple enzymes on magnetic nanoparticles by the magnetosome display system for efficient cellulose hydrolysis. Biomacromolecules 16: 3863–3868.

Hong, J., P. Gong, D. Xu, L. Dong and S. Yao. 2007. Stabilization of alpha-chymotrypsin by covalent immobilization onamine-functionalized superparamagnetic nanogel. Journal of Biotechnology 128: 597–605.

Howard, R.L., E. Abotsi, E.J. Van Rensburg and S. Howard. 2003. Lignocellulose biotechnology: Issues of bioconversion and enzyme production. African Journal of Biotechnology 2(12): 602–619.

Huang, P.J., K.L. Chang, J.F. Hsieh and S.T. Chen. 2015. Catalysis of Rice straw hydrolysis by the combination of immobilized cellulase from *Aspergillus niger* on β-cyclodextrin-Fe_3O_4 nanoparticles and ionic liquid. International Journal of Biomedical Research 2015: http://dx.doi.org/10.1155/2015/409103.

Huang, X.J., P.C. Chen, F. Huang, Y. Ou, M.R. Chen and Z.K. Xu. 2011. Immobilization of *Candida rugosa* lipase on electrospun cellulose nanofiber membrane. Journal of Molecular Catalysis B: Enzymology 70: 95–100.

Hwang, E.T. and M.B. Gu. 2013. Enzyme stabilization by nano/microsized hybrid materials. Engineering in Life Sciences 13: 49–61.

IESS 2047. 2016. Available at http://www.india energy.gov.in/bioenergy.php.

Isroi, R., S. Millati, S. Syamsiah, C. Nikklason, M.N. Cahyanton, K. Lundquist and M.J. Taherzadeh. 2011. Biological pretreatment of lignocellulosic with white rot-fungi and its applications: A review. Bioresources 6: 5226–5259.

Jafari Khorshidi, K., H. Lenjannezhadian, M. Jamalan and M. Zeinali. 2016. Preparation and characterization of nanomagnetic cross-linked cellulase aggregates for cellulose bioconversion. Journal of Chemical Technology and Biotechnology 91(2016): 539–546.

Ji, C., L.N. Nguyen, J. Hou, F.I. Hai and V. Chen. 2017. Direct immobilization of laccase on titania nanoparticles from crude enzyme extracts of *P. ostreatus* culture for micro-pollutant degradation. Separation and Purification Technology 178(7): 215–223.

Joshi, V. and A. Pandey. 1999. Biotechnology: Food Fermentation: Microbiology, Biochemistry, and Technology. Educational Publishers & Distributors.

Kang, Q., L. Appels, T. Tainwei and R. Dewil. 2014. Bioethanol from lignocellulosic biomass: Current findings determine research priorities. The Scientific World Journal. DOI: 10.1155/2014/298153.

Kantharaj, P., B. Boobalan, S. Sooriamuthu and R. Mani. 2017. Lignocellulose degrading enzymes from fungi and their industrial applications. International Journal of Research and Review 9(21): 1–12.

Kaur, J., B.S. Chandha and B.A. Kumar. 2007. Purification and characterization of β-glucosidase from *Melanocarpus* sp. MTCC3922. Electronic Journal of Biotechnology 10: 260–270.

Khindaria, A., I. Yamazaki and S.D. Aust. 1996. Stabilization of the veratryl alcohol cation radical by lignin peroxidase. Biochemistry (N Y) 35: 6418–6424.

Khoshnevisan, K., A.K. Bordbar, D. Zare, D. Davoodi, M. Noruzi, M. Barkhi and M. Tabatabaei. 2011. Immobilization of cellulase enzyme on superparamagnetic nanoparticles and determination of its activity and stability. Chemical Engineering Journal 171: 669–673.

Khoshnevisan, K., F. Vakhshiteh, M. Barkhi, H. Baharifar, E. Poor-Akbar, N. Zari, H. Stamatis and A.K. Bordbar. 2017. Immobilization of cellulase enzyme onto magnetic nanoparticles: Applications and recent advances. Molecular Catalyst 442: 66–73.

Kim, M. and D.F. Day. 2011. Composition of sugar cane, energy cane, and sweet sorghum suitable for ethanol production at Louisiana sugar mills. Journal of Industrial Microbiology and Biotechnology 38(7): 803–807.

Kim, J., J.W. Grate and P. Wang. 2008. Nanobiocatalysis and its potential applications. Trends in Biotechnology 26: 639–646.

Kim, K.H., O.K. Lee and E.Y. Lee. 2018. Nano-immobilized biocatalysts for biodiesel production from renewable and sustainable resources. Catalysts 8(2): 68. DOI: 10.3390/catal8020068.

Kocher, G.S., P. Kaur and M.S. Taggar. 2017. An overview of pretreatment processes with special reference to biological pretreatment for rice straw delignification. Journal of Biochemical Engineering 4(3): 151–163.

Kubicek, C.P., R. Messner, F. Gruber, R.L. Mach and E.M. Kubicek-Pranz. 1993. The *Trichoderma* cellulase regulatory puzzle: From the interior life of asecretory fungus. Enzyme and Microbial Technology 15: 90–99.

Kumar, A., S. Singh, R. Tiwari, R. Goel and L. Nain. 2017. Immobilization of indigenous holocellulase on iron oxide (Fe_2O_3) nanoparticles enhanced hydrolysis of alkali pretreated paddy straw. International Journal of Biological Macromolecules 96: 538–549.

Kumar, A., S. Singh and L. Nain. 2018. Magnetic nanoparticle immobilized cellulase enzyme for saccharification of paddy straw. International Journal of Current Microbiology and Applied Sciences 7(4): 881–893.

Lee, S.M., L.H. Jin, S.O. Han, H.B. Na, T. Hyeon, Y.M. Koo, J. Kim and J.H. Lee. 2009. ß-Glucosidase coating on nanofibers for improved cellulosic ethanol production. Bioprocess and Biosystem Engineering 33: 141–147.

Lee, H.V., S.B.A. Hamid and S.K. Zain. 2014. Converson of lignocellulosic biomass to nanocellulose: Structure and chemical process. The Scientific World Journal. DOI: 10.1155/2014/631013.

Leo, V.V., A.K. Passari, J.B. Joshi, V.K. Mishra, S. Uthandi, N. Ramesh and B.P. Singh. 2016. A novel triculture system (CC3) for simultaneous enzyme production and hydrolysis of common grasses through submerged fermentation. Frontiers in Microbiology 7.

Li, Y., X.Y. Wang, R.Z. Zhang, X.Y. Zhang, W. Liu, X.M. Xu and Y.W. Zhang. 2014. Molecular imprinting and immobilization of cellulase onto magnetic Fe_3O_4@SiO_2 nanoparticles. Nanoscience and Nanotechnology 14: 2931–2936.

Licht, F.O. 2006. World Ethanol Markets: The Outlook to 2015, Tunbridge Wells, Agra Europe Special Report, UK.

Lundquist, K., T.K. Kirk and W.J. Connors. 1977. Fungal degradation of kraft lignin and lignin sulphonates prepared from synthetic ^{14}C-lignins. Archives of Microbiology 112: 291–296.

Lupoi, J.S. and E.A. Smith. 2011. Evaluation of nanoparticle-immobilized cellulase for improved yield in simultaneous saccharification and fermentation reactions. Biotechnology and Bioengineering 108: 2835–2843.

Lynd, L.R., W.H. Van Zyl, J.E. McBride and M. Laser. 2005. Consolidated bioprocessing of cellulosic biomass: An update. Current Opinions in Biotechnology 16: 577–583.

Ma, D.I.N.G., M. Li, A.J. Patil and S. Mann. 2004. Fabrication of protein/silica core–shell nanoparticles by microemulsion-based molecular wrapping. Advanced Materials 16: 1838–1841.

Malherbe, S. and T.E. Cloete. 2002. Lignocellulose biodegradation: Fundamentals and applications. Reviews in Environmental Science and Biotechnology 1: 105–114.

Mandal, D., M.E. Bolander, D. Mukhopadhay, G. Sarkar and P. Mukherjee. 2005. The use of microorganism for the formation of metal nanoparticles and their application. Applied Microbiology and Biotechnology 69: 485–492.

Martin, H. 2002. Review: Lignin conversion by manganese peroxidase (MnP). Enzyme and Microbial Technology 30: 454–466.

McKendry, P. 2002. Energy production from biomass (part 1): Overview of biomass. Bioresource Technology 83(1): 37–46.

Mei, X.Y., R.H. Liu, F. Shen and H.J. Wu. 2009. Optimization of fermentation conditions for the production of ethanol from stalk juice of sweet sorghum by immobilized yeast using response surface methodology. Energy Fuels 23: 487–491.

Mester, T. and J.A. Field. 1998. Characterization of a novel manganese peroxidase-lignin peroxidase hybrid isozyme produced by *Bjerkandera* sp. strain BOS55 in the absence of manganese. The Journal of Biological Chemistry 273: 15412–1541.

Mubarak, N.M., J.R. Wong, K.W. Tan, J.N. Sahu, E.C. Abdullah, N.S. Jayakumar and P. Ganesan. 2017. Immobilization of cellulase enzyme on functionalized multiwall carbon nanotubes. Journal of Molecular Catalysis B: Enzymology 107: 124–131.

Naik, S.N., V.V. Goud, P.K. Rout and A.K. Dalai. 2010. Production of first and second generation biofuels: A comprehensive review. Renewable and Sustainable Energy Reviews 14(2): 578–597.

Nizami, A. and M. Rehan. 2018. Towards nanotechnology-based biofuel industry. Biofuel Research Journal 18: 798–799.

Pan, X., Z. Fan, W. Chen, Y. Ding, L. Luo and X. Bao. 2007. Enhanced ethanol production inside carbon nanotube reactors containing catalytic particles. Nature Materials 6: 507–511.

Patila, M., I.V. Pavlidis, A. Kouloumpis, K. Dimos, K. Spyrou, P. Katapodis, D. Gournis and H. Stamatis. 2016. Graphene oxide derivatives with variable alkyl chain length and terminal functional groups as supports for stabilization of cytochrome c. International Journal of Biological Macromolecules 84: 227–235.

Pavlidis, I.V., T. Vorhaben, D. Gournis, G.K. Papadopoulos, U.T. Bornscheuer and H. Stamatis. 2012. Regulation of catalytic behaviour of hydrolases through interactions with functionalised carbon-based nanomaterials. Journal of Nanoparticle Research 14: 842.

Placido, J. and S. Capareda. 2015. Ligninolytic enzymes: a biotechnological alternative for bioethanol production. Bioresources and Bioprocessing 2: 23. DOI: 10.1186/s40643-015-0049-5.

Pozdnyakova, N.N. 2012. Involvement of the ligninolytic system of white-rot and litter decomposing fungi in the degradation of polycylic aromatic hydrocarbons. Biotechnology Research International. DOI: 10.1155/2012/243217.

Prassad, S., A. Singh and H.C. Joshi. 2007. Ethanol as an alternative fuel from agricultural, industrial and urban residues. Resources, Conservation and Recycling 50: 1–39.

Qi, H., H. Duan, X. Wang, X. Meng, X. Yin and L. Ma. 2015. Preparation of magnetic porous terpolymer and its application in cellulase immobilization. Polymer Engineering and Science 55: 1039–1045.

Rai, M., J.C. dos Santos, M.F. Soler, P.R.F. Marcelino, L.P. Brumano, A.P. Ingle, S. Gaikwad, A. Gade and S.S. da Silva. 2016. Strategic role of nanotechnology for production of bioethanol and biodiesel. Nanotechnology Reviews 5(2): 231–250.

Reetz, M.T., A. Zonta, V. Vijayakrishnan and K. Schimossek. 1998. Entrapment of lipases in hydrophobic magnetite-containing sol-gel materials: Magnetic separation of heterogeneous biocatalysts. Journal of Molecular Catalysis A: Chemical 134: 251–258.

Ren, Y., J.G. Rivera, L. He, H. Kulkarni, D.K. Lee and P.B. Messersmith. 2011. Facile, high efficiency immobilization of lipase enzyme on magnetic iron oxide nanoparticles via a biomimetic coating. BMC Biotechnology 11(1): 63.

Rezania, S., M. Ponraj, M.F. Md. Din, A.R. Songip, F.M. Sairan and S. Chelliapan. 2015. The diverse applications of water hyacinth with main focus on sustainable energy and production for new era: An overview. Renewable and Sustainable Energy Reviews 41: 943–954.

Robak, K. and M. Balcerek. 2018. Review of second generation bioethanol production from residual biomass. Food Technology and Biotechnology 56(2): 174–187.

Robson, L.M. and G.H. Chambliss. 1989. Cellulases of bacterial origin. EMT 11: 626–644.

Roth, H.S., F.P. Schwaminger and S. Berensmeier. 2016. Immobilization of cellulase on magnetic nanocarriers. Chemistry Open 5(3): 183–187.

Ruiz-Duenas, F.J., M. Morales, E. Garcia, Y. Miki, M.J. Martinez and A.T. Martinez. 2009. Substrate oxidation sites in versatile peroxidase and other basidiomycete peroxidases. Journal of Experimental Biology 60: 441–52.

Saini, J.K., R. Saini and T. Tewari. 2015. Lignocellulosic agricultural waste as biomass feedstocks for second generation bioethanol production. Concepts and Recent Developments. 3 Biotech. 5: 337–353.

Saito, T., S. Kimura, Y. Nishiyama and A. Isogai. 2007. Cellulose nanofibers prepared by TEMPO-mediated oxidation of nativecellulose. Biomacromolecules 8: 2485–2491.

Sánchez-Ramírez, J., J.L. Martínez-Hernández, P. Segura-Ceniceros, G. López, H. Saade, M.A. Medina-Morales, R. Ramos-González, C.N. Aguilar and A. Ilyina. 2017. Cellulases immobilization on chitosan-coated magnetic nanoparticles: application for *Agave atrovirens* lignocellulosic biomass hydrolysis. Bioprocess and Biosystem Engineering 40: 9–22.

Schwaminger, S.P., P. Fraga-García, F. Selbach, F.G. Hein, E.C. Fuß, R. Surya, H.C. Roth, S.A. Blank-Shim, F.E. Wagner, S. Heissler and S. Berensmeier. 2017. Bio-nano interactions: cellulase on iron oxide nanoparticle surfaces. Adsorption 23: 281–292.

Siro, I. and D. Plackett. 2010. Microfibrillated cellulose and new nanocomposite materials: A review. Cellulose 17: 459–494.

Sette, L.D., V.M. de Oliveira and M.F.A. Rodrigues. 2008. Microbial lignocellulolytic enzymes: industrial applications and future perspectives. Microbiology Australia 29: 18–20.

Sheldon, R.A. 2007. Cross-linked enzyme aggregates (CLEAs): Stable and recyclable biocatalysts. Biochemical Society Transactions 35: 1583–1587.

Sojitra, U.V., S.S. Nadar and V.K. Rathod. 2016. A magnetic tri-enzyme nanobiocatalyst for fruit juice clarification. Food Chemistry 213: 296–305.

Srivastava, N., R. Rawat, R. Sharma, H.S. Oberoi, M. Srivastava and J. Singh. 2014. Effect of nickel-cobaltite nanoparticles on production and thermostability of cellulases from newly isolated thermotolerant *Aspergillus fumigatus* NS (Class: Eurotiomycetes). Applied Biochemistry and Biotechnology 174(3): 1092–1103.

Sun, Y. and J. Cheng. 2002. Hydrolysis of lignocellulosic materials for ethanol production: A review. Bioresource Technology 83(1): 1–11.

Sundaramoorthy, M., K. Kishi, M.H. Gold and T.L. Poulos. 1994. The crystal structure of manganese peroxidase from *Phanerochaete chrysosporium* at 2.06-A°resolution. Journal of Biological Chemistry 269: 32759–32767.

Szijártó, N., Z. Szengyel, G. Lidén and K. Réczey. 2004. Dynamics of cellulose production by glucose grown cultures of *Trichoderma reesei* Rut-C30 as a response to addition of cellulose. Paper presented at the Proceedings of the Twenty-Fifth Symposium on Biotechnology for Fuels and Chemicals Held in Breckenridge, CO. Springer Verlag: Germany, 2012, pp. 525–548.

Tang, H., J. Chen, S. Yao, L. Nie, G. Deng and Y. Kuang. 2004. Amperometric glucose biosensor based on adsorption of glucose oxidase at platinum nanoparticle-modified carbon nanotube electrode. Analytical Biochemistry 331: 89–97.

Tao, Q.L., Y. Li, Y. Shi, R.J. Liu, Y.W. Zhang and J. Guo. 2016. Application of molecular imprinted magnetic $Fe_3O_4@SiO_2$ nanoparticles for selective immobilization of cellulase. Journal of Nanoscience and Nanotechnology 16: 6055–6060.

Tuomela, M., M. Vikman, A. Hatakka and M. Itävaara. 2000. Biodegradation of lignin in a compost environment: A review. Bioresource Technology 72: 169–183.

UNEP. 2015. Biofuel roadmap for India. Available at NEP http://www.unep.org/transport/ lowcarbon/PDFs/Biofuel_Roadmap_for_India.pdf.

US EIA. 2018. Available at :https://www.naturalgasintel.com/articles/115170-energy-use-by-china-india-africa-forecast-to-surge-through-2040-says-eia#.

US EPA. 2016. Available at https://www.epa.gov/ nutrientpollution/sources-and-solutions-fossil-fuels.

Vaenzuela, R., J.F. Castro, C. Parra, J. Baeza, N. Duran and J. Freer. 2014. β-Glucosidase immobilisation on synthetic superparamagnetic magnetite nanoparticles and their application in saccharification of wheat straw and *Eucalyptus globules* pulps. Journal of Experimental Nanoscience 9: 177–185.

Valdez-Vazquez, V. and A. Sanchez. 2017. Proposal for biorefineries based on mixed cultures for lignocellulosic biofuel production: A techno-economic analysis. Biofuel, Bioproduct and Biorefinery 12(1). DOI: 10.1002/bbb.1828.

Verardi, A., I. De Bari, E. Ricca and V. Calabrò. 2011. Hydrolysis of cellulose with immobilized cellulases: process analysis and control. Proceedings of the 19th European Biomass. Conference & Exhibition, Berlin (Germany).

Verardi, A., I. De Bari, E. Ricca and V. Calabrò. 2012. Hydrolysis of lignocellulosic biomass: Current status of process and technologies and future perspectives. *In*: Lima, M.A.P. and A.P.P. Natalense (eds.). Bioethanol, In Tech Publisher, London, UK, p. 302.

Verma, M.L., R. Chaudhary, T. Tsuzuki, C.J. Barrow and M. Puri. 2013. Immobilization of β-glucosidase on a magnetic nanoparticle improves thermostability: Application in cellobiose hydrolysis. Bioresource Technology 135: 2–6.

Wheals, A.E., L.C. Basso, D.M.G. Alves and H.V. Amorim. 1999. Fuel ethanol after 25 years. Trends in Biotechnology 17: 482–87.

Wongwatanapaiboon, J., K. Kangvansaichol, V. Burapatana, R. Inochanon, P. Winayanuwattikun, T. Yongvanich and W. Chulalaksananuku. 2012. The potential of cellulosic ethanol production from grasses in Thailand. BioMed Research International 2012(303748): 1–10.

Yang, H.H., S.Q. Zhang, X.L. Chen, Z.X. Zhuang, J.G. Xu and X.R. Wang. 2004. Magnetite-containing spherical silica nanoparticles for biocatalysis and bioseparations. Analytical Chemistry 76: 1316–1321.

Yiu, H.H. and M.A. Keane. 2012. Enzyme–magnetic nanoparticle hybrids: New effective catalysts for the production of high value chemicals. Journal of Chemical Technology and Biotechnology 87(5): 583–594.

Zang, L., J. Qiu, X. Wu, W. Zhang, E. Sakai and Y. Wei. 2014. Preparation of magnetic chitosan nanoparticles as support for cellulase immobilization. Industrial and Engineering Chemistry Research 53: 533448–3454.

Zhang, D., M. Deng, H. Cao, S. Zang and S. Zhao. 2017. Laccase immobilized on magnetic nanoparticles by dopamine polymerization for 4-chlorophenol removal. Green Energy Environment 2(4): 393–400.

Zhou, T.Y., S.N. Su, M.M. Song, H.L. Nie, L.M. Zhu and C.B. White. 2009. Improving the stability of cellulase by immobilization on chitosan-coated magnetic nanoparticles modified alpha-ketoglutaric acid. 2009 3rd international conference on bioinformatics and biomedical engineering. IEEE. 10.1109/ICBBE.2009.5162941.

Zuliani, A., I. Francisco and L. Rafael. 2018. Advances in nanocatalysts design for biofuels production. ChemCatChem 10: 1968–1981.

Chapter 8

Nanoparticles in Methane Production from Anaerobic Digesters

Advantages, Challenges and Perspectives

Efraín Reyes Cruz, Lilia Ernestina Montañez Hernández, Inty Omar Hernández De Lira **and** ***Nagamani Balagurusamy****

1. Introduction

Anaerobic digestion (AD) comprises a series of processes that interact for the degradation of complex organic materials by the participation of a wide microbial consortium. During AD, the microbial metabolism allows the recovery of renewable energy in the form of biogas (mainly methane or CH_4) and an organic fertilizer rich in nutrients from organic waste. AD is divided into the next stages: hydrolysis and/or fermentation, acetogenesis and methanogenesis. At the start of the process, hydrolytic bacteria degrade highly insoluble particles, forming simple and soluble products that will be catabolized by fermentative bacteria to produce alcohols, fatty acids, CO_2 and H_2. In the next stage denoted as acetogenesis, three groups participate: obligate-hydrogen producing bacteria that generate H_2; acetate and CO_2 from the oxidation of fatty acids; syntrophic acetate oxidizing bacteria that produce H_2 and CO_2 from acetate and homoacetogenic bacteria which carry on the formation of acetate from H_2 and CO_2. Finally, in methanogenesis, CO_2, acetate and methylated C1 compounds are reduced to CH_4 by hydrogenotrophic, aceticlastic and methylotrophic methanogens, respectively. Other microbial groups present in the process might use nitrates (NO_3^-), e.g., denitrifying bacteria and dissimilative nitrate-reducing bacteria or sulfates (SO_4^{2}) like sulfate-reducing bacteria as electron acceptors. There is even a microbial group with the capacity to oxidize CH_4 to H_2 and CO_2 called anaerobic methanotrophic bacteria (Alvarado et al. 2014).

Laboratorio de Bioremediación, Facultad de Ciencias Biológicas, Universidad Autónoma de Coahuila, Torreón, C.P. 27000, Coahuila, México.

* Corresponding author: bnagamani@uadec.edu.mx

Naturally, methanogens participate in the consumption of organic matter in anaerobic environments which results in the production of CH_4 because this compound is a by-product of its energy metabolism (Casals et al. 2014, Enzmann et al. 2018). This microbial group belongs to the Archaea domain and *Euryarchaeota* phylum which is divided into seven orders (Alvarado et al. 2014). According to the environment, the methanogenic population varies, e.g., in mesophilic digesters, hydrogenotrophic and aceticlastic methanogens prevail, being *Methanosarcina* or *Methanoculleus* the dominating species, respectively (Enzmann et al. 2018).

As was mentioned before, a lot of interactions are required for the AD process. Therefore, microbial groups and their complex metabolic relations can determine the performance of an anaerobic system. Therefore, it is important to study their behavior at different operational conditions (Casals et al. 2014). One of the most common parameters that affect AD performance is the concentration of volatile fatty acids (VFAs) where the main representative of this group is acetate. For instance, it has been noted that the increase of initial acetate concentration from 20 to 200 mM extended the lag-phase time, increased the none-dissociated form of acetic acid from 0.114 to 1.444 mM and decreased CH_4 yield from 0.88 to 0.36 mole CH_4/mole acetate (Yang et al. 2015). However, low VFAs concentration can limit biogas production (Jia et al. 2017). Therefore, strategies are necessary to improve microbial activity and evade problems like reactor acidification. Among other parameters that have been modified are digester design, treatment conditions and substrate pretreatments (Alvarado et al. 2014). Nowadays, the investigations have turned toward the enhancement of the AD process with the enrichment of trace elements required for microbial metabolism. In the case of methanogenic archaea, metallic elements such as Fe, Ni, Co, Mo, Zn, Cu and Mn are important metals for their growth (Scherer et al. 1983). A novel enhancement method to fulfill this gap is the use of nanotechnology, specifically nanoparticles (NPs). Furthermore, the growing presence of NPs in everyday products, their increasing demand and their consequent disposal as waste have caused these to reach wastewater treatment plants (WWTPs) that mainly work with anaerobic digestion processes (Wang et al. 2016). All the sewage sludge produced in these plants requires to be treated and the application of AD is adequate for that. After sludge stabilization, part of the organic matter present in it is degraded, otherwise, its discharge to the environment will be damaging. Also, it is possible to obtain renewable energy in the form of CH_4 from the AD process which can be useful within WWTPs (Mu et al. 2011). To this day, the effects that NPs can trigger remain unknown. Therefore, it is necessary to study the effects of NPs in AD (Wang et al. 2016).

NPs have been applied in experiments that managed waste-activated sludge (Mu and Chen 2011, Feng et al. 2014), anaerobic granular sludge (Ambuchi et al. 2017, He et al. 2017) and cattle slurry (Abdelsalam et al. 2016, 2017, Juntupally et al. 2017).

Some studies have determined the positive effects of NPs in AD (Chen et al. 2018, Jia et al. 2017) but it also has shown that their excessive concentration can inhibit microbial activities (Abdelsalam et al. 2017, Jia et al. 2017). In addition to that, there is the possibility that NPs do not influence the performance of AD (Chen et al. 2018). Hence, it is necessary to collect information about the effects caused by NPs in AD and elucidate their participation in this process. Generally, the set of positive effects in AD process by the enrichment with NPs includes the increase of biogas and CH_4 volume (Abdelsalam et al. 2016, 2017, Chen et al. 2018), reduction of the retention time of biomass (Chen et al. 2018) and the enhancement of microbial enzymatic activity (Tian et al. 2017a). Moreover, the addition of NPs has demonstrated their capacity to maintain stability in some physicochemical parameters during the AD processes, such as ammonia nitrogen, pH and VFAs (Jia et al. 2017). Consequently, the application of NPs can be a suitable method for AD improvement. If AD technology continues progressing, it would become the main option to deal with the excessive waste generated from anthropogenic activities and the increasing energy demand by taking advantage of biomass to produce renewable energy, thus decreasing our dependence of fossil fuels and mitigating the emission of greenhouse gases (Chen et al. 2018). In this chapter, the effects of the addition of nanoparticles in AD will be addressed.

2. Trace Elements in Methanogenic Archaea

Through all AD process, a great number of enzymes are used for the degradation of organic matter. Some of them require metals for their activities, such as hydrogenases. These remarkable enzymes activate H_2, represented by Equation 1, to use it as an electron donor for downstream redox reactions.

$$H_2 \rightarrow 2e^- + 2H^+ \tag{1}$$

Since H_2 is produced and consumed extensively during AD as a way to conserve redox balance in fermentative and acetogenic bacteria, hydrogenases are key enzymes during the degradation of organic matter. H_2-forming bacteria and hydrogenotrophic methanogens require some of the three known types of hydrogenases: [NiFe]-hydrogenases, [FeFe]-hydrogenases and [Fe]-hydrogenase. Four methanogenic enzymes belong to [NiFe]-hydrogenases and another to [Fe]-hydrogenase; the latter has the capacity to function during Ni limitations (Thauer et al. 2010). Therefore, the limitation of the latter trace elements can negatively impact the AD process (Qiang et al. 2012).

Another key enzyme that impacts methane production is methyl-coenzyme M reductase (MCR). This enzyme is the one that catalyzes the formation of methane in all methanogenesis pathways, and it requires the binding of a prosthetic group, coenzyme F_{430}, which contains Ni. It has been previously demonstrated that Ni limitation greatly impacts the distribution and activity of MCR (Ermler et al. 1997, Ferry 1999, Takano et al. 2013), therefore small amounts of this metal must be present in order to guarantee methanogenic activity.

Finally, it might be possible that other trace elements are necessary for new processes that might not involve enzymes.

3. Nanoparticles

Nanotechnology science studies processes that are carried out at the molecular level, which is often referred to a nanoscale size, a thousand millionth part of a unit, e.g.,

1 nanometer = 10^{-9} m. Therefore, NPs belong to this field of study that is because their range size is approximate to or lower than 100 nm. In addition to that, NPs have physical, chemical and biological unique properties due to their size, e.g., a larger surface area to react, the possibility to present different electrical properties (quantum confinement) and uniformity as essential features. Such properties have allowed the application of nanotechnology in sciences like medicine and engineering among others. The characterization of NPs according to their size, morphology and charge have been possible with the use of specialized microscopic techniques, such as atomic force microscopy (AFM), scanning electron microscopy (SEM) and transmission electron microscopy (TEM) (Bhatia et al. 2016, Strambeanuet al. 2015). NPs elaboration can be done by physical or chemical processes. Physical methods refers to diminish the particle size of a bulk compound until it reaches the sought size, while chemical methods comprise the reduction of ions or decomposition of metal compounds both coupled with their consequent controlled aggregation (Yonezawa 2018). Hence, each characteristic of the NP can affect its interactions and behavior. For instance, cellular uptake of NPs is influenced by their shape and size (Verma and Stellacci 2010).

3.1 Enhancement of Anaerobic Digestion by NPs Enrichment

The addition of metal NPs can improve AD performance with an increase of biogas and CH_4 production and accelerating substrate consumption rate. Consequently, their application can reduce the retention time of biomass and energy requirements (Abdelsalam et al. 2016). Table 1 shows the improvement of the two aforementioned parameters by distinct NPs concentrations.

As it was mentioned earlier in the chapter, NPs have a different behavior due to their size. In order to differentiate the effect of NPs in comparison with a bulk compound in AD, biogas and CH_4 production were determined in reactors fed with

Table 1. Nanoparticles concentrations for enhancement in AD performance compared with their respective controls.

NP	Average Size (nm)	Concentration	Biogas Improvement ratio	Methane Improvement Ratio	Reference
Co	28	1 (mg/L)	1.7	2	Abdelsalam et al. 2016
Ni	17	2 (mg/L)	1.8	2.17	Abdelsalam et al. 2016
Fe_2NiO_4 (Ni)	< 100	100 (mg/L)	N/A	1.3	Chen et al. 2018
Fe	9 ± 0.3	20 (mg/L)	1.5	1.67	Abdelsalam et al. 2016
nZVI	128 ± 7.9	10 (mg/g TSS)	N/A[a]	1.2	Wang et al. 2016
Fe_2O_3	20	750 (mg/L)	1.25	1.28	Ambuchi et al. 2017
Fe_3O_4	7 ± 0.2	20 (mg/L)	1.7	2.16	Abdelsalam et al. 2016
Graphene	0.5–2	30 (mg/L) 120 (mg/L)	N/A	1.17 1.51	Tian et al. 2017a
MWCNTs	10–20	1500 mg/L	1.08	1.12	Ambuchi et al. 2017

[a] Not available.

cattle slurry and enriched with Fe NPs, magnetite (Fe_3O_4) NPs or bulk ferric chloride ($FeCl_3$) at a concentration of 10 mg/L. The enhancement of biogas production was 1.2, 1.45 and 1.64 times higher the control volume (660.1 mL/day) while CH_4 content increased 1.21, 1.63 and 1.90 times the control (336.3 mL/day) with $FeCl_3$, Fe and Fe_3O_4 NPs, respectively. Therefore, NPs boost was greater than in the bulk compound. The authors of the study suggested that the spherical shape of Fe and Fe_3O_4 NPs could influence the startup enhancement of biogas production (Abdelsalam et al. 2017). NPs with spherical shape have demonstrated to enhance reaction rate due to their ability to cross cell membrane 500% more than other NPs structures, such as rods (Verma and Stellacci 2010). Furthermore, this effect could be enhanced by the average size of iron NPs (< 10 nm). Another study by Juntupally et al. (2017) evaluated the effect of micronutrients against their respective NPs in AD from cattle slurry. It was observed that all the treatments enhanced biogas production and reduced the hydraulic retention time (HRT) compared with the control. Even so, NPs addition had the greatest effect (Juntupally et al. 2017).

NPs concentrations have a great influence on the effect that will produce in AD. To find this, researchers have studied different concentrations of the same NPs. For instance, biogas production from cattle slurry with 5, 10 and 20 mg/L of Fe NPs was almost the same, enhancing biogas volume 1.44, 1.45 and 1.45 times compared with the control (660.1 mL/day), respectively. However, CH_4 production increased proportionally to the dosage and obtained values of 1.38, 1.53 and 1.59 times the CH_4 volume of the control (336.3 mL/day) with the previous concentrations. As well, Fe_3O_4 NPs enhanced biogas production by 1.2, 1.63 and 1.64 times the control volume production (660.1 mL/day) for 5, 10 and 20 mg/L of Fe_3O_4 NPs, respectively. In a similar way as Fe NPs, CH_4 production was proportional to the dosage of Fe_3O_4 NPs, the same concentrations showed 1.82, 1.90 and 1.96 times the CH_4 volume of the control (336.3 mL/day), respectively. As the experiment advanced, substrate content decreased. On the 50th day, the highest decomposition of substrate among Fe NPs was achieved with 20 mg/L. This was observed with the decrease of total solids (TS) and volatile solids (VS) from 7.16 and 5.85% to 4.86 and 4.14%, respectively, which explained the high content of biogas and CH_4 observed. In addition, the highest concentration of Fe_3O_4 NPs (20 mg/L) aided to consume the largest amount of substrate and just remained 4.23 and 3.87% of TS and VS, respectively. Therefore, the concentration of 20 mg/L of iron NPs showed the best performance in AD and between them, Fe_3O_4 NPs had the major impact. In addition to that, their effects lasted through all the experimentation time (Abdelsalam et al. 2017).

Earlier, the same group of authors experimented with iron NPs and obtained almost equal results. Enrichment of cattle slurry with 20 mg/L of Fe or Fe_3O_4 NPs showed an enhancement in biogas volume production of 1.5 and 1.7 times the control volume (667 mL/day) and CH_4 enhancement production was 1.67 and 2.16 times the content in control (326.2 mL/day) for Fe and Fe_3O_4 NPs, respectively. Furthermore, biogas and CH_4 production started on the first day of the experiment due to the reduction in lag-phase of the microbial consortium. At the end of the experimentation time (40th day) with 20 mg/L of Fe NPs, TS and VS decreased from 7.16 and 5.85% to 5.25 and 4.5%, respectively; for Fe_3O_4 NPs, the resultant values were 4.95 and 4.15%. Even if initial concentrations of TS and VS were the same as in their later study, consumption probably decreased due to the shorter time of experimentation (10 days less). Another remarkable feature was that Fe_3O_4 NPs demonstrated the highest methane content in biogas at the startup phase (31.79% for 5 days) and reached the highest peak between 24th–26th day with approximately 79% of CH_4 content (Abdelsalam et al. 2016).

Moreover, the addition of Fe_3O_4 NPs to an inoculum can be used to accelerate CH_4 production. Paddy soil, for instance, was enriched with 20 mM of Fe_3O_4 NPs to evaluate acetate consumption rate for five generations in AD (approximately 130 days) and helped to reduce the required time to end the process by 23% to a maximum of 48% compared with the control. A similar effect was already discussed in the last paragraph (Abdelsalam et al. 2016). Thus, the CH_4 production rate in the first generation was approximately 61.2% higher compared with the control, allowing the possibility to reduce operation time. However, if cumulative CH_4 production is compared with that of the control experiment at the final stage, similar content was observed in all the treatments (21 mL). Average CH_4 yield was 0.96 moles CH_4 per 1 mole acetate, very close to the theoretical value (1 mole CH_4 per 1 mole acetate), which demonstrated that almost all acetate was converted to methane. Furthermore, different concentrations of Fe_3O_4 NPs were tested, e.g., 20, 80, 160 and 320 mM, with no significant difference in the acceleration effect. Hence, the improvement of these NPs was dose-independent when concentration exceeded 20 mM. For the next part of the experiment, the authors decided to evaluate the paddy soil enriched with Fe_3O_4 as a possible inoculum for a two-stage AD process. For this purpose, CH_4 produced from an artificial medium of ethanol and acetate, in two cycles of incubation, was measured. Their results demonstrated the production of 28.9 and 32.7 mL of CH_4 in its respective cycle compared with the 11 and 3.6 mL of CH_4 of the control (paddy soil without NPs). Therefore, an improvement of 1.5 and 5.5 times in CH_4 production was calculated. Furthermore, the same enriched paddy soil was applied in the effluents of an H_2-producing stage reactor. CH_4 production was 31 mL on day 10 of the first cycle, 37 mL on day 8 of the second cycle compared with the 28 mL on day 11 in the first cycle and 36 mL on day 18 of the second cycle. Therefore, the addition of Fe_3O_4 NPs accelerated CH_4 production and reached an enhancement of 0.2 and 1.7 times compared with the control in the first and the second cycle, respectively. Then it can be concluded that 20 mM of Fe_3O_4 NPs can be a viable option for AD improvement by the increase in CH_4 production and reduction of the process retention time. Moreover, paddy soil demonstrated to be an adequate inoculum for the AD process (Yang et al. 2015).

It is necessary to know the limit range of NPs that can change a beneficial effect into a detrimental one. In that way, only useful concentrations would be applied for further experiments. This leads to the study of the effect of large NPs concentrations on AD performance. Higher concentrations of Fe_3O_4 NPs had been employed in AD. For instance, the addition of 100 mg/L of Fe_3O_4 NPs to an anaerobic reactor treating waste showed that cumulative biogas production per gram of organic matter increased 180% compared with the control and reached a value of approximately 2,800 mL at the end of the 60th day of experimentation. After the determination of biogas composition, the authors found an increase of 8% in CH_4 compared with the control

(48.8 ± 0.2%). Therefore, the reactor reached 56.7 ± 2.1% of CH_4 content in biogas and consequently, improved cumulative CH_4 production by 234%. Furthermore, concentrations of other additional gases in biogas, e.g., CO_2 and H_2S were detected less than 1% of the total volume (Casals et al. 2014).

Similarly, in another research, the addition of 500 and 1,000 mg of nZVI in anaerobic batch reactors for sludge AD was studied. At the seventh day, a reactor with 1,000 mg/L of nZVI had the highest content of ammonia nitrogen (2,600 mg/L), probably derived from the denitrification process of organic nitrogenous material. That sudden increase helped to counter the pH decreased by VFAs accumulation and maintained pH values between 6.5–7.5. On day 10, ammonia concentration decreased to 600 mg/L and for the next few days continued at similar levels (between 600–800 mg/L). Therefore, nZVI stabilized the pH in the medium and provided an adequate environment for methanogenic activity. To understand the effect of nZVI particles better, the authors also monitored the changes in the COD, VFAs, biogas and CH_4. With 1,000 mg/L of nZVI, COD removal rate was 79.45%, higher than 500 mg/L of nZVI and the control (76.54%). Hence, the addition of nZVI promoted COD degradation. The latter effect could be attributed to the acceleration of enzymatic activity coupled with microbial growth. VFAs production with concentrations of 500 and 1,000 mg/L of nZVI NPs showed similar and stable behavior. In the acidogenic phase, until day 15, the first concentration almost reached 2,800 mg/L of VFAs while the second recorded 3,841 mg/L. After this period, methanogenic activity decreased in both values (between 800–1,200 mg/L of CH_4) and it remained stable for the rest of the experiment. This suggested that nZVI particles enhanced acid-producing bacteria that will form VFAs useful for methanogenic archaea. Another positive effect by the addition of nZVI was that cumulative biogas production increased 7.30 and 18.11% compared with the control (500 mL at 35th d) for 500 and 1,000 mg/L of nZVI NPs, respectively. After the eight-day, CH_4 content dramatically increased in all reactors. At the end of the experiment, a reactor with 1,000 mg/L of nZVI reached the higher CH_4 content with 68.75%. If control reached a maximum value of 61.82% approximately, then there was an increment of 1.11 times the CH_4 content compared with the control (Jia et al. 2017). From these results it is possible to observe that correct development of AD by the addition of nZVI particles is related to high degradation of COD and VFAs, coupled with the enhancement of biogas production and CH_4 content. In addition to that, the particles aided to maintain the physicochemical parameters stable.

To assess the effects on the parameters that previously mentioned, it also has to be taken into account the exposure time between NPs and the components of the process because the triggered effects can change according to this relationship.

In order to understand the effect of nZVI particles in AGS (Anaerobic Granular Sludge), short and long-term effect experiments were carried out with concentrations of 0, 15 and 30 mM of nZVI particles. In short-term reactors (225 hours), the addition of nZVI did not improve the degradation rate of glucose compared with the control. However, both concentrations improved the CH_4 content. With 30 mM of nZVI particles, CH_4 increased from 5.74 to 7.45 mmol, approximately 30% higher than the control. In addition to that, this dosage allowed a longer phase of propionic acid consumption and kept pH stable between the ranges of 6.9–7.3. Furthermore, in all experiments COD did not increase with nZVI addition, suggesting that NPs did not cause an important cell membrane disruption (He et al. 2017).

However, there is evidence that nZVI particles increased COD content in sludge AD (Yu et al. 2016, Jia et al. 2017). In long-term experiments (33 days), the addition of 15 or 30 mM of nZVI particles to AGS did not cause negative effects and even if the experiment was divided into four cycles, application of more nZVI was not involved because almost all Fe^0 remained trapped in the AGS (98.6%). However, the degradation of glucose and CH_4 production were similar to control after the first cycle (He et al. 2017). The lack of improvement in the rest of the cycles can be related to the structure of AGS and the behavior of microbial groups. The AGS structure contains three layers, each one inhabited by different microbial groups. External contains a heterogeneous population of fermentative and acetogenic bacteria with hydrogenotrophic methanogens; the second one has a varied microbial population and the internal layer is dominated by aceticlastic methanogens. Following this arrangement, more NPs will interact with the microorganisms present in the outer layer, but the existent pores of the granular sludge could allow the passage of nZVI and reach aceticlastic methanogens (He et al. 2017). Nevertheless, this structure can be modified with the production of extracellular polymeric substances (EPS) that are a common mechanism of microorganisms against metallic NPs. Their protective mechanism consists of the formation of dense aggregates of biomolecules, e.g., carbohydrates, DNA, lipids and proteins with a little pore size (10–50 nm) that surround the microorganism and act as a barrier to stop NPs penetration. Furthermore, the barrier can be released when it gets saturated with these particles (Tang et al. 2018). It is possible that this aggregation was the reason for the limited diffusion of nZVI particles through the pores of the AGS, resulting in less interaction with aceticlastic methanogens that inhabited the inner layer of the granule. This statement suggests that hydrogenotrophic methanogenesis was the main pathway to produce CH_4 (He et al. 2017). However, it must be highlighted that if dense aggregates continuously form in the sludge, there will be a point in which even external microorganisms would not be in contact with nZVI particles and their beneficial effect would not be achieved.

The interaction between EPS and nZVI has also demonstrated to decrease the protein and polysaccharide content in EPS. In the last cited experiment (He et al. 2017), protein and polysaccharide concentrations in the control reactor were 155.45 ± 19.65 and 23.56 ± 3.75 mg/g-VSS, respectively. The addition of 15 mM of nZVI decreased both parameters to 143.15 ± 12.79 and 15.33 ± 0.32 mg/g-VSS, respectively. The same effect occurred with 30 mM, decreasing the protein and polysaccharide content to 123.58 ± 10.21 and 15.53 ± 2.43 mg/g-VSS, respectively. Moreover, this interaction increased importantly Fe^{2+} content, being 2.48 times higher than the control experiment. It is speculated that the release of these ions would help to prevent the toxic effect of nZVI particles in the microbial consortium. However, they might precipitate with S or P present in AGS with the consequent formation of FeS or stable complexes with $P_3O_4^-$, respectively, affecting the bio-availability of these elements to microorganisms. Nevertheless, it was demonstrated in the study that both elements were not around the iron NPs clusters in an AGS slice. Instead, the surface of AGS had nZVI particles in the form of FeO(OH) or iron oxides.

The H_2 concentration in AD systems can increase when water and nZVI particles interact. This is represented in Equation 2 (Yu et al. 2016).

$$Fe^0 + H_2O \rightarrow Fe^{2+} + H_2 + 2OH^-, \Delta G^0 = -150.5 \text{ kJ mol}^{-1} \quad (2)$$

A study determined that 30 mM of nZVI produced 0.4 mmol of H_2 compared with the 0.06 mmol (approximately) of the control. The latter statement was supported by an extra experiment without inoculum which produced similar values of H_2. In addition to that, CO_2 consumption increased in treatments with nZVI compared with the control. The authors confirmed that this consumption was related to H_2 decrease and consequent CH_4 production by the stimulation of hydrogenotrophic methanogenesis (He et al. 2017).

Iron NPs have been also applied in waste anaerobic granular sludge (AGS) as well. An experiment analyzed the effect of iron NPs in AGS. The additives used were nZVI and Fe_2O_3 NPs at concentrations of 10 and 100 mg/g total soluble solids (TSS), respectively. The enhanced CH_4 production measured was 1.2 and 1.17 times the control (181.72 mL/g volatile suspended solids or g-VSS) for nZVI and Fe_2O_3 NPs respectively. Even though Fe_2O_3 NPs did not release ions due to its low solubility, there was an enhancement of the AD process (Wang et al. 2016).

Another research determined the participation of nano and micro ZVI in AD of sludge fermentation liquor (SFL). At the end of the experiment (32nd day), cumulative CH_4 with ZVI was 108.24 mL/g-VS, 46.1% higher than the control. However, the increment was lower compared with the different sizes of ZVI (μm). From the total CH_4 production with nZVI just 13.5% was produced in the first 3 days compared with the 32.9% of the control. This low yield could be caused by the high reactivity of nZVI that released more H_2 than other treatments and coupled with the consequent Fe^{2+} release that can react with P to form precipitates and limit elements uptake. These effects would suppress microbial activity. Nevertheless, cumulative CO_2 production with nZVI (4.08 mL/g-VS) was lower compared with the other treatments and the control (17.48 mL/g-VS), this factor improved biogas quality (Yu et al. 2016). Even if in this case, nZVI performance did not surpass the efficiency of micro ZVI, it has to be taken into account that the concentration used was importantly high. Indeed, 10 g/L (10,000 mg/L) was the highest nZVI concentration mentioned in this chapter and it was demonstrated that even lower concentrations have triggered negative effects (Jia et al. 2017). Regarding the initial total VFAs content in the experiment, a concentration of 32.1 mg-COD/L was measured, composed mainly of acetic and propionic acids. Treatment with nZVI increased this content and reached a peak of 610.8 mg-COD/L at day 3 as a result of cell membrane disruption with the consequent release of its components. Then, VFAs consumption recovered and CH_4 production was observed. At the end of the experiment, the residual concentration of VFAs was 64.9 mg-COD/L. Consumption of both acids suggested the favorable reductive environment created by these NPs for AD development (Yu et al. 2016).

Due to the complex interactions in AD, it is difficult to differentiate the effect of NPs in a specific stage of the process. However, it is possible to inhibit microbial groups that participate in the other stages. For instance, to study the effect of nZVI particles in the hydrolysis-acidification stage for three days, it was necessary to avoid VFAs consumption by methanogens. To get rid of them, 40 mM

2-bromoethanesulfonate was added. In that study, the addition of nZVI enhanced acid-producing stage and reached a total VFAs production of 1,082.1 mg-COD/L, a value that overpassed the production of other treatments and the control (approximately 600 mg-COD/L) when compared the VFAs behavior in complete AD with the hydrolysis-acidification process in the reactor with nZVI. The accumulation of acetic acid and low CO_2 content led the authors to suggest that VFAs consumption was due to homoacetogenic bacteria activity coupled with aceticlastic methanogenesis instead of H_2/CO_2 formation. Homoacetogenic bacteria can use CO_2 and H_2 to form acetate (represented by Equation 3) and acetoclastic methanogens consume this VFA and produce CH_4 (Yu et al. 2016):

$$4H_2 + 2CO_2 \rightarrow CH_3COO^- + H^+ + 2H_2O, \Delta G^0 = -95 \text{ kJ mol}^{-1} \quad (3)$$

To understand the enhancement by NPs addition in a better way, it is possible to relate their effects with the same compounds at a normal scale. For instance, the addition of ZVI particles has previously demonstrated to accelerate sludge digestion by decreasing the content of soluble proteins and polysaccharides. In addition to that, VFAs production with a dosage of 4 g/L of ZVI was 37.3% higher compared with no-dosage treatment. Furthermore, among total VFAs production acetate was predominant with a presence of 48.7%. The presence of ZVI particles limited propionate production due to the reductive environment created by the compound, which increased butyric-type and acetic-type fermentation. All these positive effects culminate in CH_4 production increment (Feng et al. 2014). ZVI particles can also be used to reduce azo dye which is an organic compound commonly employed in the textile industry. ZVI addition in an anaerobic reactor system for wastewater treatment recorded a higher COD and color removal (59.7 and 74.4%) compared with the control (56.8 and 49.7%). Moreover, when the temperature was decreased from 35°C to 25°C, the reactor gradually approached the color removal rate presented at a higher temperature. However, COD removal just reached 49.7%. These results suggested that ZVI particles helped to relieve stress conditions caused by low temperatures.

Another positive effect by ZVI addition was an increase in the content of CH_4 in biogas, which reached 50.4%. Instead, biogas produced by control only contained 34.7% of CH_4 and the presence of 4.4% of H_2. Moreover, HRT with ZVI reduced 12 hours of the treatment time and presented almost the same activity achieved by the control for 24 hours of operation. Furthermore, the pH regulation problems due to acidogenesis stage were minimal because of ZVI regulation as explained in Equation 4 (Zhang et al. 2011).

$$Fe^0 - 2e \rightarrow Fe^{2+} \quad (4)$$

According to the effects accounted for ZVI particles, iron NPs can act in a similar way to enhance the AD process but with the advantage of a larger surface area.

An increase of CH_4 production by the addition of iron NPs can have more explanations. As it was mentioned in previous paragraphs, Fe additives promote acetate generation, a required substrate for aceticlastic methanogenesis. Furthermore, when the same compound is consumed by syntrophic oxidizing-bacteria, they will produce CO_2 and H_2, the required substrates of hydrogenotrophic methanogens (Karakashev et al. 2006). Another means to promote this effect is the substitution of

H_2 by Fe NPs as electron donors, represented by Equations 5 and 6 (Feng et al. 2014, Yu et al. 2016).

$$CO_2 + 4Fe^0 + 8H^+ \rightarrow CH_4 + 4Fe^{2+} + 2H_2O, \Delta G^0 = -150.5 \text{ kJ mol}^{-1} \quad (5)$$

$$CO_2 + 4H_2 \rightarrow CH_4 + 2H_2O, \Delta G^0 = -131 \text{ kJ mol}^{-1} \quad (6)$$

These reactions can help to control the negative impact caused by the constant formation of VFAs and their breakage that release hydrogen ions and acidify the AD process. Thus, NPs offer a mean to regulate pH. But excessive H_2 consumption can also increase pH. Therefore, a strategy is required to control the shift of this parameter due to NPs. Another advantage shown by the equations is the improvement in biogas quality by the removal of CO_2 particles (Abdelsalam et al. 2017).

The bioavailability of trace metals can be limited by the medium. It is assumed that when iron NPs were added to an aqueous medium, they produced insoluble precipitates of ferric hydroxide. When these particles are reduced to Fe^{2+}, they become soluble and available for microorganisms (Casals et al. 2014). Therefore, the release of Fe ions (Fe^{2+}/Fe^{3+}) allows the microorganisms to assimilate them (Abdelsalam et al. 2016). A special feature of iron ions is their capacity to gain or lose electrons that make them useful cofactor in proteins. Bioavailability of ions also changes with the oxidation state of the molecule. For instance, solubility for Fe^{2+} and Fe^{3+} is 0.1 M and 10^{-18} M, respectively (Casals et al. 2014).

In addition to iron and iron oxide NPs, other experiments have tested more metals. For instance, Ni ferrite (Fe_2NiO_4) NPs were added in synthetic municipal wastewater to obtain a concentration of 100 mg/L of Ni. After 7 days of AD, it demonstrated to enhance cumulative CH_4 production 1.3 times compared with the control (50 mL) (Chen et al. 2018). Application of elemental NPs, such as Co and Ni, is another option for AD improvement. An investigation analyzed the production of biogas and CH_4 from the previously mentioned elements in concentrations of 1 and 2 mg/L, respectively. In the same order, the improvement of biogas volume was 1.7 and 1.8 times compared with the control (667 mL/day). In the same fashion, CH_4 production was highly benefited and showed an increase of 2 and 2.17 times the volume of the control (326.2 mL/day) (Abdelsalam et al. 2016). As it can be seen in other experiments, NPs accelerated process development and increased organic matter decomposition. Their beneficial effect could be related to their function as trace elements. Likewise, the addition of Fe, Ni and Co compounds has demonstrated to reduce lag phase, microbial stress and the required time to reach biogas peak. They have also stimulated specific methanogenic activity (SMA) (Krongthamchat et al. 2006).

Carbon compounds have been used as NPs in the form of graphene with the purpose of being applied in AD and other industries. Tian et al. (2017a) studied the short and long-term influence of nano-graphene in the AD process. Short-term experimentation (14 days) with batch reactors showed that CH_4 production was dependent on the dosage of nano-graphene. Therefore, 30 mg/L increased CH_4 production 1.17 times while 120 mg/L increased it 1.51 times, all compared with the control (0.22 ± 0.01 mmol-CH_4/g-VSS/d). It was possible that the activity of aceticlastic methanogens was stimulated by nano-graphene because acetate was a prevalent substrate in the medium. In the long-term experimentation with three up-flow anaerobic sludge blanket (UASB) reactors, temperature and HRT were modified to determine reactor efficiency and diverse effects were present due to these variations. In the first 13 days with a HRT of 12 hours at 20°C, all reactors showed similar biogas production. However, with the decrease of HRT to 6 hours and the temperature to 10°C (day 36 to 55), CH_4 production fell. Yet, reactors in which 30 mg/L of nano-graphene were added had the best methane production (12.8 ± 0.4 mL/g-VSS/d) with 14.3% more than control production (11.2 ± 0.6 mL/g-VSS/d). Furthermore, CO_2 production was higher compared with 120 mg/L of graphene NPs. On the other hand, the reactor enriched with 120 mg/L of nano-graphene lowered methane production to 77.6% (8.7 ± 1.2 mL/gVSS/d) of the control, which evidenced a slight inhibition caused by a long-term exposure (55 days). The average COD removal efficiency with 30 mg/L of graphene NPs decreased by approximately 38% when the temperature changed from 20°C to 10°C, similar to the control behavior. Even with those important changes in the reactor performance, pH remained stable in all treatments (Tian et al. 2017a). Therefore, according to the duration of a process, a shift from high to low nano-graphene concentration can help to reach the best performance in AD.

In relation to carbon compounds, another study evaluated the effects of either 1,500 mg/L multi-wall carbon nanotubes (MWCNTs) or 750 mg/L oxide nanoparticles (Fe_2O_3) in expanded granular sludge bed (EGSB) reactors. In this experiment, granular sludge from a beer industry was applied for AD of beet sugar industrial wastewater. The influent contained a COD of 3,000 ± 300.5 mg/L and the dominant treatment for COD degradation shift over time. In the first three days with a HRT of 48 hours, COD effluent of control, MWCNTs and Fe_2O_3 NPs reactors were 738, 837 and 1,089 mg/L, respectively. The addition of NPs may have been responsible for lower COD degradation in reactors due to the required time for microbial adaptation. On the 18th day, the reactor with Fe_2O_3 NPs demonstrated a COD removal of 95%, while the other reactors showed removal of 92%. On the 48th day, after HRT was adjusted to from 48 to 24 hours, COD removal reached 95, 94 and 89% for Fe_2O_3, control and MWCNTs reactors, respectively. Finally, after the last modification of HRT to 12 hours at the end of operation time (74th day), remaining COD concentrations were 142, 154 and 197 mg/L for MWCNTs, Fe_2O_3, and control experiments, respectively. Therefore, AD enriched with MWCNTs or Fe_2O_3 NPs and operated with short HRT enhanced COD degradation compared with the control. Either in biogas or CH_4 production, inhibition by NPs addition was absent. Similarly to COD removal, the long-time of experimentation changed the trend in CH_4 production in the reactors. For the initial 24 days, values of cumulative biogas formed were 485.7, 395.2 and 370.5 mL/g-VSS for Fe_2O_3, control and MWCNTs reactors, respectively. However, on 38th day, the Fe_2O_3 reactor produced the highest methane content (1,857 mL/g-VSS), while MWCNTs treatment shifted to second place and the control to the last with 1,345 and 1,030.4 mL/g-VSS, respectively. Accordingly to final values of biogas production were 25,144.4, 21,876.0 and 20,082.8 mL/g-VSS for Fe_2O_3, MWCNTs and control, respectively, with maximum CH_4 production of 8,374.9, 7,313.2 and 6,496.5 mL/g-VSS for the same reactors. Thus, MWCNTs and Fe_2O_3 treatments increased 12.6 and 28.9% of the CH_4 content compared with the control. Markedly, the increment of biogas and cumulative CH_4 were observed when HRT changed to 6 hours. Biogas production started to decrease after 66 days of experimentation and eight days before the end of the experiment. Moreover, the H_2 consumption evolved well while the experiment advanced and both treatments showed a similar trend as the control. Nevertheless, less H_2 was released in the reactor with Fe_2O_3 NPs, which was attributed to its fast consumption for CH_4 production. Acetic and propionic acids generation was high until day 20, especially in the presence of Fe_2O_3 NPs. After this period, the concentrations of both acids dropped to low levels (between 1–2 mg/L). Furthermore, when the 50th day of experimentation was reached, the concentration of the acids decreased even more and at day 74 the concentration was undetectable. Therefore, Fe_2O_3 treatment was the best regarding VFAs consumption, followed by MWCNTs and control reactors (Ambuchi et al. 2017).

When Mn oxides are present, they serve as electron acceptors and Mn respiration can be coupled with methanogenesis. Thus, when these compounds are added as NPs, their enhancement effect and microbial interaction can change due to the characteristics of the particles. In order to comprehend this behavior, the research studied the influence of MnO_2 NPs addition in microbial communities and the possible modifications to NPs due to this interaction. For that, concentrations of 10, 50, 100, 200 and 400 mg/g-VSS of MnO_2 NPs were added in anaerobic batch reactors inoculated with anaerobic activated sludge from an UASB reactor. The CH_4 production with the first two concentrations was almost the same as the control, while the increase of 100 mg/g-VSS recorded a slight enhancement. Improvement in CH_4 production was stronger with concentrations of 200 and 400 mg/g VSS. In contrast, with the highest concentration, the CH_4 yield rate increased 42% (38.2 ± 1.6 mL/g-VSS/d) compared with the control (26.9 ± 1.1 mL/g-VSS/d) and reached an increase of 83.3 ± 3.7 mL/g-VSS. Furthermore, the addition of NPs did not increase EPS content, instead protein and carbohydrate content remained similar or below the control (80 and 60 mg/g-VSS for carbohydrate and protein). Therefore, microbial stress did not occur in the reactors. Likewise, as the concentration of MnO_2 NPs increased, cumulative CO_2 went down. When the experiment finished (192 hours after), concentrations of 200 and 400 mg/g-VSS of MnO_2 NPs were shown to decrease CO_2 content by 25.1 ± 2.7 and 48 ± 2.2 mL/g-VSS, respectively. Consequently, CH_4 content in biogas increased from 64.7 to 72.8 and 81.8% for the same concentrations. In order to explain what caused this enhancement, it is important to know that MnO_2 NPs cannot serve directly as electron donors for CO_2 reduction. To achieve this, it is necessary to reduce Mn oxides to Mn^{2+} in order to provide electrons. In the experiment, dissolved Mn^{2+} increased when higher concentrations of NPs were added and the authors argued that the production of ions was not related to abiotic mechanisms. Instead, almost all reduction was due to microbial manganese reduction. To determine if Mn^{2+} ions were responsible for methanogenesis improvement, a complementary experiment that used a gradient of Mn^{2+} in AD was realized. However, the experiment showed that ions remained stable over time and were not responsible for methanogenesis improvement. Variation in VFAs content was also studied in relation to MnO_2 NPs. It was shown that total VFAs content increased sharply in the first 24 hours and then continued more slowly until 96 hours. In addition to that, enrichment with MnO_2 NPs lowered VFAs accumulation rates compared with the control. Cumulative acetate in reactors with more than 50 mg/g-VSS of MnO_2 NPs reached their peak at 24 hours too but such concentrations decreased immediately to zero at 144 hours. However, lower dosages of MnO_2 NPs and control reactor recorded acetate concentrations at this hour. Therefore, the degradation rate of total VFAs, especially acetate, was enhanced with 50, or more mg/g-VSS of MnO_2 NPs. This consumption rate of acetate combined with Mn^{2+} generation suggested that microbial manganese reduction required this substrate for their metabolism. Therefore, microorganisms with such metabolism can compete by substrate with acetoclastic methanogens. That would explain why at higher concentrations VFAs consumption was faster but methanogenic activity was not importantly improved. However, the authors were not sure about this statement because the reactor with 400 mg/g-VSS showed a very different pattern. Nevertheless, they hypothesized that the addition of MnO_2 NPs could increase electron flow and reduce the effect of competition (Tian et al. 2017b).

Other effects of MnO_2 NPs in AD that were studied included their effect in propionate and butyrate degradation by microbial activity. In a study by Tian et al. (2019), a concentration of 50 mg/g-VSS of MnO_2 NPs was used. The CH_4 production rate derived from propionate degradation was enhanced and showed an increment of 25.6% compared with the control, which increased from 0.43 ± 0.03 to 0.54 ± 0.02 mL/g-VSS/h. The CO_2 production reached its maximum accumulation at 120 hours followed by a decreasing period, a trend that was different from the one measured in control, in which CO_2 was always produced. With butyrate as substrate, addition of 50 mg/gVSS of MnO_2 NPs increased CH_4 production rate from 0.92 ± 0.02 mL/g-VSS/h to 1.12 ± 0.01 mL/g-VSS/h, which represented an increment of 21.7%. In addition to that, CO_2 content was less compared with the control. Furthermore, Mn respiration was measured by the changes in Mn^{+2} concentrations during AD. Reactor for propionate degradation enriched with 50 mg/g-VSS of MnO_2 NPs showed a fast increment of Mn^{2+} production during the first 48 hours until its maximum value registered at 72 hours. On the other hand, butyrate degradation with the same concentration reached its peak at 24 hours, and there was no change after that. Regarding content in biomolecules such as carbohydrates and protein, in the experiment of propionate degradation, there was an increment of carbohydrate and a decrease of protein content compared with the control. While in butyrate with the same NPs concentration, there was an increment of both compounds (24.7% and 60.3%) of carbohydrates and proteins, respectively, compared with the control. Therefore, the addition of this NPs concentration caused stress to microbial groups, enhanced acid degradation and CH_4 production as a response (Tian et al. 2019).

There are new investigations that focus on the use of bimetallic NPs. For instance, it was studied the effect of iron copper bimetallic NPs (nZVI/Cu°) at different concentrations on biogas and CH_4 production in comparison to those with the addition of nZVI particles. In the physical form of nZVI/Cu°, these NPs were chain-like clusters, each one surrounded with an iron core, and oxide shell and cooper particles randomly coated iron particles on the surface. During their application in AD, there was a relation between dissolved Fe^{2+} ions and NPs concentration. With

500 mg/L of nZVI particles, the content of Fe^{2+} released was 114.65 mg/L after 24 hours. Bimetallic NPs also showed a similar behavior at the same concentration. Ferric ions (Fe^{3+}) release were stable and low quantities. Even at 500 mg/L of nZVI NPs, the content did not surpass 55 mg/L. However, in bimetallic NPs at 1,500 mg/L, Fe^{3+} ion concentration surpassed 600 mg/L. Until day 5, the reactor with 3,000 mg/L of nZVI increased Fe^{2+} release from 99.99 mg/L to 532.58 mg/L and then decreased to 285.99 mg/L at the end of the experiment (day 14). At the same concentration, maximum produced Fe^{2+} concentration from nZVI/Cu^0 was reached at day 4 and remained stable after that, with a value of 160.72 mg/L. This was an important parameter because biogas production was correlated with Fe^{2+} ions release. Biogas cumulative production from control was 229 mL; however, reactors enriched with NPs exceeded this value, except reactors with 50 and 100 mg/L nZVI that produced 192 and 209 mL of biogas, respectively. These results were attributed to possible low Fe^{2+} ions production. Sludge treated with 1,500 mg/L of nZVI/Cu^0 demonstrated the highest production. Nevertheless, 3,000 mg/L of the same NPs decreased biogas production by 31.1%. It is possible that 3,000 mg/L of nZVI/Cu^0 NPs affected AD by fast H_2 production due to the presence of Fe^0 or by toxic divalent Cu ions formed by released Cu NPs. Cumulative biogas production from reactors with 50, 100, 250, 500, 1,500 and 3,000 of nZVI/Cu^0, NPs was respectively 337, 355, 373, 516, 693 and 447 mL. These values in percentages surpassed the control by 47.16, 55.02, 62.88, 125.33, 202.62 and 108.30%, respectively. Cumulative biogas production from 250, 500, 1,500 and 3,000 of nZVI, NPs were 1.19, 1.31, 1.73 and 2.05 times the biogas produced from the control, respectively. In reactors with nZVI particles, maximum biogas production was observed in the first six days. In contrast, treatments with nZVI/Cu^0 reached their peak production after the two first days. In all experiments, when the maximum peak was reached there was not a considerable biogas production on later days. Therefore, these experiments demonstrated the effectiveness of NZVI/Cu^0 bimetallic NPs in AD performance and the capacity to surpass nZVI particles effect with relation to biogas production. Regarding CH_4 generation, concentrations of 500, 1,500 and 3,000 mg/L nZVI stimulated CH_4 production 24 hours after the start of the experiment. By the second day of experimentation, CH_4 content in reactors containing 500, 250, 1,500 and 3,000 mg/L nZVI increased to 62%, 57%, 59% and 50%, respectively. These results indicated a reduction of lag-phase for methanogenic populations. In addition to that, reactors enriched with 500 mg/L of nZVI NPs or more, increased CH_4 content to 82% by the seventh day to the end of the experiment. These results were opposite to the values of reactors with nZVI/Cu^0 NPs since they did not reach the same values of CH_4 content. Just 500 and 1,500 mg/L of bimetallic NPs exceeded 50% in CH_4 content, being approximately 63% and 70% %, respectively. From the seventh day, until the end of the experiment, there was not important CH_4 production. Therefore, it can be suggested that methanogenesis stage was inhibited and biogas produced after this period was mainly composed of CO_2. Another effect observed by the addition of NPs was in Soluble COD content. Initial SCOD was 5,453 mg/L and increased to its maximum at day one. After NPs addition, SCOD values remained stable for the rest of the experiment. Therefore, the authors suggested that NPs allowed rapid solubilization of sludge which increased SCOD. Ranges for SCOD with nZVI/Cu^0 NPs were between 11,889

and 21,652 mg/L, higher if compared with SCOD of nZVI that ranged between 7,650 and 21,507 mg/L and with the control (9,220 mg/L). All these results suggested that a coating of Cu NPs enhance nZVI particles activity in the fermentation of sludge, coupled with better production of biogas. However, this result was observed with a specific concentration (1,500 mg/L) (Amen et al. 2018). Due to the novelty of this investigation, further research is necessary to understand and determine the potential of bimetallic NPs. In addition to that, long-term experimentations and the impact in microbial communities would be very useful. Possibly, more combinations of NPs would overcome the deficiencies observed between them and improve in a better way AD for CH_4 production, process stability and microbial activity.

In summary, there is a vast array of NPs that can be implemented to improve AD development. Among the advantages of their use is high substrate degradation that will cause the release of available substrates for biogas and CH_4 production, improvement of biogas quality, reduction in process time and physicochemical parameters stability.

3.2 Different Effects in AD

NPs that have demonstrated an improvement effect in AD can also promote inhibition or not affect under different circumstances (Table 2). The reasons for this could be the distinct enrichment concentrations in a substrate, the oxidation state of NPs ions, time of exposure, pH of the media and interaction of NPs with other compounds among others (Bożym et al. 2015).

The effect of four NPs (nZVI, Ag, Fe_2O_3 and MgO) was evaluated to determine their impact in CH_4 production in anaerobic granular sludge. With a concentration of 1 mg/g-TSS of the previous NPs, there was no significant difference in the effect of CH_4 production compared with the control (181.72 mL/g-VSS). When the sludge was subjected to higher concentrations of Ag or MgO NPs (500 mg/g-TSS), there was a deleterious effect in cumulative CH_4 production. With each concentration of NPs, the concentration of released ions changed. In this case, 3.3 mg/L and 9.8 mg/L of Ag^+ and Mg^{2+}, respectively, were suggested to cause the inhibitory effect. Therefore, a high concentration of NPs coupled with the production of free ions can affect heavily AD process (Wang et al. 2016).

It is important to verify VFAs production and behavior when NPs are added because these compounds are the precursors of CH_4 but also inhibitors of methanogenesis. Unlike other studies that only focused on the direct enrichment of NPs in wastewater sludge, there was an inquiry that compared the direct application of 50 mg/L of Cu NPs into the sludge against their addition in the system through feeding. The purpose was to mimic real conditions in wastewater treatment plants and evaluate if there existed a difference in NPs effect according to the phase they were added. Their results showed that NPs added directly to the sludge decreased maximal VFA production from 223.5 to 120 mgCOD/g-VSS. Nonetheless, resultant Cu NPs from the addition to the biological process had no impact on VFAs production (225.1 mgCOD/g-VSS). In the same way, there were differences in the content of soluble proteins and polysaccharides. In this case, NPs added initially favored sludge solubilization. Another effect caused by Cu NPs was the smaller size

Table 2. Different effects caused by NPs in AD performance.

Nanoparticle	Size (nm)	Concentration	Response	Reference
Ag	170 ± 7.9	1 mg/g TSS 500 mg/g TSS	No effect Inhibition	Wang et al. 2016
MgO	154 ± 7.9	1 mg/g TSS 500 mg/g TSS	No effect; Almost total inhibition	Wang et al. 2016
Fe_2O_3	108 ± 7.9	1 mg/g TSS	No effect	Wang et al. 2016
nZVI	128 ± 7.9 55 ± 11	1 mg/g TSS 1 mM 10 mM 30 mM	No effect 79.6% 79.6% 30.8% of cumulative CH_4 in control	Wang et al. 2016 Yang et al. 2013
Fe_2NiO_4 (Ni)	46 ± 7.9	1 mg/L 10 mg/L	No effect No effect	Chen et al. 2018
Fe_4NiO_4Zn (Ni)	89 ± 7.9	1 mg/L 10 mg/L 100 mg/L	94.4% 34.4% 1.6% of cumulative CH_4 in control	Chen et al. 2018
CuO	30	10.7 mg-Cu/L 30.2 mg-Cu/L	50% inhibition 100% inhibition CH_4 production	Luna-delRisco et al. 2011
ZnO	50-70	57.3mg-Zn/L 181mg-Zn/L	50% inhibition 100% inhibition CH_4 production	Luna-delRisco et al. 2011
TiO_2	< 25	6, 30 and 150 mg/g-TSS	No effect	Mu and Chen 2011
Al_2O_3	< 50	6, 30 and 150 mg/g-TSS	No effect	Mu and Chen 2011
SiO_2	10–20	6, 30 and 150 mg/g-TSS	No effect	Mu and Chen 2011

of sludge particles compared with the control being 106 and 194 μm, respectively. Interestingly, sludge treated directly with Cu NPs showed a similar size to the control (189 μm). If biogas and its composition had measured in the experiment, maybe a correlation of VFAs and biogas production would have been found (Chen et al. 2014).

Another research determined the impact in biogas and CH_4 production by the addition of CuO and ZnO particles in bulk and nanosize during AD of cattle manure. At 14th day, 15 mg/L of these NPs reduced cumulative biogas production by 30% compared with the control. On the other hand, bulk CuO at 120 and 240 mg/L decreased biogas content by 19% and 60%, respectively. In the same way, addition of ZnO NPs at 120 and 240 mg/L decreased cumulative biogas production by 43% and 74%, respectively, compared with the control. In the case of bulk ZnO particles with the same concentrations, there was an inhibition of 18% and 72%. The inhibition of both sizes was not significantly different. NP can be toxic to microorganisms due to the damage of metal ions in cell membranes. Thus, biogas production inhibition might occur. Exposition time also influenced the effect of NPs.

In the first six days of the experiments with ZnO, biogas values were similar to control. However, CuO NPs inhibited biogas production at the moment of addition. From 11–14 days, some reactors enriched with ZnO NPs, bulk CuO and ZnO particles with less than 120 mg/L increased biogas production. For CuO NPs, concentrations of less than 15 mg/L demonstrated the same effect. However, all reactors produced less biogas than the control. Therefore, adaptation and recuperation of AD microbial consortium and activity could not occur. To determine the required concentration of compounds for 50% and 100% of inhibition in CH_4 production in cattle manure, the effective concentration values (EC) were determined. The EC_{50} for bulk and nano CuO particles was found to be 129 and 10.7 mg Cu/L, respectively, for total inhibition 330 and 30.2 mg Cu/L were required. Regarding ZnO, for EC_{50} with bulk and nanosize particles values of 101 and 57.3 mg Zn/L were obtained, respectively, and for total inhibition 246 and 181 mg Zn/L were necessary. Therefore, CuO and ZnO NPs had a major impact on CH_4 production and caused a great inhibition compared with bulk particles. NPs toxic effect might also be related to their ability to dissolve in aqueous solutions (Luna-delRisco et al. 2011).

There are more NPs that have been applied in WAS produced from WWTPs in order to study their effects. That is the case of ZnO NPs which is a compound known to inhibit microbial groups. To prove this, a study employed dosages of 1, 30 and 150 mg/g-TSS of ZnO NPs to determine their influence in WAS. The results suggested that the dosage of ZnO NPs had a direct effect on CH_4 production. A low concentration of 1 mg/g-TSS did not show any effect in the process. However, when the concentration increased to 30 mg/g-TSS, the average CH_4 production just reached 81.7% of the control volume. Furthermore, with 150 mg/g-TSS of ZnO NPs, there was an important decrease in CH_4 production, and it only reached 24.9% of the control. When the authors investigated the concentration of the released Zn^{2+} ions, 1.2, 11.6 and 17.6 mg/L from the lower to the higher value of NPs concentrations were found. Therefore, it is possible to conclude that ZnO concentrations of > 30 mg/g-TSS damage the AD process and cause complete inhibition of AD. Furthermore, due to the large difference in the first two concentrations employed, it is still unclear what value range does not cause inhibition. Finally, another feature that can be dangerous is the accumulation of ZnO NPs on the surface of the sludge. The authors suggested that ZnO NPs might inhibit AD by interfering in the conversion of acetic acid to CH_4 (Mu and Chen 2011).

NZVI are other NPs that have demonstrated positive effects in AD. However, there is the possibility that it can affect and alter microbial activity, which would result in low CH_4 production. Therefore, the effects of different concentrations of nZVI and bulk ZVI on methanogenesis were investigated. For nZVI, the concentrations were established on 0, 1, 10 and 30 mM. In the case of bulk ZVI, the concentration was kept at 30 mM. The results showed that cumulative CH_4 production with 30 mM of ZVI reached 135 mL of CH_4, higher than control (123 mL). However, the opposite effect was observed in the case of all nZVI concentrations. They achieved low cumulative CH_4 production with values of 98, 98 and 38 mL for 1, 10 and 30 mM nZVI, respectively. Furthermore, nZVI concentrations of 10 and 30 mM produced 111 and 174 mL of H_2, respectively. Generally, H_2 is required by homoacetogenic and methanogenic activities and thus H_2 partial pressure is maintained at minimal values.

But, the high presence of this gas (from 318 to 13,025 Pa) is a warning factor for AD and can inhibit VFAs oxidation and therefore methane production. In the presence of 30 mM nZVI, there was an accumulation of VFAs which was possible due to the rupture of syntrophic association between acetogenic bacteria and methanogens. Other parameters that showed destabilization at this concentration were SCOD and pH. Methanogenic growth was slow and the SCOD increased, which might be related to cell breakage. Bulk ZVI did not show high H_2 production. Therefore, it is possible to suggest that high nZVI concentration inhibited methanogens and lead to the acidification of the AD process by increasing H_2 partial pressures (Yang et al. 2013).

Other concentrations of nZVI particles had demonstrated to inhibit the AD process. An experiment that managed concentrations of 1,500 and 2,000 mg/L of nZVI particles determined a decrease in cumulative biogas production of 27.30 and 46.45%, respectively, compared with the control. In addition to that, it demonstrated an unstable cumulative biogas production. With 1,500 mg/L of nZVI NPs, biogas production stopped at day 10 and started again on 17th day approximately. In reactor with 2,000 mg/L, biogas production stopped on day 3 and re-started on day 16 approximately. Furthermore, CH_4 content in biogas was unstable during all the process but in some days reached more than 60%. Another negative effect was that both concentrations resulted in a lower COD removal rate compared with the control, 1,500 and 2,000 mg/L obtained a removal rate of 75.37 and 74.46%, respectively (Jia et al. 2017).

Another study focused on the AD process of synthetic municipal wastewater in the presence of different Ni concentrations (0, 10 and 100 mg/L) by the addition of two magnetic nanoparticles (MNPs), Ni ferrite (Fe_2NiO_4) and Ni Zn ferrite (Fe_4NiO_4Zn). After 7 days, it was found out that Ni concentrations of 1 and 10 mg/L by Fe_4NiO_4Zn MNPs, decreased cumulative CH_4 production to 94.4 (47.2 mL) and 34.4% (17.2 mL) of the control (50 mL), respectively. Furthermore, the high concentration of Ni (100 mg/L) caused nearly total inhibition of AD (1.6% of control; 0.8 mL). In contrast, low concentrations of Ni (1 and 10 mg/L) by Fe_2NiO_4 MNPs showed no difference compared with the control (50 mL). It is important to state that high NPs concentration does not ensure that microorganisms will use them. The parameter that determines their employment is their bio-availability. For instance, 100 mg/L of Ni present with Fe_2NiO_4 displayed more bio-availability than in the reactors enriched with lower concentrations of the same NPs, indicating that the major part of NPs remained trapped in the sludge (Chen et al. 2018).

More NPs have demonstrated negative effects on AD. For instance, with addition of 400 mg/g-VSS of MnO_2 NPs,CH_4 production derived from propionate degradation, decreased to 0.27 ± 0.03 mL/g-VSS/h (62.8%) less compared with the control (0.43 ± 0.03 mL/g-VSS/h). Furthermore, CO_2 accumulation was very low, its maximum value at 120 hours was 3.0 mL/g-VSS compared with 35 mL/g-VSS of the control. Regarding butyrate degradation, the addition of 400 mg/g-VSS of MnO_2 NPs severely affected the CH_4 production rate since it just reached 6.5% of the control (0.06 ± 0.01 mL/g-VSS/h). Therefore, the latter concentration inhibited the adaptation of microbial groups and decreased EPS content in propionate degradation. In contrast, EPS concentration in butyrate degradation at high MnO_2, NPs concentration was the same as the control. Consequently, it can be concluded that EPS content can show a different behavior according to the substrate (Tian et al. 2019).

There was a study focused on the addition of TiO_2, Al_2O_3, SiO_2 and ZnO NPs in order to determine their effect in WAS AD. It was shown that concentrations of 6, 30 and 150 mg/g-TSS of TiO_2, Al_2O_3 and SiO_2 NPs did not affect CH_4 production compared with the control. The 6 mg/g-TSS of ZnO had the same behavior. However, higher concentrations of ZnO demonstrated a dosage-dependent inhibitory effect. On 18th day, the CH_4 that was formed in reactors with 30 and 150 mg/g-TSS was 99.5 and 24.5 mL/g-VSS, respectively, which represented a decrease of 22.8% and 81.1% compared with the control (129.1 mL/g-VSS). A complementary experiment to study the importance of released metal oxides ions in CH_4 production determined that ZnO NPs showed a significant amount of metal ion (Zn^{2+}) release. In contrast, the ions released by the other NPs were negligible. The concentrations of released Zn^{2+} from ZnO were 4.4, 11.6 and 17.6 mg/L for the treatments of 6, 30 and 150 mg/g-TSS, respectively. The lowest concentration of ions did not affect CH_4 generation, but the higher ones decreased production to 84.4% and 22.1% of the control. Thus, released metal oxide ions could be a key factor in CH_4 inhibition. The WAS employed in the study was mainly composed of proteins and carbohydrates (61.5% of sludge TCOD) but due to its particulate state, it was necessary to first solubilize it for further hydrolysis. Therefore, the NPs effect on solubilization was also studied. However, NPs did not produce changes in soluble protein and carbohydrate production. The authors suggested that sludge solubilization was not a result of microbial activity and for that reason NPs did not affect this parameter. The effect of NPs in the hydrolysis stage was studied in the experiment too. Bovine serum albumin (BSA) and dextran were used as protein and carbohydrate substrates, respectively. From the four NPs, just 150 mg/g-TSS of ZnO NPs affected negatively the degradation efficiencies of both biomolecules by 62.5% and 87.7%, respectively, compared with the control. These results could be related to low CH_4 production. Additionally, to study the effect of NPs in acidogenesis stage, L-glutamic acid and glucose were used as the source of amino acids and monosaccharides that will later be converted to short-chain fatty acids (SCFA). Again, higher dosages of ZnO NPs had a negative effect on the process. The concentration of 30 mg/g-TSS of ZnO NPs decreased the production of butyric and valeric acids to 87.4% and 78.2% of the control while 150 mg/g-TSS decreased the production of acetic, propionic, butyric and valeric acids to 39.3, 20.4, 59.3 and 18.5%, respectively, compared with the control. Due to this inhibition, it was determined that both ZnO concentrations produced less short-chain fatty acids compared with the control (2770 ± 65 mg COD/L), which represented 2,600 ± 70 and 1,160 ± 50 mg COD/L for the treatments of 30 and 150 mg/g-TSS NPs, respectively. Finally, CH_4 production from acetate as a substrate represented 79.9% and 20.8% of the control in reactors with 30 and 150 mg/g-TSS of ZnO NPs. Thus, the negative effect due to ZnO NPs was reaffirmed (Mu et al. 2011).

The aforementioned studies focused on the effects of individual nanosize compounds. However, in WWTPs, a mixture of NPs will likely be very complex.

Nowadays, less information about this type of experiments is available. Therefore, it is important to study and determine if the interaction of different compounds can affect wastewater treatment. It is possible that a combination of NPs in WWTPs enhances, inhibits or helps to restore AD development. This last effect was observed in a study that tested the effects of single and combined ZnO and TiO_2 NPs during fermentation of anaerobic sludge at dosages of 0, 5, 30 and 100 mg/g-TSS of ZnO and TiO_2. When ZnO NPs increased from 0 to 100 mg/g-TSS, soluble protein increased from 827.8 to 1,147.6 mg-COD/L while soluble polysaccharide content was not affected. In addition to that, the removal rate of proteins decreased indicating microbial inhibition. It was probable that protein content increased due to cell lysis. On the other hand, all concentrations of individual TiO_2 NPs did not affect protein or polysaccharides production; therefore, they did not pose a threat for the AD process. In experiments with ZnO and TiO_2, NPs combined at different concentrations; the content of soluble polysaccharides was the same compared with reactors enriched with single ZnO NPs, but there was a notable change in soluble protein content when the combination reached 100 mg/g-TSS of TiO_2 NPs. For instance, combinations of 5, 30 and 100 mg/g-TSS of ZnO NPs with 100 mg/g-TSS of TiO_2 NPs decreased soluble protein from 891.87 to 805.32, 955.80 to 830.67 and 1,147.60 to 870.56 mg-COD/L, respectively. Thus, the toxic effect of ZnO NPs was reduced with the addition of TiO_2 NPs. In addition to that, among the measured VFAs, acetic acid was predominant and the most affected by the addition of ZnO NPs. Acetate concentrations of 0, 5, 30 and 100 mg/g-TSS of ZnO NPs were 2,113.9, 2,110.1, 2,045.2 and 1,968.7 mg COD/L, respectively. Reactors with only TiO_2 NPs did not register any change again. However, a combination of 100 mg/g-TSS of these NPs with 100 mg/g-TSS of ZnO NPs produced 2,057.1 mg COD/L of acetate which was not very different from the control reactor. Thus, the authors suggested the possibility that TiO_2 NPs contributed to stabilizing the reactor from the toxicity caused by ZnO NPs in the acidification stage. It has been mentioned previously that increasing ZnO NPs concentration resulted in the higher release of Zn^{2+} (Mu et al. 2011). In control sample, Zn^{2+} concentrations in supernatant and sediment were found to be < 0.02 mg/L or 518.5 mg/kg, respectively. On the other hand, sludge reactors obtained concentrations of Zn^{2+} in supernatant and sediment from 0.02–0.1 mg/L and 469.5–12,100 mg/kg, respectively. Therefore, it was concluded that a common feature of ZnO NPs was that they remained trapped in anaerobic sludge. Furthermore, in systems with both NPs, the ones with the highest TiO_2 NPs dosage decreased the concentration of Zn^{2+}. The possible reactions that could explain such phenomena are presented in Equations 7 and 8.

$$ZnO\,(s) + 2H^{+}\,(aq) \rightarrow Zn^{2+}\,(aq) + H_2O(1) \qquad (7)$$

$$Zn^{2+}\,(aq) + TiOH(s) \rightarrow \; -TiOZn^{2+}\,(s) + H^{+}\,(aq) \qquad (8)$$

As it is shown, Zn^{2+} interacts with the surface of TiO_2 forming a complex. Therefore, less Zn^{2+} ions are less available, which reflects in a lower concentration. Consequently, in reactors that contained both NPs, acetic acid inhibition was not observed because of lack of free Zn^{2+}. This prominent and positive effect was observed to be dosage-dependent. However, the maximum concentration employed in the experiment was 100 mg/g-TSS, so more concentrations will need to be tested (Zhang et al. 2019).

In conclusion, in the same way, that NPs can trigger great enhancement effects in AD, their application can also damage the process importantly. Hence, more research is required in order to gain knowledge of the exact mechanisms by which NPs influence the AD process. Such information will be useful in the development of strategies that will help to evade or recover reactors' performance from the conditions in which NPs might unleash detrimental effects. If that happens, this technology that has proven to be very useful would continue to evolve and its application in large-scale size would be possible, benefiting greatly clean energy production by AD.

3.3 Microbial Behavior

As it was mentioned in the introductory part of this chapter, the process of AD is carried out by the participation of microbial groups that interact within a consortium; each one degrades or transforms the organic matter in simple compounds that are finally used by methanogenic archaea. However, due to the different microbial populations found in each ecosystem, it is necessary to take into count the possible variations in metabolic pathways (Karakashev et al. 2006). Therefore, a better understanding of microbial diversity is needed in order to elucidate their interactions and the possible effects of NPs in them (Figure 1).

The initial interaction of NPs with cells occurs in the plasma membrane, the permeable barrier that controls the inlet and outlet of molecules to maintain intracellular conditions. Non-polar molecules, like O_2 or CO_2, can go through this lipid bilayer without problems. On the contrary, polar molecules require a mechanism to enter the cell. Therefore, the membrane's mechanisms must work effectively to prevent the alteration of the cell. However, some nanomaterials can evade those defenses and cross the membrane, which is the case of cationic (metallic) NPs. Nonetheless, other nanomaterials like needle-shaped materials cannot penetrate it (Verma and Stellacci 2010).

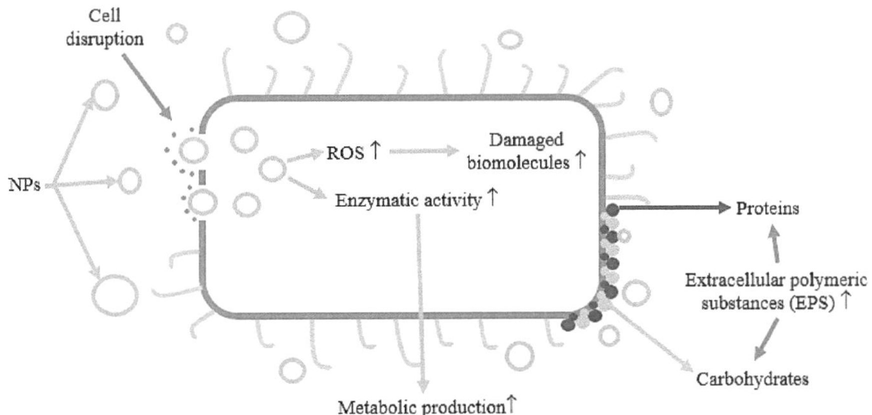

Figure 1. Microbial response to stressful conditions caused by NPs presence.

As it was described in this chapter, the interaction of methanogenic archaea with NPs can improve their performance by the reduction of their lag-phase time (microorganisms acclimation time) and the enhancement of specific methanogenic activity, that is the rate of substrate degradation to produce CH_4 expressed in g-of-substrate/g-VSS/d. The last parameter can be used to determine the impact of environmental conditions in their growth and metabolism (Krongthamchat et al. 2006). In addition to that, methanogenic population varies according to the sample composition and physicochemical parameters. *Methanobacteriales*, *Methanococcales* and *Methanomicrobiales* belong to hydrogenotrophic methanogens. On the other hand, the families *Methanosaetaceae* and *Methanosarcinaceae* comprise aceticlastic methanogens. Common methanogenic populations in wastewater sludge belong to *Methanosaetacea* (Karakashev et al. 2006). Therefore, the presence and constant growth of some of these groups can suggest the proper development of AD.

In the study of Wang et al. (2016) on AGS, the concentrations of nZVI and Fe_2O_3 NPs that improved methane production had the most active microorganisms in comparison with the other treatments. Microbial composition demonstrated the presence of α-Proteobacteria, β-Proteobacteria and Bacteroidetes phylum which may participate actively in hydrolysis and acetogenesis stages. Furthermore, the same effect was observed in *Methanosaeta* genus (now *Methanotrix*), the dominant acetate-consuming archaea and responsible for methane production (Wang et al. 2016). In another experiment, archaeal populations that were found in a bed with ZVI in bulk comprised 65.4% of microbial diversity. Coupled with this enhancement there was an increment in COD removal and less HRT (Zhang et al. 2011).

Another experiment that studied the effects of nZVI particles reported that Clostridia were the predominant bacterial class, with a relative abundance of 53.2%. This class is recognized as an important acid-forming bacteria and an H_2-producer. That could explain the high VFAs and H_2 production at the initial AD stage during the experiment. Anaerolineae class was also present in the reactor (11% approximately). At the genus level, hydrogenotrophic methanogens detected were *Methanospirillum* (13.7%), *Methanobacterium* (11.6%), *Methanoculleus* (6.2%) and *Methanobrevibacter* (3.2%). The relative abundance of *Methanosaeta* (now *Methanotrix*), an aceticlastic methanogen, was 36% approximately. In addition to that, the higher relative abundance of *Methanosarcina* genus was present in the reactor with nZVI (22.6%). This genus can perform all known methanogenic pathways, being a versatile methanogen. Therefore, the addition of 10 g/L of nZVI particles enhanced methanogenic growth and improved CH_4 production. Furthermore, the accumulation of acetate in this treatment promoted *Methanosarcina* development (Yu et al. 2016).

The addition of nZVI NPs did not affect the species richness in AGS. Reactor enriched with 30 mM of nZVI NPs showed a higher microbial richness with 1,358 OTUs compared with 15 mM (1342 OTUs) and the control (1,268 OTUs). Reactors with nZVI NPs shared 211 OTUs between control and reactor with 15 mM of nZVI NPs were 152 OTUs and 30 mM were 142 OTUs. All reactors shared 790 OTUs. The prominent among them were Proteobacteria (26.58%), Firmicutes (18.86%), Chloroflexi (10.38%) and Actinobacteria (9.24%). Phylum and class distribution between control and 15 mM nZVI were similar. However, a remarkable difference was present with 30 mM of nZVI. The changes were reflected in the percentage of the abundance of Proteobacteria, Chloroflexi and Firmicutes that decreased from 25.94% to 20.20%, 16.25% to 13.38% and 5.27% to 3.94%, respectively. However, there was an increment of Thermotogae and Euryarchaeota from 6.21% to 14.03% and 8.38% to 12.44%. Proteobacteria is an acetate producing bacteria alongside Firmicutes. In addition to that, the latter contributes to the degradation of butyrate. Archaea present in AGS belonged to Methanobacteria and Methanomicrobia classes. *Methanobacterium* genus, a hydrogenotrophic methanogen, increased its abundance from 4.14% to 10.83% with 30 mM of nZVI. In addition to that, the presence of this group decreased the presence of Chloroflexi bacteria that are H_2 consuming microorganisms. Another observation was the decreased of *Methanomicrobia* abundance, with *Methanosaeta* genus as the predominant in the class (He et al. 2017). Therefore, it is possible that the increment of nZVI enhanced hydrogenotrophic methanogenesis due to the fast H_2 production by these NPs, leaving in second place aceticlastic methanogenesis.

Among the microbial populations present in paddy soil enriched with Fe_3O_4 NPs, *Desulfovibrio* was the dominant species in the family. Desulfovibrionaceae, this sulfate reducing bacteria (SRB), is capable to produce CO_2 from the oxidation of acetate (Madigan et al. 2015) which is one of the main components in the medium. Synergistaceae, Dethiosulfovibrionaceae and Anaerolinaceae families were present too. The genera *Cloacibacillus* and *Aminobacterium* belong respectively to the first two families and it was assumed that the members are focused on amino acid degradation released from dead cells. Other observed bacterial families in paddy soil were Acholeplasmataceae, Comamonadaceae and Pseudomonadaceae. These groups consisted of the genera *Acholeplasma*, *Diaphorobacter* and *Pseudomonas*, respectively, but the role of the first two was unknown. In the case of *Pseudomonas*, they are capable to reduce Fe(III) with H_2 as an electron donor. Clostridiaceae, in which *Clostridium* was the dominant genus, are fermentative bacteria capable of oxidizing acetate in a syntrophic relation with hydrogen scavenger microorganisms such as members of Methanobacteriaceae, an archaeal family composed of hydrogenotrophic methanogens that are also found in the medium. Moreover, the presence of Rhodocyclaceae was predominant, within this family the genera *Thauera* and *Dechloromonas* were abundant and both can reduce Fe(III) using acetate. Furthermore, *Methanosaeta* was also detected (Yang et al. 2015).

An experiment that managed Fe_2O_3 NPs and MWCNTs demonstrated, in addition with the control, the presence of Euryarchaeota, Bacteroidetes, Firmicutes, Proteobacteria and Chloroflexi phylum in order of abundance. In comparison with control, a reactor with Fe_2O_3 NPs had a higher content of Euryarchaeota (63%), an archaeal phylum. However, in the same reactor, another archaeal phylum called Crenarchaeota decreased importantly in its population. In MWCNTs' reactor, Bacteroidetes (11.2%) and Firmicutes (8.4%) phylum were more abundant but only to a certain extent. At the class level, Methanomicrobia, Bacteroidia, Deltaproteobacteria, Anaerolineae and Clostridia were predominant. However, Methanobacteria, Thermoplasmata and Thermoprotei demonstrated a visible relative abundance too. Methanomicrobia and Anaerolinae classes increased in reactors with NPs but Deltaproteobacteria, Thermoprotei and Thermoplastmata abundance decreased compared with the control. In Fe_2O_3 reactor Methanomicrobia and Methanobacteriaclasses were more abundant; these are recognized for comprising hydrogenotrophic methanogens. Therefore, this can explain why faster H_2 consumption was observed during the experiment (section 3.1). At order-level Methanosarcinales, Methanomicrobiales, Bacteroidales, Anaerolinales, Clostridiales, Desulfovibrionales and Methanobacteriales predominated. The Fe_2O_3 NPs enhanced the abundance of Methanosarcinales and Methanobacteriales but decreased the content of Bacteroidales and Thermoplasmatales. Reactor with MWCNTs had a short increment in Methanomicrobiales, Methanosarcinales, Bacteroidales and Anaerolineales order. At the genus level, the reactor added with Fe_2O_3 NPs increased *Methanosaeta, Methanobacterium, Methanospaera* and *Methanolinea.* The same effect was observed in *Methanospirillum, Parabacteroides, Methanomethylovorans, Longilinea* and *Pyrolobus* in MWCNTs reactor. Both treatments decreased *Thermogymnomonas, Desulfovibrio* and *Peludibacter* proportions. In addition to the changes observed by NPs addition, the structural features of sludge were modified too. In the control reactor, microorganisms were dispersed in loose aggregates, but in both treatments, the sludge granules were densely packed. This feature was suggested to occur by the increment of EPS content as a shell to block the pass of NPs. Thus, cytotoxicity was reduced by the limited interaction between the cell membrane and NPs. In addition to that, due to the controlled interaction between microorganisms and NPs there was an enhancement in biogas, CH_4 production, adequate substrate fermentation and VFAs formation. The abundance of Bacteroidetes and Firmicutes phylum indicated the good development of the fermentation process. In addition to that, due to the highest relative abundance of *Methanosaeta* (now *Methanothrix*) genus for Fe_2O_3 reactors (39.4%) compared with MWCNTs (26.5%) and control (24.6%), higher acetate consumption was observed with the consequent enhancement of CH_4 production. Therefore, microbial activity coupled with the other results of laboratory-scale EGSB reactors enriched with 750 mg/L of Fe_2O_3 NPs and 1,500 mg/L of MWCNTs suggested the capacity to adapt a viable, continuous and well-functioning AD system with 12 hours of HRT using industrial wastewater to produce energy in CH_4 form (Ambuchi et al. 2017).

In the case of graphene NPs, they did not show an inhibitory effect in microbial populations. Instead, they enhanced aceticlastic methanogens and accelerated the substrate utilization coupled with CH_4 production. In anaerobic batch experiments with glucose as the carbon source, it was observed that concentrations of acetate decreased to 425.4 mg/L (88% of the control) when 120 mg/L of graphene NPs were added. The higher removal of acetate was linked with higher CH_4 production. Therefore, it was assumed that graphene NPs enhanced aceticlastic methanogens and thus accelerated the process. However, a complementary experiment with the addition of CH_3F, an inhibitor of aceticlastic methanogens did not show a difference in CH_4 production compared with the control. That suggested that process enhancement by graphene NPs was not only linked to acetate-consuming methanogenesis. As was explained, the common mechanism of defense demonstrated by microorganisms against stress by NPs is the production of EPS. Probably the increase of graphene concentrations led to a higher carbohydrate production and protein removal. EPS also benefited the formation of bio-granules that resisted harsh environmental conditions (Tian et al.

2017a). In addition to that, bacterial and archaeal communities were also determined. *Methanoregula* (66.70%), *Methanosaeta* (18.72%) and *Methanospirillum* (9.67%) were dominant methanogenic genera in both concentrations of graphene NPs. In the control, *Methanobacterium* (31.39%), a hydrogenotrophic methanogen, and *Methanosarcina* (4.29%) were detected. However, there was a marked decrease in *Methanosaeta* (2.25%) too. That could suggest a shift of aceticlastic to hydrogenotrophic methanogenesis. In control, *Caldisericum* (30%) predominated and *Olsenella* (13.37%) was the primary lactate producer. There was a change of bacterial community structure with the addition of graphene NPs. *Anaerolinea*, a VFAs degrader, was present (24.26% and 16.90% with 30 and 120 mg/L of graphene NPs) along with similar sequences of *Lactococcus*. Other bacterial groups present with different concentrations of graphene were *Desulfovibrio*, *Sunxiuqina*, *Pseudomonas* and *Sedimentibacter*. These groups could be related to hydrolysis, acidogenesis and acetogenesis with H_2 production. Another observation was the decrease of *Methanosaeta*, *Lactococcus* and *Anaerolinea* with the addition of 120 mg/L of nano-graphene at long-time exposure, which might explain for the decay in CH_4 production. Compared with 30 mg/L graphene NPs treatment, *Geobacter* genus was enhanced too from the 0.15% in the control passed to 1.35% and 1.64% with 30 and 120 mg/L of nano-graphene, respectively. Therefore, it is possible that direct interspecies electron transfer (DIET) was promoted between *Geobacter* and *Methanosaeta* species when graphene NPs were present (Tian et al. 2017a).

Some NPs have demonstrated a bactericidal effect at high concentrations, which is the case of Ag (Morones et al. 2005). Therefore, it was shown that a concentration of 500 mg/g-TSS of Ag and MnO NPs reduced microbial population and diversity (Wang et al. 2016).

An experiment that studied the effects of ZnO and TiO_2 NPs found that there were a total of 13 phyla on their reactors. The predominant microorganisms in all reactors belonged to the phyla Proteobacteria, Firmicutes, Bacteroidetes, Chloroflexi, Actinobacteria and Acidobacteria. Their combined abundance accounted for more than 93% in control and treatments with NPs. However, Firmicutes abundance decreased with the addition of 100 mg/g-TSS of ZnO NPs compared with the control and even in combination with 100 mg/g-TSS of TiO_2 NPs, Firmicutes population did not recover. Therefore, the authors speculated that inhibition of microbial populations that are in charge of producing VFAs would reduce and limit CH_4 production by methanogens. The dominating classes in reactors were Gammaproteobacteria, Betaproteobacteria, Clostridia, Alphaproteobacteria, Bacteroidia and Anaerolineae. Bacteroidia and Clostridia classes decreased in the presence of ZnO NPs due to the release of Zn^{2+}. Finally, it was shown that microbial populations were similar in reactors in which ZnO NPs were added but different from those detected in control and reactors with only TiO_2 NPs. Thus, the addition of ZnO NPs might change microbial consortium (Zhang et al. 2019).

The addition of ZnO NPs can cause constant production of reactive oxygen species (ROS). ROS include superoxide anion and hydrogen peroxide. The reaction of the former with iron-sulfur clusters in proteins causes the release of iron ions (Fe^{2+}) that can interact with hydrogen peroxide and form hydroxyl radicals. These radicals are dangerous due to their ability to damage DNA, proteins and lipids (Van

Acker and Coenve 2017). Therefore, the increase of ZnO concentration markedly lowers CH_4 production. It was reported that at lower ZnO concentrations (1 mg/g-TSS), 97.3% of biomass was retained in comparison with the control. However, this concentration decreased to 88.7%, when the concentration was increased to 30 mg/g-TSS. The highest depletion of biomass was observed at 150 mg/g-TSS of ZnO NPs, where it accounted for 62.4% of the control. Therefore, almost 40% of the control biomass was affected by high ZnO NPs concentrations. However, the acidification process in sludge was not influenced by the addition of ZnO NPs (Mu and Chen 2011).

The addition of NPs at high concentrations can affect the normal development of enzymes. Therefore, monitoring the enzymatic activity can help to identify microbial growth or inhibition. For instance, some important enzymes in anaerobic sludge are acetate kinase (AK), which catalyzes the conversion of acetyl-CoA to acetate and coenzyme F_{420} which is a hydrogenase that reduces carbon compounds to obtain CH_4. Enrichment with 150 mg/g-TSS of ZnO NPs decreased AK activity markedly. In addition to that, the reduction of coenzyme F_{420} activity was related to the dosage of the NP. Therefore, from low to high concentration of ZnO NPs (1, 30 and 150 mg/g-TSS), the enzymatic activity was 99.3, 89.8 and 66.2% of the activity presented in the control. These results were related to the composition of the microbial community. In the control, archaeal and bacterial populations were 39.5% and 52.6% of the total microbial populations. However, when reactors were enriched with ZnO NPs, the archaeal populations decreased to 38.6%, 27.15% and 3.5% of the total microbial population with 1, 30 and 150 mg/g-TSS. Therefore, it can be suggested that the decrease of the archaeal population reflected a lower coenzyme F_{420} activity. However, with the same concentrations, bacterial population growth was 51.3%, 60.8% and 87.4% of the total microbial population, respectively. From these results, it can be suggested that bacterial growth was enhanced by ZnO NPs and that they were more resistant to ROS (Mu and Chen 2011).

A way to understand the effect of ROS in AD was with the application of radical scavengers, as T-butanol and catalase, to study the influence of •OH and H_2O_2. A combination of 100 mg/g-TSS of ZnO NPs with one radical scavenger recovered acetic acid formation compared with the control. Nevertheless, that enhancement was less important compared with the enhancement of acetic acid by the addition of 100 mg/g-TSS of TiO_2 NPs that have demonstrated to adsorb Zn^{2+}. Therefore, it was suggested that Zn^{2+} concentration was more toxic than ROS species (Zhang et al. 2019).

Another experiment measured protease, AK and coenzyme F_{420} activity as an approach to infer microbial development. Enzymatic activity of the control experiment was not different from the values obtained with all concentrations of TiO_2, Al_2O_3 and SiO_2 NPs. Reactor with 6 mg/g-TSS of ZnO NPs entered in this statement too. Nevertheless, the on 18th day, with 30 mg/g-TSS of ZnO NPs, there was an inhibition of the coenzyme F_{420} activity of 12.9%. Furthermore, 150 mg/g-TSS of ZnO NPs decreased respectively activity of protease, AK and coenzyme F_{420} by 25.3%, 22.9% and 40.9% (Mu et al. 2011).

Similarly, other investigations determined the change of activity in AK and coenzyme F_{420}. The addition of nZVI particles modified the activity of AK, but the change was not significantly different to the control. In contrast, the activity of F_{420} increased its activity by 24.90 ± 2.14 and 26.85 ± 0.74% with 15 and 30 mM of nZVI. This improvement was demonstrated with an increment of CH_4 production compared with the control (He et al. 2017).

The enzymes used in the latter study have been also measured in reactors with graphene too. When biomass was exposed to 30 and 120 mg/L of graphene NPs, the AK activity decreased by 9.7% and 79.4%, respectively. In contrast, the activity of coenzyme F_{420} was promoted 1.82 and 2.09-folds than the control at concentrations of 30 and 120 mg/L, respectively. Therefore, more F_{420} activity resulted in faster conversion of acetate to methane. Originally, the researchers assumed that grapheme acted as an electron shuttle to promote electron transfer. To prove this they added anthraquinone-2,6-disulfonate (AQDS), extracellular quinones to investigate if these compounds could substitute graphene. However, they were not able to replicate graphene stimulation on methanogenesis. They conclude that graphene did not act as an electron shuttle in the study (Tian et al. 2017a).

Lactate dehydrogenase (LDG) activity, an indicator of cell membrane damage, has been used to measured cellular lysis by NPs. With 100 mg/g-TSS of ZnO NPs, LDH content increased importantly from the control that had approximately 10 U/L to 50 U/L. This content decreased to 32 and 39 U/L, approximately when T-butanol or catalase was added, respectively. However, in the presence of a combination of ZnO and TiO_2 at 100 mg/gTSS, the content of LDH decreased to a range between 20 and 28 U/L. It seems probable that trapping the Zn^{2+} in TiO_2 NPs helped to avoid the damage developed by ZnO NPs. In addition to that, in experiments with either of the NPs, amylase activity was not affected. A similar situation happened with protease and AK activity assays. However, with 100 mg/g-TSS of ZnO NPs, their value decreased to 77.88% and 47.39% of the control respectively. In contrast, the reactor with both NPs at 100 mg/g-TSS increased the activity of protease and AK by 10.03% and 23.68% compared with the same concentration of ZnO NPs. However, their values were lower than the control. A decrease of AK activity could be related to the low production of acetate in reactors at high concentrations of ZnO NPs (Zhang et al. 2019).

In addition to the previous enzymes, other enzymes can be used as indicators of microbial growth. For instance, an experiment determined the activities of butyrate kinase (BK), phosphotransacetylase (PTA) and phosphotransbutyrylase (PTB). The authors concluded that when the set of enzymes recorded high concentration and were constantly active, fast degradation of VFAs coupled with a high content of acetate and butyrate were achieved (Feng et al. 2014).

Another way to study microbial development is the resazurin assay. This reagent act as an oxidation-reduction indicator that gives a specific signal that can be correlated with microbial activity. Resazurin reduction to resorufin and then to dihydroresorufin can only be done by living cells and not by the medium. Therefore, it can be suggested that a higher constant reduction rate of resazurin will show a faster microbial activity in anaerobic sludge. Kinetics of resazurin reduction, with Fe_4NiO_4Zn NPs addition, showed similar results in control (0.513 ± 0.028 mM/min) and 1 and 10 mg/L of Ni. When 100 mg/L of Ni was added, the reaction was significantly slower than the other treatments (0.488 ± 0.023 mM/min). In the case of Fe_2NiO_4 NPs addition, the kinetics of 1 and 10 mg/L of Ni were similar to control (0.577 ± 0.034 mM/min). However, the addition of 100 mg/L of Ni showed a faster kinetic (0.632 ± 0.018 mM/min) than the control. Then, all Fe_4NiO_4Zn NPs concentrations reduced microbial activity, while high concentrations of Fe_2NiO_4 NPs improved CH_4 production and microbial activity even if the concentration of metals (Ni, Fe and Zn) remained between 0.04 and 0.08 mg/g-TSS in the liquid phase. With the addition of Fe_4NiO_4Zn NPs, Zn concentration remained stable at different concentrations, Ni was higher in 100 mg/L compared with 1 mg/l and Fe concentration in 100 mg/L of Ni was the highest in comparison with the other treatments (Chen et al. 2018).

The knowledge of microbial communities, their possible role in AD process, enhancement or inhibition of microorganisms coupled with their enzymatic activity changes by the addition of NPs are useful parameters to construct a better idea of how they act and could show a way to improve their performance. Therefore, a further investigation about the microbial changes it essentially in these investigations.

4. Perspectives

Inorganic NPs are a potential technology for the improvement of the AD process. The common improvements produced by their enrichment can have a great impact. But the experiments are commonly performed at laboratory-scale with a limited range of substrates and inoculum; therefore, it is necessary to investigate the effects of NPs in reactors working at a higher scale with different process configurations in order to validate the results. Furthermore, NPs synthesis require an elaborate methodology, reagents are expensive and the tests to determine NPs features require specialized equipment and specific materials which can be expensive. In addition to that, the development of a system for NPs or ions recovery is required to avoid excessive spending of compounds. Hence, more research is needed to elucidate the physical, chemical and biochemical mechanisms that NPs mediate in microbial populations, calculate associated costs, evaluate the benefits and possible drawbacks, develop strategies to mitigate detrimental effects caused by high concentrations of NPs and establish schemes that allow recuperating NPs trapped in a potential medium like anaerobic sludge. By doing so, the presence of NPs might become a useful strategy to achieve high methane yields in anaerobic reactors.

References

Abdelsalam, E., M. Samer, Y.A. Attia, M.A. Abdel-Hadi, H.E. Hassan and Y. Badr. 2016. Comparison of nanoparticles effects on biogas and methane production from anaerobic digestion of cattle dung slurry. Renewable Energy 87: 592–598.

Abdelsalam, E., M. Samer, Y.A. Attia, M.A. Abdel-Hadi, H.E. Hassan and Y. Badr. 2017. Influence of zero valent iron nanoparticles and magnetic iron oxide nanoparticles on biogas and methane production from anaerobic digestion of manure. Energy (Oxf). 120: 842–853.

Alvarado, A., L.E. Montañez-Hernández, S.L. Palacio-Molina, R. Oropeza-Navarro, M.P. Luévanos-Escareño and N. Balagurusamy. 2014. Microbial trophic interactions and mcrA gene expression in monitoring of anaerobic digesters. Frontiers in Microbiology 5: 1–14.

Ambuchi, J.J., Z. Zhang, L. Shan, D. Liang, P. Zhang and Y. Feng. 2017. Response of anaerobic granular sludge to iron oxide nanoparticles and multi-wall carbon nanotubes during beet sugar industrial wastewater treatment. Water Research 117: 87–94.

Amen, T.W.M., O. Eljamal, A.M.E. Khalil, Y. Sugihara and N. Matsunaga. 2018. Methane yield enhancement by the addition of new novel of iron and copper-iron bimetallic nanoparticles. Chemical Engineering and Processing–Process Intensification 130: 253–261.

Bhatia, S. 2016. Nanoparticles types, classification, characterization, fabrication methods and drug delivery applications. *In*: Natural Polymer Drug Delivery Systems: Nanoparticles, Plants, and Algae. Springer, Cham.

Bożym, M., I. Florczak, P. Zdanowska, J. Wojdalski and M. Klimkiewicz. 2015. An analysis of metal concentrations in food wastes for biogas production. Renewable Energy 77: 467–472.

Casals, E., R. Barrena, A. García, E. González, L. Delgado, M. Busquets-Fité, X. Font, J. Arbiol, P. Glatzel, K. Kvashnina, A. Sánchez and V. Puntes. 2014. Programmed iron oxide nanoparticles disintegration in anaerobic digesters boosts biogas production. Small 10: 2801–2808.

Chen, H., Y. Chen, X. Zheng, X. Li and J. Luo. 2014. How does the entering of copper nanoparticles into biological wastewater treatment system affect sludge treatment for VFA production. Water Research 63: 125–134.

Chen, J.L., T.W.J. Steele and D.C. Stuckey. 2018. The effect of $Fe_2NiO_4$and Fe_4NiO_4Zn magnetic nanoparticles on anaerobic digestion activity. Science of the Total Environment 642: 276–284.

Enzmann, F., F. Mayer, M. Rother and D. Holtmann. 2018. Methanogens: biochemical background and biotechnological applications. AMB Express 8: 1–22.

Ermler, U., W. Grabarse, S. Shima, M. Goubeaud and R.K. Thauer. 1997. Crystal structure of methyl-coenzyme M reductase: The key enzyme of biological methane formation. Science 278: 1457–1462.

Feng, Y., Y. Zhang, X. Quan and S. Chen. 2014. Enhanced anaerobic digestion of waste activated sludge digestion by the addition of zero valent iron. Water Research 52: 242–250.

Ferry, J.G. 1999. Enzymology of one-carbon metabolism in methanogenic pathways. FEMS Microbiology Reviews 23: 13–38.

He, C.S., P.P He, H.Y. Yang, L.L. Li, Y. Lin, Y. Mu and H.Q. Yu. 2017. Impact of zero-valent iron nanoparticles on the activity of anaerobic granular sludge: From macroscopic to microcosmic investigation. Water Research 127: 32–40.

Jia, T., A. Wang, H. Shan, Y. Liu and L. Gong. 2017. Effect of nanoscale zero-valent iron on sludge anaerobic digestion. Resources, Conservation and Recycling 127: 190–195.

Juntupally, S., S. Begum, S.K. Allu, S. Nakkasunchi, M. Madugula and G.R. Anupoju. 2017. Relative evaluation of micronutrients (MN) and its respective nanoparticles (NPs) as additives for the enhanced methane generation. Bioresource Technology 238: 290–295.

Karakashev, D., D.J. Batstone, E. Trably and I. Angelidaki. 2006. Acetate oxidation is the dominant methanogenic pathway from acetate in the absence of Methanosaetaceae. Applied and Environmental Microbiology 72: 5138–5141.

Krongthamchat, K., R. Riffat and S. Dararat. 2006. Effect of trace metals on halophilic and mixed cultures in anaerobic treatment. International Journal of Environmental Science and Technology 3: 103–112.

Luna-delRisco, M., K. Orupõld and H.C. Dubourguier. 2011. Particle-size effect of CuO and ZnO on biogas and methane production during anaerobic digestion. Journal of Hazardous Materials 189: 603–608.

Madigan, M.T., J.M. Martinko, K.S. Bender, D.H. Buckley and D.A. Stahl. 2015. Brock biology of microorganisms. Pearson, Boston.

Morones, J.R., J.L. Elechiguerra, A. Camacho, K. Holt, J.B. Kouri, J.T. Ramírez and M.J. Yacaman. 2005. The bactericidal effect of silver nanoparticles. Nanotechnology 16: 2346–2353.

Mu, H. and Y. Chen. 2011. Long-term effect of ZnO nanoparticles on waste activated sludge anaerobic digestion. Water Research 45: 5612–5620.

Mu, H., Y. Chen and N. Xiao. 2011. Effects of metal oxide nanoparticles (TiO_2, Al_2O_3, SiO_2 and ZnO) on waste activated sludge anaerobic digestion. Bioresource Technology 102: 10305–10311.

Qiang, H., D.L. Lang and Y.Y. Li. 2012. High-solid mesophilic methane fermentation of food waste with an emphasis on iron, cobalt, and nickel requirements. Bioresource Technology 103: 21–27.

Scherer, P., H. Lippert and G. Wolff. 1983. Composition of the major elements and trace elements of 10 methanogenic bacteria determined by inductively coupled plasma emission spectrometry. Biological Trace Element Research 5: 149–163.

Strambeanu, N., L. Demetrovici, D. Dragos and M. Lungu. 2015. Nanoparticles: Definition, classification and general physical properties. pp. 3–8. *In*: Lungu, M., A. Neculae, M. Bunoiu and C. Biris (eds.). Nanoparticles Promises and Risks: Characterization, Manipulation, and Potential Hazards to Humanity and the Environment. Springer International Publishing, Switzerland.

Takano, Y., M. Kaneko, J. Kahnt, H. Imachi, S. Shima and N. Ohkouchi. 2013. Detection of coenzyme F_{430} in deep sea sediments: A key molecule for biological methanogenesis. Organic Geochemistry 58: 137–140.

Tang, J., Y. Wu, S. Esquivel-Elizondo, S.J. Sørensen and B.E. Rittmann. 2018. How microbial aggregates protect against nanoparticle toxicity. Trends in Biotechnology 36: 1171–1182.

Thauer, R.K., A.-K. Kaster, M. Goenrich, M. Schick, T. Hiromoto and S. Shima. 2010. Hydrogenases from methanogenic archaea, nickel, a novel cofactor, and H_2 storage. Annual Review of Biochemistry 79(1): 507–536.

Tian, T., S. Qiao, X. Li, M. Zhang and J. Zhou. 2017a. Nano-graphene induced positive effects on methanogenesis in anaerobic digestion. Bioresource Technology 224: 41–47.

Tian, T., S. Qiao, C. Yu, Y. Tian, Y. Yang and J. Zhou. 2017b. Distinct and diverse anaerobic respiration of methanogenic community in response to MnO_2 nanoparticles in anaerobic digester sludge. Water Research 123: 206–215.

Tian, T., S. Qiao, C. Yu and J. Zhou. 2019. Effects of nano-sized MnO_2 on methanogenic propionate and butyrate degradation in anaerobic digestion. Journal of Hazardous Materials 364: 11–18.

Van Acker, H. and T. Coenye. 2017. The role of reactive oxygen species in antibiotic-mediated killing of bacteria. Trends in Microbiology 25: 456–466.

Verma, A. and F. Stellacci. 2010. Effect of surface properties on nanoparticle-cell interactions. Small 6: 12–21.

Wang, T., D. Zhang, L. Dai, Y. Chen and X. Dai. 2016. Effects of metal nanoparticles on methane production from waste-activated sludge and microorganism community shift in anaerobic granular sludge. Scientific Reports 6: 1–10.

Yang, Y., J. Guo and Z. Hu. 2013. Impact of nano zero valent iron (NZVI) on methanogenic activity and population dynamics in anaerobic digestion. Water Research 47: 6790–6800.

Yang, Z., X. Xu, R. Guo, X. Fan and X. Zhao. 2015. Accelerated methanogenesis from effluents of hydrogen-producing stage in anaerobic digestion by mixed cultures enriched with acetate and nano-sized magnetite particles. Bioresource Technology 190: 132–139.

Yonezawa, T. 2018. Preparation of metal nanoparticles and their application for materials. pp. 829–837. *In*: Naito, M., T. Yokoyama, K. Hosokawa and K. Nogi (eds.). Nanoparticle Technology Handbook. Elsevier B.V.

Yu, B., X. Huang, D. Zhang, Z. Lou, H. Yuan and N. Zhu. 2016. Response of sludge fermentation liquid and microbial community to nano zero-valent iron exposure in a mesophilic anaerobic digestion system. RSC Advances 6: 24236–24244.

Zhang, Y., Y. Jing, J. Zhang, L. Sun and X. Quan. 2011. Performance of a ZVI-UASB reactor for azo dye wastewater treatment. Journal of Chemical Technology and Biotechnology 86: 199–204.

Zhang, L., Z. Zhang, X. He, L. Zheng, S. Cheng and Z. Li. 2019. Diminished inhibitory impact of ZnO nanoparticles on anaerobic fermentation by the presence of TiO_2 nanoparticles: Phenomenon and mechanism. Science of the Total Environment 647: 313–322.

Chapter 9

Contributions of Nanomaterials in Biohydrogen Production

Santosh Kumar[1,#,*], *Rekha Kushwaha*[1,#] and *Madan L. Verma*[2]

1. Introduction

For a long time, fossil fuels have been the main source of the energy known to fulfill the global primary energy demand (Vijayaraghavan and Mohd Soom 2006). According to the Global Energy and CO_2 Status Report (International Energy Agency 2018), 70% of the world energy need is fulfilled by fossil fuels. This unfettered use of fossil fuels has caused a disastrous effect on the world's climate. In addition to the disastrous effect of fossil fuels on environmental, these are also responsible for generating economic and political concerns. Therefore, dependency on fossil fuel is now declining and increasing attention is paid to renewable sources of the energy (Balat and Kirtay 2010). A recent report on the statistical review of world energy by the British Petroleum Company (BP 2018) showed strong growth in renewable energy consumption in 2017. The consumption of renewable power grew by 17%, higher than the 10-year average. Renewable energy has thus emerged as an increasingly competitive way of meeting the needs of new power generation. The main reason for this increase is that the portfolio of renewable energy technologies, in an increasingly wide range of circumstances, is becoming cost-competitive, and in some cases, are offering investment opportunities without the need for specific economic support (IRENA 2018).

There are many forms of renewable energy such as solar, wind, hydroelectric power geothermal energy, tidal energy and so on that are now being utilized for global energy demands. Besides these, the uses of biofuels (biogas, bioethanol, biohydrogen, biodiesel, etc.) are increasing day by day. Biohydrogen is most

[1] Department of Biochemistry, 242 Christopher S. Bond Life Sciences Center, 1201 Rollins, Street, Columbia, MO 65211, USA.

[2] Department of Biotechnology, School of Basic Sciences, Indian Institute of Information Technology Una, Himachal Pradesh-177220, India.

* Corresponding author: kumarsant@missouri.edu

# Equal contribution

commonly produced from biomass and is now considered as the versatile fuel of the future. It is an important key to a sustainable world power supply. Being rich in the high energy (122 kJ/g) hydrogen gas is clean fuel and thus has a potential to replace fossil fuels (Meher Kotay and Das 2008, Vijayaraghavan and Mohd Soom 2006, Seth 2002, Han et al. 2016, Chandrasekhar et al. 2015, Van Ginkel et al. 2001, Bhatia 2014).

Production of the hydrogen from the microbial or vegetal conversion of biomass is called biohydrogen (Fernández Rodríguez et al. 2016, Güell et al. 2015, Herbert et al. 2006, Mohan et al. 2007, Jones et al. 2017, Patrick Hallenbeck 2009, Azman et al. 2016, Sui et al. 2017, Bastidas-Oyanedel and Schmidt 2018). The major biomass used for the biohydrogen production include agricultural crops and their waste products, lignocellulosic products, food wastes, various microorganism and effluents produced in the human habitat (Saratale et al. 2008, Maneeruttanarungroj et al. 2010, Guo et al. 2010, 2016, Radjaram and Saravanane 2011, Cheng et al. 2011, Chen et al. 2013, Pragya et al. 2013, Skjånes et al. 2013, Khetkorn et al. 2013, 2017, Zilouei and Taherdanak 2015, Isikgor and Becer 2015, Sriwuryandari et al. 2016, Jiang et al. 2016, Liu et al. 2017, Srivastava et al. 2017, Łukajtis et al. 2018, Diah et al. 2018, Abd-Alla et al. 2019).

Biohydrogen from the biomass is generated by various routes, such as anaerobic/ fermentation, photobiological, enzymatic and electrogenic mechanisms (Mohan and Pandey 2013, Holladay et al. 2009). Each method faced challenges in the case of energy efficiency and practicality and thus every method has both advantages as well as disadvantages when compared with the other (Chandrasekhar et al. 2015, Azwar et al. 2014, Rittmann and Herwig 2012, Nath and Das 2004). To achieve a hydrogen economy, several technical challenges (lowering the cost of hydrogen production, delivery, storage, conversion and end-use applications) have to be faced by the manufactures. Therefore, researches are continuously working on the basic and applied research to overcome these hurdles by using a multidisciplinary approach for environmentally friendly and economically feasible biohydrogen production. These approaches include advances in-process modifications, physiological manipulations by metabolic and genetic engineering. One of these technological advances includes the use of nanotechnology during the biohydrogen production (Lee et al. 2010, Yan et al. 2010, Carvalho et al. 2011, Chang et al. 2011, Ivanova et al. 2011, Lupoi and Smith 2011, Mu et al. 2011, Risco et al. 2011, García et al. 2012, Khan et al. 2012, Uygun et al. 2012, Feyzi et al. 2013, 2014, Kim et al. 2013, Sachdeva and Saroj 2013, Tao et al. 2013, Verma et al. 2013).

Nanoparticles are particles that range from 1 to 100 nanometers (nm) in size with unique optical, electronic or mechanical properties (Hübler and Osuagwu 2010, Buzea et al. 2007, Daima and Bansal 2015, Mao et al. 2012). These unique properties of the nanoparticle are very effective and are widely utilized in production of biofuels (Yang et al. 2013, Zhao et al. 2013, Casals et al. 2014, Otero-González et al. 2014, Abraham et al. 2014, Yulianti et al. 2014, Degirmenbasi et al. 2015, Carpenter et al. 2015, Abdelsalam et al. 2016, 2017a, Dalla Vecchia et al. 2016, Abdelsalam et al. 2017b, Rahmani Vahid and Haghighi 2016, Suanon et al. 2016, Bet-Moushoul et al. 2016, Alsharifi et al. 2017, Amen et al. 2017, Baskar et al. 2017, Park et al. 2018, Xie et al. 2018, Alaei et al. 2018).

Many researchers are nowadays are using these nanomaterials for improving the bioactivity of hydrogen-producing microbes to increase the yield and rate of hydrogen production as these nanoparticles act as a bridge between bulk materials and atomic or molecular structures (Klabunde 2001, Kumar Gupta et al. 2013, Das et al. 2014, Lin et al. 2016, Han et al. 2011).

Therefore, the main focus of the present chapter is to summarize the role of nanomaterial or nanoparticle during biohydrogen production.

2. Methods of Biohydrogen Production

A variety of organisms including the archaea, anaerobic and facultative aerobic bacteria, cyanobacteria and lower eukaryotes are known to produce hydrogen (Chandrasekhar et al. 2015). Biohydrogen from different biomass is produced by many methods (Gürtekin 2014, Michał 2012, Sołowski 2018, Patrick Hallenbeck and Benemann 2002) (see Figure 1). These methods include direct biophotolysis (Das et al. 2014, Das and Veziroglu 2008, Kapdan and Kargi 2006, Kosourov et al. 2002, Melis et al. 2000), indirect bio photolysis (Lindberg et al. 2012, Dutta et al. 2005, Sveshnikov et al. 1997), photo fermentation (Tsygankov et al. 1994, 1998, Fedorov et al. 1998, Zürrer and Bachofen 1979, Fascetti and Todini 1995, Francou and Vignais 1984, Kovács et al. 2006, Bélafi-Bakó et al. 2006, Melis and Happe 2001, Melis 2002, Kim et al. 2006, Kondo et al. 2006), dark fermentation (Nath et al. 2005, Van Ginkel et al. 2001, Hallenbeck and Ghosh 2009, Ntaikou et al. 2010, Taguchi et al. 1996, Chang et al. 2002, Van Niel et al. 2002, Lay 2001), two stage process-integration of dark and photo fermentation (Tao et al. 2007) and biocatalyzed electrolysis (Liu et al. 2005, Rozendal et al. 2006) (Table 1).

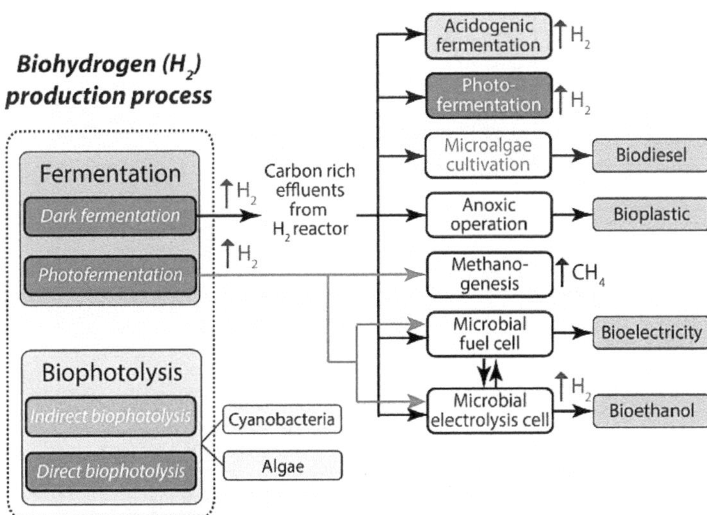

Figure 1. Schematic representation of the primary biological routes integrated with various secondary processes for effective H_2 production. Adapted from the Chandrasekhar et al. 2015. Biohydrogen production: Strategies to improve process efficiency through microbial routes. International Journal of Molecular Sciences 16(4): 8266–8293.

Table 1. Different methods of biohydrogen production.

S. No.	Methods	No. of Reaction Steps	Chemical Reaction	Examples	References
1.	Direct biophotolysis	1	$2H_2O + light \rightarrow 2H_2 + O_2$	*Scenedesmus obliquus, Chlorococcum littorale, Platymonas subcordiformis* and *Chlorella fusca*	(Das and Veziroĝlu 2002)
2.	Indirect biophotolysis	2	1. $12H_2O + 6CO_2 + light \rightarrow C_6H_{12}O_6 + 6O_2$ 2. $C_6H_{12}O_6 + 12H_2O + light \rightarrow 12H_2 + 6CO_2$	*Anabaena variabilis*	(Sveshnikov et al. 1997)
3.	Photo fermentation	1	1. $CH_3COOH + 2H_2O +$ 'light' $\rightarrow 4H_2 + 2CO_2$ OR 2. $CO + H_2O \rightarrow CO_2 + H_2$	1. Purple non-sulfur bacteria 2. Rhodospirillaceae	(Zürrer and Bachofen 1979, Francou and Vignais 1984, Tsygankov et al. 1998, 1994, Fascetti and Todini 1995, Fedorov et al. 1998, Melis and Happe 2001, Melis 2002, Bélafi-Bakó et al. 2006, Kim et al. 2006, Kondo et al. 2006, Kovács et al. 2006, Winkler et al. 2002)
4.	Two stage process with integration of dark and photo fermentation	2	1. $C_6H_{12}O_6 + 2H_2O \rightarrow 2CH_3COOH + 2CO_2 + 4H_2$ 2. $2CH_3COOH + 4H_2O \rightarrow 8H_2 + 4CO_2$	1. Dark fermentation (facultative anaerobes) 2. Photofermentation (photosynthetic bacteria)	(Pandu and Joseph 2012, Eroĝlu et al. 2006)
5.	Biocatalyzed electrolysis	1	1. Anodic Rxn$2CH_3COOH+2H_2O \rightarrow 2CO_2$ +8H+ +8e_ 2. Katot: 8H+ + 8e_ $\rightarrow 4H_2$	*Geobacter, Shewanella, Pseudomonas*	(Liu et al. 2005, Rozendal et al. 2006, Bond et al. 2002, Bond and Lovley 2003, Rabaey et al. 2003, Park et al. 1999, Kim et al. 2002, Fedorovich et al. 2009)

All the biohydrogen methods have many limitations. Low hydrogen yield as well as low conversion efficiency are two main challenges faced in biohydrogen production units. In addition, there are many limitations in each method, such as (1) in dark fermentation, final product is a mixture of hydrogen and carbon dioxide which require separation; (2) in photo fermentation, there is a need for an external light source such as solar light that is further limited by day and night cycles; (3) in-direct biophotolysis, generation of oxygen and requirement of photobioreactors are main limitations and (4) in-indirect biophotolysis, lower hydrogen yield is caused by hydrogenase(s) and requirement of an external light source mainly affects the yield (Patrick Hallenbeck 2011, Chandrasekhar et al. 2015, Pandu and Joseph 2012, Muhamad et al. 2011, Khanna and Das 2013, Benemann 1998) (Table1) (Figure 1) (Adapted from the Chandrasekhar et al. 2015).

3. Main Critical Factors Responsible for Enhancing the Biohydrogen Production

There are several main factors responsible for the fermentative hydrogen production these includes temperature, pH, nutrients (carbon, nitrogen, phosphate, metal ion, hydraulic retention time (Das and Veziroglu 2008, Das and Veziroglu 2002, Wang and Wan 2009, Vijayaraghavan and Mohd Soom 2006).

The H_2 production and the metabolic product formation are affected by temperature during fermentation process by influencing the microbial use of the substrate and the specific growth rate (Yokoyama et al. 2009, Levin et al. 2004, Lee et al. 2006, Lin et al. 2008, Yokoyama et al. 2007a,b). Temperature during the fermentation process mainly depends upon the type of the H_2-producing microorganism and the type of substrate used. Levin et al. (2004) and Yokoyama et al. (2009) reported the production of hydrogen at various temperatures from ambient 15°C to the extremely thermophilic, i.e., > 60°C. The optimal growth temperatures for *C. butyricum* and *E. cloacae* are reported to have 30 to 45°C and for *Pyrococcus furiosus* and *Caldicellulosiruptor bescii* varies from 50 to 80°C (Tang et al. 2014, Wang and Wan 2009, Yokoyama et al. 2009).

The pH is another important factor responsible for the activity of the hydrogen-producing microorganism. It regulates the hydrogenase activity of these organisms by regulating the metabolic pathway. The optimal pH of hydrogen production depends on the type of microorganism. Several researchers investigated the effect of initial pH on fermentative hydrogen production in batch mode and showed that optimum pH range varies from 4.5 to 9.0 (Fang et al. 2006, Kumar and Das 2000, Wang et al. 2005, Mu et al. 2006, Yang Mu et al. 2006, Fang and Liu 2002, Zhao and Yu 2008).

Various nutrients factors such as (C/N) ratio, metal ions and optimal concentration of phosphorous are the crucial for growth of hydrogen-producing bacteria as well as the yield of hydrogen production (Lin and Lay 2004, Wang et al. 2009, Chen et al. 2008, Redwood and Macaskie 2006). According to Lin and Lay (2004) optimum C/N ratio directly influenced the microbial growth as well as H_2 yield in mixed or pure cultures and hence affected the overall production costs. The optimum phosphorous concentrations enhance the yield of the H_2 by regulating the buffering capacity as well as regulating the energy generation in the bacterial cell (Nath and Das 2011). Various trace elements, such as Fe, Ni, Mg, Mn, Na, Zn, K, I, Co, Cu, Mo and Ca, also play an important role in the H_2 production (Srikanth and Mohan 2012, Lin and Shei 2008, Karadag and Puhakka 2010, Zhang and Shen 2006).

Besides these factors, optimum hydraulic retention time also influences the hydrogen yield as well as the selection of microorganisms. According to many researchers, optimum hydraulic retention time for different substrates varies between 8 and 14 hours (Hawkes et al. 2007, Vijaya Bhaskar et al. 2008, Ren et al. 2005, Chen et al. 2009).

4. Role of Nanoparticles During Biohydrogen Production

4.1 Nanomaterial Used During Biohydrogen Production From the Algae

The production of biohydrogen from the algal biomass has got much attention despite being in its early stage of development (Hankamer et al. 2007, Khetkorn et al. 2017). Many microalgae are used for the biohydrogen production, these include *C. reinhardtii, Phaeodactylum tricornutum, Thalassiosira pseudonana, Cyanidioschyzon merolae, Ostreococcus lucimarinus, Ostreococcus tauri, Micromonaspusilla, Fragilariopsis cylindrus, Pseudo-nitzschia, Thalassiosira rotula, Chlorella vulgaris, Dunaliella salina, Micromonas pusilla, Galdieria sulphuraria, Porphyra purpurea, Volvox carteri, Aureococcus anophageferrens, Chlorella pyrenoidosa, Chlamydomonas moewusii, Scenedesmus oblique, Anabaena variabilis, Rhodobacter sphaeroides, Enterobacter aerogenes, Clostridium butyricum, Bacillus coagulans, Clostridium acetobutylicum ATCC 824, Laminaria japonica, Gelidiumamansii* and many more (Kars et al. 2006, Voloshin et al. 2016, Greenwell et al. 2010, Winkler et al. 2002, Liu et al. 2006, Radakovits et al. 2010, Fabiano and Perego 2002, Fang et al. 2006, Zhang et al. 2006, Shi et al. 2011, Park et al. 2011, Kotay and Das 2007, Sharma and Arya 2017). Photobioreactors and open-air systems are mainly used for the algal biohydrogen. But the low biomass concentration and costly down streaming processes are the two main hurdles in the algal biohydrogen production. Therefore, researchers are continuously working on the use of new technology such as nanotechnology for better yield of algal biohydrogen production silica gel suspension, iron oxide, NiO, etc., are some of the used nanoparticles (Table 2).

Giannelli and Torzillo (2012) used the suspension of silica nanoparticles to avoid the light saturation effect for improving the growth of the microalgae (*Chlamydomonas reinhardtii*) in photobioreactor and hence the hydrogen production. These researchers also reported reactor fluid dynamics and a tri-dimensional light profile inside the photobioreactor. Zaidi et al. (2019) suggested an energy-efficient way for biohydrogen production that can be scaled up for commercial biohydrogen production by combining microwave pretreatment with iron oxide nanoparticles. Their method was able to increase the biohydrogen production by 54.71%. The increase in hydrogen production was due to the improvement in biodegradability of green algae *Enteromorpha* which can further be applied for biohydrogen production

Table 2. Role of nanomaterials in biohydrogen production from various organisms.

S. No.	Organisms	Methods	Nanoparticle	Effect on Hydrogen Yield	References
1.	*Chlamydomonas reinhardtii*	Photobioreactor	Suspension of silica	+	(Giannelli and Torzillo 2012)
2.	*Chlamydomonas reinhardtii*				
3.	*Enteromorpha*	Batch-wise anaerobic digestion	Iron oxide	+	(Zaidi et al. 2019)
4.	Anaerobic mixed bacteria dominated with *Clostridium butyricum*	Batch assays	Hematite	+	(Han et al. 2011)
5.	Anaerobic mixed bacteria dominated with *Clostridium butyricum*	Batch experiments	Gold	+	(Zhang and Shen 2007)
6.	*Rhodopseudomonas palustris*	Dark fermentation	Nano-TiO_2	+	(Zhao and Chen 2011)
7.	Anaerobic mixed bacteria dominated with *Clostridium butyricum*	Batch experiments	Mesoporous magnetic Fe_3O_4	+	(Zhao et al. 2011)
8.	Anaerobic mixed bacteria dominated with *Clostridium butyricum*	Batch experiments	Silver nanoparticles	+	(Zhao et al. 2013)
9.	*Clostridium butyricum* strain	Dark fermentation	Encapsulated metallic (Pd, Ag and Cu) or metallic oxide (FexOy) nanoparticles	No effect on the yield	(Beckers et al. 2013)
10.	Mixed culture	Batch fermentation	Nickle	+	(Mullai et al. 2013)
11.	Anaerobic mixed Culture	Dark fermentation	Fe, Zn, Mg, Al hydrotalcite	+	(Wimonsong et al. 2013)
12.	*Enterobacter cloacae*	Dark fermentation	Iron oxide	+	(Mohanraj et al. 2014)
13.	*Enterobacter cloacae*	Dark fermentation	Palladium	+	(Sundaresan Mohanraj et al. 2014)
14.	Anaerobic mixed Culture	Dark fermentation	Fe-Zn-Mg-Al-O hydrotalcites supported Au	+	(Wimonsong et al. 2014)
15.	*Clostridium* and *Rhodopseudomonas palustris*	The two-step process of sequential dark-photo fermentation	Maghemite	+	(Nasr et al. 2015)

Table 2 contd. ...

...Table 2 contd.

S. No.	Organisms	Methods	Nanoparticle	Effect on Hydrogen Yield	References
16.	Anaerobic sludge	Batch fermentation test	Nickel oxide and hematite	+	(Gadhe et al. 2015a)
17.	Anaerobic sludge	Batch fermentation test	Nickel oxide and hematite	+	(Gadhe et al. 2015b)
18.	*Clostridium beijerinckii* NCIMB8052	Dark fermentation	magnetite		(Seelert et al. 2015)
19.	Anaerobic sludge	Dark fermentation	Nano-titanium dioxide		(Jafari and Zilouei 2016)
20.	Anaerobic sludge	Dark fermentation	Fe^0, Ni^0		(Taherdanak et al. 2016, Taherdanak et al. 2015)
21.	*Clostridium pasteurianum*	Dark fermentation	Nano-titanium dioxide and magnetic hematite	+	(Hsieh et al. 2016)
22.	*Enterobacter aerogenes*	Dark fermentation	Ferric oxide	+	(R. Lin et al. 2016)
23.	Mixed culture	Thermophilic mixed fermentation	Iron(II) oxide and	+	(Engliman et al. 2017)

from all biomass with resistant cell walls. Zaidi et al. (2019) also studied the kinetic parameters of the reaction by using a modified Gompertz and Logistic function model (Table 2).

4.2 Nanomaterial Used During Biohydrogen Production from Various Bacteria

Biohydrogen can be produced by many anaerobes, facultative anaerobes, aerobes, methyotrophs and photosynthetic bacteria (Nandi and Sengupta 1998, Gray and Gest 1965, Stephen et al. 2017, Cabrol et al. 2017, Dutta et al. 2005, Puyol et al. 2018, Wang et al. 2012, Lambert and Smith 1977, Masukawa et al. 2001).

Zhang and Shen (2007) enhanced the biohydrogen production by using gold nanoparticles as a nanocatalyst. They also studied the effect of the size of gold nanoparticles on the biohydrogen production. The studied nanoparticles improved the bioactivity of hydrogen-producing microbes. The bioactivity of the microbes was dependent on the size of nanoparticles. These authors obtained the maximum cumulative yield of hydrogen by 5 nm gold nanoparticles. This increase in hydrogen production was related to the surface effect and the quantum size effect of the nanoparticles (Belloni et al. 1991, Chen et al. 1996) (Table 2).

Zhao and Chen (2011) studied the effect of the nano-titanium oxide in enhancing the photo-fermentative hydrogen production from the dark fermentation liquid of waste activated sludge. They reported a 46.1% increase in biohydrogen production by adding 100 mg/L nano TiO_2. This increase was due to the improved decomposition of protein and polysaccharide to small-molecule organic compounds by TiO_2. Titanium oxide is responsible for the promotion of growth and nitrogenase activity of photosynthetic bacteria but it declines the hydrogenase activity.

Han et al. (2011) studied the effects of hematite nanoparticles concentration (0–1600 mg/L) and initial pH (4.0–10.0) on hydrogen production in batch assays. Besides hydrogen production, they also studied the degradation efficiency, metabolites distribution and the bacteria morphology. For this study, these authors used the sucrose-fed anaerobic mixed bacteria and reported a 66.1% increase in hydrogen yield at pH 6.0 and 200 mg/L hematite nanoparticle. Transmission electron microscopic analysis by Han et al. (2011) reported an increase in the length and decline in the width of the bacteria under the effect of hematite nanoparticles. Thus, hematite nanoparticles not only improved the hydrogen production but also modified the bacterial growth as well as their metabolites distribution (Table 2).

Iron is known to have an indispensable role in the activity of the hydrogen-producing activity of microbes (Zhang et al. 2005, Dabrock et al. 1992, Junelles et al. 1988). Zhao et al. (2011) studied the catalytic effect of mesoporous magnetic Fe_3O_4 nanoparticles on biohydrogen production from the mixed culture dominated by *Clostridium butyricum* obtained from the aeration basin of the local municipal wastewater treatment plant. Mixed anaerobic culture was shocked by the high strength of alkaline influent (10 $g \cdot L^{-1}$ sodium carbonate) and lost its bioactivity of hydrogen production. To recover the hydrogenase activity of the alkaline treated culture, these authors used the mesoporous Fe_3O_4 nanoparticles and ferrous ions as activators and enhanced the hydrogen yield by 44% than that of the blank. Thus, these authors also explained the role of these Fe_3O_4 nanoparticles as an activator when the bioactivity of hydrogen-producing bacteria was inhibited. Zhao et al. (2013) further studied the effect and biocompatibility of silver nanoparticles on glucose-fed mixed hydrogen-producing bacterial culture (rich in *Clostridium butyricum*) growing under three experimental conditions. In the first experiment these authors added the activated culture in different concentrations of nanoparticles, in the second experiment alkaline pretreated culture was added in varying concentrations of inorganic ammonium and in the third experiment alkaline pretreated culture was added in different concentration of nanoparticles as well the inorganic ammonium concentration. These authors further concluded that silver nanoparticles help in enhancing the acidogenesis, and fermentative hydrogen production silver nanoparticles could not only increase the hydrogen yield but also reduced the lag phase for hydrogen production simultaneously (Table 2).

Beckers et al. (2013) studied the improving effect of nanometre-sized metallic (Pd, Ag and Cu) or metallic oxide (FexOy) nanoparticles encapsulated in porous silica on fermentative biohydrogen production by *Clostridium butyricum*. One-step sol-gel process (Lambert et al. 2004, Heinrichs et al. 2008, Pohaku Mitchell et al. 2012) was applied to obtain the nanocatalyst (nanoparticles + SiO_2 = catalyst). These nanoparticles increased the hydrogen production by 38% in culture containing iron oxide. The studied nanoparticles showed an increase in the hydrogen volume and rate of production but yields and metabolic pathways were not affected by them. To monitor the hydrogen production and metabolites in the cultures, Beckers et al. (2013) applied the Gompertz model volumetric production curves (Table 2).

Mullai et al. (2013) investigated the effect of initial glucose concentration, pH and nickel nanoparticles concentration on biohydrogen production using anaerobic microflora in batch culture. They further optimized biohydrogen production by employing a response surface methodology with a central composite design. This study reported a significant linear and interactive effect of initial glucose concentration and nickel nanoparticles concentration on biohydrogen production. Maximum biohydrogen yield of 2.54 mol of hydrogen/mol of glucose was achieved with an initial glucose concentration of 14.01 g/L at an initial pH of 5.61 and nickel nanoparticles concentration of 5.67 mg/L. With nickel nanoparticles, biohydrogen production was increased by 22.71%. The response surface methodology by Mullai et al. (2013) illustrated the significant impacts of initial glucose concentration and nickel nanoparticles concentration on biohydrogen production both individually and interactively (Table 2).

Mohanraj et al. (2014) enhanced the hydrogen yield by using synthesized palladium nanoparticles as well as the $PdCl_2$ on glucose using *Enterobacter cloacae* and mixed culture. They synthesized the palladium nanoparticles from $PdCl_2$ using *Coriandrum sativum* leaf extract which is an environmentally friendly method, i.e., green synthesis of palladium nanoparticles. The $PdCl_2$ has an inhibitory effect on the fermentative hydrogen production using *Enterobacter cloacae* and mixed culture, whereas Pd nanoparticles enhance the fermentative hydrogen production using mixed culture. This study resulted in confirmation of the effect of palladium nanoparticles on the metabolic pathway toward the high yield of hydrogen (Vavilin et al. 1995). In another study, Mohanraj et al. (2014) studied the influence of photosynthesized iron oxide nanoparticles and ferrous iron on fermentative hydrogen production using *Enterobacter cloacae* and reported the highest hydrogen yield but no change in the metabolic pathway was reported. These authors reported maximum hydrogen yields with glucose (2.07 ± 0.07 mol H_2/mol) and with sucrose (5.44 ± 0.27 mol H_2/mol) when 125 mg/L and 200 mg/L of iron oxide nanoparticles were used in the experiment as compared to the iron oxide supplement experiments. This increase in hydrogen production can be explained by enhancing the ferredoxin activity in fermentative hydrogen production (Gorrell 1985) (Table 2).

Wimonsong et al. (2014) fused four different Fe-Zn-Mg-Al-O hydrotalcites supported gold nanocatalysts synthesized by the incipient impregnation method. They further characterized these nanocatalysts by X-ray photoelectron spectroscopy as well as scanning electron microscopy. The use of these nanocatalysts enhances the biohydrogen yield as well as of biohydrogen production. The maximum hydrogen yield was obtained by the Au/Zn-Mg-Al hydrotalcites, and calcined Zn-Mg-Al hydrotalcites did not enhance the biohydrogen production. This improvement in hydrogen production was due to the combined effects of the metallic gold on the support surface containing Zn which acts as the active sites for hydrogenase enzyme and the gold catalyst serve as electron sinks at active sites (Chang et al. 2006, Wang and Wan 2009) (Table 2).

Nasr et al. (2015) used maghemite nanoparticles for biohydrogen production. Maghemite nanoparticles enhanced hydrogen production as well as yield by promoting the anaerobic conversion of starch wastewater into biohydrogen. Wimonsong et al. (2013) studied the activity as well as categorized the Fe-Zn-Mg-Al hydrotalcites for hydrogen production in batch tests using sucrose-fed anaerobic mixed culture. Maximum hydrogen yield was obtained by the Mg-Al and minimum yield by Fe-Mg-Al hydrotalcites (Table 2).

Gadhe et al. (2015) evaluated the potentials of hematite (Fe_2O_3) and nickel oxide (NiO) nanoparticles (separately as well as a combine) on biohydrogen production. Chemical precipitation method followed by thermal decomposition was used for the synthesis of NiO, and Fe_2O_3 (Darezereshki et al. 2012) nanoparticles. Synthesized nanoparticles were further used for the enhancement of biohydrogen production from complex distillery wastewater. The highest hydrogen yield was obtained for the batch test of complex distillery wastewater with addition of Fe_2O_3 (200 mg/L) and NiO NP (5 mg/L). These authors (Gadhe et al. 2015b) also reported a similar increase in the biohydrogen yield from dairy wastewater with Fe_2O_3 plus NiO nanoparticles. This increase in the biohydrogen production by combined hematite (Fe_2O_3) and nickel oxide (NiO) can be explained by the increased activity of various enzymes involved in the hydrogen production pathways such as ferredoxin oxidoreductase, ferredoxin and hydrogenase enzyme (Patrick Hallenbeck 2009, Johnson et al. 2014, Mohanraj et al. 2014, Zhang and Shen 2007). The increase in hydrogen yield was explained by the enhanced activity of ferredoxin oxidoreductase and hydrogenase due to Fe_2O_3 plus NiO nanoparticles interface (Table 2).

Seelert et al. (2015) improved the biohydrogen production by using *Clostridium beijerinckii* immobilized with magnetite nanoparticles. To promote the bacterial attachment, magnetite nanoparticles were functionalized with chitosan and alginic acid polyelectrolytes by using a layer-by-layer method.

Jafari and Zilouei (2016) enhanced the biohydrogen production by 127% from the sugarcane bagasse by using nano-titanium dioxide (nanoTiO_2) under ultraviolet irradiation (UV) followed by dilute sulfuric acid hydrolysis in consecutive dark fermentation and anaerobic digestion. This increase in hydrogen production was due to the effects of nano-TiO_2 pretreatment on the destruction of surface morphology and the reduction of crystallinity (Table 2).

Taherdanak et al. (2015) studied the effects of Fe^0 and Ni^0 nanoparticles on mesophilic dark hydrogen fermentation from starch using heat-shock pretreated anaerobic sludge in batch reactors and reported starch and Fe^0 nanoparticles are a most important factor affecting the hydrogen production. In another study, these authors (Taherdanak et al. 2016) compared the effect of Fe^0 and Ni^0 nanoparticles (NPs) with Fe^{2+} and Ni^{2+} ions on mesophilic dark hydrogen fermentation from glucose and reported 55% increase in hydrogen yield with Ni^{2+} and lowest with the Ni^0 nanoparticles.

Hsieh et al. (2016) studied the enhanced effect of titanium dioxide (TiO_2) nano-metals, magnetic hematite and soluble iron (II, III) nanoparticle addition on biohydrogen production by *Clostridium pasteurianum*. Hsieh et al. (2016) also evaluated gene expression and growth activity. To further explore the possible stimulation mechanism, these authors measured the enzyme activity and cell growth through a reverse-transcription polymerase chain reaction. The significant stimulation effect of hematite nanoparticles on hydrogen production was reported by them, but they did not find any change in the hydrogen yield. Interestingly, titanium dioxide nanoparticles have not shown any significant stimulation effect on biohydrogen production. This is due to the lower electrical and thermal conductivities of TiO_2 (Li et al. 2011) (Table 2).

Lin et al. (2016) studied the effects of ferric oxide nanoparticles on hydrogen fermentation from glucose and pretreated cassava starch using *Enterobacter aerogenes*. These authors also evaluated the effect of these nanoparticles in terms of hydrogen yield, production rate and metabolites distribution and reported 17.0% and 63.1% increase in the hydrogen yield in glucose and pretreated starch, respectively, by use of 200 mg/L ferric oxide nanoparticles. They also studied the effect of ferric oxide nanoparticles on the morphology of *Enterobacter aerogenes* cells by using scanning electron microscopy and the transmission electron microscopy. The transmission electron microscopic image of *E. aerogenes* showed cellular internalization of ferric oxide nanoparticles. The scanning electron microscopy showed the aggregates on the surfaces of bacterial cells derived from the nanowires related proteins in response to the addition of ferric oxide nanoparticles. Nanowires are extracellular appendages suggested as pathways for electron transfer in microorganisms (El-Naggar et al. 2010) (Table 2).

Engliman et al. (2017) investigated the impact of initial pH, metal oxide and concentration of nanoparticles on biohydrogen production in batch assays. For this study, they used glucose-fed anaerobic mixed bacteria in the thermophilic condition of 60°C and also used the metal oxide nanoparticles of iron (II) oxide and nickel oxide. These two-metal oxide nanoparticles were able to enhance the hydrogen yield by 34.38% and 5.47% than the control. They optimized pH 5.5 for higher production of hydrogen without nanoparticles. In combination experiment, nanoparticles (either iron(II) oxide or nickel oxide) as well and pH 5.5 of maximum hydrogen yield 1.92 mol H_2/mol glucose and 51% hydrogen content were obtained with optimal iron(II) oxide concentration of 50 mg/L. In this study, the researchers were able to recover nanoparticles without any loss after the experiment. Besides hydrogen production, Engliman et al. (2017) also studied the characteristic of the fermentation operations, such as metabolites distribution, sugar consumption and cell growth (Table 2).

5. Conclusion

Biohydrogen is a clean, efficient and sustainable energy option for the future and can be called as an energy carrier of the future. To be a marketable fuel, biohydrogen production should be cost-effective and effortlessly available. For this, there is a need for the advancement of technology for scaling up the production as well as lowering the cost. Nanotechnologists can play an important role in this regard by designing new nanomaterial for efficient production, storage, delivery and the end-use of biohydrogen.

6. Disclosure of Potential Conflicts of Interest

Authors declare no conflict of interest

References

Abd-Alla, M.H., F.A. Gabra, A.W. Danial and A.M. Abdel-Wahab. 2019. Enhancement of biohydrogen production from sustainable orange peel wastes using enterobacter species isolated from domestic wastewater. International Journal of Energy Research 43(1): 391–404.

Abdelsalam, E., M. Samer, Y.A. Attia, M.A. Abdel-Hadi, H.E. Hassan and Y. Badr. 2016. Comparison of nanoparticles effects on biogas and methane production from anaerobic digestion of cattle dung slurry. Renewable Energy 87: 592–598.

Abdelsalam, E., M. Samer, Y.A. Attia, M.A. Abdel-Hadi, H.E. Hassan and Y. Badr. 2017a. Influence of zero valent iron nanoparticles and magnetic iron oxide nanoparticles on biogas and methane production from anaerobic digestion of manure. Energy 120: 842–853.

Abdelsalam, E., M. Samer, Y.A. Attia, M.A. Abdel-Hadi, H.E. Hassan and Y. Badr. 2017b. Effects of Co and Ni nanoparticles on biogas and methane production from anaerobic digestion of slurry. Energy Conversion and Management 141: 108–119.

Abraham, R.E., M.L. Verma, C.J. Barrow and M. Puri. 2014. Suitability of magnetic nanoparticle immobilised cellulases in enhancing enzymatic saccharification of pretreated hemp biomass. Biotechnology for Biofuels 7(1): 1–12.

Alaei, S., M. Haghighi, J. Toghiani and B. Rahmani Vahid. 2018. Magnetic and reusable MgO/ $MgFe_2O_4$ Nanocatalyst for biodiesel production from sunflower oil: Influence of fuel ratio in combustion synthesis on catalytic properties and performance. Industrial Crops and Products 117(September): 322–332.

Alsharifi, M., H. Znad, S. Hena and M. Ang. 2017. Biodiesel production from canola oil using novel Li/TiO_2 as a heterogeneous catalyst prepared via impregnation method. Renewable Energy 114: 1077–1089.

Amen, T.W.M., O. Eljamal, A.M.E. Khalil and N. Matsunaga. 2017. Biochemical methane potential enhancement of domestic sludge digestion by adding pristine iron nanoparticles and iron nanoparticles coated zeolite compositions. Journal of Environmental Chemical Engineering 5(5): 5002–5013.

Azman, N.F., P. Abdeshahian, A. Kadier, N.K. Nasser Al-Shorgani, N.K.M. Salih, I. Lananan, A.A. Hamid and M.S. Kalil. 2016. Biohydrogen production from de-oiled rice bran as sustainable feedstock in fermentative process. International Journal of Hydrogen Energy 41(1): 145–156.

Azwar, M.Y., M.A. Hussain and A.K. Abdul-Wahab. 2014. Development of biohydrogen production by photobiological, fermentation and electrochemical processes: A review. Renewable and Sustainable Energy Reviews 31: 158–173.

Balat, H. and E. Kirtay. 2010. Hydrogen from biomass—present scenario and future prospects. International Journal of Hydrogen Energy 35(14): 7416–7426.

Baskar, G., A. Gurugulladevi, T. Nishanthini, R. Aiswarya and K. Tamilarasan. 2017. Optimization and kinetics of biodiesel production from mahua oil using manganese doped zinc oxide nanocatalyst. Renewable Energy 103: 641–646.

Bastidas-Oyanedel, J.R. and J.E. Schmidt. 2018. Increasing profits in food waste biorefinery-a techno-economic analysis. Energies 11(6).

Beckers, L., S. Hiligsmann, S.D. Lambert, B. Heinrichs and P. Thonart. 2013. Improving effect of metal and oxide nanoparticles encapsulated in porous silica on fermentative biohydrogen production by *Clostridium butyricum*. Bioresource Technology 133: 109–117.

Bélafi-Bakó, K., D. Búcsú, Z. Pientka, B. Bálint, Z. Herbel, K.L. Kovács and M. Wessling. 2006. Integration of biohydrogen fermentation and gas separation processes to recover and enrich hydrogen. International Journal of Hydrogen Energy 31(11): 1490–1495.

Belloni, J., M. Mostafavi, J.L. Marignier and J. Amblard. 1991. Quantum size effects and photographic development. Journal of Imaging Science 35(2): 68–74.

Benemann, J.R. 1998. The technology of biohydrogen. pp. 19–30. *In*: Zaborsky, O.R., J.R. Benemann, T. Matsunaga, J. Miyake and A. San Pietro (eds.). BioHydrogen. Boston, MA: Springer US.

Bet-Moushoul, E., K. Farhadi, Y. Mansourpanah, A.M. Nikbakht, R. Molaei and M. Forough. 2016. Application of CaO-Based/Au nanoparticles as heterogeneous nanocatalysts in biodiesel production. Fuel 164: 119–127.

Bhatia, S.C. 2014. Biohydrogen. pp. 627–644. *In*: Bhatia, S.C. (ed.). Advanced Renewable Energy Systems. Woodhead Publishing India.

Bond, D. and D. Lovley. 2003. Electricity production by geobacter sulfurreducens attached to electrodes. Applied and Environmental Microbiology 69(3): 1548–1555.

BP. 2018. 67th Edition Contents is one of the Most Widely Respected. Statistical Review of World Energy.

Buzea, C., I.I. Pacheco and K. Robbie. 2007. Nanomaterials and nanoparticles: Sources and toxicity. Biointerphases 2(4): MR17–MR71.

Cabrol, L., A. Marone, E. Tapia-Venegas, J.P. Steyer, G. Ruiz-Filippi and E. Trably. 2017. Microbial ecology of fermentative hydrogen producing bioprocesses: Useful insights for driving the ecosystem function. FEMS Microbiology Reviews 41(2): 158–181.

Carpenter, A.W., S.N. Laughton and M.R. Wiesner. 2015. Enhanced biogas production from nanoscale zero valent iron-amended anaerobic bioreactors. Environmental Engineering Science 32(8): 647–655.

Carvalho, M.S., R.A. Lacerda, J.P.B. Leão, J.D. Scholten, B.A.D. Neto and P.A.Z. Suarez. 2011. *In situ* generated palladium nanoparticles in imidazolium-based ionic liquids: A versatile medium for an efficient and selective partial biodiesel hydrogenation. Catalysis Science and Technology 1(3): 480–488.

Casals, E., R. Barrena, A. García, E. González, L. Delgado, M. Busquets-Fité, X. Font, J. Arbiol, P. Glatzel, K. Kvashnina, A. Sánchez and V. Puntes. 2014. Programmed iron oxide nanoparticles disintegration in anaerobic digesters boosts biogas production. Small 10(14): 2801–2808.

Chandrasekhar, K., Y.J. Lee and D.W. Lee. 2015. Biohydrogen production: Strategies to improve process efficiency through microbial routes. International Journal of Molecular Sciences 16(4): 8266–8293.

Chang, F.W., H.Y. Yu, L.S. Roselin, H.C. Yang and T.C. Ou. 2006. Hydrogen production by partial oxidation of methanol over gold catalysts supported on TiO_2-MOx (M = Fe, Co, Zn) composite oxides. Applied Catalysis A: General 302(2): 157–167.

Chang, J.S., K.S. Lee and P.J. Lin. 2002. Biohydrogen production with fixed-bed bioreactors. In International Journal of Hydrogen Energy 27: 1167–1174. Pergamon.

Chang, R.H.Y., J. Jang and K.C.W. Wu. 2011. Cellulase immobilized mesoporous silica nanocatalysts for efficient cellulose-to-glucose conversion. Green Chemistry 13(10): 2844–2850.

Chen, R., Y.Z. Wang, Q. Liao, X. Zhu and T.F. Xu. 2013. Hydrolysates of lignocellulosic materials for biohydrogen production. BMB Reports 46(5): 244–251.

Chen, S.-D., K.-S. Lee, Y.-C. Lo, W.-M. Chen, J.-F. Wu, C.-Y. Lin and J.-S. Chang. 2008. Batch and continuous biohydrogen production from starch hydrolysate by clostridium species. International Journal of Hydrogen Energy 33(7): 1803–1812.

Chen, W.-H., S. Sung and S.-Y. Chen. 2009. Biological hydrogen production in an anaerobic sequencing batch reactor: PH and cyclic duration effects. International Journal of Hydrogen Energy 34(1): 227–234.

Chen, Z.J., X.M. Ou, F.Q. Tang and L. Jiang. 1996. Effect of nanometer particles on the adsorbability and enzymatic activity of glucose oxidase. Colloids and Surfaces B: Biointerfaces 7(3-4) (September 16): 173–179.

Cheng, C.L., Y.C. Lo, K.S. Lee, D.J. Lee, C.Y. Lin and J.S. Chang. 2011. Biohydrogen production from lignocellulosic feedstock. Bioresource Technology 102(18): 8514–8523.

Dabrock, B., H. Bahl and G. Gottschalk. 1992. Parameters affecting solvent production by Clostridium pasteurianum. Applied and Environmental Microbiology 58(4): 1233–1239.

Daima, H.K. and V. Bansal. 2015. Influence of physicochemical properties of nanomaterials on their antibacterial applications. pp. 151–166. *In*: Rai, M. and K. Kon (eds.). Nanotechnology in Diagnosis, Treatment and Prophylaxis of Infectious Diseases. Academic Press.

Dalla Vecchia, C., A. Mattioli, D. Bolzonella and E. Palma. 2016. Impact of magnetite nanoparticles supplementation on the anaerobic digestion of food wastes: batch and continuous-flow investigations. Chemical Engineering Transactions 49: 1–6.

Darezereshki, E., F. Bakhtiari and M. Alizadeh. 2012. Direct thermal decomposition synthesis and characterization. Materials Science in Semiconductor Processing 15(1): 91–97.

Das, D. and T. Veziroĝlu. 2002. Hydrogen production by biological processes: A survey of literature. International Journal of Hydrogen Energy 26(January 1): 13–28.

Das, D. and T.N. Veziroglu. 2008. Advances in biological hydrogen production processes. International Journal of Hydrogen Energy 33(21)(November 1): 6046–6057.

Das, D., N. Khanna and C.N. Dasgupta. 2014. Biohydrogen production: Fundamentals and technology advances. Biohydrogen Production: Fundamentals and Technology Advances. Boca Raton, Florida, United States: CRC CRC Press.

Degirmenbasi, N., S. Coskun, N. Boz and D.M. Kalyon. 2015. Biodiesel Synthesis from canola oil via heterogeneous catalysis using functionalized CaO nanoparticles. Fuel 153: 620–627.

Diah, K., P. Joni, S. Endang and H. Sumi. 2018. Biohydrogen production through separate hydrolysis and fermentation and simultaneous saccharification and fermentation of empty fruit bunch of palm oil. Research Journal of Chemistry and Environment 22(II): 193–197.

Dutta, D., D. De, S. Chaudhuri and S.K. Bhattacharya. 2005. Hydrogen production by cyanobacteria. Microbial Cell Factories 4(1): 36.

El-Naggar, M.Y., G. Wanger, K.M. Leung, T.D. Yuzvinsky, G. Southam, J. Yang, W.M. Lau, K.H. Nealson and Y.A. Gorby. 2010. Electrical transport along bacterial nanowires from shewanella oneidensis MR-1. Proceedings of the National Academy of Sciences 107(42): 18127–18131.

Engliman, N.S., P.M. Abdul, S.Y. Wu and J.M. Jahim. 2017. Influence of iron(II) oxide nanoparticle on biohydrogen production in thermophilic mixed fermentation. International Journal of Hydrogen Energy 42(45): 27482–27493.

Eroĝlu, E., I. Eroĝlu, U. Gündüz, L. Türker and M. Yücel. 2006. Biological hydrogen production from olive mill wastewater with two-stage processes. International Journal of Hydrogen Energy 31(11)(September 1): 1527–1535.

Fabiano, B. and P. Perego. 2002. Thermodynamic study and optimization of hydrogen production by enterobacter aerogenes. International Journal of Hydrogen Energy 27(2): 149–156.

Fang, H.H.P., H. Zhu and T. Zhang. 2006. Phototrophic hydrogen production from glucose by pure and co-cultures of Clostridium butyricum and Rhodobacter sphaeroides. International Journal of Hydrogen Energy 31.

Fang, Herbert H.P. and H. Liu. 2002. Effect of PH on hydrogen production from glucose by a mixed culture. Bioresource Technology 82(March 1): 87–93.

Fang, Herbert H.P., C. Li and T. Zhang. 2006. Acidophilic biohydrogen production from rice slurry. International Journal of Hydrogen Energy 31(6): 683–692.

Fascetti, E. and O. Todini. 1995. Rhodobacter sphaeroides RV cultivation and hydrogen production in a one- and two-stage chemostat. Applied Microbiology and Biotechnology 44(3-4)(December): 300–305.

Fedorov, A., A. Tsygankov, K.K. Rao and D.O. Hall. 1998. Hydrogen photoproduction by rhodobacter sphaeroides immobilised on polyurethane foam. Biotechnology Letters 20: 1007–1009.

Fedorovich, V., M.C. Knighton, E. Pagaling, F.B. Ward, A. Free and I. Goryanin. 2009. Novel electrochemically active bacterium phylogenetically related to arcobacter butzleri, isolated from a microbial fuel cell. Applied and Environmental Microbiology 75(23): 7326–7334.

Fernández Rodríguez, C., E.J. Martínez Torres, A. Morán Palao and X. Gómez Barrios. 2016. Procesos Biológicos Para El Tratamiento de Lactosuero Con Producción de Biogás e Hidrógeno. Revisión Bibliográfica. Revista Investigación, Optimización y Nuevos Procesos En Ingeniería 29(1)(June 1): 47–62.

Feyzi, M., A. Hassankhani and H.R. Rafiee. 2013. Preparation and characterization of $Cs/Al/Fe_3O_4$ nanocatalysts for biodiesel production. Energy Conversion and Management 71: 62–68.

Feyzi, M., L. Nourozi and M. Zakarianezhad. 2014. Preparation and characterization of magnetic $CsH_2PW12O40/Fe\text{-}SiO_2$ nanocatalysts for biodiesel production. Materials Research Bulletin 60: 412–420.

Francou, N. and P.M. Vignais. 1984. Hydrogen production by rhodopseudomonas capsulata cells entrapped in carrageenan beads. Biotechnology Letters 6(October 10): 639–644.

Gadhe, A., S.S. Sonawane and M.N. Varma. 2015a. Influence of nickel and hematite nanoparticle powder on the production of biohydrogen from complex distillery wastewater in batch fermentation. International Journal of Hydrogen Energy 40(34): 10734–10743.

Gadhe, A., S.S. Sonawane and M.N. Varma. 2015b. Enhancement effect of hematite and nickel nanoparticles on biohydrogen production from dairy wastewater. International Journal of Hydrogen Energy 40(April 13): 4502–4511.

García, A., L. Delgado, J.A. Torà, E. Casals, E. González, V. Puntes, X. Font, J. Carrera and A. Sánchez. 2012. Effect of cerium dioxide, titanium dioxide, silver, and gold nanoparticles on the activity of microbial communities intended in wastewater treatment. Journal of Hazardous Materials 199-200: 64–72.

Giannelli, L. and G. Torzillo. 2012. Hydrogen production with the microalga chlamydomonas reinhardtii grown in a compact tubular photobioreactor immersed in a scattering light nanoparticle suspension. International Journal of Hydrogen Energy 37(22): 16951–16961.

Gorrell, T.E. 1985. Effect of culture medium iron content on the biochemical composition and metabolism of trichomonas vaginalis. Journal of Bacteriology 161(3): 1228–1230.

Gray, C.T. and H. Gest. 1965. Biological formation of molecular hydrogen. Science 148(3667): 186–192.

Greenwell, H.C., L.M.L. Laurens, R.J. Shields, R.W. Lovitt and K.J. Flynn. 2010. Placing microalgae on the biofuels priority list: A review of the technological challenges. Journal of the Royal Society, Interface/the Royal Society 7(46): 703–26.

Güell, E.J., B.T. Maru, R.J. Chimentão, F. Gispert-Guirado, M. Constantí and F. Medina. 2015. Combined heterogeneous catalysis and dark fermentation systems for the conversion of cellulose into biohydrogen. Biochemical Engineering Journal 101: 209–219.

Guo, W., Z. He and J. Li. 2016. The development of hydrogen production from lignocellulosic biomass: Pretreatment and process. pp. 250–255. *In*: The 2016 International Conference on Advances in Energy, Environment and Chemical Science (AEECS 2016). Singapore.

Guo, X.M., E. Trably, E. Latrille, H. Carrre and J.P. Steyer. 2010. Hydrogen production from agricultural waste by dark fermentation: A review. International Journal of Hydrogen Energy 35(19): 10660–10673.

Gürtekin, E. 2014. Biological hydrogen production methods. pp. 463–471. *In*: 2nd International Symposium on Environment and Morality. Adıyaman–Turkey.

Hallenbeck, P.C. and D. Ghosh. 2009. Advances in fermentative biohydrogen production: The way forward? Trends in Biotechnology 27(5): 287–97.

Hallenbeck, Patrick C. and J.R. Benemann. 2002. Biological hydrogen production; fundamentals and limiting processes. International Journal of Hydrogen Energy 27(11-12)(November 1): 1185–1193.

Hallenbeck, Patrick C. 2009. Fermentative hydrogen production: principles, progress, and prognosis. International Journal of Hydrogen Energy 34(17): 7379–7389.

Hallenbeck, Patrick C. 2011. Microbial paths to renewable hydrogen production. Biofuels 2(3): 285–302.

Han, H., M. Cui, L. Wei, H. Yang and J. Shen. 2011. Enhancement effect of hematite nanoparticles on fermentative hydrogen production. Bioresource Technology 102(17): 7903–7909.

Han, W., Y. Yan, Y. Shi, J. Gu, J. Tang and H. Zhao. 2016. Biohydrogen production from enzymatic hydrolysis of food waste in batch and continuous systems. Scientific Reports 6(December): 1–9.

Hankamer, B., F. Lehr, J. Rupprecht, J.H. Mussgnug, C. Posten and O. Kruse. 2007. Photosynthetic biomass and H_2 production by green algae: From bioengineering to bioreactor scale-up. Physiologia Plantarum 131(1): 10–21.

Hawkes, F.R., I. Hussy, G. Kyazze, R. Dinsdale and D.L. Hawkes. 2007. Continuous dark fermentative hydrogen production by mesophilic microflora: Principles and progress. International Journal of Hydrogen Energy 32(2): 172–184.

Heinrichs, B., L. Rebbouh, J.W. Geus, S. Lambert, H.C.L. Abbenhuis, F. Grandjean, G.J. Long, J.P. Pirard and R.A. van Santen. 2008. Iron(III) species dispersed in porous silica through sol-gel chemistry. Journal of Non-Crystalline Solids 354(2-9): 665–672.

Holladay, J.D., J. Hu, D.L. King and Y. Wang. 2009. An overview of hydrogen production technologies. Catalysis Today 139(4): 244–260.

Hsieh, P.H., Y.C. Lai, K.Y. Chen and C.H. Hung. 2016. Explore the possible effect of TiO_2 and magnetic hematite nanoparticle addition on biohydrogen production by clostridium pasteurianum based on gene expression measurements. International Journal of Hydrogen Energy 41(46): 21685–21691.

Hübler, A.W. and O. Osuagwu. 2010. Digital quantum batteries: Energy and information storage in nanovacuum tube arrays. Complexity 15(5): 48–55.

International Energy Agency. 2018. Global Energy and CO_2 Status Report 2017. https://www.iea.org/publications/freepublications/publication/GECO2017.pdf.

IRENA. 2018. Power Generation Costs in 2017. International Renewable Energy Agency. Vol. Abu Dhabi.

Isikgor, F.H. and C.R. Becer. 2015. Lignocellulosic biomass: A sustainable platform for the production of bio-based chemicals and polymers. Polymer Chemistry 6(25): 4497–4559.

Ivanova, V., P. Petrova and J. Hristov. 2011. Application in the ethanol fermentation of immobilized yeast cells in matrix of alginate/magnetic nanoparticles, on chitosan-magnetite microparticles and cellulose-coated magnetic nanoparticles. International Review of Chemical Engineering 3(2): 289–299.

Jafari, O. and H. Zilouei. 2016. Enhanced biohydrogen and subsequent biomethane production from sugarcane bagasse using nano-titanium dioxide pretreatment. Bioresource Technology 214: 670–678.

Jiang, D., Z. Fang, S.X. Chin, X.F. Tian and T.C. Su. 2016. Biohydrogen production from hydrolysates of selected tropical biomass wastes with clostridium butyricum. Scientific Reports 6(June): 1–11.

Johnson, B.J., W. Russ Algar, A.P. Malanoski, M.G. Ancona and I.L. Medintz. 2014. Understanding enzymatic acceleration at nanoparticle interfaces: approaches and challenges. Nano Today. Elsevier Ltd.

Jones, R.J., J. Massanet-Nicolau, M.J.J. Mulder, G. Premier, R. Dinsdale and A. Guwy. 2017. Increased biohydrogen yields, volatile fatty acid production and substrate utilisation rates via the electrodialysis of a continually fed sucrose fermenter. Bioresource Technology 229: 46–52.

Junelles, A.M., R. Janati-Idrissi, H. Petitdemange and R. Gay. 1988. Iron effect on acetone-butanol fermentation. Current Microbiology 17(September 5): 299–303.

Kapdan, I.K. and F. Kargi. 2006. Bio-hydrogen production from waste materials. Enzyme and Microbial Technology. Elsevier.

Karadag, D. and J.A. Puhakka. 2010. Enhancement of anaerobic hydrogen production by iron and nickel. International Journal of Hydrogen Energy 35(16): 8554–8560.

Kars, G., U. Gündüz, M. Yücel, L. Türker and İ. Eroglu. 2006. Hydrogen production and transcriptional analysis of NifD, NifK and HupS genes in rhodobacter sphaeroides O.U.001 grown in media with different concentrations of molybdenum and iron. International Journal of Hydrogen Energy 31(11)(September 1): 1536–1544.

Khan, M.J., Q. Husain and A. Azam. 2012. Immobilization of porcine pancreatic α-amylase on magnetic Fe_2O_3 nanoparticles: Applications to the hydrolysis of starch. Biotechnology and Bioprocess Engineering 17(2): 377–384.

Khanna, N. and D. Das. 2013. Biohydrogen production by dark fermentation. Wiley Interdisciplinary Reviews: Energy and Environment 2(4): 401–421.

Khetkorn, W., N. Khanna, A. Incharoensakdi and P. Lindblad. 2013. Metabolic and genetic engineering of cyanobacteria for enhanced hydrogen production. Biofuels 4(5): 535–561.

Khetkorn, W., R.P. Rastogi, A. Incharoensakdi, P. Lindblad, D. Madamwar, A. Pandey and C. Larroche. 2017. Microalgal hydrogen production—a review. Bioresource Technology 243: 1194–1206.

Kim, D.H., S.K. Han, S.H. Kim and H.S. Shin. 2006. Effect of gas sparging on continuous fermentative hydrogen production. International Journal of Hydrogen Energy 31(15): 2158–2169.

Kim, H.J., H.S. Park, M.S. Hyun, I.S. Chang, M. Kim and B.H. Kim. 2002. A mediator-less microbial fuel cell using a metal reducing bacterium, shewanella putrefaciens. Enzyme and Microbial Technology 30(2)(February 14): 145–152.

Kim, M., H.S. Lee, S.J. Yoo, Y.S. Youn, Y.H. Shin and Y.W. Lee. 2013. Simultaneous synthesis of biodiesel and zinc oxide nanoparticles using supercritical methanol. Fuel 109: 279–284.

Klabunde, K.J. 2001. Nanoscale materials in chemistry. pp. 11–292. *In*: Kenneth J. Klabunde (ed.). Nanoscale Materials in Chemistry, New York: Wiley-Interscience.

Kondo, T., T. Wakayama and J. Miyake. 2006. Efficient hydrogen production using a multi-layered photobioreactor and a photosynthetic bacterium mutant with reduced pigment. International Journal of Hydrogen Energy 31(11): 1522–1526.

Kosourov, S., A. Tsygankov, M. Seibert and M.L. Ghirardi. 2002. Sustained hydrogen photoproduction by chlamydomonas reinhardtii: effects of culture parameters. Biotechnology and Bioengineering 78(7): 731–740.

Kotay, S.M. and D. Das. 2007. Microbial hydrogen production with bacillus coagulans IIT-BT S1 isolated from anaerobic sewage sludge. Bioresource Technol. 98.

Kovács, K.L., G. Maróti and G. Rákhely. 2006. A novel approach for biohydrogen production. International Journal of Hydrogen Energy 31(11)(September 1): 1460–1468.

Kumar Gupta, S., S. Kumari, K. Reddy and F. Bux. 2013. Trends in biohydrogen production: Major challenges and state-of-the-art developments. Environmental Technology 34(13-14)(July 1): 1653–1670.

Kumar, N. and D. Das. 2000. Enhancement of hydrogen production by enterobacter cloacae IIT-BT 08. Process Biochem. 35.

Lambert, G.R. and G.D. Smith. 1977. Hydrogen formation by marine blue-green algae. FEBS Letters 83.

Lambert, S., C. Alié, J.P. Pirard and B. Heinrichs. 2004. Study of Textural properties and nucleation phenomenon in Pd/SiO_2, Ag/SiO_2 and Cu/SiO_2 cogelled xerogel catalysts. Journal of Non-Crystalline Solids 342(1-3): 70–81.

Lay, J.J. 2001. Biohydrogen generation by mesophilic anaerobic fermentation of microcrystalline cellulose. Biotechnology and Bioengineering 74(4): 280–287.

Lee, K.-S., P.-J. Lin and J.-S. Chang. 2006. Temperature effects on biohydrogen production in a granular sludge bed induced by activated carbon carriers. International Journal of Hydrogen Energy 31(4): 465–472.

Lee, S.M., L.H. Jin, J.H. Kim, S.O. Han, H. Bin Na, T. Hyeon, Y.M. Koo, J. Kim and J.H. Lee. 2010. β-glucosidase coating on polymer nanofibers for improved cellulosic ethanol production. Bioprocess and Biosystems Engineering 33(1): 141–147.

Levin, D.B., L. Pitt and M. Love. 2004. Biohydrogen production: Prospects and limitations to practical application. International Journal of Hydrogen Energy 29(2): 173–185.

Li, C. and H.H.P. Fang. 2007. Fermentative hydrogen production from wastewater and solid wastes by mixed cultures. Critical Reviews in Environmental Science and Technology 37(1): 1–39.

Li, M., M.E. Noriega-Trevino, N. Nino-Martinez, C. Marambio-Jones, J. Wang, R. Damoiseaux, F. Ruiz and E.M.V. Hoek. 2011. Synergistic bactericidal activity of Ag-TiO_2 nanoparticles in both light and dark conditions. Environmental Science and Technology 45(20): 8989–8995.

Lin, C.Y. and C.H. Lay. 2004. Carbon/Nitrogen-ratio effect on fermentative hydrogen production by mixed microflora. International Journal of Hydrogen Energy 29(1): 41–45.

Lin, C.-Y. and S.-H. Shei. 2008. Heavy metal effects on fermentative hydrogen production using natural mixed microflora. International Journal of Hydrogen Energy 33(2): 587–593.

Lin, C.-Y., C.-C. Wu and C.-H. Hung. 2008. Temperature effects on fermentative hydrogen production from xylose using mixed anaerobic cultures. International Journal of Hydrogen Energy 33(1): 43–50.

Lin, R., J. Cheng, L. Ding, W. Song, M. Liu, J. Zhou and K. Cen. 2016. Enhanced dark hydrogen fermentation by addition of ferric oxide nanoparticles using enterobacter aerogenes. Bioresource Technology 207(May 1): 213–219.

Lindberg, P., E. Devine, K. Stensjö and P. Lindblad. 2012. HupW protease specifically required for processing of the catalytic subunit of the uptake hydrogenase in the cyanobacterium nostoc sp. strain PCC 7120. Applied and Environmental Microbiology 78(January 1): 273–276.

Liu, H., S. Grot and B.E. Logan. 2005. Electrochemically assisted microbial production of hydrogen from acetate. Environmental Science & Technology 39(11): 4317–4320.

Liu, J., V.E. Bukatin and A.A. Tsygankov. 2006. Light energy conversion into H_2 by anabaena variabilis mutant PK84 dense cultures exposed to nitrogen limitations. International Journal of Hydrogen Energy 31(11): 1591–1596.

Liu, S., C.Y. Wang, L.L. Yin, W.Z. Li, Z.J. Wang and L.N. Luo. 2017. Optimization of hydrogen production from agricultural wastes using mixture design. International Journal of Agricultural and Biological Engineering 10(3): 246–254.

Łukajtis, R., I. Hołowacz, K. Kucharska, M. Glinka, P. Rybarczyk, A. Przyjazny and M. Kamiński. 2018. Hydrogen production from biomass using dark fermentation. Renewable and Sustainable Energy Reviews 91(April): 665–694.

Lupoi, J.S. and E.A. Smith. 2011. Evaluation of nanoparticle-immobilized cellulase for improved ethanol yield in simultaneous saccharification and fermentation reactions. Biotechnology and Bioengineering 108(12): 2835–2843.

Maneeruttanarungroj, C., P. Lindblad and A. Incharoensakdi. 2010. A newly isolated green alga, tetraspora sp. CU2551, from Thailand with efficient hydrogen production. International Journal of Hydrogen Energy 35(24): 13193–13199.

Mao, S.S., S. Shen and L. Guo. 2012. Nanomaterials for renewable hydrogen production, storage and utilization. Progress in Natural Science: Materials International 22(6): 522–534.

Masukawa, H., K. Nakamura, M. Mochimaru and H. Sakurai. 2001. Photobiological hydrogen production and nitrogenase activity in some heterocystous cyanobacteria. pp. 63–66. *In*: Miyake, J., T. Matsunaga and A.S. Pietro (eds.). Biohydrogen II. Oxford: Pergamon.

Meher Kotay, S. and D. Das. 2008. Biohydrogen as a renewable energy resource—prospects and potentials. International Journal of Hydrogen Energy 33(January 1): 258–263.

Melis, A., L. Zhang, M. Forestier, M.L. Ghirardi and M. Seibert. 2000. Sustained photobiological hydrogen gas production upon reversible inactivation of oxygen evolution in the green alga *Chlamydomonas reinhardtii*. Plant Physiology 122(1): 127–136.

Melis, A. and T. Happe. 2001. Hydrogen production. green algae as a source of energy. Plant Physiology 127: 740–748.

Melis, A. 2002. Green alga hydrogen production: progress, challenges and prospects. International Journal of Hydrogen Energy 12: 1–12.

Michał, M. 2012. Biological methods for obtaining hydrogen. Chemik 66(8): 827–834.

Mohan, S.V., G. Mohanakrishna, S. Veer Raghavulu and P.N. Sarma. 2007. Enhancing biohydrogen production from chemical wastewater treatment in anaerobic sequencing batch biofilm reactor (AnSBBR) by bioaugmenting with selectively enriched kanamycin resistant anaerobic mixed consortia. International Journal of Hydrogen Energy 32(15) SPEC. ISS.: 3284–3292.

Mohan, S.V. and A. Pandey. 2013. Biohydrogen production: An introduction. pp. 1–24. *In*: Pandey, A., J.-S. Chang, P.C. Hallenbecka and C. Larroche (eds.). Biohydrogen. Elsevier B.V.

Mohanraj, S., S. Kodhaiyolii, M. Rengasamy and V. Pugalenthi. 2014. Phytosynthesized iron oxide nanoparticles and ferrous iron on fermentative hydrogen production using enterobacter cloacae: Evaluation and comparison of the effects. International Journal of Hydrogen Energy 39(23): 11920–11929.

Mohanraj, Sundaresan, K. Anbalagan, S. Kodhaiyolii and V. Pugalenthi. 2014. Comparative evaluation of fermentative hydrogen production using enterobacter cloacae and mixed culture: Effect of Pd (II) ion and phytogenic palladium nanoparticles. Journal of Biotechnology 192(Part A): 87–95.

Mu, H., Y. Chen and N. Xiao. 2011. Effects of metal oxide nanoparticles (TiO_2, Al_2O_3, SiO_2 and ZnO) on waste activated sludge anaerobic digestion. Bioresource Technology 102(22): 10305–10311.

Mu, Y., G. Wang and H.-Q. Yu. 2006. Response surface methodological analysis on biohydrogen production by enriched anaerobic cultures. Enzyme Microb. Technol. 38.

Mu, Yang, H.-Q. Yu and Y. Wang. 2006. The role of PH in the termentative H_2 production from an acidogenic granule-based reactor. Chemosphere 64(June 3): 350—358.

Muhamad, N.S., N.A. Johan, M.H. Isa and S.R.M. Kutty. 2011. Biohydrogen production using dark and photo fermentation: A mini review. pp. 1–9. *In*: 2011 National Postgraduate Conference—Energy and Sustainability: Exploring the Innovative Minds, NPC 2011, IEEE.

Mullai, P., M.K. Yogeswari and K. Sridevi. 2013. Optimisation and enhancement of biohydrogen production using nickel nanoparticles—a novel approach. Bioresource Technology 141: 212–219.

Nandi, R. and S. Sengupta. 1998. Microbial production of hydrogen: An overview. Crit. Rev. Microbiol. 24.

Nasr, M., A. Tawfik, S. Ookawara, M. Suzuki, S. Kumari and F. Bux. 2015. Continuous biohydrogen production from starch wastewater via sequential dark-photo fermentation with emphasize on maghemite nanoparticles. Journal of Industrial and Engineering Chemistry 21: 500–506.

Nath, K. and D. Das. 2004. Biohydrogen production as a potential energy resource—present state-of-art. Journal of Scientific and Industrial Research 63(9): 729–738.

Nath, K., A. Kumar and D. Das. 2005. Hydrogen production by rhodobacter sphaeroides strain O.U.001 using spent media of enterobacter cloacae strain DM11. Applied Microbiology and Biotechnology 68(4): 533–541.

Nath, K. and D. Das. 2011. Modeling and optimization of fermentative hydrogen production. Bioresource Technology 102(September 18): 8569–8581.

Ntaikou, I., G. Antonopoulou and G. Lyberatos. 2010. Biohydrogen production from biomass and wastes via dark fermentation: A review. Waste and Biomass Valorization 1(March 1): 21–39.

Otero-González, L., J.A. Field and R. Sierra-Alvarez. 2014. Inhibition of anaerobic wastewater treatment after long-term exposure to low levels of CuO nanoparticles. Water Research 58: 160–168.

Pandu, K. and S. Joseph. 2012. Comparisons and limitations of biohydrogen production processes: A review. International Journal of Advances in Engineering & Technology 2(1): 2231–1963.

Park, D.H., P.K. Shin, I.S. Chang, H.J. Kim, W. Habermann, A.M. Biotechnology, R.M. Allen et al. 1999. Mediator-Less Biofuel Cell. Us Patent.

Park, H.J., A.J. Driscoll and P.A. Johnson. 2018. The development and evaluation of β-glucosidase immobilized magnetic nanoparticles as recoverable biocatalysts. Biochemical Engineering Journal 133: 66–73.

Park, J.H., J.J. Yoon, H.D. Park, Y.J. Kim, D.J. Lim and S.H. Kim. 2011. Feasibility of biohydrogen production from gelidium amansii. International Journal of Hydrogen Energy 36(21): 13997–14003.

Pohaku Mitchell, K.K., A. Liberman, A.C. Kummel and W.C. Trogler. 2012. Iron(III)-doped, silica nanoshells: A biodegradable form of silica. Journal of the American Chemical Society 134(34)(August): 13997–14003.

Pragya, N., K.K. Pandey and P.K. Sahoo. 2013. A review on harvesting, oil extraction and biofuels production technologies from microalgae. Renewable and Sustainable Energy Reviews 24: 159–171.

Puyol, D., F. Martinez, A. Esteve-Nuñez, C. Manchon, I.A. Vasiliadou, A. Berná and J.A. Melero. 2018. Biological and bioelectrochemical systems for hydrogen production and carbon fixation using purple phototrophic bacteria. Frontiers in Energy Research 6(November): 1–12.

R Bond, D., D. Holmes, L.M. Tender and D.R. Lovley. 2002. Electrode-reducing microorganisms that harvest energy from marine sediments. Science 295: 483–485.

Rabaey, K., G. Lissens, S.D. Siciliano and W. Verstraete. 2003. A microbial fuel cell capable of converting glucose to electricity at high rate and efficiency. Biotechnology Letters 25(18)(September): 1531–1535.

Radakovits, R., R.E. Jinkerson, A. Darzins and M.C. Posewitz. 2010. Genetic engineering of algae for enhanced biofuel production. Eukaryotic Cell 9(4): 486–501.

Radjaram, B. and R. Saravanane. 2011. Start up study of UASB reactor treating press mud for biohydrogen production. Biomass and Bioenergy 35(7): 2721–2728.

Rahmani Vahid, B. and M. Haghighi. 2016. Urea-nitrate combustion synthesis of $MgO/MgAl_2O_4$ nanocatalyst used in biodiesel production from sunflower oil: Influence of fuel ratio on catalytic properties and performance. Energy Conversion and Management 126: 362–372.

Redwood, M.D. and L.E. Macaskie. 2006. A two-stage, two-organism process for biohydrogen from glucose. Int. J. Hydrogen Energ. 31.

Ren, N., Z. Chen, A. Wang and D. Hu. 2005. Removal of organic pollutants and analysis of MLSS–COD removal relationship at different HRTs in a submerged membrane bioreactor. International Biodeterioration & Biodegradation 55(4): 279–284.

Risco, M.L., K. Orupõld and H.C. Dubourguier. 2011. Particle-Size effect of CuO and ZnO on biogas and methane production during anaerobic digestion. Journal of Hazardous Materials 189(1-2): 603–608.

Rittmann, S. and C. Herwig. 2012. A comprehensive and quantitative review of dark fermentative biohydrogen production. Microbial Cell Factories 11(August 1): 115.

Rozendal, R.A., H.V.M. Hamelers, G.J.W. Euverink, S.J. Metz and C.J.N. Buisman. 2006. Principle and perspectives of hydrogen production through biocatalyzed electrolysis. International Journal of Hydrogen Energy 31(12): 1632–1640.

Sachdeva, H. and R. Saroj. 2013. ZnO nanoparticles as an efficient, heterogeneous, reusable, and ecofriendly catalyst for four-component one-pot green synthesis of pyranopyrazole derivatives in water. The Scientific World Journal 2013: 680671.

Saratale, G.D., S. Der Chen, Y.C. Lo, R.G. Saratale and J.S. Chang. 2008. Outlook of biohydrogen production from lignocellulosic feedstock using dark fermentation—A review. Journal of Scientific and Industrial Research 67(11): 962–979.

Seelert, T., D. Ghosh and V. Yargeau. 2015. Improving biohydrogen production using clostridium beijerinckii immobilized with magnetite nanoparticles. Applied Microbiology and Biotechnology 99(9): 4107–4116.

Seth, D. 2002. Hydrogen futures: Toward a sustainable energy system. International Journal of Hydrogen Energy 27(3): 235–264.

Sharma, A. and S.K. Arya. 2017. Hydrogen from algal biomass: A review of production process. Biotechnology Reports 15(February): 63–69.

Shi, X., K.W. Jung, D.H. Kim, Y.T. Ahn and H.S. Shin. 2011. Direct fermentation of laminaria japonica for biohydrogen production by anaerobic mixed cultures. International Journal of Hydrogen Energy 36(10): 5857–5864.

Skjånes, K., C. Rebours and P. Lindblad. 2013. Potential for green microalgae to produce hydrogen, pharmaceuticals and other high value products in a combined process. Critical Reviews in Biotechnology 33(2)(June 1): 172–215.

Sołowski, G. 2018. Biohydrogen production-sources and methods: A review. International Journal of Bioprocessing and Biotechniques. Vol. 2018.

Srikanth, S. and S.V. Mohan. 2012. Regulatory function of divalent cations in controlling the acidogenic biohydrogen production process. RSC Adv. 2(16): 6576–6589.

Srivastava, N., M. Srivastava, P.K. Mishra, P. Singh, H. Pandey and P.W. Ramteke. 2017. Nanoparticles for biofuels production from lignocellulosic waste. pp. 263–278. *In*: Ranjan, S., N. Dasgupta and E. Lichtfouse (eds.). Nanoscience in Food and Agriculture 4. Cham: Springer International Publishing.

Sriwuryandari, L., E.A. Priantoro, N. Sintawardani, J.T. Astuti, D. Nilawati, A.M.H. Putri, Mamat, S. Sentana and T. Sembiring. 2016. The organic agricultural waste as a basic source of biohydrogen production. In AIP Conference Proceedings 1711: 080002-1-6.

Stephen, A.J., S.A. Archer, R.L. Orozco and L.E. Macaskie. 2017. Advances and bottlenecks in microbial hydrogen production. Microbial Biotechnology 10(September 5): 1120–1127.

Suanon, F., Q. Sun, D. Mama, J. Li, B. Dimon and C.P. Yu. 2016. Effect of nanoscale zero-valent iron and magnetite (Fe_3O_4) on the fate of metals during anaerobic digestion of sludge. Water Research 88: 897–903.

Sui, H., J. Dong, M. Wu, X. Li, R. Zhang and G. Wu. 2017. Continuous hydrogen production by dark fermentation in a foam SiC ceramic packed up-flow anaerobic sludge blanket reactor. Canadian Journal of Chemical Engineering 95(1): 62–68.

Sveshnikov, D.A., N.V. Sveshnikova, K.K. Rao and D.O. Hall. 1997. Hydrogen metabolism of mutant forms of anabaena variabilis in continuous cultures and under nutritional stress. FEBS Microbiol. Lett. 147.

Taguchi, F., K. Hasegawa, T. Saito-Taki and K. Hara. 1996. Simultaneous production of xylanase and hydrogen using xylan in batch culture of clostridium sp. strain X53. Journal of Fermentation and Bioengineering 81(2)(January 1): 178–180.

Taherdanak, M., H. Zilouei and K. Karimi. 2015. Investigating the effects of iron and nickel nanoparticles on dark hydrogen fermentation from starch using central composite design. International Journal of Hydrogen Energy 40(38): 12956–12963.

Taherdanak, M., H. Zilouei and K. Karimi. 2016. The effects of Fe0 and Ni0 nanoparticles versus Fe^{2+} and Ni^{2+} ions on dark hydrogen fermentation. International Journal of Hydrogen Energy 41(1): 167–173.

Tang, D., Y. Dong, N. Guo, L. Li and H. Ren. 2014. Metabolomic analysis of the polyphenols in germinating mung beans (vigna radiata) seeds and sprouts. Journal of the Science of Food and Agriculture 94: 1639–1647.

Tao, G., Z. Hua, Z. Gao, Y. Zhu, Y. Chen, Z. Shu, L. Zhang and J. Shi. 2013. KF-loaded mesoporous Mg-Fe Bi-metal oxides: High performance transesterification catalysts for biodiesel production. Chemical Communications 49(73): 8006–8008.

Tao, Y., Y. Chen, Y. Wu, Y. He and Z. Zhou. 2007. High hydrogen yield from a two-step process of dark- and photo-fermentation of sucrose. International Journal of Hydrogen Energy 32(2): 200–206.

Tsygankov, A.A., A.S. Fedorov, T.V. Laurinavichene, I.N. Gogotov, K.K. Rao and D.O. Hall. 1998. Actual and potential rates of hydrogen photoproduction by continuous culture of the purple non-sulphur bacterium rhodobacter capsulatus. Applied Microbiology and Biotechnology 49(1): 102–107.

Tsygankov, Anatoly A., Y. Hirata, M. Miyake, Y. Asada and J. Miyake. 1994. Photobioreactor with photosynthetic bacteria immobilized on porous glass for hydrogen photoproduction. Journal of Fermentation and Bioengineering 77(5)(January 1): 575–578.

Uygun, D.A., N. Ozutuk, A. Akgol and A. Denizli. 2012. Novel magnetic nanoparticles for the hydrolysis of starch with bacillus licheniformis a-amylase. Journal of Applied Polymer Science 123: 2574–581.

Van Ginkel, S., S. Sung and J.J. Lay. 2001. Biohydrogen production as a function of PH and Substrate concentration. Environmental Science and Technology 35(24): 4726–4730.

Van Niel, E.W.J., M.A.W. Budde, G. De Haas, F.J. Van der Wal, P.A.M. Claassen and A.J.M. Stams. 2002. Distinctive properties of high hydrogen producing extreme thermophiles, caldicellulosiruptor saccharolyticus and thermotoga elfii. In International Journal of Hydrogen Energy 27: 1391–1398.

Vavilin, V.A., S.V. Rytow and L.Y. Lokshina. 1995. Modelling hydrogen partial pressure change as a result of competition between the butyric and propionic groups of acidogenic bacteria. Bioresource Technology 54(2)(January 1): 171–177.

Verma, M.L., R. Chaudhary, T. Tsuzuki, C.J. Barrow and M. Puri. 2013. Immobilization of β-glucosidase on a magnetic nanoparticle improves thermostability: Application in cellobiose hydrolysis. Bioresource Technology 135: 2–6.

Vijaya Bhaskar, Y., S. Venkata Mohan and P.N. Sarma. 2008. Effect of Substrate loading rate of chemical wastewater on fermentative biohydrogen production in biofilm configured sequencing batch reactor. Bioresource Technology 99(15)(October): 6941–6948.

Vijayaraghavan, K. and M.A. Mohd Soom. 2006. Trends in bio-hydrogen generation—A review. Environmental Sciences 3(4): 255–271.

Voloshin, R.A., M.V. Rodionova, S.K. Zharmukhamedov, T. Nejat Veziroglu and S.I. Allakhverdiev. 2016. Review: Biofuel production from plant and algal biomass. International Journal of Hydrogen Energy 41(39): 17257–17273.

Wang, A.-J., G.-L. Cao and W.-Z. Liu. 2012. Biohydrogen production from anaerobic fermentation. pp. 143–163. *In*: Bai, F.-W., C.-G. Liu, H. Huang and G.T. Tsao (eds.). Biotechnology in China III: Biofuels and Bioenergy. Berlin, Heidelberg: Springer Berlin Heidelberg.

Wang, B., W. Wan and J. Wang. 2009. Effect of ammonia concentration on fermentative hydrogen production by mixed cultures. Bioresource Technology 100(3)(February): 1211–1213.

Wang, G., Y. Mu and H.-Q. Yu. 2005. Response surface analysis to evaluate the influence of PH, temperature and substrate concentration on the acidogenesis of sucrose-rich wastewater. Biochemical Engineering Journal 23(2): 175–184.

Wang, J. and W. Wan. 2009. Factors influencing fermentative hydrogen production: A review. Int. J. Hydrogen Energ. 34: 799–811.

Wimonsong, P., J. Llorca and R. Nitisoravut. 2013. Catalytic activity and characterization of Fe-Zn-Mg-Al hydrotalcites in biohydrogen production. International Journal of Hydrogen Energy 38(25): 10284–10292.

Wimonsong, P., R. Nitisoravut and J. Llorca. 2014. Application of Fe-Zn-Mg-Al-O hydrotalcites supported Au as active nano-catalyst for fermentative hydrogen production. Chemical Engineering Journal 253: 148–154.

Winkler, M., A. Hemschemeier, C. Gotor, A. Melis and T. Happe. 2002. [Fe]-hydrogenases in green algae: Photo-fermentation and hydrogen evolution under sulfur deprivation. International Journal of Hydrogen Energy 27(11-12): 1431–1439.

Xie, W., Y. Han and H. Wang. 2018. Magnetic Fe_3O_4/MCM-41 composite-supported sodium silicate as heterogeneous catalysts for biodiesel production. Renewable Energy 125: 675–681.

Yan, S., S. Mohan, C. Dimaggio, M. Kim, K.Y.S. Ng and S.O. Salley. 2010. Long term activity of modified ZnO nanoparticles for transesterification. Fuel 89(10): 2844–2852.

Yang, Y., J. Guo and Z. Hu. 2013. Impact of Nano Zero Valent Iron (NZVI) on methanogenic activity and population dynamics in anaerobic digestion. Water Research 47(17): 6790–6800.

Yokoyama, H., M. Waki, A. Ogino, H. Ohmori and Y. Tanaka. 2007a. Hydrogen fermentation properties of undiluted cow dung. Journal of Bioscience and Bioengineering 104(July 1): 82–85.

Yokoyama, H., M. Waki, N. Moriya, T. Yasuda, Y. Tanaka and K. Haga. 2007b. Effect of fermentation temperature on hydrogen production from cow waste slurry by using anaerobic microflora within the slurry. Applied Microbiology and Biotechnology 74(2)(February): 474–483.

Yokoyama, H., H. Ohmori, M. Waki, A. Ogino and Y. Tanaka. 2009. Continuous hydrogen production from glucose by using extreme thermophilic anaerobic microflora. Journal of Bioscience and Bioengineering 107(1): 64–66.

Yulianti, C.K., D. Hartanto, T.E. Purbaningtias, Y. Chisaki, A.A. Jalil, C.K.N.L.C.K. Hitam and D. Prasetyoko. 2014. Synthesis ofCaOZnO nanoparticles catalyst and its application in transesterification of refined palm oil. Bulletin of Chemical Reaction Engineering and Catalysis 9(2): 100–110.

Zaidi, A., R. Feng, A. Malik, S. Khan, Y. Shi, A. Bhutta and A. Shah. 2019. Combining microwave pretreatment with iron oxide nanoparticles enhanced biogas and hydrogen yield from green algae. Processes 7(1): 24.

Zhang, H., M.A. Bruns and B.E. Logan. 2006. Biological hydrogen production by clostridium acetobutylicum in an unsaturated flow reactor. Water Res. 40.

Zhang, Y., G. Liu and J. Shen. 2005. Hydrogen Production in batch culture of mixed bacteria with sucrose under different iron concentrations. International Journal of Hydrogen Energy 30(8): 855–860.

Zhang, Y. and J. Shen. 2006. Effect of temperature and iron concentration on the growth and hydrogen production of mixed bacteria. International Journal of Hydrogen Energy 31(4): 441–446.

Zhang, Y. and J. Shen. 2007. Enhancement effect of gold nanoparticles on biohydrogen production from artificial wastewater. International Journal of Hydrogen Energy 32(1): 17–23.

Zhao, L., Z. Qiu and S.M. Stagg-Williams. 2013. Transesterification of canola oil catalyzed by nanopowder calcium oxide. Fuel Processing Technology 114: 154–162.

Zhao, Q.-B. and H.-Q. Yu. 2008. Fermentative H_2 production in an upflow anaerobic sludge blanket reactor at various PH values. Bioresource Technology 99(5)(March): 1353–1358.

Zhao, W., J. Zhao, G.D. Chen, R. Feng, J. Yang, Y.F. Zhao, Q. Wei, B. Du and Y.F. Zhang. 2011. Anaerobic biohydrogen production by the mixed culture with mesoporous Fe_3O_4 nanoparticles activation. Advanced Materials Research 306-307: 1528–1531.

Zhao, W., Y. Zhang, B. Du, D. Wei, Q. Wei and Y. Zhao. 2013. Enhancement effect of silver nanoparticles on fermentative biohydrogen production using mixed bacteria. Bioresource Technology 142(August 1): 240–245.

Zhao, Y. and Y. Chen. 2011. Nano-TiO_2 enhanced photofermentative hydrogen produced from the dark fermentation liquid of waste activated sludge. Environmental Science and Technology 45(19): 8589–8595.

Zilouei, H. and M. Taherdanak. 2015. Biohydrogen from lignocellulosic wastes. pp. 253–288. *In*: Karimi, K. (ed.). Lignocellulose-Based Bioproducts. Cham: Springer International Publishing.

Zürrer, H. and R. Bachofen. 1979. Hydrogen production by the photosynthetic bacterium rhodospirillum rubrum. Applied and Environmental Microbiology 37(May 5): 789–793.

Chapter 10

Nanobiotechnological Solutions for Sustainable Bioenergy Production

Madan L. Verma,[1,*] ***Kaushal Kishor,***[2] ***B.S. Dhanya,***[3] ***Jatinder S. Randhawa***[4] **and** ***Asim K. Jana***[5]

1. Introduction

Global warming is a world-known challenge and its existence is primarily due to fossil fuel combustion (Demirbas 2010). The emission data indicates a high amount of carbon dioxide is released due to the burning of fossil fuels for various purposes such as transportation, heat and electricity (Maria 2014). Many strategies have been developed to minimize the emanation of carbon dioxide. Some of the strategies include capturing carbon dioxide at the source level, removing carbon dioxide with the aid of microbes, minimizing the use of fossil fuels and using predominant alternative sources such as wind, hydro, solar or biomass (Subhadra and Edwards 2010). Biomass, a principal source of energy, contributes about 10% of world energy supply (Saidur et al. 2011). Replacing fossil fuel with biomass is environmentally feasible and by 2050, it is expected to meet about 38% of the global energy supply (Demirbas 2000).

The fuel originated by the biological carbon fixation process is predicted to be biofuel. First-generation biofuels are bioethanol and biodiesel produced from feedstock (Antunes et al. 2014). The feedstocks for biodiesel production are animal fats and vegetable oils and for bioethanol production, sucrose and starch are the feedstocks (Silva and Chandel 2014). Biochemical methods (transesterification

[1] Department of Biotechnology, School of Basic Sciences, Indian Institute of Information Technology Una, Himachal Pradesh-177220, India.

[2] Technology Research and Advisory, Aranca Pvt Ltd, Mumbai-400076, India.

[3] Department of Biotechnology, Udaya School of Engineering, Nagercoil-629204, India.

[4] Department of Environmental Science and Technology, School of Environment and Earth Sciences, Central University of Punjab, Bathinda, 151001.

[5] Department of Biotechnology, Dr. B. R. Ambedkar National Institute of Technology, Jalandhar-144011.

* Corresponding author: madanverma@gmail.com

and hydrolysis), thermochemical methods (pyrolysis, liquefaction, combustion and gasification) and microbiological methods (anaerobic digestion) are the conventionally used technologies for producing biofuel from feedstock (Hahn-Hagerdal et al. 2006). The sage of feedstock is a limitation to be concerned about, and hence the second-generation biofuel is availed from non-food feedstocks such as agricultural and wood waste (Eggert and Greaker 2014). Beyond this, third-generation biofuel exists that is generated using microbial biomass (Ho et al. 2014). The challenges associated with the third-generation biofuel productions are its high cost and less energy-efficiency (Patumsawad 2011).

Nanobiotechnology is approved as an alternative technology for producing biofuel. It is an interdisciplinary study covering the core principles and related applications in the field of science and technology. The integration of nanotechnology and biotechnology has made tremendous nanobiocatalytic developments. Research is ongoing to construct novel nanobiocatalytic systems (Verma et al. 2013). Fabrication of novel catalyst and its application by using nanomaterials are considered to be phenomenal due to nanometric properties, such as high surface area, pollutant-free and reusability (Zhang et al. 2010). For attaining sustainable biofuel and bioenergy production, different kinds of nanomaterials are used. Nanomaterials include nanofiber, nanosheets, nanoparticles, nanoporous, nanocomposites and nanotubes. The specific property of nanomaterial, high surface-area-to-volume ratio favors more reaction kinetics and increased capacity for enzyme loading, by which biocatalytic effectiveness is augmented for industry-oriented applications (Verma et al. 2013).

Nanoparticles are tiny molecules that show quantum effects. Improvising the working parameters of the diesel engine or compression ignition is necessary with the available biofuels. Efforts have been initiated to incorporate nanoadditives to biofuels for inculcating additional properties. The potential nanoadditives include ceria, carbon nanotubes, alumina, aluminum, etc. The characteristic features of nanoadditives, i.e., high rate of burning, minimal delay in the ignition, increased number of cetanes, high catalytic and reaction specificity add to the advantages of its usage in biofuels. Particularly, nanoadditives minimize the effect of hazardous gas pollutants such as carbon monoxide, hydrocarbon and nitrogen oxides (Sadhik Basha 2016).

Another application of nanotechnology is in the domain of developing nanocatalyst. Nanocatalyst ameliorates the biomass conversion process. It renders green friendly, environmental process by inducing high catalytic changes with more selectivity and thus, enhancing the production of biofuels. Nanosized catalysts hold number of active sites when compared to the conventionally used catalysts (Chan and Tanksale 2014).

Current technology depends upon the nanomaterials/nanoparticles in the field of engineering too (Sadhik Basha 2015b). Rheological issues such as settling, clogging, abrasion and friction can be tackled by nanoparticles (Sadhik Basha and Anand 2014). Nanoparticles have been proven with powerful thermal, physical and chemical properties (Sadhik Basha 2015b). Hence, nanoparticles are demanded to ameliorate the fuel properties (Sadhik Basha and Anand 2015a). Brake thermal efficiency, combustion improvement and minimized emission of harmful gases can be attained with the aid of nanotechnology.

The application of nanoscience and technology and its development show the ability to intensify the production cycle and further concentrates on conserving natural resource. An intense number of studies is available on the involvement of nanoparticles on bioenergy and biofuel production (Ahmad and Sardar 2014). The present chapter is concentrated on various strategies involved in the nanobiotechnological solutions and the impact of nanomaterials on bioenergy production.

2. Nanomaterials Types and Their Association in Biofuels Production

Nanomaterials have gained wide importance in the field of biofuel production. Many types of nanomaterials are used for achieving the desired result and overcoming the disadvantages felt in the conventional techniques.

2.1 Types of Nanomaterials

Different types of nanomaterials are used in the process of biofuel production as shown in Figure 1.

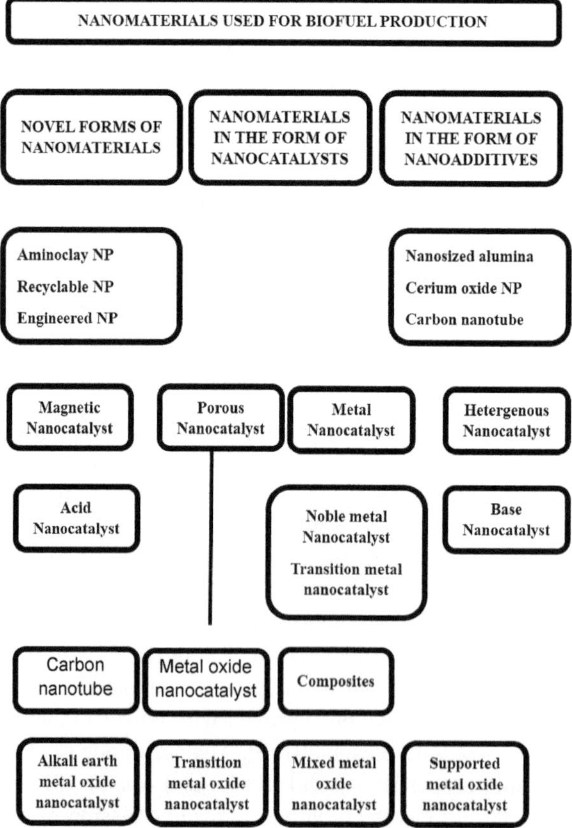

Figure 1. Nanomaterials used for biofuel production.

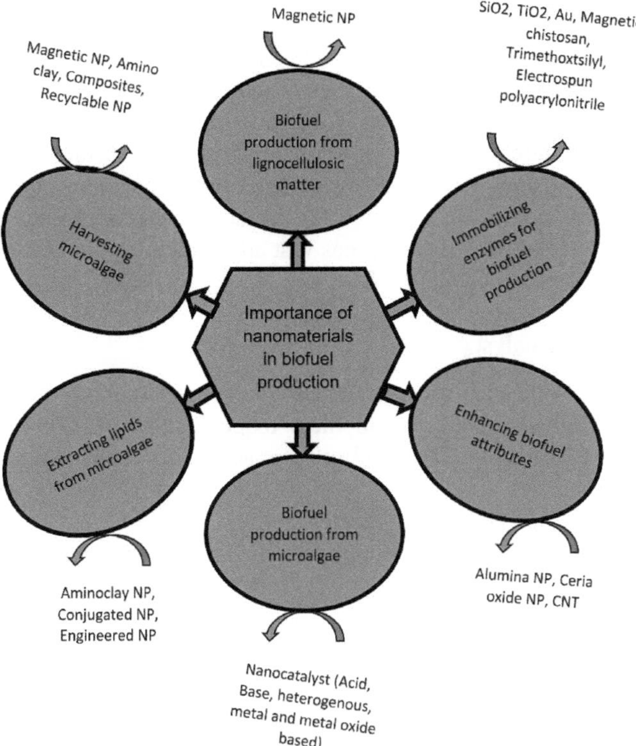

Figure 2. Importance of nanomaterials in biofuel production.

2.1.1 Nanomaterials in the Form of Nanocatalysts

Nanocatalysts are active nanomaterials with high catalytic property and particle size of less than 100 nm. It is also known as a soluble heterogeneous catalyst or semi-heterogeneous catalyst. More active sites are visible when the particle size is reduced to nanoscale and in turn, causes the turnover frequency to be increased. Nanoparticles with tetrahedral shapes show innumerable corners and edges and hence active participation in reaction. Also, the surface-area-to-volume ratio is high in nanoparticles. All these properties induce greater importance for nanocatalysts. High selectivity, reusability, activity and durability are the requirements for an effective nanocatalyst for biofuel production (Lee and Juan 2016). It can be classified broadly into the following types:

- Magnetic nanocatalyst
- Acid nanocatalyst
- Base nanocatalyst
- Heterogenous nanocatalyst
- Nanoparticle catalyst
- Porous nanocatalyst
- Novel forms of nanocatalyst

2.1.1.1 Magnetic Nanocatalyst

Magnetic nanoforms are widely used as a potent catalyst, and it finds application in biomedical, engineering, biotechnology, environmental science and material science. Nickel alloy, platinum alloy, iron alloy, cobalt alloy and various types of metal oxides act as a catalyst (Laurent et al. 2008). An increased ratio of surface-area-to-volume, the capability to incorporate other compounds, good quantum capacity, unique catalytic property and easy recovery after use are the characteristics of magnetic nanoparticles (Nicolas et al. 2014). It has vast applications in bioethanol production and harvesting microalgae, e.g., Fe_3O_4.

2.1.1.2 Acid Nanocatalyst

Acid nanocatalysts include nanocrystalline zeolite (e.g., beta and ZSM-5), sulfonated carbon-based catalyst (MBC) and niobium oxide (CBMM's HY-340), molecular sieve zeolite A (Chen et al. 2011) and sulfonic acid functionalized platelet SBA-15 (Chen et al. 2011).

2.1.1.3 Base Nanocatalyst (Al_2O_3 supported MgO and Al_2O_3 supported CaMgO).

2.1.1.4 Heterogeneous Nanocatalyst (Ni/ HZSM-5)

2.1.1.5 Metal Nanocatalyst

- Noble metal nanocatalyst (Pt and Pd)
- Transition metal nanocatalyst (Ni, Co, Cu, K, Ca and Na)

2.1.1.6 Porous Nanocatalyst

- Metal oxide (TiO_2, SiO_2)
- Mesoporous particles (SBA-15)
- Carbon (carbon nanotube, graphene)
- Zeolite (ZSM-5)
- Composites (Al_2O_3-MCM-41)

2.1.1.6.1 Metal Oxide Nanocatalysts

Metal oxide nanocatalysts exist with more active sites and are used in pyrolysis/ gasification process. In addition to this, the recovery and reusability are high with metal oxide nanocatalysts. It can be categorized into 4 types as follows:

- **Alkali Earth Metal Oxide Nanocatalyst**

  CaO NP and MgO NP belong to the category of alkali earth metal oxide nanocatalyst, and it is used in transesterification reaction of soybean oil (Reddy et al. 2006, Wang and Yang 2007).

- **Transition Metal Oxide Nanocatalyst**

  During transesterification reaction in fatty acids, transition metal oxides such as nanosulfated TiO_2 are applied and the FAME yield is predicted to be 98% (Hosseini-Sarvari and Sodagar 2013).

- **Mixed Metal Oxide Nanocatalyst**

  In palm oil and soybean oil transesterification, to produce biodiesel, mixed metal nanocatalyst such as TiO_2-ZnO (Madhuvilakku and Piraman 2013) and iron cadmium, oxide/iron tin oxide (Alves et al. 2014) is used.

- **Supported Metal Oxide Nanocatalyst**

  Nanocatalyst such as KF/CaO-MgO (Wang et al. 2009b), CsNanoMgO (Montero et al. 2010), KF/CaO (Wen et al. 2010), CaO/Fe_3O_4 (Liu et al. 2010), Li/CaO (Kaur and Ali 2011), Cs/Al/Fe_2O_3 (Feyzi et al. 2013), Cs-MgO (Woodford et al. 2014) and copper doped ZnO (Gurunathan and Ravi 2015a) are some of the examples for supported metal oxide nanocatalyst.

2.1.1.6.2 Mesoporous Nanocatalysts

The diffusional limitations cause the formation of coke, leading to the loss of catalytic activity and hence mesoporous nanocatalysts of size between 2 and 50 nm are desired. SBA-15, MCM-41/Al and MCM-41/Fe are mesoporous nanocatalysts (Lee and Juan 2016).

2.1.1.6.3 Carbon Nanotube

Methods such as laser ablation, arc discharge and chemical vapor deposition are used to synthesize carbon nanotube. It is made up of graphite sheets that are cylindrically aligned with high biocompatibility and high surface area. It is used for immobilizing the enzyme by the conjugation process (Feng and Ji 2011, Shi et al. 2007). Mechanical, biocompatible, structural and thermal property of carbon nanotube is unique. Due to its high surface area, a high amount of enzymes can be loaded and there will be low diffusion resistance. The three-dimensional elective area is high for carbon nanotubes and hence it is widely used in the field of biofuel production. Particularly, porosity and conductivity are also high (Holzinger 2012).

2.1.1.6.4 Composites

- Aminoclay TiO_2 composites
- PVP/Fe_3O_4 composite

2.1.1.7 Other Types of Nanocatalyst

- Organometallic nanocatalyst
- Nanoalloy catalyst

2.1.2 Nanomaterials in the Form of Nanoadditives

- Nanosized alumina: It is spherical in shape, equipped with a more surface-area-to-volume ratio and at high temperature, the stability is pronounced more.
- Cerium oxide nanoparticle
- Carbon nanotube

2.1.3 Novel Forms of Nanomaterials

- Aminoclay NP, e.g., magnesium aminoclay
- Recyclable NP, e.g., stearic acid coated Fe_3O_4-ZnO nanocomposites
- Engineered NP, e.g., alkyl-grafted Fe_3O_4@SiO_2 nanoparticle

2.2 Importance of Nanomaterials in Biofuel Production

Table 1 lists out the types of nanomaterials used for biofuel production.

2.2.1 Nanomaterials for Biofuel Production from Lignocellulosic Matters

Bioethanol is produced from lignocellulosic matters by immobilizing enzymes such as hemicellulases and cellulases on magnetic nanoparticles. By applying magnetic fields, these magnetic particles can be recovered and reused (Alftren and Hobley 2013). In addition to the application of enzyme immobilization support, magnetic nanoparticles can also be used for photo-oxidation, inductive heating and hydrogenation (Govan and Gunko 2014). Immobilizing cellulases using magnetic nanoparticles increases the yield of reaction and large-scale immobilization is also possible with seldom use of any surfactants. The property of superparamagnetism exists in magnetic nanoparticles and it is advantageous since it avoids the residual magnetism on the removal of the extrinsic magnetic field (Suh 2009). Metal oxides or metals are applied for immobilizing enzymes on magnetic nanoparticles. Ti, CoPt, FePt, Fe_3O_4, Mg, Co, Ni and FeCO are the commonly used metals and metal oxides. Fe_3O_4 nanoparticles are used for immobilizing enzymes. Before subjecting to immobilization, the Fe_3O_4 nanoparticles are activated with carbodiimide (Jordan et al. 2011).

Biofuel production from lignocellulosic biomass using nanomaterials can be categorized as the green nanotechnology field. Cobalt nanoparticle is used as a catalyst to synthesize biodiesel and bioethanol from *Camelia sinensis* (spent tea-solid waste) (Mahmood and Hussain 2010). Cellulase enzyme is immobilized with silica nanoparticles to obtain a high amount of bioethanol from cellulose by undergoing alternative solid-state fermentation and saccharification steps (Lupoi and Smith 2011). The efficiency of bioethanol production is increased by immobilizing cellulose enzyme retrieved from *Aspergillus fumigatus* on manganese dioxide nanoparticles (Cherian et al. 2015). From the biomass of *Sesbania aculeate*, a high amount of bioethanol is retrieved using magnetic nanoparticles (Baskar et al. 2016).

2.2.2 Nanomaterials for Immobilizing Enzymes in Biofuel Production

The nanoparticles Fe_3O_4, electrospun polyacrylonitrile, SiO_2, TiO_2, magnetic-chitosan, trimethoxysilyl, silica, chitosan-coated Fe_3O_4, acrylamide-bisacrylamide and gold are vehemently used for immobilizing cellulose (Gao and Kyratzis 2008), lipase (Sakai et al. 2010), xylanase (Dhiman et al. 2012), cellulose (Ahmad and Sardar 2014), laccases (Fang et al. 2009), hemicellulases beta-D-xylosidae (Hegedus et al. 2011), laccases (Wang et al. 2010), lipase (Kuo et al. 2012), beta-D-mannosidase (Hegedus et al. 2011) and lipase (Wang et al. 2011), respectively.

Table 1. Types of nanomaterials for biofuel production.

Nanomaterials	Application	Source of Biofuel	Type of Biofuel	Reference
Cobalt NP	Nanobiocatalyst	*Camellia sinensis*	Bioethanol, Biodiesel	Mahmood and Hussain 2010
CaO	Nanocatalyst	Soyabean oil	Biodiesel	Reddy et al. 2006
MnO_2 NP	Nanobiocatalyst	Sugarcane leaves and jack fruit waste	Bioethanol	Cherian et al. 2015
MgO	Nanocatalyst	Soyabean oil	Biodiesel	Wang and Yang 2007
		Rapeseed and sunflower oils		Verziu et al. 2008
CsNanoMgO	Nanocatalyst	Tributyrin	Biodiesel	Montero et al. 2010
Silica NP	Nanobiocatalyst	Cellulose	Bioethanol	Lupoi and Smith 2011
KF/CaO-MgO	Nanocatalyst	Rapeseed oil	Biodiesel	Wang et al. 2009a
Magnetic NP	Nanobiocatalyst	*Sesbania aculeate*	Bioethanol	Baskar et al. 2016
KF/CaO	Nanocatalyst	Chinese tallow seed oil	Biodiesel	Wen et al. 2010
Nanosulfated TiO_2	Nanocatalyst	Fatty acids	Biodiesel	Hosseini Sarvari and Sodagar 2013
Nanocrystalline lithium-impregnated with Calcium oxide	Nanocatalyst	Karanja and Jatropha oils	Biodiesel	Kaur and Ali 2011
CaO nanoparticle	Nanocatalyst		Biodiesel	Luz Martinez et al. 2010
KF/CaO-Fe_3O_4	Nanocatalyst	Stillingia oil	Biodiesel	Hu et al. 2011
Potassium bitartrate	Nanocatalyst	Soyabean oil	Biodiesel	Qiu et al. 2011
TiO_2-ZnO	Nanocatalyst	Palm oil	Biodiesel	Madhuvilakku and Piraman 2013
KF/CaONiO	Nanocatalyst	WCO	Biodiesel	Kaur and Ali 2014
Iron/Cadmium	Nanocatalyst	Soybean oil	Biodiesel	Alves et al. 2014
Cs/Al/Fe_3O_4	Nanocatalyst	Soybean oil	Biodiesel	Feyzi et al. 2013
Cs promoted MgO	Nanocatalyst	Olive oil	Biodiesel	Woodford et al. 2014
Iron/tin oxides	Nanocatalyst	Macauba oil	Biodiesel	Alves et al. 2014
Ferric manganese-doped tungstated zirconia naoparticles	Nanocatalyst	Waste cooking oil	Biodiesel	Alhassan et al. 2015a
Cs-Ca/SiO_2-TiO_2	Nanocatalyst	Blend refined vegetable oil	Biodiesel	Feyzi and Shahbazi 2015
Ferric manganese-doped sulfated zirconia naoparticles	Nanocatalyst	Waste cooking oil	Biodiesel	Alhassan et al. 2015a
CZO	Nanocatalyst	Waste cooking oil	Biodiesel	Gurunathan and Ravi 2015a

2.2.3 Nanomaterials for Enhancing Biofuel Attributes

The quality of fuel is increased, and the problem of engine starting in cold weather can be overcome by nanoparticles. In earlier days, for improving ignition, alkyl nitrate is added which is quite corrosive and toxic. Nanoadditives are incorporated into the biofuels for surpassing the challenges (Imdadul et al. 2015). In nanoadditive combined biodiesel, the evaporation rate is minimum. Alumina nanoparticles act as a catalyst and it is worth the high surface-area-volume ratio (Mimani and Patil 2001). It is more stable at high oxidation temperature. Ceria oxide nanoparticles are mixed with jatropha biodiesel and the brake thermal efficiency is found to be higher with minimized hazardous gas emissions (Sajith et al. 2010). Carbon nanotubes are used to boost up the fuel quality (Sadhik Basha and Anand 2014).

2.2.4 Nanomaterials for Harvesting Microalgae

Harvesting is significant in a microalgal biorefinery. Conventional harvesting techniques are filtration, floatation, immobilization, centrifugation, sedimentation, magnetophoretic separation and electrophoresis (Wang et al. 2015). Considering the cost of the process, reliability, consumption of energy, toxic effect to the environment and end concentration of microalgae using the conventional methods, nanoparticles are chosen for enhancing the efficiency of harvesting (Lee et al. 2015b). The nanoparticles used are magnetic nanoparticles, aminoclay nanoparticles, conjugated nanoparticles, and recyclable nanoparticles.

2.2.4.1 Magnetic Nanoparticles

Harvesting with the aid of magnetic nanoparticles is rapid and scalable (Borlido et al. 2013), e.g., Fe_3O_4 magnetic nanoparticle. Polydiallyl dimethyl ammonium chloride coated Fe_3O_4 magnetic nanoparticle, polyacrylamide modified Fe_3O_4 magnetic nanocomposites (Wang et al. 2014), $BaFe_{12}O_{19}$ nanoparticle functionalized with APTES, (Seo et al. 2014) and chitosan-Fe_3O_4 nanoparticle (Lee et al. 2013a) are highly productive in the harvesting process. Magnetic nanoparticle such as Fe_3O_4 along with PEI resulted in a high capacity for harvesting (Hu et al. 2014).

2.2.4.2 Aminoclay Nanoparticles

Aminoclays, an amine rich organophyllosilicate, constitutes metal cations and phyllosilicate sheets (Farooq et al. 2013). Mg-aminoclay and humic acid are used together for efficient extraction with cost reduction (Lee et al. 2014b). Mg-aminoclay/nanoscale zero-valent iron composite is applied for harvesting *Chlorella* sp. It is prepared by subjecting the ferric ions to the reduction process by sodium borohydride in the presence of aminoclay (Lee et al. 2014b). Aminoclay with Mg^{2+} and Fe^{3+} shows high harvesting efficiency (Farooq et al. 2013).

2.2.4.3 Conjugated Nanoparticles

Reducing the cost of harvesting using nanoparticles can be facilitated by combined nanoparticles. Aminoclay TiO_2 composites, PVP/Fe_3O_4 composites and triazabicyclodecene functionalized Fe_3O_4@ silica nanoparticles (Chiang et al. 2015) are conjugated nanoparticles. PVP/Fe_3O_4 composite is a lipophilic, cationic and magnetic nanoparticle, enabling the extraction of intracellular lipids (Seo et al. 2015). Aminoclay TiO_2 composite assists both in cell disruption and flocculation steps (Lee et al. 2014a).

2.2.4.4 Recyclable Nanoparticles

While using nanoparticles, its reusability must be a major concern since it is environmentally pleasing too. The used nanoparticle such as magnetic nanoparticle can be detached from the algal flocs using a water-nonpolar organic solvent (Lee et al. 2015a). Stearic acid-coated Fe_3O_4-ZnO nanocomposites are detached using irradiation with ultraviolet rays (Ge et al. 2015b). In addition to this, the sonication method is used to detach magnetic nanoparticles (Ge et al. 2015a). The pH is directly linked with the detachment efficiency. The detachment effect is pronounced more in APTES-BaFe12O19 when the pH is about 12 (Seo et al. 2014).

2.2.5 Nanomaterials for Extracting Lipids from Micro-algae

Microalgae consist of cell walls which are rigid structures that make the cell disruption process harder. Conventional methods for cell disruption such as solvent extraction method and mechanical methods (expeller press, ultrasonication, bead beating and microwave) face many disadvantages, and it is not economically viable too (Kim et al. 2016). As a substitute, nanoparticles are used for the lipid extraction process.

2.2.5.1 Aminoclay Nanoparticles

During the time of microalgal harvesting, flocculants are added for maximum lipid extraction, fatty acid methyl ester productivity and content. The added flocculants are aminoclay nanoparticles. AI-APTES clay, Mg-N_3 clay, Mg-APTES clay and Ca-APTES clay are some of the examples of aminoclay nanoparticles (Lee et al. 2013b). Aminoclays destabilize the cell walls of microalgae and cause maximum extraction of lipids. Fame productivity is high for AI-APTES clay.

2.2.5.2 Conjugated Nanoparticles

It includes aminoclay conjugated with TiO_2, which is a powerful cell-disruption agent caused by the mechanism of UV-irradiated TiO_2 photocatalytic reaction (Lee et al. 2014a).

2.2.5.3 Engineered Nanoparticles

These nanoparticles are engineered with various types of nanomaterials that are effective in disrupting the cell walls for enhanced extraction of lipid contents. The examples are enzyme functionalized nanoparticle and surfactant functionalized nanoparticle. The biotoxic and antimicrobial nature of surfactants inhibits the growth of microalgae (Mohareb et al. 2015). CTAB (etyl Trimethyl Ammonium

Bromide) is a cationic surfactant that is used for extracting lipid from *Chlorella* sp. and the extraction efficiency is found to be maximum. Also, the percentage of saturated fatty acids and monounsaturated fatty acids formed are high with CTAB that is up to the marked standard of biodiesel (Coward et al. 2014). Enzymes, such as beta glucosidases and cellulases, hydrolyze the glucan biomolecule to glucose for obtaining maximum lipid extraction from *Chlorella vulgaris* (Cho et al. 2013). Other enzymes used are lysozymes (Taher et al. 2014), i.e., lipase immobilized on alkyl-grafted Fe_3O_4@SiO_2 nanoparticle (Tran et al. 2013).

2.2.6 Nanomaterials for Biofuel Production from Microalgae

Due to the rapid rate of growth and increased the content of lipid, microalgae are accepted as a source of biofuels and they have the capacity to sequestrate and absorb carbon dioxide at a high rate (Praveenkumar et al. 2014). In biofuel generation from microalgae, the most common challenges include enzymatic control, improvisation of enzyme loading capacity, storing the bioenergy products, biofuel separation and purification part and increasing the performance of bioreactor. Various factors such as high yield protein, lipid and carbohydrate algal strain selection and reactor designing influence biofuel generation. Nanotechnology is relevant for tackling the limitations and used to design new bioreactors. Nanoparticles of multifunctional nature act as a tool for future developments in the area of extraction, conversion and harvesting (Wang et al. 2015). Nanoparticles enhance the productivity of total microalgal biomass, increase the extraction efficiency of lipids leading to higher productivity of biodiesel and on the whole minimize the economic parameters.

2.2.6.1 Nanocatalyst

Nanocatalyst is rich with more selectivity, high stability, excellent activity, easy recovery and high energy efficacy (Lee and Juan 2017). A high amount of tar is generated during the process of gasification in algal biomass conversion. This leads to corrosion, piping blockage and engine failure (Nordgreen 2011). To minimize this tar, the nano-based catalyst is used. Nanocrystalline zeolite is specific with more surface area that paves the acid sites to be accessed by the oil extracted from microalgae. Examples of nanocrystalline zeolites include beta and ZSM-5 (Carrero et al. 2011). ZSM-5 has narrow pores, and the FAME yield is high for the beta. Sulfonated carbon-based catalyst is a solid acid catalyst with -COOH, -SO_3H and phenolic-OH functional groups. The efficiency of this catalyst is 60% and is high when compared to the conventional solid acid catalysts such as niobic acid, amberlyst-15 and nafion NR50 (Sani et al. 2013). MBC catalyst is one of the examples for sulfonated carbon-based catalysts, and this catalyst is highly efficient in increasing the transesterification yield. Niobium oxide is used as a catalyst in hydroesterification process with 94.27% FAME yield, e.g., CBMM's HY-340.

The transesterification process is a rapid using base nanocatalyst (Lee et al. 2015a), such as Al_2O_3 supported MgO, Al_2O_3 supported CaO, Al_2O_3 supported CaMgO, Mg-Zr catalyst and TBD-Fe_3O_4@SiO_2 (Chiang et al. 2015). The conversion yield is high with base nanocatalyst than with the pure CaO or MgO (Umdu et al. 2009). The fame yield is high for Al_2O_3 supported CaMgO.

Heterogenous nanocatalyst is used for producing greener biodiesel, such as HZSM-5 (Li and Savage 2013), Ni/HZSM-5 (Peng et al. 2012a), Ni/HBeta (Peng et al. 2012a), Ni/ZrO_2 (Peng et al. 2012b), aminopropyl-Ni-mesoporous silica nanoparticle (Kandel et al. 2013), FE-mesoporous silica nanoparticle (Kandel et al. 2014). For maximum conversion of tar into producer gas (in algal biomass), nickel-based nanocatalysts are effective (Anis and Zainal 2011). Also, maximum hydrogen gas is generated in biofuel generation from algal biomass by utilizing nickel-based catalysts (Sinag 2012). Potassium-based nanocatalyst is effective in the gasification of char (Nzihou et al. 2013). Calcium-based nanocatalyst maximizes the algal biomass conversion efficiency (Nzihou et al. 2013). Sodium-based nanocatalyst maximizes the standard of gas production (Nzihou et al. 2013). Hybrid nanocatalyst includes combined NiO with alumina (metal carrier) (Li et al. 2008). Organometallic nanocatalyst is used in algal biomass gasification with a high yield of hydrogen (Aladi et al. 2010). Nanoalloy catalyst is used in algal biomass gasification. In the presence of nanoalloy catalyst, the conversion efficiency of algal biomass is enhanced at minimal gasification temperature, such as $Ni_3Cu(SiO_2)_6$. Nanosized nickel oxide catalyst gains wide application in biomass gasification. Also, it is used as a catalyst in the algal biomass pyrolysis step. Low temperature is required during pyrolysis in the presence of a catalyst for achieving maximum efficiency, enhancing the gaseous product's quality and minimizing the formation of tar (Li et al. 2008). Nanosized zinc oxide catalyst is used for gasification in biofuel generation from algal biomass. There are specifically used for less temperature (300°C) water-gas shift mechanism. The shift reaction is accelerated due to a high surface area of the nanoparticles.

Nanosized tin oxide catalyst is also applied in biofuel production from algae during gasification and used in high temperature (400–500°C) water-gas shift mechanism (Sinag et al. 2011). Electrospun material is a productive tool for microalgal conversion procedure (Verdugo et al. 2014). Mesoporous materials act as a powerful aid for microalgal conversion action (Bautista et al. 2015).

3. Biodiesel Production by Lipase Immobilized on Nanomaterial

Enzymatic transesterification is an attractive and efficient method for biodiesel production for the generation of high purity biodiesel production. Lipase is employed to convert fatty acids and triglycerides into biodiesel. The free lipase system shows low activities due to the inhibitory effect of free methanol. Also, enzyme recovery is poor in the case of employing free lipases. In order to reuse and reduce cost, lipase can be immobilized on suitable substrates. These complexes have many-fold higher activities in comparison with free lipases. Furthermore, employing nanomaterial-based support provides higher surface area, improved enzyme loading and stability to pH, temperature, solvents and inhibitors. Various investigators immobilized lipase on nanomaterial by establishing chemical (covalent linkage between the amino acid residue of lipase and an active group of the support) and physical (adsorption and entrapment) linkages between lipase and nanomaterial. Most of these investigators incorporated magnetic material such as iron oxides particles for the ease of separation of these complexes after one reaction cycle under the influence of applied magnetic field and enabling repeated uses. Furthermore, in order to protect these magnetic iron oxides from oxidation, often core/shell structured, coating and surface functionalization strategies are employed. Silica, carbon, noble metals, metal oxides and chitosan are employed to cover these iron magnetic nanomaterials. Non-magnetic immobilized complexes such as silver nanoparticles (Dumri and Anh 2014) and silica nanoparticles (Babaki et al. 2014, Kalantari et al. 2013) have also been employed. Such materials are often recovered via centrifugation. At present, operational stability, recovery and loss of activities during reuse of these complexes are major challenges to be addressed in order to make them commercially successful.

Wang et al. (2009) immobilized lipases onto magnetic amino-functionalized magnetic nanoparticles (Fe_3O_4) for biodiesel production. Three different lipases, such as porcine pancreas, Candida rugosa and Pseudomonas cepacia lipase, were immobilized onto the amino-functionalized Fe_3O_4 nanoparticles using glutaraldehyde as a linker. The activities of these immobilized porcine pancreas lipase, Candida rugosa lipase and Pseudomonas cepacia lipases are 95.7, 70.1 and 82.6% of free lipases, respectively. These immobilized enzymes complexes were applicable to continuous production methods with a high conversion rate, reuse stability and easy recovery by using an external magnetic field. In the same year, Xie and Ma (2009, 2010) employed magnetic Fe_3O_4 nanoparticles treated with (3-aminopropyl) triethoxysilane to immobilize *T. lanuginosa* lipase using glutaraldehyde as a coupling reagent. The immobilized enzyme exhibited good pH tolerance and thermostability. The soybean oil was used as a substrate for transesterification. The maximum activity reported was above 90% at a temperature 50°C and pH ~ 7.5. Sakai et al. (2010) studied the transesterification of rapeseed oil in batch and flow-through reactors by employing lipase (*P. cepacia*) physically-adsorbed onto electrospun polyacrylonitrile nanofibers. The conversion rate after 24 hours was 80%. Dussan et al. (2010) immobilized *C. rugosa* lipase on iron magnetic nanostructures and studied the reactive extraction method. The reactive extraction directly produced 77% biodiesel using lower ethanol/triolein ratios. Wang et al. (2011) studied the performance of the lipase-nanoporous gold biocomposite for biodiesel production. They observed excellent stability and activity in the transesterification of soybean oil. The composite activity was above 90% in the first two-cycle of the run. Wang et al. (2011) also studied the performance of the packed-bed reactor system employing using lipase-Fe_3O_4 nanoparticle biocomposite for transesterification of soybean oil. The conversion rate in the four-packed-bed reactor was 100% after 24 hours. The conversion rate was 75% even after 240 hours. In comparison to a single packed bed reactor, the four-packed-bed-reactor offered better activity and stability owing to longer residence time and lower product inhibition. Lipase (*P. cepacia*) covalently immobilized on polyacrylonitrile nanofibrous membrane activated through amidination for transesterification of soybean oil. The conversion rate under optimized condition was 90% after 24 hours. Tran et al. (2012) immobilized lipase (*Burkholderia* sp.) on ferric silica nanocomposite for transesterification of olive oil with methanol to produce fatty acid methyl esters. The conversion rate was 92% after 30 hours. The better reuse and tolerance to methanol and water by lipase (*Burkholderia* sp.) was also reported in this study. In another study, a hybrid-nanosphere composed of L-α- phosphatidylcholine (lioposome), silica nanoparticle and lipase from *Rhizomucor miehei* were employed to carry out transesterification of triolein by methanol (Macario et al. 2013). The biodiesel yield was 89% after 3 hours at 37°C and the solvent to oil ratio was 6:1. Yu et al. (2013) employed lipase (*Pseudomonas cepacia*) immobilized onto magnetic nanoparticles (Fe_3O_4 with a spinel structure) for the synthesis of biodiesel from waste cooking oil. The immobilized enzymes showed excellent storage stability and reusability. Under the optimized conditions, the conversion rate was 79%. Kalantari et al. (2013) covalently immobilized lipase from *Pseudomonas cepacia* on magnetic silica nanocomposite particles using glutaraldehyde as a coupling agent. The immobilized enzyme exhibited high enzymatic activities, high stability and easy recovery in the presence of an external magnetic field. Furthermore, the mesoporous structure of silica was found to more superior to nonporous silica and pore-expanded mesoporous silica structures in terms of stability and activity. Babaki et al. (2014) studied immobilized enzymes for the production of biodiesel from canola oil. The investigators employed lipase from three different microorganisms, e.g., *Candida antarctica*, *Thermomyces lanuginosus* and *Rhizomucor miehei*. These lipases are covalently attached to epoxy-functionalized mesoporous silica (SBA-15). In comparison to the chemically and physically immobilized enzyme, the epoxy-functionalized mesoporous silica showed better stability and catalytic activity. The lipase (*Thermomyces lanuginosus*) showed better activity (> 94%) even after 20 cycles of use. Dumri and Anh (2014) employed covalently attached lipase on silver nanoparticles using adhesive polydopamine for the conversion of soybean oil into biodiesel. The complex yield 95% of conversion at 40°C after six hours. Xie and Wang (2014) covalently immobilized lipase (*Candida rugosa*) on Fe_3O_4/Poly (styrene-methacrylic acid) magnetic microsphere as a biocatalyst and carried out biodiesel production from soybean oil. The immobilized lipase exhibited better pH and thermal stability. The maximum bioconversion rate obtained was 86% at 35°C for 24 hours with a methanol-to-oil molar ratio of 4:1. Andrade et al. (2016) investigated biodiesel production from soybean oil using lipase (*Pseudomonas cepacia*) covalently attached to magnetite nanoparticles coated with a thin film of polydopamine. The yield of the immobilized enzyme was 97% at 37°C which fell below 60% after five cycles of use. In a different approach, Chen et al. (2016) demonstrated a method for attaching magnetic Fe_3O_4 nanoparticles to lipase immobilized on the acrylic resin for producing biodiesel from soybean oil. However, the maximum FAME yield reported in this studied was on the lower side (44%). Karimi (2016) synthesized mesoporous superparamagnetic iron oxide nanoparticles SPION-silica core-shell nanoparticles and grafted aldehyde group in order to immobilize lipase (*B. cepacia*) for enzymatic transesterification of waste cooking oil. The optimized yield was 91% after 35 hours. Mukherjee and Gupta (2016) tested lipase from (*Thermomyces lanuginosus*) precipitated over the clusters of Fe_3O_4 nanoparticles for the production of biodiesel from soybean oil.

Thangaraj et al. (2017) immobilized lipases on Fe_3O_4@SiO_2 magnetic nanoparticles functionalized with 3-amino-propyl-triethoxysilane and 3-mercapto-propyl-trimethoxysilane. The immobilized enzyme with low ratios of tetraethyl orthosilicate produced better results. EL-Batal et al. (2016) immobilized lipase (*Aspergillus niger*) on barium ferrite magnetic nanoparticles for biodiesel production from waste cooking oil. Several parameters methanol/oil molar ratio, reaction temperature, shaking, reaction time and enzyme concentration were studied. The studied reported that complex can be employed for transesterification for five cycles without any treatment or washing with affordable yield. Fan et al. (2016) employed superparamagnetic multi-walled carbon nanotubes (mMWCNTs) with iron oxide to covalently immobilized *Burkholderia cepacia* lipase using polyamidoamine (PAMAM) dendrimers. The maximum biodiesel conversion rate was 92.8% and retained 90% of its original activity even after 20 cycles of repeated use. Li et al. (2017) developed a novel hetero-functional carrier in order to lower activity loss and improve the operational stability of the immobilized lipase. They immobilized lipase (*Burkholderia cepacia*) on a matrix prepared using 2,3-epoxy-propyl-trimethylammonium chloride with epoxy and quaternary ammonium group and glutaraldehyde grafted onto aminated magnetic nanoparticles. The matrix showed improved operational stability, better reusability and higher esters as compared with nanoparticle modified with (3-aminopropyl) triethoxysilane and glutaraldehyde. Kalantari et al. (2018) synthesized benzene bridged dendritic mesoporous organosilica nanoparticles (BDMONs) having enriched benzene groups in pore channel walls. The lipase was covalently immobilized on these nanoparticles using aldehyde functionalized particle solution. The nanobiocatalyst showed a 93% conversion in the first cycle of the transesterification of corn oil to produce biodiesel and retained 94% of initial activity after five cycles of reuses.

4. Bioethanol Production by Cellulase Immobilized on Nanomaterial

Saccharification is a process of conversion of cellulose into glucose which is further used by fermentative microorganisms to produce ethanol. The enzymes such as cellulases and glucosidases are employed to breakdown glycosidic bonds in cellulosic materials to produce glucose. However, the cost of the enzyme is a key issue in employing them at industrial scales. Employing enzymes as a free system often leads to its loss in the absence of recovery steps in solvent processing. Therefore, in order to reuse these enzymes and reduce cost, researchers are trying to immobilize enzymes on suitable substrates. The immobilized enzyme offers better thermal and pH stabilities, solvent and inhibitors tolerance as well as reusability. The nanomaterials as immobilization support materials offer high surface area, better enzyme loading capacity and lower substrate inhibitions. Similar to lipase immobilized complexes, these enzyme-nanomaterials conjugate can be recovered via applying magnetic particles, centrifugation or filtration.

Cellulase immobilization on nanomaterials began last decade with initial works from Liao et al. (2008) (Polyvinyl Alcohol/Fe_2O_3 Nanoparticles), Zhou et al. (2009) (chitosan-coated magnetic nanoparticles modified with α-ketoglutaric acid) and Jordan (2009) (polystyrene-coated particles containing a Fe_3O_4 core). Lee et al (2010) employed beta-glucosidase immobilized on polystyrene and polystyrene-co-maleic anhydride nanofiber by relieving the product inhibition of cellobiose. The magnetic nanoparticles were added to nanofibers to enhance reusability. Das et al.

(2011) immobilized cellobiase enzyme on sol-gel routed mesoporous silica tetraethyl orthosilicate. The porosity of the complex was controlled using fructose during the sol-gel step. The immobilized cellobiase system showed hydrolysis efficiency of up to 74% and 81% in comparison to hydrolysis with free cellobiase enzyme. Cho et al. (2012) performed coimmobilization of three cellulases (endo-glucanase, exo-glucanase and β-glucosidase) on gold nanoparticle and gold-doped magnetic silica nanoparticles. The gold-doped silica nanoparticle (Au-MSNP) reported higher pH and thermal stability which was due to the covalent bonding of cellulase and nanoparticle. Also, Au-MSNP has better activity in reusability. Lupoi and Smith (2013) immobilized cellulase (*T. viride*) on silica nanoparticle (40 nm) using physical adsorption. The complex produced 1.6 times more glucose than free cellulase after 96 hours at pH 4.8 and 35°C. Ungurean et al. (2013) immobilized cellulase from *Trichoderma reesei* by sol-gel encapsulation. The cellulase was entrapped by sol-gel using silane precursors (binary or ternary mixtures of tetra-methoxysilane with methyl-trimethoxysilane and phenyl-trimethoxysilane). The entrapped enzyme displayed high temperature and pH stability, lower inhibition and multiple reuses. Gokhale et al. (2013) covalently immobilized cellulase on magnetoresponsive graphene nano-supports. This support was polyacrylic acid (PAA) coated maghemite-magnetite graphene scaffolds. The hydrophilic carboxylic groups of polyacrylic acid helped to interact with an iron oxide layer of nanoparticle, while 1-ethyl-3-(3'-dimethylaminopropyl) carbodiimide-hydrochloride spacer molecule helped to interact with cellulase. The cellulase immobilized on this nanosystem replicated the activity of free enzyme as much as 90–95% at the optimum temperature of at 60°C and a pH of 5.1. Abraham et al. (2014) covalently immobilized cellulase (*Trichoderma reesei*) on zinc-doped magnetite nanoparticles using glutaraldehyde. The CMC and pretreated hemp substrate were employed as the substrate. The complex showed superior stability at a higher temperature (80°C), storage stability (45 days) and retained 50% of its activity after five cycles of reuse. The complex hydrolyzed 93% of hemp pre-treated biomass while in the case of CMC the hydrolysis rate was 83% after 48 hours. Zhang et al. (2014) covalently immobilized cellulase on Fe_3O_4-chitosan nanoparticles using glutaraldehyde. The immobilized enzyme was able to maintain 50% of its initial activity after 10 cycles. The optimum pH and temperature were 5 and 50°C, respectively, with the activity of nearly 5 IU/mg of cellulase. Khorshidi et al. (2014) immobilized cross-linked cellulase aggregates on the amine-functionalized Fe_3O_4@ silica core-shell magnetic nanoparticles. The enzyme complex exhibited higher temperatures and pH tolerance. Huang et al. (2015) immobilized cellulase (*A. niger*) on β-Cyclodextrin-Fe_3O_4 nanoparticles by silanization and reductive amidation. The immobilization was carried out via coating with 3-aminopropyltriethoxysilane, aldehyde-functionalization and cyanoborohydride activation. The cyanoborohydride triggers reductive amidation to convert Schiff's base to a carbon-nitrogen single bond. This immobilized enzyme complex was able to retain 80% activity after 16 cycles of reuse. The activity of this complex further increased after employing ionic liquid such as 1-butyl-3-methylimidazolium chloride, 1-butyl-3-methylimidazolium hydrogen sulfate and 1-ethyl-3-methylimidazolium diethyl phosphate. In another study carried out by Zhang et al. (2015), cellulase was immobilized on functionalized magnetic nanospheres which were functionalized using three different silanes obtained using co-condensation of tetraethylorthosilicate with three different amino-silanes: 3-(2-aminoethylamino propyl)-triethoxysilane, 3-(2-aminoethylamino propyl)-trimethoxysilane and 3-aminopropyltriethoxysilane. The functionalization of nanospheres resulted in higher loading capacity. Among these silanes, 3-(2-aminoethylamino propyl)-trimethoxysilane functionalized nanospheres displayed the highest loading capacity (enzyme immobilization), temperature and pH stability. Xu et al. (2016) studied the performance of cellulase immobilized on PEGylated graphene oxide nanosheets in butyl-3-methylimidazolium chloride ([Bmim][Cl]). For this purpose, investigators grafted 4-armed polyethyleneglycol (PEG)-amine stars on graphene oxide via amide formation. The PEGylated graphene oxide nanosheets were employed to immobilize cellulase using glutaraldehyde. The PEGylated GO-cellulase was able to retain 61% of the initial activity in 25% (w/v) of butyl-3-methylimidazolium chloride as compared to free cellulase system where only 2% of initial activity was observed. There was 30 times greater tolerance of ionic liquid, such as butyl-3-methylimidazolium chloride, which was achieved using immobilization in comparison to the free enzyme system. Grewal et al. (2017) immobilized cellulase (*Trichoderma reesei*) on two nanomatrices, i.e., magnetic and silica nanoparticles. These nanobioconjugates displayed higher V_{max}, pH and temperature stability. Furthermore, these nanobioconjugates were stable in ionic liquid 1-ethyl-3-methylimidazoliumacetate, which is used as a solvent to dissolve cellulosic biomass. A simultaneous fermentation of treated wheat straw and sugarcane bagasse to produce ethanol was also carried out using *S. pastorianus* and one-pot method. Califano et al. (2018) immobilized β-glucosidase on wrinkled silica nanoparticles with central-radial pore structure and a hierarchical trimodal micro-/mesoporous pore-size distribution via adsorption. The complex has lower apparent K_M value as compared with a free enzyme which may be due to higher enzyme-substrate affinity. Qi et al. (2018) performed immobilization of cellulase on a magnetic core-shell metal-organic framework (MOF) material, i.e., Poly-(sodium 4-styrenesulfonate) surface modified Fe_2O_3-UIO-66-NH2. The complex exhibited high protein loading efficiency of (126.2 g/g support), high enzyme activity recovery, pH and temperature stability and high inhibitors (formic acid and vanillin) tolerance. Ahmad and Khare (2018) functionalized MWCNTs to immobilize cellulase (*Aspergillus niger*) via carbodiimide coupling using N-ethyl-N-(3-dimethylaminopropyl)carbodiimide hydrochloride in the presence of N-hydroxysuccinimide. The bionanoconjugates displayed catalytic efficiency, better affinity, improved pH and thermal stability and reusability up to 10 times without much loss in its activity.

5. Impact of Nanoparticles in Biohydrogen Production

This section of the chapter reviews the recent development in the application of nanoparticles for the production of biohydrogen and microalgae harvesting. It mainly emphasizes the role of nanoparticles in enhancing the production of biohydrogen from various pathways of production of biohydrogen. Nanomaterials can be used to improve the production of biofuels like biohydrogen (Ivanova et al. 2009, Srivastava et al. 2015). In biohydrogen, nanoparticles have been studied as catalysts and reactive materials.

Inspecting the future scenario, the production of energy from biofuels is considered to be a potential possibility to swap the usage of fossil fuels. Besides the widespread usage and restricted resources of fossil fuels for energy making, globally researchers are exploring numerous ways for upgrading energy production from biofuels. The application of nanomaterials among various other approaches is considered important in the upgrading of biofuels production (Manzanera et al. 2008, Choedkiatsaku et al. 2011). As per the studies by Patel et al. (2016), Kim et al. (2016) and Otari et al. (2016), nanoparticles (NPs) are being used more for applications like biosensors, protein immobilization and biofuels production. The engrossment of the nanomaterials gives extra value to the process of biofuel production by lowering the costs and constructive impacts on the environment.

Biohydrogen is one of the recorded clean sources of energy due to high energy content, i.e., approximately 142 MJ/kg, which is about 2.75 times more than other fuels such as hydrocarbon and petroleum (Arimi et al. 2015, Kumar et al. 2016). At present, about 96% of the global H_2 production is principally from fossil fuels, viz., 48%, 30%, 18% and 4% from natural gas, oil, coal and electrolysis of water, respectively (Das et al. 2014, Sivagurunathan et al. 2016). Biohydrogen produces via various production pathways such as biophotolysis, photo-fermentation and dark-fermentation, photo-dark combined fermentation and microbial electrolysis (MEC) processes (Azwar et al. 2014). Biohydrogen production is a complex process and relies on various factors like operational condition, nature of substrates, inorganic nutrients including metal ions, etc. (Hallenbeck 2005, Ferchichi et al. 2005, Wang and Wan 2008, 2009). Still, due to certain limitations, biohydrogen production is not yet feasible.

In photosynthetic H_2 production, photosynthetic bacteria converts solar energy into H_2. The practical usability of H_2 production is mainly affected by the requirement of expensive bioreactor design, poor efficiency of light transfer and slow rates of hydrogen production (Das et al. 2014). Except for the experimental design, microorganisms play a substantial part in biohydrogen production process, but dark-fermentation biohydrogen production method provides higher yields; in addition to that, it requires simpler reactor operation and diverse range of feedstock that makes them a striking method for the sustainable and clean hydrogen production (Rajhi et al. 2016, Wang and Yin 2018). But the usage of these kinds of feedstocks becomes unsatisfactory because of lesser hydrogen production (4.4–37.8 mL/g-dry grass). So, it is necessary to boost the yield of hydrogen using efficient production methods. The swift development of nanomaterials has unlocked a new research approach for the improvement of biohydrogen process efficacy by stimulating the bioactivity of the microorganisms due to its selective physicochemical properties. Microbes' stimulation activity is mediated by the electron-electron transport direction of the favorable metabolite (Lin et al. 2016). Nath et al. (2015) reported the application of zero-valent iron nanoparticles (F^0NPs) in improving the process of biohydrogen fermentation. As projected by the works of Beckers et al. (2013), microorganisms may be profited by nanoparticles in anaerobic conditions because electron transfer is more appropriate to acceptors.

The addition of Fe^0NPs effectively accelerates the biohydrogen production in hydrogen fermentation process through two different processes. Firstly, Fe^0NPs in the hydrogen fermentation process eliminates the unwanted oxygen from the system and amplifies the activity of oxygen-sensitive hydrogenase enzyme. Secondly, Fe^0NPs acts as a reductive material and can reduce the oxidation-reduction potential in the system and thus providing a more apposite environment for the growth of fermentative bacteria along with facilitating the synthesis of hydrogenase enzyme in the system. Yang and Wang (2018) in their study reported that the activity of enzymes could be promoted by Fe^0NPs, and the yield of hydrogen was 73.1% higher than the control and the lag time along with total fermentation time which was also shortened.

Buzea et al. (2007) reported that for growth and metabolism of microbes and a trace amount of metal co-factors like nickel, potassium, zinc and iron are essential. Under aerobic conditions, H_2 production can be significantly influenced by NPs due to the efficient transfer of electrons (Beckers et al. 2013). The application of nanoparticles benefits the processes such as H_2 yield, energy or conversion efficiency, hydrogen production rate and capital input (Bunker et al. 2011). The positive effect of various NPs, including gold (Au), silver (Ag), copper (Cu), iron (Fe), nickel (Ni), titanium (Ti), silica (SiO_2), palladium (Pd), carbon nanotubes (CNTs), activated carbon and composite were detected in biohydrogen production (Mohanraj et al. 2014, 2016, Elreedy et al. 2017). Han et al. (2011) in their study used hematite nanoparticles and reported an improved biohydrogen production. Zhang and Shen (2007) reported that the addition of gold nanoparticle (AuNPs) improves the efficiency of biohydrogen fermentation significantly, and the hydrogen yield increases by 36.3% as compared to the control. Zhao et al. (2013) reported that silver nanoparticles (AgNPs) may also advance the process of the fermentative route and conversion efficiency of the substrate to biohydrogen production. As concluded by the researchers, bacterial cell immobilization on the nanoparticles could be the reason for enhanced biohydrogen production. In another study by Lower et al. (2001), robust affinity was identified in the distance range of a few hundred nanometers between *Shewanella oneidensis* and goethite (α-FeOOH), which increased by 2–2.5 times in anaerobic condition.

Silica has been well recognized as biocompatible support toward proteins and microorganism (Giannelli and Torzillo 2012, Patel et al. 2014). Venkta Mohan et al. (2008) established the efficient use of mesoporous SiO_2 particles in biohydrogen production using mixed consortia from the chemical wastewater of common effluent treatment plants. Zhao and Chen (2011) assessed the effects of TiO_2 nanoparticles (TiO_2NPs) on the activity and growth of H_2 production enzymes (nitrogenase and H_2-uptake) for the photo-fermentative process by *Rhodopseudomonas palustris* from dark-fermentative effluent. The H_2 production and nitrogenase activity were significantly enhanced over the control in the presence of TiO_2 (100 mg/L). Likewise, in another study, Pandey et al. (2015) reported that *Rhodopseudomonas sphaeroides* NMBL-02 showed enhancement of 1.7 and 1.9 times in the average H_2 production rate and duration as compared with control using 60 mg/L of TiO_2NPs under the photo-fermentative conditions.

Ni^{2+} ions are well recognized for improving the yield of H_2 production by enhancing the catalytic activity of hydrogenases enzyme. Taherdanak et al. (2016)

have demonstrated the influence of Ni^{2+} ions as well as Ni nanoparticles (NiNPs) on H_2 production by anaerobic sludge from glucose as feed. NiNPs up to 2.5 mg/L showed significant results as 0.9% higher H_2 production yield than the control. In contrast, Mullai et al. (2013) showed a 22.7% enhancement with a maximum H_2 production yield of 2.54 mol/mol glucose by anaerobic sludge in the presence of 5.7 mg/L of NiNPs. Likewise, Gadhe et al. (2015) reported an upsurge in H_2 production with industrial wastewater sources like dairy and molasses using NiNPs.

The individual effect of Fe and Ni metals has been widely reported for their positive effects on the activity of the hydrogenases (Taherdanak et al. 2016). Therefore, the combined effect of Fe and NiNPs was seen to improve the H_2 production by anaerobic sludge using different concentrations (0–50 mg/L) of Fe and Ni NPs mixture sludge from starch as feed. The maximum H_2 production of 150 L/kg VS was detected at Fe and Ni concentration of 37.5 and 37.5 mg/L, respectively. Here, an upgrading of nearly 200% in H_2 yield was observed as compared to controls. Interestingly, individual NPs (25 mg/L) showed maximum H_2 production of 66.8 and 61 L/kg VS for Fe and Ni, respectively (Taherdanak et al. 2015).

Gadhe et al. (2015) conducted a study in the presence of Fe_2O_3 (50 mg/L) and NiO (10 mg/L) and maximum H_2 production of 17.2 mol/kg in dairy wastewater were observed. The co-addition of Fe_2O_3 and NiO enhanced H_2 yield by 27% compared to controls. Significantly, it also declined from 3.6 to 2.8 h in the lag phase of H_2 production. On the other hand, maximum H_2 production was observed to be 8.83 mol/kg COD using Fe_2O_3 (200 mg/L) and NiO (5 mg/ L) NPs from complex distillery wastewater.

It can be, thus, concluded that despite the number of exertions made to advance the biological hydrogen production using a variety of nanomaterials, biofuels production in this area of research is still struggling. Additionally, more stages are required toward the large-scale production of biohydrogen (fermentative or photo fermentative) through the nanomaterials-based approach.

Indeed, nanoparticles' usage has led to an improved biohydrogen production rate and yield, the latter being the major. Viewed from a strictly scientific position, irrespective of the frequency of the reactions and size of the bioreactors used when testing the nanoparticles, nanoscience holds grave optimism in making the biohydrogen 'fuel economy' a forthcoming reality. Yet, the transition from the scale of 'nanoscience' to a mature 'nanobiotechnology' as adequately pertinent and applicable to the production of biohydrogen from stabilized biomass types has to be well understood and manifested thereafter designed taking a number of technical and economic considerations. This advancement will naturally facilitate more research, the results of which will be boosting the development of new routes for emboldening the application of nanobiotechnology in biohydrogen production. The transition from laboratory-scale biohydrogen production to streamlined large-scale production of biohydrogen is imperative, at least, to provide a wholesome and sustainable foundation for the 'biohydrogen green economy'.

6. Microalgae as a Biofuel

Microalgae can be used as alternative biofuel feedstock compared to other oil yielding plants or crops such as sunflower, soybeans, palm, corn and rapeseed because microalgae have many advantages including extensive lipid content. It has a high photosynthetic rate, there is no requirement of extra farmland and it can easily grow in wastewater (Andersson et al. 2014, Feng et al. 2016, Zhu et al. 2016, Prasad et al. 2017). The biofuels derived from microalgae faces many challenges for being commercialized; the major issue is poor economics (Huntley et al. 2015, Zhu et al. 2016, Chew et al. 2017).

6.1 Problems Associated with Microalgae Harvesting

One of the major issues associated with the harvesting process of microalgae is its high cost, it is energy demanding and requires huge time. Bestowing to several reports, the cost of harvesting the microalgal biomass and its dying process accounts for about 20–30% of the total expense of microalgal production (Wang et al. 2016, Zhu et al. 2016). The high cost of microalgae harvesting and drying is because of its small size ranging from 2–20 μm diameter and its low density ranging between 0.5–5 g dry W/L of growth medium (Agbakpe et al. 2014). For the harvesting of microalgae, the common methods include sedimentation, flocculation, flotation, filtration, centrifugation and also the combination of flotation and flocculation (Pragya et al. 2013, Rawat et al. 2013, Tasić et al. 2016, Zhu et al. 2017). Although these methods efficiently harvest microalgal with a high concentration of biomass slurry, these methods are still expensive and time-consuming (Zhu et al. 2016). Therefore, new methods need to be explored to overcome the problems associated with them.

6.2 Nanotechnology Applications in Microalgae Harvesting

Xu et al. (2011) reported the harvesting of microalgae biomass using bare iron oxide nanoparticles. In general, magnetic nanoparticles (MNPs) have been gaining much attention since the last 20 years due to its magnetic property. Nevertheless, robust growth in research on magnetic agents as tools for harvesting microalgae has been newly commenced by considering the advantages of such materials (Lim et al. 2012). Recently, magnetophoretic harvesting of microalgae using functionalized magnetic nanoparticles (MNPs) has arisen, its benefits include energy and time-saving process (Prochazkova et al. 2013, Verma et al. 2016). The mechanism behind the harvesting of microalgae biomass using magnetic nanoparticles and the microalgae harvesting process is illustrated in Figure 3. The positively charged magnetic nanoparticles attached with negatively charged microalgal cells and further attached microalgal cells with magnetic nanoparticles can be separated from the suspension medium using an external magnetic field. Lee et al. (2015) enabled 98.5% harvesting efficiency by preparing magnetic nanoflocculant functionalized dually by organosilane compound with a dosage of 1.6 g nanoflocculant per gram of cells. Furthermore, Wang et al. (2016) coated amino-rich polyamidoamine dendrimer on the surfaces of magnetic iron oxide nanoparticles with a dose of 80 mg/L in order to harvest *Chlorella* sp.

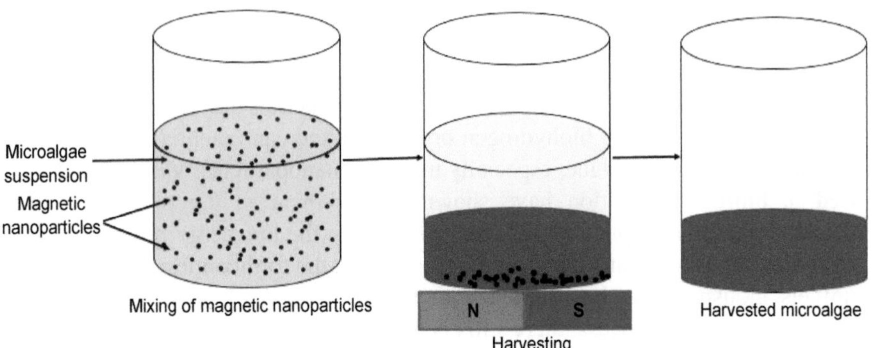

Figure 3. Harvesting process of microalgae biomass using magnetic nanoparticles (adopted from Zhu et al. 2017).

The harvesting efficiency of 95% was achieved within 2 minutes of time duration and at pH 8. In yet another study, 3-aminopropyl triethoxysilane-grafted $BaFe_{12}O_{19}$ was applied on magnetic nanoparticles to harvest *Chlorella* sp. (Seo et al. 2014) and harvest efficiency of 99% was achieved within 2–3 minutes in an external magnetic field. Similarly, tri-functional (cationic, magnetic, and lipophilic) carbon microparticles were implanted with iron oxide and then coated materials were utilized to harvest *Chlorella* sp. by Seo et al. (2015), who were able to achieve a harvesting efficiency of nearly 99%. The magnetic flocculant prepared with iron oxide and 0.1 mg/mL concentration of cationic polyacrylamide yielded harvesting efficiency of more than 95% in 10 minutes using 25 mg/L for *Botryococcus braunii* and 120 mg/L for *Chlorella ellipsoidea* (Wang et al. 2014).

Numerous studies on the topic have already been testified, yet this process still needs further extensive exploration. Moreover, the direct application of bare magnetic particles can help in the harvesting process by evading complex surface modification through coating or functionalization. Zhu et al. (2017) harvested the *Chlorella vulgaris* using yttrium iron oxide ($Y_3Fe_5O_{12}$) and iron oxide nanoparticles (Fe_3O_4). The $Y_3Fe_5O_{12}$ was found to be more efficiently harvested with 90% algal biomass than Fe_3O_4 nanoparticles. Fraga-García et al. (2018) utilized bare iron oxide magnetic nanoparticles in the harvesting of microalgae and could achieve 95% efficiency for the species *Scenedesmus ovalternus* and *Chlorella vulgaris.*

Ferro-ferric oxide altered with chitosan showed the removal of algae in freshwater (Liu et al. 2009). This magnetic polymer used was able to remove 99% of algal cells, which is more effective compared to chitosan or Fe_3O_4 only. It was found that the high algal removal effectiveness of magnetic polymer was because of the assistance between chitosan and Fe_3O_4 particles. Chitosan enhances the function of netting and linking in the flocculating algal cells, whereas the Fe_3O_4 separates the flocculated algal cells from water using a magnetic field. The network connection of chitosan and the magnetic Fe_3O_4 agglomerates the cells. After agglomeration, the algae cells retained their original shape and showed that flocculant could not destroy the cells. This method has been successfully pragmatic in the pilot-scale in Chaohu Lake, China.

7. Conclusion

This chapter critically reviewed the impact of nanomaterials in the field of bioenergy production. It is concluded that the nanomaterials play an important role in biodiesel, bioethanol and biohydrogen production and can potentially contribute to the harvesting of microalgae, especially magnetic nanoparticles. Various feedstocks used in biofuel production have shown enhancement in the production using nanomaterials. The application of nanomaterials in the bioenergy production helps to provide economical and cleaner energy source in the forthcoming years. Although biofuels industries can be profited with the application of nanomaterials in the biofuels production process, there still needs to be more research to be carried out in this area. Therefore, to study the effect of nanoparticles on biofuel production, a more focused study is needed that may play a substantial role in the commercialization of biofuels.

References

Abraham, R.E., M.L. Verma, C.J. Barrow and M. Puri. 2014. Suitability of magnetic nanoparticle immobilised cellulases in enhancing enzymatic saccharification of pretreated hemp biomass. Biotechnology for Biofuels 7(1): 90.

Agbakpe, M., S. Ge, W. Zhang, X. Zhang and P. Kobylarz. 2014. Algae harvesting for biofuel production: Influences of UV irradiation and polyethylenimine (PEI) coating on bacterial biocoagulation. Bioresource Technology 166: 266–272.

Ahmad, R. and M. Sardar. 2014. Immobilization of cellulase on TiO_2 nanoparticles by physical and covalent methods: A comparative study. Indian Journal of Biochemistry and Biophysics 51: 314–20.

Ahmad, R. and S.K. Khare. 2018. Immobilization of Aspergillus niger cellulase on multiwall carbon nanotubes for cellulose hydrolysis. Bioresource Technology 252: 72–75.

Alftren, J. and T.J. Hobley. 2013. Covalent immobilization of β-glucosidase on magnetic particles for lignocellulose hydrolysis. Applied Biochemistry and Biotechnology 169: 2076–2087.

Alhassan, F.H., U. Rashid and Y.H. Taufiq-Yap. 2015a. Synthesis of waste cooking oil-based biodiesel via effectual recyclable bi-functional Fe_2O_3-MnO-SO_4/ZrO_2 nanoparticle solid catalyst. Fuel 142: 38–45.

Alhassan, F.H., U. Rashid and Y.H. Taufiq-Yap. 2015b. Synthesis of waste cooking oil based biodiesel via ferric-manganese promoted molybdenum oxide/zirconia nanoparticle solid acid catalyst: influence of ferric and manganese dopants. Journal of Oleo Science 64(5): 505–514.

Alves, M.B., F. Medeiros, M.H. Sousa, J.C. Rubim and P.A. Suarez. 2014. Cadmium and tin magnetic nanocatalysts useful for biodiesel production. Journal of the Brazilian Chemical Society 25(12): 2304–2313.

Andersson, V., S.B. Viklund, R. Hackl, M. Karlsson and T. Berntsson. 2014. Algae-based biofuel production as part of an industrial cluster. Biomass and Bioenergy 71: 113–124.

Andrade, M.F., A.L. Parussulo, C.G. Netto, L.H. Andrade and H.E. Toma. 2016. Lipase immobilized on polydopamine-coated magnetite nanoparticles for biodiesel production from soybean oil. Biofuel Research Journal 3(2): 403–409.

Anis, S. and Z.A. Zainal. 2011. Tar reduction in biomass producer gas via mechanical, catalytic and thermal methods: A review. Renewable and Sustainable Energy Reviews 15: 2355–2377.

Antunes, F.A.F., A.K. Chandel, T.S.S. Milessi, J.C. Santos, C.A. Rosa and S.S. Da Silva. 2014. Bioethanol production from sugarcane bagasse by a novel Brazilian pentose fermenting yeast *Scheffersomyces shehatae* UFMG-HM 52.2: Evaluation of fermentation medium. International Journal of Chemical Engineering Article ID 180681: 1–8.

Aradi, A., J. Roos and T.C. Jao. 2010. Nanoparticle catalyst compounds and/or volatile organometallic compounds and method of using the same for biomass gasification. US 20100299990A.

Arimi, M.M., J. Knodel, A. Kiprop, S.S. Namango, Y. Zhang and S.U. Geißen. 2015. Strategies for improvement of biohydrogen production from organic-rich wastewater: A review. Biomass and Bioenergy 75: 101–118.

Azwar, M.Y., M.A. Hussain and A.K. Abdul-Wahab. 2014. Development of biohydrogen production by photobiological, fermentation and electrochemical processes: A review. Renewable and Sustainable Energy Reviews 31: 158–173.

Babaki, M., M. Yousefi, Z. Habibi, J. Brask and M. Mohammadi. 2015. Preparation of highly reusable biocatalysts by immobilization of lipases on epoxy-functionalized silica for production of biodiesel from canola oil. Biochemical Engineering Journal 101: 23–31.

Baskar, G., R.N. Kumar, X.H. Melvin, R. Aiswarya and S. Soumya. 2016. Sesbania aculeate biomass hydrolysis using magnetic nanobiocomposite of cellulase for bioethanol production. Renewable Energy 98: 23–28.

Bautista, L.F., G. Vicente, A. Mendoza, S. Gonzalez and V. Morales. 2015. Enzymatic production of biodiesel from *Nannochloropsis gaditana* microalgae using immobilized lipases in mesoporous materials. Energy Fuels 29: 4981–4989.

Beckers, L., S. Hiligsmann, S.D. Lambert, B. Heinrichs and P. Thonart. 2013. Improving effect of metal and oxide nanoparticles encapsulated in porous silica on fermentative biohydrogen production by *Clostridium butyricum*. Bioresource Technology 133: 109–117.

Borlido, L., A. Azevedo, A. Roque and M.R. Aires-Barros. 2013. Magnetic separations in biotechnology. Biotechnology Advances 31(8): 1374–1385.

Bunker, C.E. and M.J. Smith. 2011. Nanoparticles for hydrogen generation. J. Mat. Chem. 21: 12173–80.

Buzea, C., Pacheco II and K. Robbie. 2007. Nanomaterials and nanoparticles: Sources and toxicity. Biointerphases 2: 17–77.

Califano, V., F. Sannino, A. Costantini, J. Avossa, S. Cimino and A. Aronne. 2018. Wrinkled silica nanoparticles: efficient matrix for β-glucosidase immobilization. The Journal of Physical Chemistry C 122(15): 8373–8379.

Carrero, A., G. Vicente, R. Rodriguez, M. Linares and G.L. Del Peso. 2011. Hierarchical zeolites as catalysts for biodiesel production from *Nannochloropsis* microalga oil. Catalysis Today 167(1): 148–153.

Chan, F.L. and A. Tanksale. 2014. Review of recent developments in Ni-based catalysts for biomass gasification. Renewable and Sustainable Energy Reviews 38: 428–438.

Chen, S.Y., T. Yokoi, C.Y. Tang, L.Y. Jang, T. Tatsumi, J.C. Chan and S. Cheng. 2011. Sulfonic acid functionalized platelet SBA-15 materials as efficient catalysts for biodiesel synthesis. Green Chemistry 13(10): 2920–2930.

Chen, C.T., S. Dutta, Z.Y. Wang, J.E. Chen, T. Ahamad, S.M. Alshehri, Y. Yamauchi, Y.F. Lee and K.C.W. Wu. 2016. An unique approach of applying magnetic nanoparticles attached commercial lipase acrylic resin for biodiesel production. Catalysis Today 278: 330–334.

Cherian, E., M. Dharmendirakumar and G. Baskar. 2015. Immobilization of cellulase onto MnO_2 nanoparticles for bioethanol production by enhanced hydrolysis of agricultural waste. Chinese Journal of Catalysis 36(8): 1223–1229.

Chew, K.W., J.Y. Yap, P.L. Show, N.H. Suan, J.C. Juan, T.C. Ling, D.J. Lee and J.S. Chang. 2017. Microalgae biorefinery: High value products perspectives. Bioresource Technology 229: 53–62.

Chiang, Y.D., S. Dutta, C.T. Chen, Y.T. Huang, K.S. Lin, J.C. Wu, N. Suzuki, Y. Yamauchi and K.C.W. Wu 2015. Functionalized Fe_3O_4@ silica core–shell nanoparticles as microalgae harvester and catalyst for biodiesel production. ChemSusChem 8(5): 789–794.

Cho, E.J., S. Jung, H.J. Kim, Y.G. Lee, K.C. Nam, H.J. Lee and H.J. Bae. 2012. Co-immobilization of three cellulases on Au-doped magnetic silica nanoparticles for the degradation of cellulose. Chemical Communications 48(6): 886–888.

Choedkiatsaku, N.K., W. Kiatkittipona, N. Laosiripojana and S. Assabumrungrat. 2011. Patent review on biodiesel production process. Recent Patents on Chemical Engineering 4: 265–279.

Coward, T., J.G. Lee and G.S. Caldwell. 2014. Harvesting microalgae by CTAB-aided foam flotation increases lipid recovery and improves fatty acid methyl ester characteristics. Biomass Bioenergy 67: 354–362.

Das, D., N. Khanna and C.N. Dasgupta. 2014. Biohydrogen Production: Fundamentals and Technology Advances. CRC Press, Boca Raton, Florida, United States.

Das, S., D. Berke-Schlessel, H.F. Ji, J. McDonough and Y. Wei. 2011. Enzymatic hydrolysis of biomass with recyclable use of cellobiase enzyme immobilized in sol–gel routed mesoporous silica. Journal of Molecular Catalysis B: Enzymatic 70(1-2): 49–54.

Demirbas, A. 2000. Biomass resources for energy and chemical industry. Energy Conversion and Management 5: 21–45.

Demirbas, A. 2010. Use of algae as biofuel sources. Energy Conversion and Management 51: 2738–2749.

Dhiman, S.S., S.S. Jagtap, M. Jeya, J.R. Haw, Y.C. Kang and J.K. Lee. 2012. Immobilization of Pholiota adiposa xylanase onto SiO nanoparticles and its application for production of xylooligosaccharides. Biotechnology Letters 34(17): 1307–1313.

Dumri, K. and D. Hung Anh. 2014. Immobilization of lipase on silver nanoparticles via adhesive polydopamine for biodiesel production. Enzyme Research, 2014.

Dussan, K.J., C.A. Cardona, O.H. Giraldo, L.F. Gutiérrez and V.H. Pérez. 2010. Analysis of a reactive extraction process for biodiesel production using a lipase immobilized on magnetic nanostructures. Bioresource Technology 101(24): 9542–9549.

Eggert, H. and M. Greaker. 2014. Promoting second generation biofuels: Does the first generation pave the road? Energies 7: 4430–4445.

El-Batal, A., A. Farrag, M. Elsayed and A. El-Khawaga. 2016. Biodiesel production by Aspergillus niger lipase immobilized on barium ferrite magnetic nanoparticles. Bioengineering 3(2): 14.

Elreedy, A., E. Ibrahim, N. Hassan, A. El-Dissouky, M. Fujji, C. Yoshimura and A. Tawfik. 2017. Nickel-graphene nanocomposite as a novel supplement for enhancement of biohydrogen production from industrial wastewater containing mono-ethylene glycol. Energy Conversion and Management 140: 133–144.

Fan, Y., F. Su, K. Li, C. Ke and Y. Yan. 2017. Carbon nanotube filled with magnetic iron oxide and modified with polyamidoamine dendrimers for immobilizing lipase toward application in biodiesel production. Scientific Reports 7: 45643.

Fang, H., J. Huang, L. Ding, M. Li and Z. Chen. 2009. Preparation of magnetic chitosan nanoparticles and immobilization of laccase. Journal of Wuhan University of Technology-Material Science Edition 24(1): 42–47.

Farooq, W., Y.C. Lee, J.I. Han, C.H. Darpito, M. Choi and J.W. Yang. 2013. Efficient microalgae harvesting by organo-building blocks of nanoclays. Green Chemistry 15(3): 749–755.

Feng, P., L. Zhu, X. Qin and Z. Li. 2016. Water footprint of biodiesel production from microalgae cultivated in photobioreactors. Journal of Environmental Engineering 142: 04016067.

Feng, W. and P. Ji. 2011. Enzymes immobilized on carbon nanotubes. Biotechnology Advances 29: 889–895.

Ferchichi, M., E. Crabbe, W. Hintz, G.H. Gil and A. Almadidy. 2005. Influence of culture parameters on biological hydrogen production by *Clostridium saccharoperbutylacetonicum* ATCC 27021. World Journal of Microbiology and Biotechnology 21: 855 862.

Feyzi, M., A. Hassankhani and H.R. Rafiee. 2013. Preparation and characterization of $Cs/Al/Fe_3O_4$ nanocatalysts for biodiesel production. Energy Conversion and Management 71: 62–68.

Feyzi, M. and E. Shahbazi. 2015. Catalytic performance and characterization of $Cs–Ca/SiO_2–TiO_2$ nanocatalysts for biodiesel production. Journal of Molecular Catalysis A: Chemical 404: 131–138.

Fraga-García, P., P. Kubbutat, M. Brammen, S. Schwaminger and S. Berensmeier. 2018. Bare iron oxide nanoparticles for magnetic harvesting of microalgae: from interaction behavior to process realization. Nanomaterials 8(5): 292.

Gadhe, A., S.S. Sonawane and M.N. Varma. 2015. Enhancement effect of hematite and nickel nanoparticles on biohydrogen production from dairy wastewater. International Journal of Hydrogen Energy 40: 4502–4511.

Gao, Y. and I. Kyratzis. 2008. Covalent immobilization of proteins on carbon nanotubes using the cross-linker 1-ethyl-3-(3-dimethylaminopropyl) carbodiimide-A critical assessment. Bioconjugate Chemistry 19(10): 1945–1950.

Ge, S., M. Agbakpe, Z. Wu, L. Kuang, W. Zhang and X. Wang. 2015a. Influences of surface coating, UV irradiation and magnetic field on the algae removal using magnetite nanoparticles. Environmental Science & Technology 49(2): 1190–1196.

Ge, S., M. Agbakpe, Z. Wu, L. Kuang, W. Zhang and X. Wang. 2015b. Recovering magnetic Fe_3O_4-ZnO nanocomposites from algal biomass based on hydrophobicity shift under UV irradiation. ACS Applied Materials & Interfaces 7(21): 11677–11682.

Giannelli, L. and G. Torzillo. 2012. Hydrogen production with the microalga *Chlamydomonas reinhardtii* grown in a compact tubular photobioreactor immersed in a scattering light nanoparticle suspension. International Journal of Hydrogen Energy 37: 16951–16961.

Gokhale, A.A., J. Lu and I. Lee. 2013. Immobilization of cellulase on magnetoresponsive graphene nano-supports. Journal of Molecular Catalysis B: Enzymatic 90: 76–86.

Govan, J. and Y.K. Gunko. 2014. Recent advances in the application of magnetic nanoparticles as a support for homogeneous catalysts. Nanomaterials 4: 222–224.

Grewal, J., R. Ahmad and S.K. Khare. 2017. Development of cellulase-nanoconjugates with enhanced ionic liquid and thermal stability for *in situ* lignocellulose saccharification. Bioresource Technology 242: 236–243.

Gurunathan, B. and A. Ravi. 2015a. Biodiesel production from waste cooking oil using copper doped zinc oxide nanocomposite as heterogeneous catalyst. Bioresource Technology 188: 124–127.

Gurunathan, B. and A. Ravi. 2015b. Process optimization and kinetics of biodiesel production from neem oil using copper doped zinc oxide heterogeneous nanocatalyst. Bioresource Technology 190: 424–428.

Hahn-Hagerdal, B., M. Galbe, M.F. Gorwa-Grauslund, G. Lidén and G. Zacchi. 2006. Bio-ethanol the fuel of tomorrow from the residues of today. Trends in Biotechnology 24(12): 549–556.

Hallenbeck, P.C. 2005. Fundamentals of the fermentative production of hydrogen. Water Science and Technology 52: 21–29.

Han, H., M. Cui, L. Wei, H. Yang and J. Shen. 2011. Enhancement effect of hematite nanoparticles on fermentative hydrogen production. Bioresource Technology 102: 7903–7909.

Hegedus, I., E. Nagy, J. Kukolya, T. Barna and C.A. Fekete. 2011. Cellulase and hemicellulose enzymes as single molecular nanobiocomposites. Hungarian Journal of Industrial Chemistry Veszprem 39(3): 341–349.

Ho, S.H., X. Ye, T. Hasunuma, J.S. Chang and A. Kondo. 2014. Perspectives on engineering strategies for improving biofuel production from microalgae: A critical review. Biotechnology Advances 32: 1448–1459.

Holzinger, M., A.L. Goff and S. Cosnier. 2012. Carbon nanotube/enzyme biofuel cells. Electrochimica Acta 82: 179–190.

Hosseini-Sarvari, M. and E. Sodagar. 2013. Esterification of free fatty acids (Biodiesel) using nano sulfated-titania as catalyst in solvent-free conditions. Comptes Rendus Chimie 16(3): 229–238.

Hu, S., Y. Guan, Y. Wang, Y. Wang and H. Han. 2011. Nano-magnetic catalyst KF/CaO–Fe_3O_4 for biodiesel production. Applied Energy 88(8): 2685–2690.

Hu, Y.R., C. Guo, F. Wang, S.K. Wang, F. Pan and C.Z. Liu. 2014. Improvement of microalgae harvesting by magnetic nanocomposites coated with polyethylenimine. Chemical Engineering Journal 242: 341–347.

Huang, P.J., K.L. Chang, J.F. Hsieh and S.T. Chen. 2015. Catalysis of rice straw hydrolysis by the combination of immobilized cellulase from Aspergillus niger on β-cyclodextrin-Fe_3O_4 nanoparticles and ionic liquid. BioMed. Research International. 2015: 409103. Doi: 10.1155/2015/409103.

Huntley, M.E., Z.I. Johnson, S.L. Brown, D.L. Sills, L. Gerber, I. Archibald, S.C. Machesky, J. Granados, C. Beal and C.H. Greene. 2015. Demonstrated large-scale production of marine microalgae for fuels and feed. Algal Research 10: 249–265.

Imdadul, H.K., H.H. Masjuki, M.A. Kalam, N.W.M. Zulkifli, M.M. Rashed, H.K. Rashedul, I.M. Monirul and M.H. Mosarof. 2015. A comprehensive review on the assessment of fuel additive effects on combustion behavior in CI engine fuelled with diesel biodiesel blends. RSC Advances 5(83): 67541–67567.

Ivanova, V., P. Petrova and J. Hristov. 2011. Application in the ethanol fermentation of immobilized yeast cells in matrix of alginate/magnetic nanoparticles, on chitosan-magnetite microparticles and cellulose-coated magnetic nanoparticles. International Review of Chemical Engineering 3: 289–299.

Jafari Khorshidi, K., H. Lenjannezhadian, M. Jamalan and M. Zeinali. 2014. Preparation and characterization of nanomagnetic cross-linked cellulase aggregates for cellulose bioconversion. Journal of Chemical Technology & Biotechnology 91(2): 539–546.

Jordan, J. 2009, Efficiency of cellulase enzyme immobilized on magnetic nanoparticles. LSU Master's Theses, 2910.

Jordan, J., C.S.S. Kumar and C. Theegala. 2011. Preparation and characterization of cellulase-bound magnetite nanoparticles. Journal of Molecular Catalysis B: Enzymatic 68(2): 139–146.

Kalantari, M., M. Kazemeini and A. Arpanaei. 2013. Evaluation of biodiesel production using lipase immobilized on magnetic silica nanocomposite particles of various structures. Biochemical Engineering Journal 79: 267–273.

Kalantari, M., M. Yu, M. Jambhrunkar, Y. Liu, Y. Yang, X. Huang and C. Yu. 2018. Designed synthesis of organosilica nanoparticles for enzymatic biodiesel production. Materials Chemistry Frontiers 2(7): 1334–1342.

Kandel, K., C. Frederickson, E.A. Smith, Y.J. Lee and I.I. Slowing. 2013. Bifunctional adsorbent-catalytic nanoparticles for the refining of renewable feedstocks. ACS Catalysis 3(12): 2750–2758.

Kandel, K., J.W. Anderegg, N.C. Nelson, U. Chaudhary and I.I. Slowing. 2014. Supported iron nanoparticles for the hydrodeoxygenation of microalgal oil to green diesel. Journal of Catalysis 314: 142–148.

Karimi, M. 2016. Immobilization of lipase onto mesoporous magnetic nanoparticles for enzymatic synthesis of biodiesel. Biocatalysis and Agricultural Biotechnology 8: 182–188.

Kaur, M. and A. Ali. 2011. Lithium ion impregnated calcium oxide as nano catalyst for the biodiesel production from karanja and jatropha oils. Renewable Energy 36(11): 2866–2871.

Kaur, M. and A. Ali. 2014. Potassium fluoride impregnated CaO/NiO: an efficient heterogeneous catalyst for transesterification of waste cottonseed oil. European Journal of Lipid Science and Technology 116(1): 80–8.

Kim, D.Y., D. Vijayan, R. Praveenkumar, J.I. Han, K. Lee, J.Y. Park, W.S. Chang, J.S. Lee and Y.K. Oh. 2016. Cell-wall disruption and lipid/astaxanthin extraction from microalgae: *Chlorella* and *Haematococcus*. Bioresource Technology 199: 300–310.

Kumar, G., A. Mudhoo, P. Sivagurunathan, D. Nagarajan, A. Ghimire and C.H. Lay. 2016. Recent insights into the cell immobilization technology applied for dark fermentative hydrogen production. Bioresource Technology 219: 725–37.

Kuo, C.H., Y.C. Liu, C.M.J. Chang, J.H. Chen, C. Chang and C.J. Shieh. 2012. Optimum conditions for lipase immobilization on chitosan-coated Fe_3O_4 nanoparticles. Carbohydrate Polymers 87(4): 2538–2545.

Laurent, S., D. Forge, M. Port, A. Roch, C. Robic, L. Vander Elst and R.N. Muller. 2008. Magnetic iron oxide nanoparticles: Synthesis, stabilization, vectorization, physicochemical characterizations, and biological applications. Chemical Reviews 108: 2064–2110.

Lee, H.V. and J.C. Juan. 2016. Nanocatalysis for the conversion of nonedible biomass to biogasoline via deoxygeneation reaction. Nanotechnology for Bioenergy and Biofuel Production 301–323.

Lee, K., S.Y. Lee, J.G. Na, S.G. Jeon, R. Praveenkumar, D.M. Kim, W.S. Chang and Y.K. Oh. 2013a. Magnetophoretic harvesting of oleaginous *Chlorella* sp. by using biocompatible chitosan/ magnetic nanoparticle composites. Bioresource Technology 149: 575–578.

Lee, K., J.G. Na, J.Y. Seo, T.S. Shim, B. Kim, R. Praveenkumar, J.Y. Park, Y.K. Oh and S.G. Jeon. 2015a. Magnetic-nanoflocculant-assisted water–nonpolar solvent interface sieve for microalgae harvesting. ACS Applied Materials & Interfaces 7(33): 18336–18343.

Lee, Y.C., Y.S. Huh, W. Farooq, J. Chung, J.I. Han, H.J. Shin, S.H. Jeong, J.S. Lee, Y.K. Oh and J.Y. Park. 2013b. Lipid extractions from docosahexaenoic acid (DHA)-rich and oleaginous *Chlorella* sp. biomasses by organic-nanoclays. Bioresource Technology 137: 74–81.

Lee, Y.C., H.U. Lee, K. Lee, B. Kim, S.Y. Lee, M.H. Choi, W. Farooq, J.S. Choi, J.Y. Park, J. Lee, and Y.K. Oh. 2014a. Aminoclay-conjugated TiO_2 synthesis for simultaneous harvesting and wet-disruption of oleaginous *Chlorella* sp. Chemical Engineering Journal 245: 143–149.

Lee, Y.C., K. Lee, Y. Hwang, H.R. Andersen, B. Kim, S.Y. Lee, M.H. Choi, J.Y. Park, Y.K. Han, Y.K. Oh and Y.S. Huh. 2014b. Aminoclay-templated nanoscale zero-valent iron (nZVI) synthesis for efficient harvesting of oleaginous microalga, *Chlorella* sp. KR-1. RSC Advances 4(8): 4122–4127.

Lee, Y.C., S.Y. Oh, H.U. Lee, B. Kim, S.Y. Lee, M.H. Choi, G.W. Lee, J.Y. Park, Y.K. Oh, T. Ryu and Y.K. Han. 2014c. Aminoclay-induced humic acid flocculation for efficient harvesting of oleaginous *Chlorella* sp. Bioresource Technology 153: 365–369.

Lee, Y.C., K. Lee and Y.K. Oh. 2015b. Recent nanoparticle engineering advances in microalgal cultivation and harvesting processes of biodiesel production: A review. Bioresource Technology 184: 63–72.

Lee, S.M., L.H. Jin, J.H. Kim, S.O. Han, H.B. Na, T. Hyeon, Y.M. Koo, J. Kim and J.H. Lee. 2010. β-Glucosidase coating on polymer nanofibers for improved cellulosic ethanol production. Bioprocess and Biosystems Engineering 33(1): 141.

Li, J., R. Yan, B. Xiao, D.T. Liang and L. Du. 2008. Development of nano-NiO/Al_2O_3 catalyst to be used for tar removal in biomass gasification. Environmental Science & Technology 42: 6224–6229.

Li, Z. and P.E. Savage. 2013. Feedstocks for fuels and chemicals from algae: treatment of crude bio-oil over HZSM-5. Algal Research 2(2): 154–16.

Li, K., Y. Fan, Y. He, L. Zeng, X. Han and Y. Yan. 2017. Burkholderia cepacia lipase immobilized on heterofunctional magnetic nanoparticles and its application in biodiesel synthesis. Scientific Reports 7(1): 16473.

Liao, H.D., L. Yuan, C.Y. Tong, Y.H. Zhu, D. Li and X.M. Liu. 2008. Immobilization of cellulase based on polyvinyl alcohol/Fe_2O_3 nanoparticles. Chemical Journal of Chinese Universities 8: 1564–1568.

Lim, J.K., D.C.J. Chieh, S.A. Jalak, P.Y. Toh, N.H.M. Yasin, B.W. Ng and A.L. Ahmad. 2012. Rapid magnetophoretic separation of microalgae. Small 8: 1683–1692.

Lin, R., J. Cheng, L. Ding, W. Song, M. Liu and J. Zhou. 2016. Enhanced dark hydrogen fermentation by addition of ferric oxide nanoparticles using *Enterobacter aerogenes*. Bioresource Technology 207: 213–9.

Liu, C., P. Lv, Z. Yuan, F. Yan and W. Luo. 2010. The nanometer magnetic solid base catalyst for production of biodiesel. Renewable Energy 35(7): 1531–1536.

Liu, D., F. Li and B. Zhang. 2009. Removal of algal blooms in freshwater using magnetic polymer. Water Science and Technology 59: 1085–1091.

Lower, S.K., M.F. Jr. Hochella and T.J. Beveridge. 2001. Bacterial recognition of mineralsurfaces: nanoscale interactions between *Shewanella* and α-FeOOH. Science 292: 1360–1363.

Lupoi, J.S. and E.A. Smith. 2011. Evaluation of nanoparticle-immobilized cellulase for improved ethanol yield in simultaneous saccharification and fermentation reactions. Biotechnology and Bioengineering 108(12): 2835–2843.

Luz Martinez, S., R. Romero, J.C. Lopez, A. Romero, V. Sanchez Mendieta and R. Natividad. 2010. Preparation and characterization of CaO nanoparticles/NaX zeolite catalysts for the transesterification of sunflower oil. Industrial & Engineering Chemistry Research 50(5): 2665–2670.

Macario, A., F.R.A.N.C.E.S.C.A. Verri, U. Diaz, A. Corma and G. Giordano. 2013. Pure silica nanoparticles for liposome/lipase system encapsulation: Application in biodiesel production. Catalysis Today 204: 148–155.

Madhuvilakku, R. and S. Piraman. 2013. Biodiesel synthesis by TiO_2–ZnO mixed oxide nanocatalyst catalyzed palm oil transesterification process. Bioresource Technology 150: 55–59.

Mahmood, T. and S.T. Hussain. 2010. Nanobiotechnology for the production of biofuels from spent tea. African Journal of Biotechnology 9(6): 858–868.

Manzanera, M., M.M.L. Molina and L.J. Gonzalez. 2008. Biodiesel: an alternative fuel. Recent Patents in Biotechnology 2: 25–34.

Maria, V.H. 2014. CO_2 emissions from fuel combustion highlights. https://www.iea.org/publications/freepublications/publication/CO_2EmissionsFromFuelCombustionHighlights2014.pdf.

Mimani, T. and K.C. Patil. 2001. Solution combustion synthase of nanoscale oxide and their composites. Materials Physics and Mechanics (Russia) 4: 134–137.

Mohanraj, S., K. Anbalagan, S. Kodhaiyolii and V. Pugalenthi. 2014. Comparative evaluation of fermentative hydrogen production using *Enterobacter cloacae* and mixed culture: effect of Pd(II) ion and phytogenic palladium nanoparticles. Journal of Biotechnology 192: 87–95.

Mohanraj, S., K. Anbalagan, P. Rajaguru and V. Pugalenthi. 2016. Effects of phytogenic copper nanoparticles on fermentative hydrogen production by Enterobacter cloacae and *Clostridium acetobutylicum*. International Journal of Hydrogen Energy 41: 10639–10645.

Mohareb, R.M., A.M. Badawi, M.R.N. El-Din, N.A. Fatthalah and M.R. Mahrous. 2015. Synthesis and characterization of cationic surfactants based on N-hexamethylenetetramine as active microfouling agents. Journal of Surfactants and Detergents 18(3): 529–535.

Montero, J.M., K. Wilson and A.F. Lee. 2010. Cs promoted triglyceride transesterification over MgO nanocatalysts. Topics in Catalysis 53(11-12): 737–745.

Mukherjee, J. and M.N. Gupta. 2016. Dual bioimprinting of *Thermomyces lanuginosus* lipase for synthesis of biodiesel. Biotechnology Reports 10: 38–43.

Mullai, P., M.K. Yogeswari and K. Sridevi. 2013. Optimisation and enhancement of biohydrogen production using nickel nanoparticles—a novel approach. Bioresource Technology 141: 212–219.

Nath, D., A.K. Manhar, K. Gupta, D. Saikia, S.K. Das, M. Mandal. 2015. Phytosynthesized iron nanoparticles: effects on fermentative hydrogen production by *Enterobacter cloacae* DH-89. Bulletin of Materials Science 38: 1533–1538.

Nicolas, P., V. Lassalle and M. Ferreira. 2014. Development of a magnetic biocatalyst useful for the synthesis of ethyloleate. Bioprocess and Biosystems Engineering 37: 585–591.

Nordgreen, T. 2011. Iron-based materials as tar cracking catalyst in waste gasification. Ph.D. Thesis, Department of Chemical Engineering and Technology Chemical Technology, KTH-Royal Institute of Technology, SE-100 44 Stockholm, Sweden.

Nzihou, A., B. Stanmore and P. Sharrock. 2013. A review of catalysts for the gasification of biomass char, with some reference to coal. Energy 58: 305–317.

Pandey, A., K. Gupta and A. Pandey. 2015. Effect of nanosized TiO_2 on photofermentation by *Rhodobacter sphaeroides* NMBL-02. Biomass Bioenergy 72: 273–279.

Patel, S.K.S., V.C. Kalia, J.H. Choi, J.R. Haw, I.W. Kim and J.K. Lee. 2014. Immobilization of laccase on SiO_2 nanocarriers improves its stability and reusability. Journal of Microbiology and Biotechnology 24: 639–647.

Patel, S.K.S., J.-H. Jeong, S. Mehariya, S.V. Otari, B. Madan, J.R. Haw, J.-K. Lee, L. Zhang and I.-W. Kim. 2016. Production of methanol from methane by encapsulated *Methylosinus sporium*. Journal of Microbiology and Biotechnology 26: 2098–2105.

Patumsawad, S. 2011. Second Generation biofuels: Technical challenge and R and D opportunity in Thailand. Journal of Sustainable Energy and Environment 47–50.

Peng, B., Y. Yao, C. Zhao and J.A. Lercher. 2012a. Towards quantitative conversion of microalgae oil to diesel-range alkanes with bifunctional catalysts. Angewandte Chemie 124(9): 2114–2117.

Peng, B., X. Yuan, C. Zhao and J.A. Lercher. 2012b. Stabilizing catalytic pathways via redundancy: selective reduction of microalgae oil to alkanes. Journal of the American Chemical Society 134(22): 9400–9405.

Pragya, N., K.K. Pandey and P.K. Sahoo. 2013. A review on harvesting, oil extraction and biofuels production technologies from microalgae. Renewable and Sustainable Energy Reviews 24: 159–171.

Prasad, M.S.V., A.K. Varma, P. Kumari and P. Mondal. 2017. Production of lipid-containing microalgal biomass and simultaneous removal of nitrate and phosphate from synthetic. Environmental Technology 39: 669–681.

Praveenkumar, R., B. Kim, E. Choi, K. Lee, S. Cho, J.S. Hyun, J.Y. Park, Y.C. Lee, H.U. Lee, J.S. Lee and Y.K. Oh. 2014a. Mixotrophic cultivation of oleaginous *Chlorella* sp. KR-1 mediated by actual coal fired flue gas for biodiesel production. Bioprocess and Biosystems Engineering 37(10): 2083–2094.

Prochazkova, G., N. Podolova, I. Safarik, V. Zachleder and T. Branyik. 2013. Physicochemical approach to freshwater microalgae harvesting with magnetic particles. Colloids and Surfaces B: Biointerfaces 112: 213–218.

Qi, B., J. Luo and Y. Wan. 2018. Immobilization of cellulase on a core-shell structured metal-organic framework composites: Better inhibitors tolerance and easier recycling. Bioresource Technology 268: 577–582.

Qiu, F., Y. Li, D. Yang, X. Li and P. Sun. 2011. Heterogeneous solid base nanocatalyst: preparation, characterization and application in biodiesel production. Bioresource Technology 102(5): 4150–4156.

Rajhi, H., D. Puyol, M.C. Martinez, E.E. Diaz and J.L. Sanz. 2016. Vacuum promotes metabolic shifts and increases biogenic hydrogen production in dark fermentation systems. Frontiers of Environmental Science & Engineering 10: 513–521.

Rawat, I., R.R. Kumar, T. Mutanda and F. Bux. 2013. Biodiesel from microalgae: A critical evaluation from laboratory to large scale production. Applied Energy 103: 444–467.

Reddy, C., V. Reddy, R. Oshel and J.G. Verkade. 2006. Room-temperature conversion of soybean oil and poultry fat to biodiesel catalyzed by nanocrystalline calcium oxides. Energy Fuels 20(3): 1310–1314.

Sadhik Basha, J. and R.B. Anand. 2014. Performance, emission and combustion characteristics of a diesel engine using carbon nanotubes blended *Jatropha* methyl esters emulsions. Alexandria Engineering Journal 53: 259–273.

Sadhik Basha, J. 2015a. Preparation of water-biodiesel emulsion fuels with CNT & Alumina nanoadditives and their impact on the diesel engine operation. SAE Technical Paper 2015-01-0904.

Sadhik Basha, J. 2015b. Chemical functionalization of carbon nanomaterials: chemistry and applications, Chap. 23. Taylor & Francis, Boca Raton, FL.

Sadhik Basha, S.J. 2016. Impact of nanoaddictive blended biodiesel fuels in diesel engines. Green Chemistry and Sustainable Technology, 325–339.

Saidur, R., E.A. Abdelaziz, A. Demirbas, M.S. Hossain and S. Mekhilef. 2011. A review on biomass as a fuel for boilers. Renewable and Sustainable Energy Reviews 15: 2262–2289.

Sajith, V., C.B. Sobhan and G.P. Peterson. 2010. Experimental investigations on the effects of cerium oxide nanoparticle fuel additives on biodiesel. *In*: Advances in Mechanical Engineering. Article ID 581407, 6 p.

Sakai, S., Y. Liu, T. Yamaguchi, R. Watanabe, M. Kawabe and K. Kawakami. 2010. Production of butyl-biodiesel using lipase physically-adsorbed onto electrospun polyacrylonitrile fibers. Bioresource Technology 101(19): 7344–7349.

Sakai, S., Y. Liu, T. Yamaguchi, R. Watanabe, M. Kawabe and K. Kawakami. 2010. Immobilization of Pseudomonas cepacia lipase onto electrospun polyacrylonitrile fibers through physical adsorption and application to transesterification in nonaqueous solvent. Biotechnology Letters 32(8): 1059–1062.

Sani, Y.M., W.M.A.W. Daud and A.A. Aziz. 2013. Solid acid-catalyzed biodiesel production from microalgal oil—the dual advantage. Journal of Environmental Chemical Engineering 1(3): 113–121.

Seo, J.Y., K. Lee, S.Y. Lee, S.G. Jeon, J.G. Na, Y.K. Oh and S.B. Park. 2014. Effect of barium ferrite particle size on detachment efficiency in magnetophoretic harvesting of oleaginous *Chlorella* sp. Bioresource Technology 152: 562–566.

Seo, J.Y., K. Lee, R. Praveenkumar, B. Kim, S.Y. Lee, Y.K. Oh and S.B. Park. 2015. Tri-functionality of Fe_3O_4-embedded carbon microparticles in microalgae harvesting. Chemical Engineering Journal 280: 206–214.

Shi, Q., D. Yang, Y. Su, J. Li, Z. Jiang, Y. Jiang and W. Yuan. 2007. Covalent functionalization of multi-walled carbon nanotubes by lipase. Journal of Nanoparticle Research 9: 1205–1210.

Silva, S.S. and A.K. Chandel. 2014. Biofuels in Brazil. Fundamental Aspects, Recent Developments and Future Perspectives. Switzerland: Springer International Publishing.

Sinag, A. 2012. Catalysts in thermochemical biomass conversion. pp. 187–197. *In*: Baskar, S., C. Baskar and R.S. Dhillon (eds.). Biomass Conversion. Springer, Berlin.

Sivagurunathan, P., G. Kumar, S.H. Kim, T. Kobayashi, K.Q. Xu and W. Guo. 2016. Enhancement strategies for hydrogen production from wastewater: A review. Current Organic Chemistry 20: 2744–2752.

Srivastava, G., S. Roy and A.M. Kayastha. 2015. Immobilisation of fenugreek β-amylase on chitosan/PVP blend and chitosan coated PVC beads: A comparative study. Food Chemistry 172: 844–51.

Subhadra, B. and M. Edwards. 2010. An integrated renewable energy park approach for algal biofuel production in United States. Energy Policy 38: 4897–4902.

Suh, W.H., K.S. Suslick, G.D. Stucky and Y.H. Suh. 2009. Nanotechnology, nanotoxicology and neuroscience. Progress in Neurobiology 87: 133–170.

Taher, H., S. Al-Zuhair, A.H. Al-Marzouqi, Y. Haik and M. Farid. 2014. Effective extraction of microalgae lipids from wet biomass for biodiesel production. Biomass Bioenergy 66: 159–167.

Taherdanak, M., H. Zilouei and K. Karimi. 2015. Investigating the effects of iron and nickel nanoparticles on dark hydrogen fermentation from starch using central composite design. International Journal of Hydrogen Energy 40: 12956–12963.

Taherdanak, M., H. Zilouei and K. Karimi. 2016. The effects of Feo and Nio nanoparticles versus Fe^{2+} and Ni^{2+} ions on dark hydrogen fermentation. International Journal of Hydrogen Energy 41: 167 173.

Tasić, M.B., L.F.R. Pinto, B.C. Klein, V.B. Veljković and R.M. Filho. 2016. *Botryococcus braunii* for biodiesel production. Renewable and Sustainable Energy Reviews 64: 260–270.

Thangaraj, B., Z. Jia, L. Dai, D. Liu and W. Du. 2017. Lipase NS81006 immobilized on functionalized ferric-silica magnetic nanoparticles for biodiesel production. Biofuels, pp. 1–9.

Tran, D.T., C.L. Chen and J.S. Chang. 2012. Immobilization of Burkholderia sp. lipase on a ferric silica nanocomposite for biodiesel production. Journal of Biotechnology 158(3): 112–119.

Tran, D.T., B.H. Le, D.J. Lee, C.L. Chen, H.Y. Wang and J.S. Chang. 2013. Microalgae harvesting and subsequent biodiesel conversion. Bioresource Technology 140: 179–186.

Umdu, E.S., M. Tuncer and E. Seker. 2009. Transesterification of *Nannochloropsis oculata* microalga's lipid to biodiesel on Al_2O_3 supported CaO and MgO catalysts. Bioresource Technology 100(11): 2828–2831.

Ungurean, M., C. Paul and F. Peter. 2013. Cellulase immobilized by sol–gel entrapment for efficient hydrolysis of cellulose. Bioprocess and Biosystems Engineering 36(10): 1327–1338.

Venkata Mohan, S., G. Mohanakrishna, S.S. Reddy, B.D. Raju, R.K.S. Rao and P.N. Sarma. 2008. Self-immobilization of acidogenic mixed consortia on mesoporous material (SBA-15) and activated carbon to enhance fermentative hydrogen production. International Journal of Hydrogen Energy 33: 6133–6142.

Verdugo, M., L.T. Lim and M. Rubilar. 2014. Electrospun protein concentrate fibers from microalgae residual biomass. Journal of Polymers and the Environment 22: 373–383.

Verma, M.L., C.J. Barrow and M. Puri. 2013. Nanobiotechnology as a novel paradigm for enzyme immobilization and stabilization with potential applications in biodiesel production. Applied Microbiology and Biotechnology 97: 23–39.

Verma, M.L., M. Puri and C.J. Barrow. 2016. Recent trends in nanomaterials immobilised enzymes for biofuel production. Critical Reviews in Biotechnology 36(1): 108–119. .

Verziu, M., B. Cojocaru, J. Hu, R. Richards, C. Ciuculescu, P. Filip and V.I. Parvulescu. 2008. Sunflower and rapeseed oil transesterification to biodiesel over different nanocrystalline MgO catalysts. Green Chemistry 10(4): 373–38.

Wang, F., C. Guo, L.R. Yang and C.Z. Liu. 2010. Magnetic mesoporous silica nanoparticles: Fabrication and their laccase immobilization performance. Bioresource Technology 101(23): 8931–8935.

Wang, J.L. and W. Wan. 2008. Effect of Fe^{2+} concentrations on fermentative hydrogen production by mixed cultures. International Journal of Hydrogen Energy 33: 1215–1220.

Wang, J.L. and W. Wan. 2009. Factors influencing fermentative hydrogen production: a review. International Journal of Hydrogen Energy 34: 799–811.

Wang, J.L. and Y.N. Yin. 2018. Fermentative hydrogen production using various biomass-based materials as feedstock. Renewable and Sustainable Energy Reviews 92: 284–306.

Wang, L. and J. Yang. 2007. Transesterification of soybean oil with nano-MgO or not in supercritical and subcritical methanol. Fuel 86(3): 328–33.

Wang, S.K., F. Wang, Y.R. Hu, A.R. Stiles, C. Guo and C.Z. Liu. 2014. Magnetic flocculant for high efficiency harvesting of microalgal cells. ACS Applied Materials & Interfaces 6(1): 109–115.

Wang, S.K., F. Wang, Y.R. Hu, A.R. Stiles, C. Guo and C.Z. Liu. 2014. Magnetic flocculant for high efficiency harvesting of microalgal cells. ACS Applied Materials & Interfaces 6: 109–115.

Wang, S.K., A.R. Stiles, C. Guo and C.Z. Liu. 2015. Harvesting microalgae by magnetic separation: a review. Algal Research 9: 178–185.

Wang, T., W.L. Yang, Y. Hong and Y.L. Hou. 2016. Magnetic nanoparticles grafted with aminoriched dendrimer as magnetic flocculant for efficient harvesting of oleaginous. Chemical Engineering Journal 297: 304–314.

Wang, X., P. Dou, P. Zhao, C. Zhao, Y. Ding and P. Xu. 2009. Immobilization of lipases onto magnetic Fe_3O_4 nanoparticles for application in biodiesel production. ChemSusChem: Chemistry & Sustainability Energy & Materials 2(10): 947–950.

Wang, X., X. Liu, X. Yan, P. Zhao, Y. Ding and P. Xu. 2011. Enzyme-nanoporous gold biocomposite: excellent biocatalyst with improved biocatalytic performance and stability. PLoS One 6(9): e24207.

Wang, X., X. Liu, C. Zhao, Y. Ding and P. Xu. 2011. Biodiesel production in packed-bed reactors using lipase–nanoparticle biocomposite. Bioresource Technology 102(10): 6352–6355.

Wang, X.H., H.P. Chen, X.J. Ding, H.P. Yang, S.H. Zhang and Y.Q. Shen. 2009a. Properties of gas and char from microwave pyrolysis of pine sawdust. BioResources 4(3): 946–959.

Wang, Y., S.Y. Hu, Y.P. Guan, L.B. Wen and H.Y. Han. 2009b. Preparation of mesoporous nanosized KF/CaO–MgO catalyst and its application for biodiesel production by transesterification. Catalysis Letters 131(3-4): 574–578.

Wen, L., Y. Wang, D. Lu, S. Hu and H. Han. 2010. Preparation of KF/CaO nanocatalyst and its application in biodiesel production from Chinese tallow seed oil. Fuel 89(9): 2267–2271.

Woodford, J.J., C. Parlett, J.P. Dacquin, G. Cibin, A. Dent, J. Montero, K. Wilson and A.F. Lee. 2014. Identifying the active phase in Cs-promoted MgO nanocatalysts for triglyceride transesterification. Journal of Chemical Technology & Biotechnology 89(1): 73–80.

Xie, W. and N. Ma. 2009. Immobilized lipase on Fe_3O_4 nanoparticles as biocatalyst for biodiesel production. Energy & Fuels 23(3): 1347–1353.

Xie, W. and N. Ma. 2010. Enzymatic transesterification of soybean oil by using immobilized lipase on magnetic nano-particles. Biomass and Bioenergy 34(6): 890–896.

Xie, W. and J. Wang. 2014. Enzymatic production of biodiesel from soybean oil by using immobilized lipase on Fe_3O_4/poly (styrene-methacrylic acid) magnetic microsphere as a biocatalyst. Energy & Fuels 28(4): 2624–2631.

Xu, L., C. Guo, F. Wang, S. Zheng and C.Z. Liu. 2011. A simple and rapid harvesting method for microalgae by *in situ* magnetic separation. Bioresource Technology 102: 10047–10051.

Xu, J., Z. Sheng, X. Wang, X. Liu, J. Xia, P. Xiong and B. He. 2016. Enhancement in ionic liquid tolerance of cellulase immobilized on PEGylated graphene oxide nanosheets: application in saccharification of lignocellulose. Bioresource Technology 200: 1060–1064.

Yang, G. and J.L. Wang. 2018a. Kinetics and microbial community analysis for hydrogen production using raw grass inoculated with different pretreated mixed culture. Bioresource Technology 247: 954–962.

Yu, C.Y., L.Y. Huang, I. Kuan and S.L. Lee. 2013. Optimized production of biodiesel from waste cooking oil by lipase immobilized on magnetic nanoparticles. International Journal of Molecular Sciences 14(12): 24074–24086.

Zhang, L., C.C. Xu and P. Champagne. 2010. Overview of recent advances in thermo-chemical conversion of biomass. Energy Conversion and Management 51(5): 969–982.

Zhang, W., J. Qiu, H. Feng, L. Zang and E. Sakai. 2015. Increase in stability of cellulase immobilized on functionalized magnetic nanospheres. Journal of Magnetism and Magnetic Materials 375: 117–123.

Zhang, Y. and J. Shen. 2007. Enhancement effect of gold nanoparticles on biohydrogen production from artificial wastewater. International Journal of Hydrogen Energy 32: 17–23.

Zhao, W., Y. Zhang, B. Dua, D. Wei, Q. Wei and Y. Zhao. 2013. Enhancement effect of silver nanoparticles on fermentative biohydrogen production using mixed bacteria. Bioresource Technology 142: 240–245.

Zhao, Y. and Y. Chen. 2011. Nano-TiO_2 enhanced photofermentative hydrogen produced from the dark fermentation liquid of waste activated sludge. Environmental Science & Technology 45: 8589–8595.

Zhou, Y.T., S.N. Su, M.M. Song, H.L. Nie, L.M. Zhu and C. Branford-White. 2009, June. Improving the stability of cellulase by immobilization on chitosan-coated magnetic nanoparticles modified alpha-ketoglutaric acid. pp. 1–4. *In*: 2009 3rd International Conference on Bioinformatics and Biomedical Engineering, IEEE.

Zhu, L. and E. Hiltunen. 2016. Application of livestock waste compost to cultivate microalgae for bioproducts production: A feasible framework. Renewable and Sustainable Energy Reviews 54: 1285–1290.

Zhu, L., Y.K. Nugroho, S.R. Shakeel, Z. Li, B. Martinkauppi and E. Hiltunen. 2017. Using microalgae to produce liquid transportation biodiesel: What is next? Renewable and Sustainable Energy Reviews 78: 391–400.

Zhu, L.D., E. Hiltunen and Z. Li. 2017. Using magnetic materials to harvest microalgal biomass: evaluation of harvesting and detachment efficiency. Environmental Technology 40: 1006–1012.

Chapter 11

Nanobiotechnology Advances in Bioreactors for Biodiesel Production

***Bhaskar Birru*,[1,*] *P. Shalini*[2] and *Madan L. Verma*[3]**

1. Introduction

The global demand for energy has increased drastically along with the rapid increase in population and urbanization. The energy consumption is proportionally increasing with high economic growth. The worldwide trend resonates the manufacturing and servicing sector growth is increasing every year. The overall development in these two sectors pushes the growth index, which subsequently leads to higher energy consumption as per the reports of International Energy Outlook 2018 (IEO 2018). The energy consumption over the world is shown to increase from 1990 to 2040 in Figure 1. All countries over the world are classified into two categories based on their membership in the Organization for Economic and Cooperation Development (OECD). It was reported that population and economic growth rapidly growing in non-OECD countries had resulted in higher energy consumption. The sector-wise consumption of the world showed that the industrial sector accounts for higher energy consumption compared to transportation, building and other end-use sectors. As per the reports of IEO 2018, the energy consumption will increase drastically in all the sectors from 2020 to 2040. The fossil fuel depletion throughout the world is questing for alternative energy sources which primarily include energy obtained from water and wind, solar energy and biomass-derived energy. This has led to a focus on the third-generation of biofuels from algae biomass.

[1] Department of Biosciences and Bioengineering, Indian Institute of Technology Guwahati, Assam-781039, India.

[2] Department of Chemical Engineering, National Institute of Technology Warangal, Telangana-506004, India.

[3] Department of Biotechnology, School of Basic Sciences, Indian Institute of Information Technology Una, Himachal Pradesh-177220, India.

* Corresponding author: bhaskar2314@gmail.com

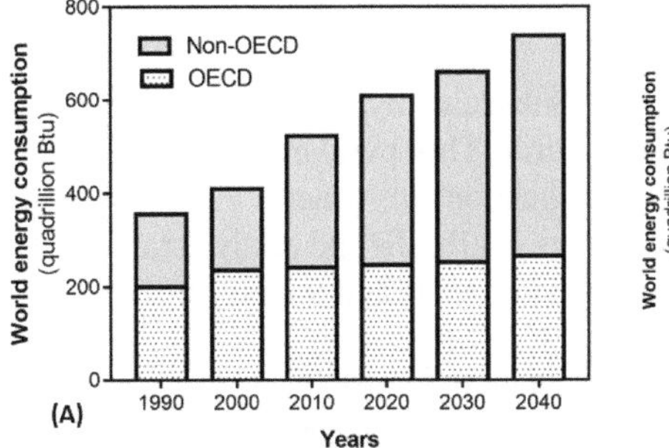

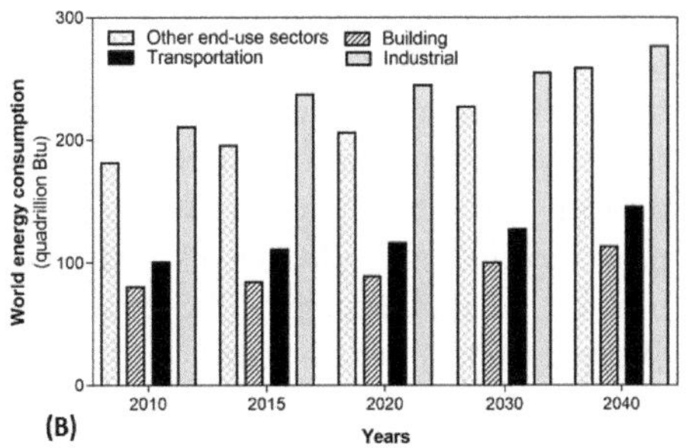

Figure 1. Global energy consumption trend for different time (A) Non-OECD and OECD energy consumption; (B) Sector-wise energy consumption. (The data reproduced using the statistics from the below mentioned website (https://www.eia.gov/pressroom/presentations/capuano_07242018.pdf)).

The biological agents, such as microorganisms and plants, were playing a pivotal role in biofuel production. Photosynthetic microorganisms were found to be the potential candidatures for biofuel production. Thus, significant research had taken place on biofuel production from algae biomass. However, developing a sustainable, cost-effective and safe method for bioenergy production from biological agents is quite essential. Nanotechnology is a branch of science found to be prominent in addressing the new challenges in the field of science and engineering. This branch of science deals with the fabrication and use of nanoscale materials for different applications (Verma et al. 2013). Numerous nanoscale materials have been used for biological applications. Mostly, the importance of nanotechnology resulted in driving the research in the field of bioenergy and biofuel production on par with different branches of science and engineering. Thus, the use of nanotechnology for biological application emerged as a nanobiotechnology (Verma et al. 2016). Nanomaterials have the ability to recognize the biological molecule and also improve the rate of reaction. The functionalization of nanomaterial is the crucial step to make it a functional material for real-time applications. Numerous nanomaterials have been developed and tested as biocatalysts. Nanobiotechnology applied for biofuels production from algae was intended to reduce the greenhouse gas emission which was found to be higher in the case of conventional approach for biofuel production (Ahmadi 2018). Nanocatalyst was used to generate the electricity by the breakdown of methane into carbon and hydrogen. Many studies have been conducted on nanocatalysts for biodiesel production from edible and non-edible oils. The introduction of nanocatalyst is highly potential and a novel approach to enhance the efficiency of biodiesel production.

The lower greenhouse gas emission is achievable with the use of biodiesel as a renewable energy resource and it is derived from fatty acids and oil. To increase the yield of biodiesel, fatty acid, lipid synthesis and metabolic engineering approach are inevitable. The initial step in the fat synthesis is the conversion of actyl-CoA into Malonyl-CoA, catalyzed by acetyl-CoA carboxylase. The soxidation of NADPH requires fat synthesis, which occurs in the cytosol. The pyruvate dehydrogenase (PDH) and fatty acid oxidation pathways generate acetyl-CoA in mitochondria; it cannot be transported into the cytosol. Kerb's cycle produces citrate in mitochondria, which is transported into the cytoplasm. Citrate breaks down into acetyl-CoA and oxalo acetate (OAA); this is catalyzed by ATP citrate lyase. The OAA converts into malate by malate dehydrogenase enzyme and NAD^+. Subsequently, pyruvate is synthesized from malate and then it enters into mitochondria. This pyruvate is also produced through the glycolytic pathway. Transesterification of triacyl glycerol (TAG) produces biodiesel. TAG serves as energy storage in all cells and easily catabolized to provide metabolic energy. TAG contains three fatty acids and the glycerol molecule (Vuppaladadiyam et al. 2018).

The present chapter discusses on microalgae cultivation, harvesting, biodiesel production, properties of biodiesel, bioreactors developed for biodiesel production, the concept of green building for self-sustained infrastructure, characterization techniques for biodiesel analysis, nanobiocatalysts for biodiesel production and microstructured devices for biodiesel production. The need for microscale technology for biodiesel production is the priority of this chapter and discusses in detail the bioreactors used for biodiesel production. It also briefly addresses the variety of microreactors used in biodiesel production, and their advantages and disadvantages are summarized. The importance of an integrated microscale system for biodiesel production and purification and their role in large scale production are presented in this chapter.

2. Nutritional Modes of Microalgae and Substrates for Microalgae Growth and Lipid Production (CO_2 and Wastewater)

Microalgae biomass is a potential feedstock for biodiesel production over other microbial biomass due to its higher biomass productivity and accumulation of lipid content (Lam and Lee 2012, Leong et al. 2018, Chen et al. 2018). Microalgae can survive in three nutritional modes: autotrophic, heterotrophic and mixotrophic. However, it has the ability to adapt and switch the metabolism according to any kind of nutritional mode. Autotrophic microalgae can generate energy and synthesize the essential molecules for cellular sustainability by using sunlight, CO_2 and H_2O. In contrast to this, heterotrophs' primary energy source is organic carbon substrate produced by autotrophs. Heterotrophic cultivation is advantageous over autotrophic because of their survival in the absence of sunlight, effective monitoring of the cultivation and the cost-effective biomass harvesting. Besides, this cultivation aids for wastewater treatment along with lipid synthesis which is a quite suitable approach to find out the solutions for the current environmental challenges.

In the case of mixotrophic cultivation, autotrophic and heterotrophic mechanism jointly aids the microalgae survival and growth and here CO_2 and organic carbon substrate are used for metabolism. It was reported that mixotrophic cultivation enhanced the growth rate of microalgae compared to autotrophic and heterotrophic cultures (Perez-garcia et al. 2010, Mohan et al. 2011, Devi et al. 2013). Microalgae biomass production using autotrophic conditions (light, CO_2, inorganic salts, water and optimal temperature of 20–30°C) is not economically viable. The best cost-effective strategy is using wastewater for nutrients, available sunlight and atmospheric CO_2

(Mohan et al. 2011). Also, the substrates in wastewater can be degraded and the same can be used as a carbon intake by the microalgae for survival and growth.

3. Cultivation Systems

The cultivation system highly influences the growth rate and cellular content of microalgae due to the change in nutrient availability and energy source. Besides, it also determines microalgae biomass production. The cultivation systems include open pond, closed, dark and offshore cultivation have chosen for algal culture on a larger scale (Wayne et al. 2018). The selection of cultivation is dependent on the desired product, algal strain, cost estimation and nutrient source. Numerous cultivation systems have been reported so far that are mainly categorized into two systems: open and closed cultivation systems (Klinthong et al. 2015).

3.1 Open Pond Cultivation Systems

Open ponds were the best choice for microalgae cultivation on a larger scale, easy to monitor the cultivation and are cost-effective. It can be done in two ways: natural water and artificial water systems. Natural water sources including ponds, lakes and lagoons can be used for open system cultivation. In the case of artificial water cultivation, containers, tanks and ponds were developed for algal cultivation. The area, shape and types (agitated and inclined) were chosen based on the product interest and its application (Klinthong et al. 2015). Unstirred, raceway and circular ponds are being used for cultivation and are shown in Figure 2. Unstirred ponds are the potential substitute for cultivation over raceway and circular ponds because their construction is easier and economical and have larger-scale cultivation. However, microalgal growth is found to be lower due to the lack of competitive growth in the presence of bacteria and protozoal natural habitats. Also, it has been reported that it is limited to some microalgal species (Chaumont 1993). The rotator is provided in circular ponds for agitation. Higher power consumption and construction cost and inadequate CO_2 supply are the major limitations of circular ponds. These can

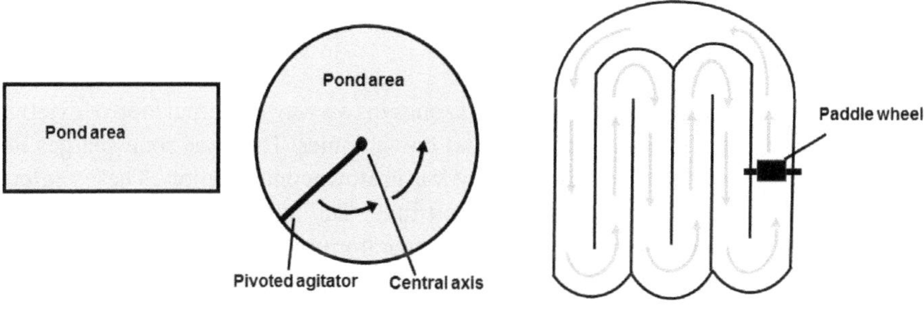

Figure 2. Different types of open pond systems for microalgae cultivation. Adapted from (Sreekumar et al. 2016) with permission of AIP Publishing © 2016.

mitigate by the raceway pond open cultivation system. In this, a paddlewheel is used to mix the algal culture and water. The limitations of open pond systems are uncontrolled process parameters, such as temperature, lighting, CO_2, evaporation and getting a chance of having contaminated bacterial or algal growth. To address these limitations, closed cultivation systems using photobioreactors have been introduced for algal cultivation. However, the closed system is expensive and scale-up is difficult compared to open pond systems. Closed cultivation system gives higher productivity, employs cloned microalgal culture, has long term culture maintenance and aids in higher surface-to-volume ratios (Hallmann 2016).

3.2 Closed Cultivation Systems

The disadvantages of an open system are addressed by introducing a closed cultivation system. This system provides controlled conditions such as the percentage of CO_2, light and required area to aid the survival and growth of microalgae (Singh and Sharma 2012). The closed system reduces the risk of contamination from bacteria, unwanted algae and protozoa. Also, they provide a controlled temperature to govern the water evaporation and overcome the loss of carbon dioxide. Different types of photobioreactors have been developed so far for microalgal cultivation, such as vertical (tubular), air-lift, bubble column, flat panel, helical-type and stirred-tank photobioreactors (Aci et al. 2017).

3.2.1 Vertical Tubular Bioreactor

The vertical bioreactor was the first photobioreactor developed for a closed cultivation system in the year 1950 (Chaumont 1993). Initially, this reactor model was fabricated using a small diameter glass tube (< 10 cm) which was produced for the fluorescent bulb industry (Miyamoto et al. 1988). The transparent tube allows light to penetrate the reactor and aids for the photosynthetic process of microalgae. The mixing culture is provided from the air sparger placed at the bottom of the reactor. The culture mixing and the CO_2 and O_2 gas transfer are efficient in this reactor system (Singh and Sharma 2012). The vertical tubular reactors are classified into two types: airlift and bubble column reactor.

3.2.1.1 Airlift Photobioreactor

Airlift reactors possess two interconnected zones in two ways: internal loop or external loop. These two zones are called riser and down comer. The riser zone sparges the gas for mixing and there is no gas provision in downcomer region. These regions are segregated by the split cylinder or draft tube. The internal loop bioreactor can be classified into two reactors based on the separator used in the reactor, namely the internal loop split and concentric tube reactor (Wayne et al. 2018). Two distinct tubes were used for separating the riser and downcomer and this can be called as external loop airlift reactor. There is no agitator provided in the reactor for mixing and the only bubble generated through sparger aids the mixing of culture. The performance of these reactors depends on the residence time of gas which highly influences the gas-liquid mass transfer, mixing, heat transfer and turbulence. The advantage of this reactor model is a circular air pattern, and the flashlight effect can be introduced for algal cells (Singh and Sharma 2012).

3.2.2 Bubble Column Reactor

The height of bubble column reactors is twice greater than its diameter. The advantage of this reactor is a larger surface-area-to-volume ratio, effective heat and mass transfer, homogenous culture environment and cost-effectiveness. The sparger generates bubbles in the reactor that helps in mixing and CO_2 mass transfer. Large-scale culture of microalgae cannot be achieved with this reactor system. The external light provided in this reactor and the gas flow rate determines the photosynthetic efficiency. The increasing in gas flow rate supports shortening light and dark cycles and this leads to enhanced photosynthesis by algal cells (Miron et al. 1999). The flat panel was used for breaking up the bubbles and redistribution of coalesced bubbles in the scale-up bubble column reactor (Janssen et al. 2002).

3.2.3 Flat Panel Reactor

Flat panel reactor is a simple design in which two transparent glass rectangular sheets are connected in cascade to expose the light. It also provides a larger surface-area-to-volume ratio, and the design can be modified easily for scale-up. The increasing mixing rate corresponds to an increase in the gas exchange rate in the reactor (CO_2-in and O_2-out) and flashing effect, and this leads to a higher yield of microalgal biomass (Klinthong et al. 2015). The advantage of this reactor system is suitable for the immobilization of algae, indoor and outdoor cultivation, easier cleaning and is economical. The design of the flat panel reactor is modified to increase the mixing rate and minimize shear stress and cell adhesion on the walls.

3.2.4 Helical-Type Bioreactor

A transparent and flexible tube of smaller diameter is arranged in a helical structure and the degassing unit is attached to it. The produced oxygen during the process is removed with the degassing unit. The CO_2 and culture medium circulate in the tube and a transparent glass tube allows light for the photosynthetic process (Singh and Sharma 2012). A centrifugal pump is used to circulate the culture medium in this reactor. The higher energy consumption for the centrifugal pump and the shear stress generated by the centrifugal forces are the limitation for scaling up the reactor. The efficiency of the reactor depends on mixing time, the residence time of the culture and the length of the tube. Microalgae harvesting and cleaning of the reactor is highly challenging. A number of light-harvesting units are required in this reactor for large scale algal culture (Aci et al. 2017). Due to these disadvantages, this reactor has not been used for commercial use to date.

3.2.5 Stirred Tank Bioreactor

Stirred tank bioreactor is a widely accepted conventional reactor for microalgae cultivation and agitation is provided by the impeller. To reduce the vortex, baffles were used in the reactor. The CO_2 was provided from the bottom of the reactor in the form of enriched air bubbles which was used as a carbon source for algal growth. The illuminating devices such as optical fibers and fluorescent lamps were introduced into the reactor model to use it as a photobioreactor (Wayne et al. 2018). The mixing pattern in the reactor is affected by the optical fibers, and this is considered one of the disadvantages of using them for lighting. This reactor provides minimal surface-area-to-volume ratio which leads to a reduction of lighting effect. Besides, microalgae biomass productivity regulates the composition of the culture medium. It was reported that higher salt concentration contained in the medium showed lower biomass productivity and induced lipid synthesis which led to higher lipid accumulation (Hong et al. 2016).

3.2.6 Photobioreactors in Urban Ecology and Aesthetics

Environmental sustainability and urban ecology principles are integrated to develop urban infrastructure by introducing various novel photo bioreactor-assisted technologies. This paves the way for the invention of new designs for cultivating algae and other biomass in urban infrastructure for energy production. The ideal way of developing urban architecture should integrate solar energy and urban ecology, which leads to the development of aesthetic designs. The living organisms' inclusion in urban infrastructure would improve aesthetic value, which is intended to come up with novel architecture designs for biomass cultivation (Elrayies 2018). Photobioreactor (PBR) facades in various modes such as flat panel, vertical, closed and tubular can be used in aesthetic designs. Recently, the new building architectures were introduced (1) Bio-Intelligent Quotient (BIQ) building, (2) process zero concept building and (3) the algae green loop tower (Elrayies 2018). These architectures were constructed by integrating PBR facades for microalgae cultivation, harvesting microalgae and energy conversion. The architecture of this building includes the supply of nutrients and CO_2, governing the temperature and liquid culture circulation, harvesting of biomass and biomass conversion into desired energy. The overall outline of green building and electricity generation is shown in Figure 3. The BIQ building produces electricity from methane, which is produced by algal biomass. This building is equipped with the flat plate PBR facades for algae cultivation. The overall energy required for the maintenance of the building is expected to meet by the microalgae cultivation in BIQ building (Kim and Ph 2013). This BIQ building can produce heat and biomass, which is subsequently used for the generation of electricity. The algae green loop tower building was in cylindrical architecture and hence the helical PBR was used for algae cultivation. The process zero concept building was developed to reduce the emissions from the building. This building holds the tubular PBR which absorbs sunlight and CO_2 and subsequently produces O_2 and lipids from algal biomass. Interestingly, the process zero concept building is designed in such a way to utilize wastewater generated from the building as a nutrient source for algal cultivation. The successful green buildings for bioenergy generation is the reason for the greater attention of research in the field of the green future building (Chan et al. 2009). The concept of this building aims to reduce greenhouse gas emission, save energy, ensure wastewater treatment and produce biofuel and release of oxygen. It can be seen in Figure 4, the wastewater is

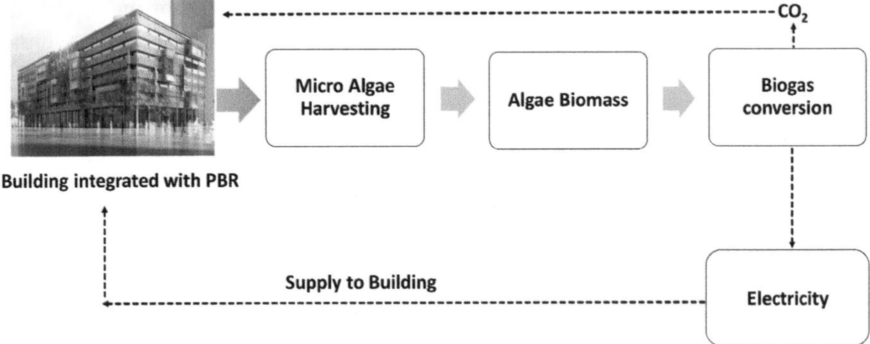

Figure 3. Outline of green buildings for bioenergy production.

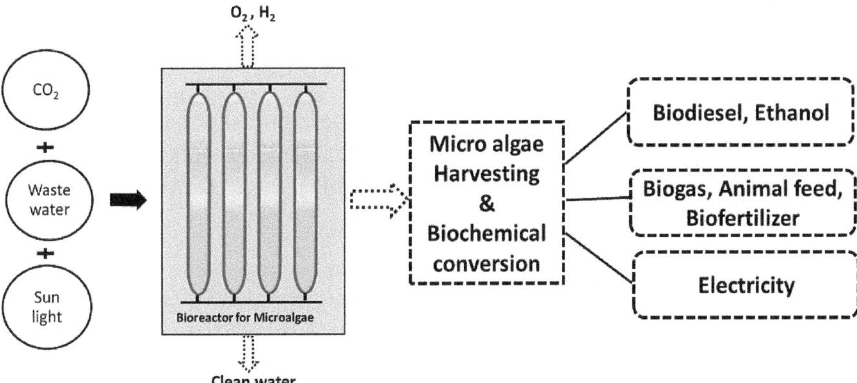

Figure 4. Schematic representation of microalgae cultivation using PBR facades for useful product synthesis.

provided as a nutrient; PBR in the building captures the sunlight and CO_2 released from industries that potentially favor microalgae cultivation. The exposure of light and light shading are considerable challenges for urban-integrated PBR buildings for algal biomass cultivation. The engineering of PBR has been provided various design of PBRs to improve the efficiency, wherein the designing aspects considered to achieve the properties include optimal light exposure, lower consumption of power, make suitable for different algal strains and contaminant free (Elrayies 2018). The efficiency of energy generation, saving and minimizing cost were the major challenges in developing urban infrastructures with PBR facades. Biogas generation plant inclusion, hydrothermal conversion and energy storage are the vital factors to be included in the design of architecture. This would contribute to developing a sustainable building without having any technical lagging.

4. Downstream Processing

The sequential steps in biodiesel production from microalgae are cultivation, harvest, extraction and conversion. The extraction process is involved in a variety of downstream processing techniques, which attributes mostly 60% of the production cost (Kim et al. 2013). The microalgal cell wall consists of complex carbohydrates and glycoproteins. Lipid extraction from microalgal biomass is a crucial step in the downstream process and subsequently, the transesterification of lipids produces biodiesel. The detailed extraction methods are presented in this section.

4.1 Cell Disruption Methods

The cell disruption is the initial step in the extraction process after harvesting microalgae biomass. The cell disruption methods can be classified into chemical, biological and mechanical methods. Chemical treatments and osmotic shock could be used as a chemical method. Various enzymes are involved in the extraction process in the degradation of complex glycoproteins and carbohydrates. The mechanical methods include ultra-sonication, microwave, electroporation, bead beating and high-pressure homogenization (Günerken et al. 2015).

4.1.1 Chemical Methods

The chemicals (acids, alkali and surfactants) or osmotic shock degrades the cell wall. The energy consumption is lower in chemical methods compared to physical methods where energy is required to apply mechanical force (Kim et al. 2013).

4.1.1.1 Chemical Treatment

Protamine, lysine polymers, polymyxin, polycationic peptides and cationic detergents had previously been used for cell wall disruption. These chemicals enhance the permeability ability of the cells leads to cell wall rupture (Günerken et al. 2015). Various acids and alkalis were used to microalgae cell wall hydrolysis. In acid and alkali treatments, temperature plays a vital role in cell disruption and it was reported that higher temperature (60°C) resulted in higher cell wall disruption compared to lower temperatures (120°C) (Harun and Danquah 2011, Halim et al. 2012, Günerken et al. 2015). However, mild temperature and low concentration of chemicals were suggested for cell disruption. Acid and alkali treatment could have degraded the proteins and pigments. Thus, the combination of any other technique with acid or alkali treatment would be a potential strategy for microalgae biorefinery.

4.1.1.2 Osmotic Shock

Osmotic shock facilitates cell disruption through the change in salt concentration which causes the imbalance of osmotic pressure between the exterior and interior of the cell. Cell disruption can be done in two ways, i.e., hyper-osmotic stress and hypo-osmotic stress (Halim et al. 2012). In the case of hyper-osmotic stress, a higher concentration in the exterior of the cell allows the cellular fluids to diffuse outside of the cell which leads to cell wall disruption. On contrary to this, hypo-osmotic stress attributes lower salt concentration in the exterior of the cell leading to the flow of water into the cell which aids for cell disruption (Kim et al. 2013). Hypo-osmotic stress requires a large amount of water for industrial-scale applications. Thus, hyper-osmotic stress is a reliable technique for microalgae cell disruption. Osmotic shock is cost-effective and is an easy process (Prabakaran and Ravindran 2011).

4.1.2 Biological Methods

Biological methods mostly employ the enzymes for cell digestion. The enzymes can regulate the biological process and are also the most abundantly available. The reaction conditions are mild and the enzymes are highly specific in action. The damage of chemical linkage occurs in mechanical methods, and the undesired reaction during the chemical treatment makes enzymatic-assisted cell disruption advantageous. To degrade the proteins and carbohydrates, mixture enzyme area is a potential approach. Enzymes are costly, so the biological approach is expensive compared to chemical and mechanical methods (Günerken et al. 2015).

4.1.3 Mechanical Methods

Physical forces are applied directly to cell disruption in mechanical methods. Various mechanical methods including ultrasonication, microwave, electroporation, bead milling and high-pressure homogenization have been used for cell disruption (Aminul et al. 2017).

5. Characterization of Biodiesel Including Analytical Techniques for Biodiesel Estimation

The quality of biodiesel can be assessed by its characterization. Viscosity, density, calorific value, oxidation stability, flash point, cloud point and cetane number are the vital fuel properties considered for biodiesel characterization (Mahmudul et al. 2017). The standard properties of biodiesel were depicted in Table 1. Viscosity determines the flow of fuel which is pivotal in deciding the efficiency of combustion and high viscosity leads to delay in combustion due to poor flow of fuel in the engine.

Table 1. Standard properties of biodiesel.

Biodiesel Property	**Biodiesel**	**Diesel**	**Reference**
Kinematic viscosity (mm^2/s) at 40°C	1.9 to 6.0	1.3 to 4.1	Mahmudul et al. 2017
Density at 20°C (kg/L)	875 to 900	75 to 840	Franjo et al. 2018
Flash point (°C)	100 to 170	60 to 80	Joshi and Pegg 2007
Moisture content (%)	Max. 0.05%	0.161	Miljic et al. 2020
Specific gravity at 15.5°C	0.88	0.85	
Ash content (%, w/w)	< 0.02	0.008 to 0.01	
Iodine value (I^2 g/100 g)	< 120	–	
Cloud point (°C)	–3 to 12	–15 to 5	
Glycerol content (%)	< 0.02	–	
Pour point (°C)	–15 to 10	–35 to –15	

The calorific value determines the power output of the engine for combustion. The biodiesel calorific values varied with source and it was proved that the calorific value of biodiesel was lower than diesel fuel. High calorific values could improve engine efficiency by producing higher heat during the combustion process (Ashraful et al. 2014). Flash point can be used to evaluate the flammability and it is influenced by the chemical composition of biodiesel. The ignition and combustion quality of the fuel depends on the cetane number, which is found to be higher in biodiesel than diesel fuel (Krishna and Mallikarjuna 2013). Given this, the characterization of biodiesel is a vital step to evaluate its quality. The analytical methods employed to characterize the biodiesel are categorized into two methods: chromatographic and spectroscopic. Besides, physical properties-based methods were not significant and hence chromatographic and spectroscopic methods will be discussed further.

5.1 Chromatographic Techniques for Biodiesel Analysis

Gas chromatography (GC), thin-layer chromatography (TLC), gel permeation chromatography (GPC) and high-performance liquid chromatography have been used for the analysis of biodiesel. However, mostly chromatographic methods restricted to identify the methyl esters are presented in the biodiesel. TLC with flame iodization technique was used for transesterification in the year 1984 (Freedman et al. 1984). The individual components of biodiesel cannot be identified using chromatography approach and it would contribute to assessing the class of compounds. Thus, chromatographic techniques are widely being used for biodiesel analysis by taking consideration of ester class of compounds. Various analytical techniques and nanoparticles used for biodiesel production are tabulated in Table 2.

5.1.1 Gas Chromatography

The minor components of biodiesel can be identified accurately using GC which is the major consideration for employing GC for analysis than other chromatographic techniques. The accuracy of GC results relied on the aging of samples and standards, baseline drift and overlapping signals. Initially, flame iodization detector was used mostly in GC, and now mass spectrophotometer is widely being used (Knothe 2006).

5.1.2 HPLC

HPLC can also be used for biodiesel characterization. However, it is used less compared to GC. Sample analysis is very easy and consumes less time than GC. Interestingly, this technique can be employed to analyze biodiesel from a variety of feedstock and especially it is highly potential for blend analysis. Though many detectors have used for HPLC, evaporating light scattering detections has been the most used detector for analyzing biodiesel (Trathnigg and Mittelbach 1990). The first report on the HPLC method for governing transesterification was done with a density detection approach which identified the mono, di, tri glycerides and methyl esters of biodiesel (Trathnigg and Mittelbach 1990).

Table 2. Various nanoparticles and analytical techniques used in biodiesel production and analysis, respectively.

Catalyst	Feedstock	Analytical Technique/ Method	Reference
Titanium dioxide (TiO_2)	Waste olive oil	NA	Mihankhah et al. 2018
Magnetic Fe_3O_4/MCM-41 composite	Soybean oil	ASTM D-6751	Xie et al. 2018
MgO	Sunflower oil	GC-FID	Yousefi et al. 2018
SO_4/Fe-Al-TiO_2	Waste cooking oil	NA	Gardy et al. 2018
Oil-palm empty fruit bunches (OP-EFB)	Palm oil	GC-MS	Lai et al. 2018
Amino-functionalized magnetic nanoparticles (APTES-Fe_3O_4)	Rapeseed oil	GC-FID	Miao et al. 2018
La_2O_3–ZrO_2	Canola oil	NA	Salinas et al. 2018
Sodium disilicate	Rapeseed oil	GC-FID	Ghaffari and Behzad 2018
MgO nanoparticles	Waste cooking oil	GC	Ranjan et al. 2018
MgO/$MgFe_2O_4$	Sunflower oil	GC	Alaei et al. 2018
Sulfonated multiwall carbon nanotube	Sunflower oil and waste chicken fat	NA	Shokuhi et al. 2018
Nickel-doped zinc oxide	Castor oil	NA	Baskar et al. 2018
Carbonated alumina	Canola oil	GC-FID	Nayebzadeh et al. 2018
Metallo-stannosilicate	Edible, non-edible and waste oils	GC	Antonio et al. 2018
Magnetic Fe_3O_4/Au-lipase	Waste tomato	GC–MS	Sarno and Iuliano 2018
Fe_3O_4-Al_2O_3	Waste cooking oil	GC–MS	Bayat et al. 2018
Fe_3O_4-lipase	Coconut oil	GC–MS	Cubides-roman et al. 2017
Hydroxyapatite-encapsulated γ-Fe_2O_3	Soybean oil	ASTM D-6751	Xie et al. 2017
Ferric-Manganese Doped Tungstated Zirconia	Waste cooking oil	GC-FID	Alhassan et al. 2015
Nanotube TiO_2	Waste olive oil	^{1}H NMR	Ghaffari et al. 2015

NA: Not available; GC: Gas Chromatography; FID-Flame ionization detector; GC-MS: Gas chromatograph with mass spectrometer; ^{1}HNMR: ^{1}H Nuclear Magnetic Resonance Spectroscopy.

5.2 Spectroscopic Techniques

Infrared and nuclear magnetic resonance (NMR) techniques are used for the characterization of biodiesel. NMR can determine the yield of biodiesel after the transesterification process by analyzing the various range of peaks in parts per million. Both biodiesel and triglycerides details in biodiesel can be figured out using IR spectroscopy analysis. However, the quality of biodiesel can be assessed based on physical and chemical properties (Knothe 2006).

6. Nanobiotechnology Advances in Bioreactors

Enzyme-based biodiesel production has major challenges that are (i) high cost of enzymes, (ii) solvents may inactivate the enzymes and (iii) large-scale production is difficult. The immobilization method is a potential approach and is well suited for scaling up the production along with the economic viability. The advantages of the immobilized enzyme method for biodiesel production are easy product recovery, enzyme stability in adverse environments of temperature and organic solvents and pH (Biswas 2019). Thus, the nanomaterial-based immobilization method is quite an efficient approach for biodiesel production. The immobilization method has proposed novel reactors for industrial production. Many bioreactors have been introduced for biodiesel production using enzyme immobilization (Ingle et al. 2018).

Nanomaterials are used more compared to conventional materials for enzyme immobilization-assisted biofuel production because of the higher enzyme loading, large surface-area-to-volume ratio and enhanced biocatalytic activity (Verma et al. 2016). The physical property of nanomaterials and enzymes aids in assessing the number of enzyme molecules attached to the supporting material. The size of the supporting matrix affects the enzyme loading and it was proved that higher enzyme loading was attained using nanomaterial compared to macromaterial. This has led to an increase in enzyme activity, which makes nanoparticles vital players for higher yield. The mass transfer limitations were mitigated with nanoparticle matrixes which were also the reason for improved enzyme activity. Poly(acrylonitrile-co-2-hydroxyethyl methacrylate) electrospun nanofibers were used in a continuous flow reactor for hydrolysis conversion (Huang et al. 2008). In the case of magnetic nanoparticle, the magnetic field can be used to easily recover the nanoparticles from the reaction volume instead of using higher centrifugal forces for separation. The operational enzyme stability could be improved and minimize the shear stress and process cost.

Various reactors and nanomaterials have been developed and also being tested for biodiesel production. Nano-biocatalyst (lipase-Fe_3O_4 nanoparticle) in a packed bed reactor system was used for biodiesel production from soybean oil. Four-packed bed reactor system resulted in a higher yield of biodiesel compared to a single packed bed bioreactor system using lipase-Fe_3O_4 nanoparticle-mediated methanolysis (Wang et al. 2011). Immobilized *Pseudomonas cepacia* lipase (PCL) on Fe_3O_4 nanoparticle in the reactor resulted in a higher yield of biodiesel compared to the flask level (Wang et al. 2009). Various nanoparticles such as carbon, graphene, TiO_2, ZnO, Fe_3O_4, SnO_2 and fullerene have been used for biodiesel production. Nanofibers' usage for immobilized enzymes has been introduced to a variety of bioreactors. PCL lipase immobilized on PAN nanofibers was studied for butyl biodiesel production in batch and continuous mode. The advantage of the nanoparticles over conventional carriers is the larger surface-area-to-volume ratio and that the immobilization is easier. Especially, magnetic nanoparticles have gained interest because of easy separation from the reaction volume. The understanding of nanomaterial properties and applications is inevitable for efficient use in biodiesel production. This has led to the invention of novel bioreactor designs to provide a favorable condition for enhancing enzymatic activity and stability. This leads to an increase in biodiesel production in terms of quality and quantity. The novel bioreactors are (1) four-packed bed reactor system, (2) magnetically stabilized fluidized bed reactor (MSFBR) and (3) microreactors.

6.1 Four-Packed Bed Reactor

A packed bed reactor is made with a glass column with the following dimension: inner radius of 1.6 cm and height 20 cm as shown in Figure 5. A 40 g of packing material can be adjusted to pack in the reactor (Wang et al. 2011). Lipase nanoparticle composite was used in this reactor and loaded on the cotton. Efficient enzyme immobilization and suitable reactor are the important factors to be considered for biodiesel production. This packed bed bioreactor is developed to enhance the biodiesel production in continuous operation. All four-packed bed reactors are connected in a series way and also a single-packed bed reactor can be used for transesterification. The reactor model can be chosen according to the required application. A combination of magnetic Fe_3O_4 and lipase nanoparticles was used in this bioreactor system to check the efficiency of this reactor for biodiesel production. Four mixtures of components were soybean oil, n-hexane, distilled water and methanol in varied ratios. The mixture was kept in a closed tank with the agitating system. Then, the reaction mixture was fed to the reactors using a peristaltic pump. The flow rate is usually optimized before the experimental start-up. The biochemical reactions take place in packed bed bioreactors due to the presence of lipase nanoparticle biocomposites. The concentration of methyl esters is estimated by collecting the effluent liquid from the outlet collection tank. The enzyme loading and flow rate are key determining variables on the yield of

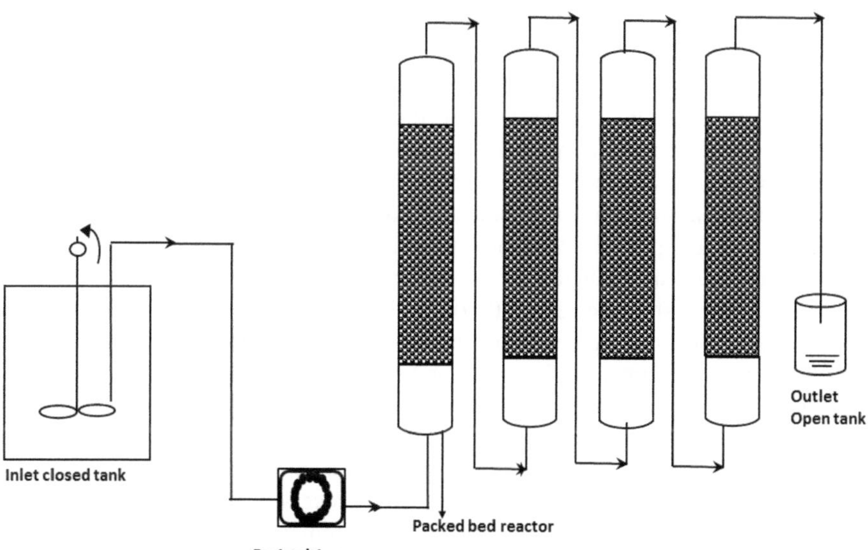

Figure 5. Four packed bed reactor packed with lipase nanoparticle composites for biodiesel production. Adapted from (Wang et al. 2011) with permission from ©2011 Elsevier.

biodiesel. A single-packed bed reactor can be used to optimize the flow rate. The flow rate affects the transesterification process. The optimal flow rate will be considered in four packed bed reactor system.

6.2 Magnetically Stabilized Fluidized Bed Reactor

Magnetically stabilized fluidized bed reactor (MSFBR) reactor is efficiently coupled with the magnetic field and whole cell or immobilized enzyme. The magnetic field is provided externally in this reactor that supports the stability of enzyme, and also transesterification reaction can be enhanced. This reactor has numerous advantages over convention FBR that are effective in solid mixing and transportation, the magnetic particle is stable and they can minimize the pressure drop. Besides these advantages, this reactor can be operated at higher fluid velocities including countercurrent. The higher mass transfer, low-pressure drop, avoiding solid mixing, easy separation and countercurrent fluid velocity are the advantages of MSFBR (Chen et al. 2017). This reactor is also capable to use for large scale biodiesel production. Immobilized *Pseudomonas mendocina* cells on Fe_3O_4-sodium alginate was used in MFBR for biodiesel production.

A glass column, peristaltic pump, water bath and reaction mixture tank are connected using a tygon tube or silicon tube. Glass column is used to keep the magnetic microspheres-immobilized lipase. The magnetic field is given using cooper wire coils, and the controlled magnetic field can be provided by adjusting DC power. The schematic representation of MSFBR reactor is shown in Figure 6. The water bath is provided to govern the temperature during the biochemical conversion.

6.3 Microreactors

Biodiesel production using conventional methods has shortcomings, for instance, high energy requisite, not economically viable process, longer residence time and lower process efficiency. The purification process on a large-scale using conventional methods and wastewater disposal into the environment is quite challenging (Madhawan et al. 2018). Thus, microstructured devices are found to be promising candidature to address the conventional method challenges and also provide logistical solutions for enhancement of the available technology and provide advanced technologies. The concept of microstructured devices ascended to improve the reaction catalysis, reduction of the required energy for a reaction and improved mass and heat transfer (Wen et al. 2009). The heat and mass transfer can be improved by providing good mixing conditions in the reactor which leads to the reduction of reaction time. The amounts of reagents and solutions for desired reaction and waste production during the process are minimal in microreactors compared to large-scale reactors. This microreactor can be relocated into the desired place and very easy to handle the reactions in nano to the microscale (Gkantzou et al. 2018). Especially, the use of microreactors for enzymes mediated process is more advantageous. The enzyme kinetic studies, enzyme specificity, optimizing biocatalysis conditions, process economic analysis before scaling up and enzyme inhibitor identification

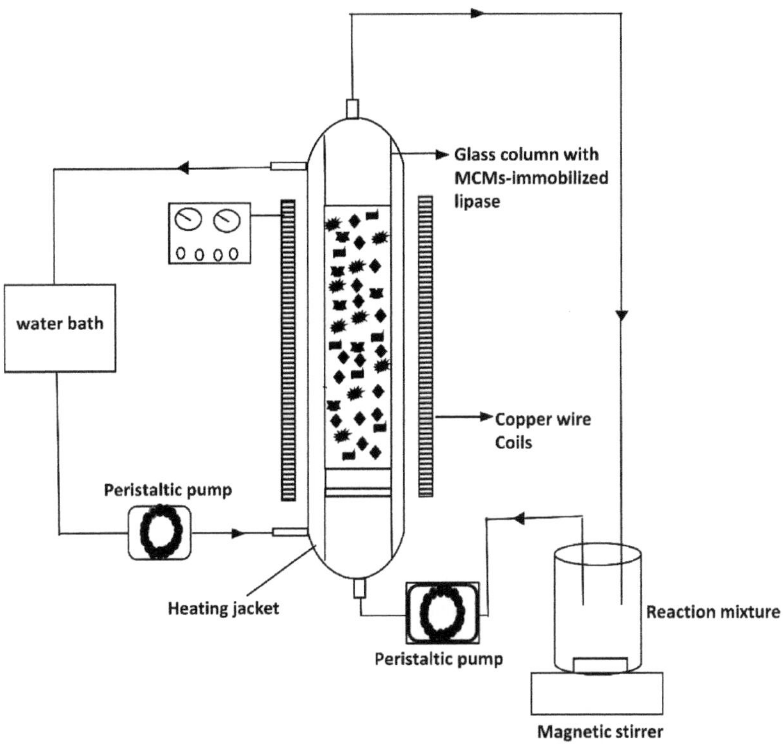

Figure 6. Schematic representation of a magnetically stabilized fluidized bed reactor for Biodiesel production. Adapted from (Gui-xionget al. 2013) with permission ©2013 Springer Nature.

studies can be carried out using microreactors. The controlled microenvironment in terms of pressure, temperature and concentration of reactants can be provided in a microreactor could increase the rate of reaction and in turn quantity and quality of the product can be enhanced. The analytical techniques namely GC, GC-MS, and HPLC can be integrated with the outlet of microreactor to improve the efficiency of the process. Numerous types of sensors can be arranged in the reactor set up to govern the temperature, pressure, dissolved oxygen and substrate and product concentrations.

Immobilized enzymes used as biocatalyst have numerous advantages such as stability of the enzyme in adverse chemical condition, enhanced the enzymatic activity, substrate utilization improved, cost-effective strategy for industrial application, easy purification procedure, higher yield of product and reuse of enzyme for biochemical conversions (Verma et al. 2016). In the microreactor scale, the product yield is significantly higher due to the good functional activity of the enzyme in microcapillary channels. The mass transfer limitations have mitigated by using microreactors, which is the reason for driving the substrates to react with enzymes (Gkantzou et al. 2018).

6.3.1 Microreactors for Biocatalysis

Microfluidic devices or microcapillary reactor uses the microchannels for the enzymatic reaction. Various materials such as polymers, glass and silicon were used for the fabrication of microreactors. GC and LC parts also can be used for microreactor fabrication. Silicon and polymers possess the light transmission ability, which is the reason for the ease of using these reactors for analyzing the product with the help of optical techniques. The use of transparent materials for microreactor fabrication enables the continuous monitoring of biochemical reactions. Thus, polydimethylsiloxane (PDMS) and poly methyl methacrylate (PMMA) are preferred in the fabrication of microreactors. Both these materials are biocompatible, light transparent and require surface modification for improving the wettability properties (Kecskemeti and Gaspar 2018).

The 3D printing was widely used to fabricate the microreactors in millimeter to centimeter size. Computer aided design (CAD) was used to design the microreactors and then printing can be done to fabricate the reactors. Through 3D printing, reproducibility and reliability can be achieved. The 3D printed molds can be used to make the PDMS-based microreactors. Interestingly, autoclavable modular microreactors also can be fabricated using 3D printing technology. Biocompatible and photo curable resigns are also used for microdevices fabrication. These microreactors are suitable to use for free and immobilized enzyme-based reactions. However, the current research majorly drove to use the immobilized enzyme microreactor. Usually, these microreactors can be operated with multi-phases and hence it is also called a multi-phase microreactor. Microreactors classified into three categories based on the microchannel design are (i) open-tubular, (ii) monolithic and (iii) packed-bed microreactor.

Open-tubular microreactors consist of microcapillary tubes that are arranged in different ways. The immobilized enzyme is placed in the inner walls of the microcapillary tube. The surface area available in the microcapillary tube is very less, which is the reason why the lower amount of enzyme loading is possible (Kolb 2013). The enzyme-substrate biochemical reaction occurs at a low reaction rate. Thus, the designing of open-tubular microreactors is quite challenging to increase the surface-area-to-volume ratio by adjusting the tube distance which improves the diffusion. Also, the inclusion of layers in addition to the existing capillary inner wall could minimize flow resistance (Kecskemeti and Gaspar 2018).

Monolithic microreactors possess remarkable properties such as the higher number available of functional groups, improved enzyme active site accessibility to substrates and solvent stable. Micropores and mesopores warrant the availability of higher surface area for the enzymatic reaction, and good mass transfer with short diffusion path are made a potential candidature for enzymatic reactions. The overall porosity, pore dimensions and surface chemistry determine the efficiency of these reactors. Organic, inorganic or hybrid polymer materials can be used in fabrication. The fabrication of a monolithic microreactor is complex.

Packed-bed reactor handling is easy compared to monolithic and open tubular microreactors. This reactor provides a higher surface area compared to both enzymatic microreactors. The procedure of simple packing and sample volume are interesting considerations for the packed-bed reactor (Chen et al. 2017). Various types of micromaterials or nanomaterials have been used for immobilization and then packed in microreactors for using different applications. Initially, the immobilization of the enzyme is carried out separately before placing it in the reactor system. The enzyme specificity, performance of enzyme and catalytic properties have to evaluate offline and then process parameters for enhanced enzymatic conversion can be monitored online mode in the reactor system. It was reported that the packed-bed reactor facilitates higher biomass conversion by capturing higher substrate, resulted in higher yield compared to open-tubular reactors (Miljic et al. 2020). The higher surface-area-to-volume ratio is the reason for effective enzymatic conversion in the packed-bed reactor, whereas the conversion rate is lower found in open-tubular reactor (Boehm et al. 2013). Nanomaterials used for immobilization are the best way to maximize the surface area, improves enzymatic reaction and higher enzyme loading. Currently, nanocarrier usage in enzymatic microreactors is a promising field of research.

Usually, a microreactor system consists of inlet, outlet chambers and fluid supply, connected with suitable silicon or tygon tubes. The fluid flow rate monitored using a syringe or peristaltic pumps. The samples can be analyzed offline or online mode using analytical techniques.

6.3.2 Magnetic Microreactors

The microreactor is integrated with a magnetic field called a magnetic microreactor. Magnetic particles in nanoscale or microscale are used either as an additive to biocatalyst or alone for immobilization. The shear forces and pressure exerted in the reactors could make the reactors unstable for enzymatic conversion. This challenge can be addressed by using an external magnetic field to regulate the biocatalyst function inside the reactor, which makes the magnetic microreactor system ideal to run the reactions using whole cell or enzyme (Gui-xiong et al. 2013). The use of an external magnetic field helps in easy product recovery and minimizes the downstream process cost. The enzymatic reaction rate is usually higher due to the employed magnetic field favors the dynamic stirring and minimal cluster formation (Gkantzou et al. 2018). Proper mixing provides a higher surface area for substrate interaction, resulting in a higher conversion rate.

6.4 Microreactors for Biodiesel Production

Microreactor technology is the only one possible way to deal with all existed problems associated with biodiesel production and purification due to the effective heat and mass transfer, provided large surface-area-to-volume ratio and shorter diffusion track (Franjo et al. 2018). Microreactors are widely used in the field of biotechnology, chemistry, pharma industries, biomedical engineering and nanotechnology for a variety of applications. In particular, organic synthesis has already been produced a significant outcome using microreactors compared to conventional reactor systems. The recent advent of studies on microreactors mediated biodiesel production found to be a potential alternative for conventional production and also suggested that catalyst-coated microreactors are more beneficial (Schürer et al. 2014). The dimensions of the reactor reduced in the case of microstructured reactors could minimize the operational and development of technology cost. The design aspect of microstructured devices for biodiesel production was found to simple and easier. The invention of various devices was noticed and classified into three categories, such as microtubular microreactors, membrane microreactors, and microstructured microreactors.

6.4.1 Microtubular Microreactors

Microtube reactors considered to be simple in view of the design aspect. Microfluidic channels possessed in this reactor with a width size of nm to mm and length of cm to m. The schematic diagram of the microreactor set-up is shown in Figure 7. This reactor designed in such a way to provide the reactants and catalysts separately. The feeding of reactants and catalyst can be done using syringe pumps. This reactor was tested for biodiesel production. These reactors were found to be used for multichemical reactions, efficient phase separation and heat exchange. Various microdevices were developed such as microtube reactors, membrane microreactor and microstructured reactors for biodiesel production (Franjo et al. 2018). All microchannels can be fabricated either in a horizontal or vertical manner. However, channels were designed in a parallel way. To provide mixing in the reactor, the design can be tailored by including mixing channels. The transesterification reaction can be completed in a shorter time using a microtube reactor compared to the laboratory-scale batch reactor. Capillary microreactors also come under microtubular microreactors. Initially, capillary reactors were used for biodiesel production. The overall production of biodiesel was improved by using a microtube reactor which confirmed intensification of the process was achieved due to thorough mixing of oil with methanol and also avoiding the mass transfer limitations. These membrane microreactors were used to intensify the process by shortening the overall reaction time and enhancing production volume. The reaction rate accelerated in microreactors

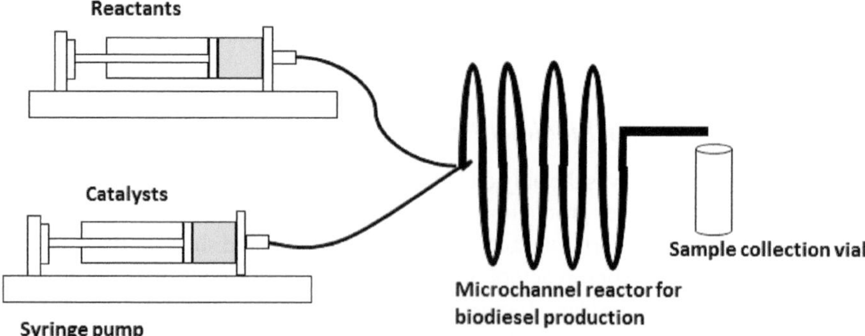

Figure 7. Schematic representation of the microreactor for biodiesel production. Adapted from (Franjo et al. 2018) with permission ©2018 Springer Nature.

due to thousands of microchannels in the reactor helping in effective mass transfer and also providing high surface-area-to-volume ratio. The material of the reactor was chosen by considering its stability toward chemical reactants and the solvents used in biodiesel production and its cost-effectiveness. Various feedstocks for biodiesel production were studied using microtubular reactors that are tabulated in Table 3.

6.4.2 Membrane Microreactor

Membrane microreactor is highly potential for biodiesel production compared to other convention reactors; also, the advantages of membrane microreactor and microreactor are integrated into one reactor and hence it is called membrane microreactor. The availability of larger surface area runs at a lower temperature, good mass and heat transfer and less enzyme loading capacity are the advantages of membrane microreactor (Machsun et al. 2010). Optimal conditions for higher yield can be attained easily in this reactor. Membrane microreactors are designed in two ways: (i) tubular and (ii) plate. Microchannels were imprinted in a plate along with catalyst deposition, and the membrane is fixed at the plate backside. Otherwise, the membrane coupled with the catalyst is placed into microchannel directly. The design possesses plate shape and hence it is called plate type membrane microreactor (Franjo et al. 2018). The occurrence of reaction and separation can be achieved in this reactor simultaneously. The microchannels for enzyme loading and membrane for separation possess tubular shape. A membrane microreactor holds an enzyme used for catalysis. The membrane in the reactor separates the product during transesterification reaction and also controls the mixing of oil with methanol/ethanol. The acid/alkali catalysts are used for biodiesel production from various feedstocks. Membrane selection depends on the type of catalyst, which is the most significant step in the overall process (Tan and Li 2013). Lipase enzyme catalyst supported membranes are also used for developing micromembrane reactors for biodiesel production. Ceramic carbon and polymeric membranes have been widely used in membrane microreactors. Ceramic membranes are thermal, chemical and mechanically stable compared to polymeric membranes. However, polymeric membranes are cheaper and production on a large scale is economically viable (Madhawan et al. 2018). In the case of biocatalytic membranes, polyethersulfone has been used for developing the layer-by-layer membrane reactor. Membrane microreactors have been studied for biodiesel production from various feedstocks such as palm, sunflower, waste cooking, rapeseed, vegetable oil and cottonseed oil (Table 3). The reaction rate in membrane microreaction is usually higher because of good mixing in reactor, continuous removal of unreacted oil and product during the continuous transesterification process.

6.4.3 Microstructured Reactors

Microstructured reactors are efficient for phase separation, chemical conversion, and heat transfer. For mixing reactants in this reactor, different plates can be used. Based on the configuration, microplate reactor and oscillatory baffled microreactor were

Table 3. Different types of microreactors for biodiesel production.

Feedstock	Type of Microreactor	Reaction in Reactor	Catalyst	Reference
Soybean oil	Microtubular reactor	Transesterification of oil with methanol	Calcined demineralized water plant sedimentation	Aghel et al. 2017
Vegetable oil	Microtubular reactor	Esterification of oleic acid with methanol	Supercritical CO_2	Quitain et al. 2018
Sunflower oil	Microtubular reactor	Transesterification of oil with ethanol	Sodium hydroxide	Santana et al. 2016
Waste cooking oil and Sunflower oil	Microtubular reactor	Transesterification of oil with water and/or oleic acid	Potassium hydroxide	Guan et al. 2010
Soybean oil for	Enzyme microreactor	Transesterification of oil with methanol	Lipase enzyme	Bi et al. 2017
Waste cooking oil	Microstructured reactors and oscillatory baffle reactors	Transesterification and esterification of oil	Potassium hydroxide or Sulfuric acid	Mazubert et al. 2014
Microalgae oil	Microreactors with micromixers Y, T, and Q-shaped mixers	Transesterification of glyceryl trioleate	Sulfuric acid	Liu et al. 2015
Soybean oil	Zig-zag microreactor	Transesterification of oil with methanol	Alkali	Wen et al. 2009
Microalgae oil	Membrane microreactor	Hydrodeoxygenation	Presulfided NiMo/γ-Al_2O_3	Zhou and Lawal 2014
Triolein	Membrane microreactor	Transesterification of triolein	Lipase	Machsun et al. 2010
Canola oil	Membrane microreactor	Transesterification of canola oil	Sulfuric acid	Dubé et al. 2007
Soybean oil and	Fixed-bed membrane reactor Wei	Transesterification of oil with methanol	Monolithic	Xu et al. 2015
Palm oil	Packed-bed membrane reactor	Transesterification of Palm oil and methanol	Potassium hydroxide/ activated carbon	Baroutian et al. 2011
Soybean	Microtubular reactor	Transesterification of oil with ethanol	Catalyst free	Da Silva et al. 2010
Vegetable oil	Oscillatory flow reactor	Transesterification of oil and methanol	Sodium hydroxide	Harvey et al. 2003
Palm oil	Microtubular reactor	Transesterification of Palm oil and methanol	Potassium hydroxide	Kaewchada et al. 2016

Table 3 contd. ...

...Table 3 contd.

Feedstock	Type of Microreactor	Reaction in Reactor	Catalyst	Reference
Waste cooking oil	Microtubular reactor	Transesterification of oil with methanol	Potassium hydroxide	Tanawannapong et al. 2013
Rapeseed oil and cottonseed oil	Capillary microreactor	Transesterification of oil with methanol	Potassium hydroxide	Sun et al. 2008
Cottonseed oil	Microstructured reactor	Transesterification of cottonseed oil with methanol	Acid catalyzed reaction	Sun et al. 2010

evolved. These reactors were investigated for the transesterification of waste cooking oil. In the case of the microplate reactor, the fluids were mixed in the channels. Oscillatory baffles were fixed in a tube-shaped reactor for mixing immiscible fluids (Franjo et al. 2018). The residence time of reaction depends on the length and diameter of the reactor. The radial mixing occurs in this reactor due to oscillatory flow. The flow rate is not fixed in this reactor, and oscillatory flow is maintained in the reactor. There are no limitations regarding heat and mass transfer. Thus, the reaction rate is comparatively very efficient over other reactor models. In order to mix well in the reactor, various structures can be incorporated into the reactor such as nozzle obstacles, zig-zag flow obstacles, microchannels or micromixers (Franjo et al. 2018). Circular obstacles were used for enhancing the transesterification of sunflower oil in the microreactor and reported that a higher conversion rate was achieved using microreactor (Santana et al. 2018). Various micromixers such as slit and rectangular inter-digital, T, Q, Y and J-shaped mixers have been used in a microstructured reactor setup for efficient mixing (Liu et al. 2015). The catalyst to methanol ratio is a significant parameter in biodiesel production. Thus, mostly all reactor studies are considered to optimize the molar ratio of methanol and catalyst irrespective of production volume. Microreactor optimization studies also choose the molar ratio of methanol and oil for obtaining optimal proportion for improved biodiesel production (Klinthong et al. 2015). The addition of magnetic particles and disruption of flow for mixing would enhance the transesterification process which yielded higher biodiesel production. Packed-bed based microreactors were also investigated for biodiesel production, many studies were proved that microreactors with packed-bed resulting in enhanced transesterification of microlipids (Ghaffari and Behzad 2018). The improved mass transfer rate was noticed in this reactor and the catalyst which was developed to use in the large-scale packed-bed reactor also can be used to study in the microscale reactor. These are the major advantages of the reactor. The process efficiency and economics evaluation could be easy with the help of a microstructured reactor. The pressure drops developed in this reactor during the transesterification would be higher and this is a critical parameter to be considered in developing the packed-bed reactor at micro-scale level (Gui-xiong et al. 2013). However, temperature monitoring in this reactor is quite easy.

6.4.4 Considerable Factors for Microreactors-Assisted Biodiesel Production

Molar ratio methanol to oil, reaction temperature, size of the microchannel, residence time, mixing, the catalyst used in the reaction and the type of microreactor are the important process variables that affect the biodiesel production in microreactor (Madhawan et al. 2018).

The molar ratio of alcohol to triglyceride is an important parameter during the transesterification reaction. The higher amount of alcohol is required to drive the reaction toward the formation of methyl esters during transesterification. However, a higher amount of alcohol creates huge trouble in product recovery. Many studies have been conducted to find out the ideal molar ratio of alcohol to triglyceride from 6:1 to 48:1 for improved biodiesel production. The maximum conversion of oils into biodiesel was noticed at 12:1 molar ratio, whereas 6:1 resulted in the lowest production. The reversible transesterification of oils takes place above 12:1 molar ratio leads to the poor conversion of oil into biodiesel (Santana et al. 2016).

Microchannel size is also one of the important parameters in deciding the yield of biodiesel. The reduction of microchannel size maximizes the conversion rate. The significant rise in pressure drop occurs due to the smaller microchannel size. The operation of the microreactor is very difficult because of the higher pressure drop and increases the production cost (Madhawan et al. 2018).

Residence time in microreactors is usually lower compared to conventional reactors. The prolonged residence time in microreactors is the possible reason for the higher yield of fatty acid methyl esters. Capillary reactors are proposed to enhance residence time. The higher yield of biodiesel production was noticed in capillary reactors (Santana et al. 2016).

Reaction temperature affects the transesterification of oils. Mostly, methanol was used in the reaction mixture. The boiling point of methanol is 64.7°C. The temperature above 64°C resulted in the evaporation of methanol and resulted in the poor conversion of triglycerides into biodiesel. It was reported that an increase in temperature from 30 to 60°C resulted in an increasing trend of methyl ester yield (Sun et al. 2008).

Mass transfer in the reactor highly depends on the mixing mechanism. This is a crucial parameter affects the yield of methyl esters. Methanol and oil are two immiscible liquids and always separates each other due to the existed higher interfacial forces between two immiscible phases. The different mixers have been used for good mixing in order to drive the reaction rate effectively. The T, J, Q and Y-shaped zig-zag flow mixers were used for efficient mixing, which attributes for maximizing the yield of methyl esters production during the reaction (Liu et al. 2015). Usually, the esterification and dehydrooxygenation reactions take place in these microreactors for biodiesel production.

7. Integration of Production and Purification Setups in a Microreactor for Biodiesel: Pros and Cons

Biodiesel production has been studied using various microreactors. Instead of having setups for biodiesel production and purification separately, the integration of both steps in one setup is likely to be more interesting. The integrated system is hypothesized to maximize the yield for commercial use. Waste edible oils can be converted into biodiesel by using a different catalyst. This is an ecofriendly approach and efficiently, products and unreacted reactants can be recovered simultaneously. For laboratory and industrial scale, integrated systems have developed for production and separation. However, integrated systems for small-scale are not yet fully elucidated for enzyme catalysis esterification reaction. As the cost of enzymes is high, process intensification of biocatalysis reaction for biodiesel production is essential. The ideal integrated system should consist of waste oil conversion, production of lipase, biodiesel production using lipase catalysis, recovery catalyst, purification of biodiesel and removal of unreacted reactants (Klinthong et al. 2015). Biodiesel production and purification shall take place in the integrated small-scale reactor setup. This integrated system is supposed to be used for lipase-assisted transesterification of edible or non-edible oils. The enzyme recovery is easier and can be used the same enzyme again in recycling for biodiesel production in this integrated system. Biodiesel will be characterized using analytical techniques after purification. Analytical instruments can be connected to the integrated system to accelerate the characterization of biodiesel during the production cycle. The major advantages of these integrated systems are easiness to transport and operate, process economics can be optimized and building efficient systems for large-scale is possible only by understanding the mechanism at a smaller scale (Franjo et al. 2018).

However, there is a shift in parameters and accuracy cannot be maintained when scaled up to the laboratory and industrial scale. The disadvantages of the small-scale integrated system are running at the lowest flow rate, higher pressure drop and residence time are shorter. Because of these factors, large-scale production is not achievable. This system can be adopted for small household, restaurant and farms. However, microchip production is still expensive and needs to come up with a cost-effective microchip manufacturing to make a sustainable integrated system for biodiesel production and purification (Mazubert et al. 2014). Mostly, waste cooking oil is used for biodiesel production. The separation of large particles from waster cooking oil is inevitable to make the flow in the microreactor well. The removal of leftover food from cooking oil could increase the overall cost of the process. The enzyme lifetime needs to be prolonged to make the process economically viable (Miljic et al. 2020). The novel reactors' development is essential, and residence time should be improved for improved reaction rate. Enzymes are highly expensive. In order to make a cost-effective system, the crude enzyme can be used for biodiesel production. However, enzyme purity and loading will influence the transesterification process and result in poor production of biodiesel. The reaction should drive to maximize the esterification process to maximize the yield of methyl esters production, which depends on the molar ratio of alcohol to triglycerides. The connecting larger number of microchips in the integrated system would be an easy way for increasing the yield of biodiesel. The mobile integrated system can be arranged at the substrate and raw material production site. This system can be relocated easily. By this approach, transportation costs can be avoided. Sustainable technology for commercial use is not yet developed so far; however, the research on microreactors for biodiesel production is rapidly going on, especially for microalgal biodiesel production. The detailed investigations with novel reactors would definitely introduce the microreactors for laboratory and industrial-scale biodiesel production. Given the details of integrated systems on a smaller scale for biodiesel production with advantages and disadvantages, these systems can contribute to understanding the process efficiency and economics at the microlevel and in turn help design the reactor set up for larger-scale production.

8. Conclusions and Future Perspectives

This chapter critically discusses various steps of biodiesel production from microalgae. It includes microalgae cultivation, harvesting, transesterification of oils into biodiesel, nanomaterials for immobilization and advances in the bioreactors for nanobiotechnology application for biodiesel production. The research studies proved that nanomaterials are highly potential for maximizing the biodiesel production. The green future building concept is also discussed to provide the outline of significant improvement of algal technology and scale-up the process for biofuel production. Microreactors' concept has to be investigated using biocatalysts. To date, significant research has been proved that microreactors enhanced the chemical conversion in small-scale. This microreactor concept would derive to develop an efficient biochemical conversion for biodiesel production.

References

Aci, F.G., E. Molina, A. Reis, G. Torzillo, G.C. Zittelli, C. Sepu and Sesto Fiorentino. 2017. Photobioreactors for the production of microalgae. pp. 1–44. *In*: Microalgae-Based Biofuels and Bioproducts. Elsevier Ltd.

Aghel, Babak, Majid Mohadesi, Sasan Sahraei and Mehrdad Shariatifar. 2017. New heterogeneous process for continuous biodiesel production in microreactors. Canadian Journal of Chemical Engineering 95(7): 1280–87.

Ahmadi, Mohammad H. 2018. Renewable energy harvesting with the application of nanotechnology: A review. International Journal of Energy Research, 1–24.

Alaei, Shervin, Mohammad Haghighi, Javad Toghiani and Behgam Rahmani. 2018. Industrial crops & products magnetic and reusable $MgO/MgFe_2O_4$ nanocatalyst for biodiesel production from sun flower oil: Influence of fuel ratio in combustion synthesis on catalytic properties and performance. Industrial Crops & Products 117: 322–32.

Alhassan, Fatah H., Umer Rashid, Robiah Yunus, Kamaliah Sirat, M. Ibrahim and Yun Hin Taufiqyap. 2015. Synthesis of ferric–manganese doped tungstated zirconia nanoparticles as heterogeneous solid superacid catalyst for biodiesel production from waste cooking oil synthesis of ferric–manganese doped tungstated zirconia nanoparticles as heterogeneous soli. International Journal of Green Energy 12(9): 987–94.

Aminul, Muhammad, Kirsten Heimann and Richard J. Brown. 2017. Microalgae biodiesel: Current status and future needs for engine performance and emissions. Renewable and Sustainable Energy Reviews 79: 1160–70.

Antonio, Danilo, Oscar A.N. Santisteban Adriano De Vasconcellos, Alex Silva, Donato A.G. Aranda, Marcus Vinicius, Christian Jaeger and José G. Nery. 2018. Metallo-stannosilicate heterogeneous catalyst for biodiesel production using edible, non-edible and waste oils as feedstock. Journal of Environmental Chemical Engineering 6(4): 5488–97.

Ashraful, A.M., H.H. Masjuki, M.A. Kalam, I.M. Rizwanul Fattah, S. Imtenan, S.A. Shahir and H.M. Mobarak. 2014. Production and comparison of fuel properties, engine performance, and emission characteristics of biodiesel from various non-edible vegetable oils: A review. Energy Conversion and Management 80: 202–28.

Baroutian, Saeid, Mohamed K. Aroua, Abdul Aziz A. Raman and Nik M.N. Sulaiman. 2011. A packed bed membrane reactor for production of biodiesel using activated carbon supported catalyst. Bioresource Technology 102(2): 1095–1102.

Baskar, G., I. Aberna Ebenezer Selvakumari and R. Aiswarya. 2018. Biodiesel production from castor oil using heterogeneous Ni doped ZnO nanocatalyst. Bioresource Technology 250: 793–98.

Bayat, Arash, Majid Baghdadi and Gholamreza Nabi Bidhendi. 2018. Tailored magnetic nano-alumina as an efficient catalyst for transesterification of waste cooking oil: Optimization of biodiesel production using response surface methodology. Energy Conversion and Management 177: 395–405.

Bi, Yicheng, Hua Zhou, Honghua Jia and Ping Wei. 2017. A flow-through enzymatic microreactor immobilizing lipase based on layer-by-layer method for biosynthetic process: Catalyzing the transesterification of soybean oil for fatty acid methyl ester production. Process Biochemistry 54: 73–80.

Biswas, Anindita. 2019. Nanotechnology in biofuels production: A Novel Approach for Processing and Production of Bioenergy.

Boehm, Christian R., Paul S. Freemont and Oscar Ces. 2013. Design of a prototype flow microreactor for synthetic biology *in vitro*. Lab on a Chip 70: 3426–32.

Chan, Edwin H.W., Queena K. Qian and Patrick T.I. Lam. 2009. The market for green building in developed asian cities—the perspectives of building designers. Energy Policy 37: 3061–70.

Chaumont, Daniel. 1993. Biotechnology of algal biomass production: A review of systems for outdoor mass culture. Journal of Applied Phycology, 593–604.

Chen, Guanyi, Jing Liu, Jingang Yao, Yun Qi and Beibei Yan. 2017. Biodiesel production from waste cooking oil in a magnetically fluidized bed reactor using whole-cell biocatalysts. Energy Conversion and Management 138: 556–64.

Chen, Jiaxin, Ji Li, Wenyi Dong, Xiaolei Zhang, Rajeshwar D. Tyagi, Patrick Drogui and Rao Y. Surampalli. 2018. The potential of microalgae in biodiesel production. Renewable and Sustainable Energy Reviews 90: 336–46.

Cubides-roman, Diana C., Victor Haber, Heizir F. De Castro, Carlos E. Orrego, Oscar H. Giraldo, Euripedes Garcia and Geraldo F. David. 2017. Ethyl esters (biodiesel) production by Pseudomonas fluorescens lipase immobilized on chitosan with magnetic properties in a bioreactor assisted by electromagnetic field. Fuel 196: 481–87.

Devi, M. Prathima, Y.V. Swamy and S. Venkata Mohan. 2013. Nutritional mode influences lipid accumulation in microalgae with the function of carbon sequestration and nutrient supplementation. Bioresource Technology 142: 278–86.

Dubé, M.A., A.Y. Tremblay and J. Liu. 2007. Biodiesel production using a membrane reactor. Bioresource Technology 98(3): 639–47.

Elrayies, Ghada Mohammad. 2018. Microalgae: Prospects for greener future buildings. Renewable and Sustainable Energy Reviews 81: 1175–91.

Franjo, Mladen, Š. Anita and Bruno Zeli. 2018. Microstructured devices for biodiesel production by transesterification. Biomass Conversion and Biorefinery, 1005–20.

Freedman, B., E.H. Pryde, T.L. Mounts and Northern Regional. 1984. Variables affecting the yields of fatty esters from transesterified vegetable oils. Journal of the American Oil Chemists Society 61(10): 1638–43.

Gardy, Jabbar, Amin Osatiashtiani, Oscar Céspedes, Ali Hassanpour, Xiaojun Lai, Adam F. Lee, Karen Wilson and Mohammad Rehan. 2018. A magnetically separable SO_4/Fe-Al-TiO_2 solid acid catalyst for biodiesel production from waste cooking oil. Applied Catalysis B: Environmental 234: 268–78.

Ghaffari, Abolfazl and Mahdi Behzad. 2018. Facile synthesis of layered sodium disilicates as efficient and recoverable nanocatalysts for biodiesel production from rapeseed oil. Advanced Powder Technology 29(5): 1265–71.

Ghaffari, Nader, Sharifah Bee, Abd Hamid and Taraneh Mihankhah. 2015. Eco-friendly biodiesel production from waste olive oil by transesterification using nano-tube TiO_2. International Conference of Social Science, Medicine and Nursing (SSMN-2015), 161–63.

Gkantzou, Elena, Michaela Patila and Haralambos Stamatis. 2018. Magnetic microreactors with immobilized enzymes—from assemblage to contemporary applications. Catalysts 8(7): 282.

Guan, Guoqing, Marion Teshima, Chie Sato, Sung Mo Son, Muhammad Faisal Irfan and Katsuki Kusakabe. 2010. Two-phase flow behavior in microtube reactors during biodiesel production from waste cooking oil. Environmental and Energy Engineering 56(5): 1383–90.

Gui-xiong Zhou, Guan-yi Chen and Bei-bei Yan. 2013. Biodiesel production in a magnetically-stabilized, fluidized bed reactor with an immobilized lipase in magnetic chitosan microspheres. Biotechnology Letters 36: 63–68.

Günerken, E., E.D. Hondt, M.H.M. Eppink, L. Garcia-gonzalez, K. Elst and R.H. Wijffels. 2015. Cell disruption for microalgae biorefineries. Biotechnology Advances 33: 243–60.

Halim, Ronald, Michael K. Danquah and Paul A. Webley. 2012. Extraction of oil from microalgae for biodiesel production: A review. Biotechnology Advances 30: 709–32.

Hallmann, Armin. 2016. Algae biotechnology-green cell-factories on the rise. Current Biotechnology 4: 389–415.

Harun, Razif and Michael K. Danquah. 2011. Influence of acid pre-treatment on microalgal biomass for bioethanol production. Process Biochemistry 46: 304–9.

Harvey, Adam P., Malcolm R. Mackley and Thomas Seliger. 2003. Process intensification of biodiesel production using a continuous oscillatory flow reactor. Journal of Chemical Technology and Biotechnology 78: 338–41.

Hong, Chung, Chun-yen Chen, Pau Loke, Tau Chuan, Hon Loong, Duu-jong Lee and Jo-shu Chang. 2016. Strategies for Enhancing lipid production from indigenous microalgae isolates. Journal of the Taiwan Institute of Chemical Engineers 63: 189–94.

Huang, Xiao Jun, An Guo Yu and Zhi Kang Xu. 2008. Covalent immobilization of lipase from *Candida rugosa* onto poly(Acrylonitrile-Co-2-Hydroxyethyl Methacrylate) electrospun fibrous membranes for potential bioreactor application. Bioresource Technology 99(13): 5459–65.

Ingle, Avinash P., Priti Paralikar, Silvio Silverio and Mahendra Rai. 2018. Nanotechnology-based developments in biofuel production: Current trends and applications. pp. 289–305. *In*: Sustainable Biotechnology-Enzymatic Resources of Renewable Energy.

Janssen, Marcel, Johannes Tramper, Luuc R. Mur and H. Wijffels. 2002. Enclosed outdoor photobioreactors: Light regime, photosynthetic efficiency, scale-up, and future prospects. Biotechnology and Bioengineering 81: 193–210.

Joshi, Rushang M. and Michael J. Pegg. 2007. Flow properties of biodiesel fuel blends at low temperatures. Fuel 86: 143–51.

Kaewchada, Amaraporn, Siriluck Pungchaicharn and Attasak Jaree. 2016. Transesterification of palm oil in a microtube reactor. Canadian Journal of Chemical Engineering 94(5): 859–64.

Kecskemeti, Adam and Attila Gaspar. 2018. Particle-based immobilized enzymatic reactors in micro fluidic chips. Talanta 180: 211–28.

Kim, Jungmin, Gursong Yoo, Hansol Lee, Juntaek Lim, Kyochan Kim, Chul Woong, Min S. Park and Ji-won Yang. 2013. Methods of downstream processing for the production of biodiesel from microalgae. Biotechnology Advances 31: 862–76.

Kim, Kyoung-hee and D. Ph. 2013. Beyond green: Growing algae facade. In ARCC 2013. The Visibility of Research Sustainability: Visualization Sustainability and Performance, pp. 500–505.

Klinthong, Worasaung, Yi-hung Yang, Chih-hung Huang and Chung-sung Tan. 2015. A review: Microalgae and their applications in CO_2 capture and renewable energy. Aerosol and Air Quality Research 15: 712–42.

Knothe, Gerhard. 2006. Analyzing Biodiesel: Standards and Other Methods 8(10): 823–33.

Kolb, Gunther. 2013. Review: Microstructured reactors for distributed and renewable production of fuels and electrical energy. Chemical Engineering & Processing: Process Intensification 65: 1–44.

Krishna, B. Murali and J.M. Mallikarjuna. 2009. Properties and performance of cotton seed oil-diesel blends as a fuel for compression ignition engines properties and performance of cotton seed oil-diesel. Journal of Renewable and Sustainable Energy 023106.

Lai, C.M.T., H.B. Chua, Z. Helwani and W. Fatra. 2018. IOP Conf. Series: Materials science and engineering. In solid catalyst nanoparticles derived from Oil-Palm Empty Fruit Bunches (OP-EFB) as a renewable catalyst for biodiesel production solid catalyst nanoparticles derived from Oil-Palm Empty Fruit Bunches (OP-EFB) as a Renewable Catalyst for Biodiesel Prod.

Lam, Man Kee and Keat Teong Lee. 2012. Microalgae biofuels: A critical review of issues, problems and the way forward. Biotechnology Advances 30: 673–90.

Leong, Wai-hong, Jun-wei Lim, Man-kee Lam, Yoshimitsu Uemura and Yeek-chia Ho. 2018. Third generation biofuels: A nutritional perspective in enhancing microbial lipid production. Renewable and Sustainable Energy Reviews 91: 950–61.

Liu, Jiao, Yadong Chu, Xupeng Cao, Yuchao Zhao, Hua Xie and Song Xue. 2015. Rapid transesterification of micro-amount of lipids from microalgae via a micro-mixer reactor. Biotechnology for Biofuels 8(1): 4–11.

Machsun, Achmadin Luthfi, Misri Gozan, Mohammad Nasikin, Siswa Setyahadi and Young Je Yoo. 2010. Membrane microreactor in biocatalytic transesterification of triolein for biodiesel production. Biotechnology and Bioprocess Engineering 15(6): 911–16.

Madhawan, Akansha, Arzoo Arora, Jyoti Das, Arindam Kuila and Vinay Sharma. 2018. Microreactor technology for biodiesel production: A review. Biomass Conversion and Biorefinery 8(2): 485–96.

Mahmudul, H.M., F.Y. Hagos, R. Mamat, A. Abdul Adam, W.F.W. Ishak and R. Alenezi. 2017. Production, characterization and performance of biodiesel as an alternative fuel in diesel engines—a review. Renewable and Sustainable Energy Reviews 72: 497–509.

Mazubert, Alex, Joelle Aubin, Sébastien Elgue and Martine Poux. 2014. Intensification of waste cooking oil transformation by transesterification and esterification reactions in oscillatory baffled and microstructured reactors for biodiesel production. Green Processing and Synthesis 3(6): 419–29.

Miao, Changlin, Lingmei Yang, Zhongming Wang, Wen Luo and Huiwen Li. 2018. Lipase immobilization on amino-silane modi Fi Ed superparamagnetic Fe_3O_4 nanoparticles as biocatalyst for biodiesel production. Fuel 224: 774–82.

Mihankhah, Taraneh, Mohammad Delnavaz and Nader Ghaffari Khaligh. 2018. Application of TiO_2 nanoparticles for eco-friendly biodiesel production from waste olive oil. International Journal of Green Energy 15: 69–75.

Miljic, Goran, Marina Tišma, Smitha Sundaram, Volker Hessel and Sandra Budz. 2017. Is there a future for enzymatic biodiesel industrial production in microreactors? Applied Energy 201: 124–34.

Miron, Asterio Sanchez, Francisco Garcia Comacho, Emilio Molina Grima and Yusuf Chisti. 1999. Comparative evaluation of compact photobioreactors for large-scale monoculture of microalgae. Journal of Biotechnology 70: 249–70.

Miyamoto, K., O. Wable and J.R. Benemann. 1988. Vertical tubuiar reactor for microalgae cultivation. Biotechnology Letters 10: 703–8.

Mohan, S. Venkata, M. Prathima Devi, G. Mohanakrishna, N. Amarnath, M. Lenin Babu and P.N. Sarma. 2011. Bioresource technology potential of mixed microalgae to harness biodiesel from ecological water-bodies with simultaneous treatment. Bioresource Technology 102: 1109–17.

Nayebzadeh, Hamed, Mohammad Haghighi and Naser Saghatoleslami. 2018. Fabrication of carbonated alumina doped by calcium oxide via microwave combustion method used as nanocatalyst in biodiesel production: Influence of carbon source type. Energy Conversion and Management 171: 566–75.

Perez-garcia, Octavio, Froylan M.E. Escalante, E. Luz and Yoav Bashan. 2010. Heterotrophic cultures of microalgae: Metabolism and potential products. Water Research 45(1): 11–36.

Prabakaran, P. and A.D. Ravindran. 2011. A comparative study on effective cell disruption methods for lipid extraction from microalgae. Letters in Applied Microbiology 53: 150–54.

Qiu, Fengxian, Yihuai Li, Dongya Yang, Xiaohua Li and Ping Sun. 2011. Heterogeneous solid base nanocatalyst: Preparation, characterization and application in biodiesel production. Bioresource Technology 102: 4150–56.

Quitain, Armando T., Elaine G. Mission, Yoshifumi Sumigawa and Mitsuru Sasaki. 2018. Supercritical carbon dioxide-mediated esterification in a microfluidic reactor. Chemical Engineering and Processing: Process Intensification 123: 168–73.

Ranjan, Alok, S.S. Dawn, J. Jayaprabakar, N. Nirmala, K. Saikiran and S. Sai Sriram. 2018. Experimental investigation on effect of MgO nanoparticles on cold flow properties, performance, emission and combustion characteristics of waste cooking oil biodiesel. Fuel 220: 780–91.

Salinas, Daniela, Catherine Sepúlveda, Néstor Escalona, J.L. Gfierro and Gina Pecchi. 2018. Sol-gel La_2O_3–ZrO_2 mixed oxide catalysts for biodiesel production. Journal of Energy Chemistry 27(2): 565–72.

Santana, Harrson S., Deborah S. Tortola, Érika M. Reis, João L. Silva and Osvaldir P. Taranto. 2016. Transesterification reaction of sunflower oil and ethanol for biodiesel synthesis in microchannel reactor: Experimental and simulation studies. Chemical Engineering Journal 302: 752–62.

Santana, Harrson S., João L. Silva, Deborah S. Tortola and Osvaldir P. Taranto. 2018. Transesterification of sunflower oil in microchannels with circular obstructions. Chinese Journal of Chemical Engineering 26(4): 852–63.

Sarno, Maria and Mariagrazia Iuliano. 2018. Highly active and stable Fe_3O_4/Au nanoparticles supporting lipase catalyst for biodiesel production from waste tomato. Applied Surface Science 474: 135–146.

Schürer, Jochen, Richard Thiele, Ole Wiborg and Athanassios Ziogas. 2014. Synthesis of biodiesel in microstructured reactors under supercritical reaction conditions. Chemical Engineering Transactions 37: 541–46.

Shokuhi, Ali, Rad Mahtab, Hoseini Nia, Fatemeh Ardestani and Hamed Nayebzadeh. 2018. Esterification of waste chicken fat: Sulfonated MWCNT toward biodiesel production. Waste and Biomass Valorization 9(4): 591–99.

Silva, Camila Da, Fernanda De Castilhos, J. Vladimir Oliveira and Lucio Cardozo Filho. 2010. Continuous production of soybean biodiesel with compressed ethanol in a microtube reactor. Fuel Processing Technology 91(10): 1274–81.

Singh, R.N. and Shaishav Sharma. 2012. Development of suitable photobioreactor for algae production—a review. Renewable and Sustainable Energy Reviews 16(4): 2347–53.

Sreekumar, Nidhin, M.S. Giri Nandagopal, Aneesh Vasudevan, Rahul Antony and N. Selvaraju. 2016. Marine microalgal culturing in open pond systems for biodiesel production—critical parameters. Journal of Renewable and Sustainable Energy 8(2).

Sun, Juan, Jingxi Ju, Lei Ji, Lixiong Zhang and Nanping Xu. 2008. Synthesis of biodiesel in capillary microreactors. Industrial and Engineering Chemistry Research 47(5): 1398–1403.

Sun, Peiyong, Juan Sun, Jianfeng Yao, Lixiong Zhang and Nanping Xu. 2010. Continuous production of biodiesel from high acid value oils in microstructured reactor by acid-catalyzed reactions. Chemical Engineering Journal 162(1): 364–70.

Tan, Xiaoyao and K. Li. 2013. Membrane microreactors for catalytic reactions. Journal of Chemical Technology and Biotechnology 88(10): 1771–79.

Tanawannapong, Yuttapong, Amaraporn Kaewchada and Attasak Jaree. 2013. Biodiesel production from waste cooking oil in a microtube reactor. Journal of Industrial and Engineering Chemistry 19(1): 37–41.

Trathnigg, Bernd and Martin Mittelbach. 1990. Analysis of triglyceride methanolysis mixtures using isocratic hplc with density detection. Journal of Liquid Chromatography 13: 95–102.

Verma, Madan L., Munish Puri, Colin J. Barrow, Madan L. Verma, Munish Puri and Colin J. Barrow. 2016. Recent trends in nanomaterials immobilised enzymes for biofuel production recent trends in nanomaterials immobilised enzymes for biofuel production. Critical Reviews in Biotechnology 36: 108–119.

Verma, Madan Lal, Colin J. Barrow and Munish Puri. 2013. Nanobiotechnology as a novel paradigm for enzyme immobilisation and stabilisation with potential applications in biodiesel production. Applied Microbiology and Biotechnology 97: 23–39.

Vuppaladadiyam, Arun K., Pepijn Prinsen, Abdul Raheem, Rafael Luque and Ming Zhao. 2018. Microalgae cultivation and metabolites production. Biofuels Bioproducts and Biorefining 12: 304–24.

Wang, Xia, Peipei Dou, Peng Zhao, Chuanming Zhao, Yi Ding and Ping Xu. 2009. Immobilization of lipases onto magnetic Fe_3O_4 nanoparticles for application in biodiesel production. Chemistry & Sustainability Energy & Materials 10: 947–50.

Wang, Xia, Xueying Liu, Chuanming Zhao, Yi Ding and Ping Xu. 2011. Biodiesel production in packed-bed reactors using lipase—nanoparticle biocomposite. Bioresource Technology 102(10): 6352–55.

Wayne, Kit, Shir Reen, Pau Loke, Yee Jiun and Tau Chuan. 2018. Effects of water culture medium, cultivation systems and growth modes for microalgae cultivation: A review. Journal of the Taiwan Institute of Chemical Engineers 91: 332–44.

Wen, Zhenzhong, Xinhai Yu, Shan Tung Tu, Jinyue Yan and Erik Dahlquist. 2009. Intensification of biodiesel synthesis using zigzag micro-channel reactors. Bioresource Technology 100(12): 3054–60.

Xie, Wenlei, Yuxiang Han and Shuangna Tai. 2017. Biodiesel production using biguanide-functionalized hydroxyapatite-encapsulated-γ-Fe_2O_3 nanoparticles. Fuel 210: 83–90.

Xie, Wenlei, Yuxiang Han and Hongyan Wang. 2018. Magnetic Fe_3O_4/MCM-41 composite-supported sodium silicate as heterogeneous catalysts for biodiesel production. Renewable Energy 125: 675–81.

Xu, Wei, Lijing Gao and Guomin Xiao. 2015. Biodiesel production optimization using monolithic catalyst in a fixed-bed membrane reactor. Fuel 159: 484–90.

Yousefi, Sina, Mohammad Haghighi and Behgam Rahmani. 2018. Facile and efficient microwave combustion fabrication of Mg-spinel as support for MgO nanocatalyst used in biodiesel production from sunflower oil: Fuel type approach. Chemical Engineering Research and Design 138: 506–18.

Zhou, Lin and Adeniyi Lawal. 2014. Evaluation of presulfided NiMo/γ-Al_2O_3 for hydrodeoxygenation of microalgae oil to produce green diesel. Energy and Fuels 29: 262–272.

Chapter 12

Impact of Nanotechnology in Biorefineries

An Overview

Rekha Kushwaha,[1,#,*] *Santosh Kumar*,[1,#] *Balraj Singh Gill*,[2] *Navgeet*[3] and *Madan Lal Verma*[4]

1. Introduction

With a growing global population, there is a high rise in global demand for food, energy and water. To meet the growing energy demand of the world, the concept of the biorefinery is one of the promising technological solutions (Zhu 2015). Biorefinery produces a spectrum of marketable products (food, feed, materials and chemicals) and energy (fuels, power and/or heat). Presently, the focus of biorefinery systems is on the production of transportation biofuel. Biomass from forestry, agriculture, aquaculture and residues from industry can be used in a biorefinery. A biorefinery is not a new concept, many of the traditional biomass converting technologies such as the sugar, starch and pulp and paper industry can be (partly) considered as biorefineries.

The term "biorefinery" appeared in 1990 in response to various industry trends. These industrial trends include increased awareness in industry of the need to use biomass resources in a more rational way both economically and environmentally, increased attention to the production of starch for energy applications growing interest in upgrading additional low-quality lignocellulosic biomass to valuable merchandise, a perceived need to develop more high-value products and diversify the product mix

[1] Department of Biochemistry, University of Missouri Columbia MO (USA)-65211.
[2] Department of Higher Education, Himachal Pradesh, Shimla, India.
[3] Department of Biotechnology, Kanya MahaVidyalaya, Jalandhar, Punjab, India.
[4] Department of Biotechnology, School of Basic Sciences, Indian Institute of Information Technology Una, Himachal Pradesh-177220, India.
* Corresponding author: kushwahar@missouri.edu
# Equal contribution

in order to meet global competition and, in some cases, utilize an excess of biomass (especially in the pulp and paper industry). The International Energy Agency (IEA) Bioenergy Task 42 and NERL (National Renewable Energy Laboratory) defined the biorefinery. According to IEA Bioenergy Task 42, the biorefinery is co-production of fuels, chemicals power and material from biomass (IEA Bioenergy Task 42 Biorefinery 2013) and biorefining is sustainable processing of biomass into a spectrum of marketable merchandise and energy (Bell et al. 2014). The definition by NERL says "biorefinery is a facility that integrates biomass conversion processes and equipment to produce fuels, power and chemicals from biomass" (Berntsson et al. 2014). There are several types of biorefinery classifications based on many factors, such as the technical implementation status (conventional and advanced biorefineries; first, second and third-generation biorefineries); type of raw materials used (full-culture biorefineries, oleochemical biorefineries and lignocellulose biorefineries); technological implementation status (conventional and advanced biorefineries; first, second, and third-generation biorefineries); type of raw materials used (whole crop biorefineries, oleochemical biorefineries, lignocellulosic feedstock biorefineries, green biorefineries and marine biorefineries); type of main intermediates produced (syngas platform biorefineries and sugar platform biorefineries) and the main type of conversion processes applied (thermochemical biorefineries, biochemical biorefineries and two-platform concept biorefineries) (Kamm and Kamm 2004a, Kamm and Kamm 2004b, van Ree et al. 2007, Benedé 2014).

In order to make broad biorefinery area more accessible for different stakeholders, to improve the overall understanding of the advantages of biorefinery processing over single-product processes and to accelerate the final market implementation of these concepts into global bio-based economies, there is still a need of a clear system to classify the different biorefinery concepts.

IEA Bioenergy Task 42 classified biorefineries based on platforms, products, feedstocks, and processes. The naming of a biorefinery system consists of the number and name of the platform(s), product(s), feedstock(s) and optionally the processes involved (Bell et al. 2014). The platforms are intermediates that can connect different biorefinery systems and their processes (e.g., C5/C6 sugars, syngas and biogas). Platforms can also be the final product. The number of involved platforms is an indication of the system complexity (Anon 2017). The feedstock is of two kinds: (i) 'energy crops' from agriculture or first-generation products and (ii) 'biomass residues' also known as second-generation products. First-generation feedstocks use sugar-rich or starch-rich crops (sugar cane, sugar beet or sweet sorghum store large amounts of saccharose) which can easily be extracted from the plant material for subsequent fermentation to ethanol or bio-based chemicals (The Royal Society 2008, Bell et al. 2014, Naik et al. 2010, Eijck et al. 2014, Eijck et al. 2014, Hahn-Hagerdal et al. 2006). The utilization of readily available food and energy crops for the production of bio-derived liquid fuels in the first-generation biorefinery is responsible for significantly higher pressure on agriculture (Stöcker 2008). Therefore, to solve this problem, the second-generation of the biorefinery is being developed to utilize all types of non-food biomass, including waste biomass, waste food and lignocelluloses materials. Second-generation feedstocks include agricultural biomass processing residual non-food parts of current crops or other non-food sources, such as straw, bark, wood chips from forest residues, used cooking oils, waste streams from biomass. Lignocellulosic biomass, jatropha oil, microalgae unicellular photo-organisms and heterotrophic organisms and animal wastes are a few examples of the second-generation feedstocks (Naik et al. 2010, Eijck and Henny 2008, Eijck et al. 2014, Chisti 2008, Greenwell et al. 2010, Smith et al. 2010, Alina Mariana Balu et al. 2012, Fitz Patrick et al. 2010). The structure of non-food lignocellulosic biomaterials is complex and contains relatively inert components like crystalline cellulose, but the conversion of lignocellulosic biomaterial into fuels and chemicals in a continuous process is extremely difficult and requires low-cost advance technology (Shuttleworth et al. 2014, Kleinert and Barth 2008, Kobayashi et al. 2010).

Four conversion-processing techniques depending on the feedstock and the desired output are applied in a biorefinery. These techniques include biochemical (e.g., fermentation and enzymatic conversion), thermochemical (e.g., gasification and pyrolysis), chemical (e.g., acid hydrolysis, synthesis and esterification) and mechanical processes (e.g., fractionation, pressing and size reduction) (Anon 2017). These methods allow for the processing of waste feedstock with resulting output can be classed in high-volume low-value products (fuels) and higher value lower volume chemicals (waxes, succinic acid, sorbitol and glycerol) (Shuttleworth et al. 2014).

There are many pros and cons associated with these techniques, such as biochemical technology; although it is very energy-efficient, it is also time-consuming and leaves lignin as waste. Thermochemical processing uses heat to break down the starting material and thus is responsible for the production of a wide range of chemicals and fuels (Bridgwater 2006, Mohan et al. 2006). But high capital input, significant energy demand and high operational costs are the drawbacks associated with thermochemical processing. Compared to biochemical processing, thermochemical processing is less selective but is still highly advantageous as it poses restrictions on the type of feedstock (Gomez et al. 2008, Sun and Cheng 2002, Bridgwater 2006). Another technique used in biorefineries is pyrolysis, but the pyrolysis products face challenges in terms of fuel applications due to unfavorable properties, such as high acidity, high water and alkali metal contents of the resulting products making them difficult to exploit as fuel (Mohan et al. 2006).

Therefore, these processing methods have been modified to improve the product as well as the yield of the desired product with low inputs and wastes. A combination of biorefinery techniques with nanotechnology is one such example. This combination not only maximizes the range of attainable products but also the individual yields (Balu et al. 2011, West and Halas 2000).

Nanotechnology, since its introduction by Nobel laureate Richard P. Feynman during his well-famous 1959 lecture "There's Plenty of Room at the Bottom" (Feynman 1960), has gone through various revolutionary developments in the field of science and technology (Khan et al. 2017). Nanotechnology is science, engineering and technology conducted at the nanoscale, i.e., 1 to 100 nanometers. National Nanotechnology Initiative defines "nanotechnology as the manipulation of matter with at least one dimension sized from one to 100 nanometers". In other words, nanotechnology produces material of various types at the nanoscale level called nanoparticles. Nanoparticles (NPs) are a wide class of materials that include particulate substances that have one dimension less than 100 nm at least (Laurent et al. 2010). There are different kinds of nanoparticles and these include carbon-based metal, ceramics, semiconductor, polymeric and lipid-based nanoparticles (Bhatia 2016, Khan et al. 2017, Astefanei et al. 2015, Rosarin and Mirunalini 2011, Sigmund et al. 2006, Ali et al. 2017, Khan et al. 2017, Mansha et al. 2017, Campani et al. 2018). These nanoparticles are composed of the three-layer surface layer (functionalized with a variety of small molecules, metal ions, surfactants and polymers), the shell layer (chemically different material from the core) and core is the central portion of the nanoparticle (Shin et al. 2016). Nanoparticles have a surrounding interfacial layer. These nanoparticles because of their tailored porosity and surface chemistry, high specific surface area, large pore volume, and high thermal, chemical and mechanical stabilities are getting much attention in various fields (Christian et al. 2008, Frenkel et al. 2011, Büscher et al. 2004, Malik et al. 2002, Mori and Hegmann 2016). Therefore, being used in drug delivery, chemical and biological sensing, gas sensing, CO_2 capturing and other related applications (Barrak et al. 2016, Ramacharyulu et al. 2015, Mansha et al. 2016, Rawal and Kaur 2013, Ullah et al. 2017, Ganesh et al. 2017, Lee et al. 2011, Shaalan et al. 2016, Rizvi and Saleh 2017, Jong and Borm 2008, Kumar et al. 2017, Rajiv et al. 2014, Thomas et al. 2015) Figure 1.

The use of these nanoparticles as heterogeneous catalysts in the biorefinery has a great potential to produce biofuel. These nanoparticles catalyze and influence the nature of the products and their distribution by modifying the chemical processes and thus making the process more efficient and robust (Xue et al. 2018, Christian et al. 2008, West and Halas 2000, Balu et al. 2011). Thus, the use of nanotechnology and nanomaterials in the biorefinery has emerged as a promising tool to improve biofuel production (Serrano et al. 2009, Sekhon 2014).

Researchers are continuously working to design and fabricate nanomaterials as biofuel and energy-related applications (Nizami and Rehan 2018). Therefore, this chapter summarizes the previous studies carried out on the use of these nanoparticles

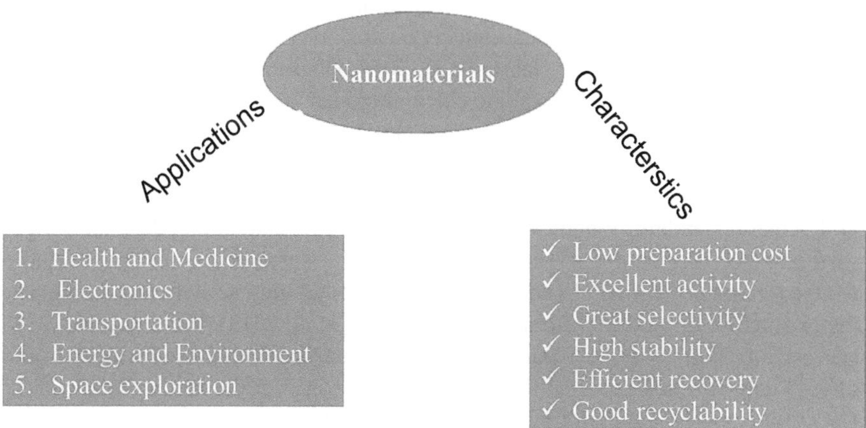

Figure 1. Applications and characteristics of the nanomaterials.

as a heterogeneous catalyst in various biorefineries, including biodiesel and bioethanol production in first and second-generation refineries, biogas biorefineries as well as the microalgal biorefineries to enhance yield, increased catalysts recycling and better product selectivity. Also, it gives a brief overview of the future aspects of nanotechnology in a biorefinery.

2. Nanotechnology and Biodiesel Production

Biodiesel is an oxygenated biofuel, also known as fatty acid alkyl esters. Due to biodegradability, bio-renewable nature, very low sulfur content and toxicity, low volatility/flammability, good transport and storage properties, higher cetane number and its salutary atmospheric CO_2 balance make biodiesel an advantageous alternative to fossil diesel fuel (Reddy et al. 2006, Toda et al. 2005, Canakci and Gerpen 2003). It is obtained through the transesterification reaction of triglycerides from vegetable oils and animal fats with short-chain alcohol (i.e., methanol or ethanol) in the presence of a catalyst (homogenous or heterogeneous) with glycerol as a product (Botero et al. 2017). Various kind of transesterification processes involved in the biodiesel production includes alkaline-catalyzed, acid-catalyzed, enzyme-catalyzed, ultrasonic irradiation and supercritical transesterification process (Math et al. 2010, Freedman et al. 1986, Marchetti et al. 2007, Bibalan and Sadrameli 2012, Ranganathan et al. 2008, Oda et al. 2005, Liu et al. 2012, Sarkar and Bhattacharyya 2012, Helwani et al. 2009, Chitra et al. 2005). Nowadays, different kinds of nanoparticles are being used as a catalyst during the conversion process.

2.1 Nanocatalyst Used in First-Generation Biorefinery

Biodiesel produced by soybean, sunflower, canola, rapeseed oil, palm oil or other vegetable oils is also considered to be a first-generation biofuel. In this section, we summarize the various nanocatalyst used in the transesterification process while producing biodiesel from food crops (Table 1).

Due to its high catalytic activity, low-cost and high-base strength calcium oxide (CaO) catalyst has been studied extensively for transesterification reactions by many researchers (Liu et al. 2008, Kamata and Sugahara 2017, Venkatesh et al. 2018, Reddy et al. 2006, Zhao et al. 2013, Hassan et al. 2018, Degirmenbasi et al. 2015). Reddy et al. (2006) used the nanocrystalline calcium oxides as catalysts for the production of environmentally compatible biodiesel fuel with high yield (99% of conversion) at room temperature by using soybean oil and poultry fat as a raw material. Obtained biodiesel exhibited a sulfated ash value of 0.020–0.004%, which meets the ASTM D-874 diesel standard. When two kinds of nanopowder calcium oxides, i.e., a higher surface area (HSA nano-CaO) and a moderate surface area (nano-CaO) were studied for the transesterification of canola oil, they exhibited high activity and stability for the production of biodiesel from canola oil (Zhao et al. 2013). Yulianti et al. (2014) successfully synthesized CaOZnO nanoparticle catalysts with Ca to Zn atomic ratios of 0.08 and 0.25 and reported that conversion of triglyceride, methyl ester yield of $CaOZnO_{0.08}$ are higher than $CaOZnO_{0.25}$ due to the increased surface area of smaller crystal.

Table 1. Nanocatalyst used in first generation biorefinery.

S. No.	Nanocatalyst Used	Yield (%)	Plant Source	Reference
1.	CaO	99	Soybean oil	(Reddy et al. 2006)
2.	CaO	99.85	Canola oil	(Zhao et al. 2013)
3.	CaO	97.67	Canola oil	(Degirmenbasi et al. 2015)
4.	CaO-based/Au	89–97	Sunflower oil	(Bet-Moushoul et al. 2016)
5.	CaOZnO	93.5	Refined Palm oil	(Yulianti et al. 2014)
6.	Synthesized mesoporous 30KF/m-$CaAl_4$(700)-700-3	98	Soybean oil	(Tao et al. 2013)
7.	Calcinated KF-impregnated nanoparticles of γ-Al2O3	97.7	Vegetable oil	(Boz et al. 2009)
8.	MgO	99	Soybean oil	(Wang and Yang 2007)
9.	1. Nanosheets (MgO(I)) 2. Conventionally prepared (MgO(II)) 3. Aerogel prepared (MgO(III)	23–98	Vegetable oil (Soybean oil and Rapeseed oil)	(Verziu et al. 2008)
10.	$MgO/MgAl_2O_4$	95	Sunflower oil	(Rahmani Vahid and Haghighi 2016)
11.	Palladium	–	Soybean oil	(Carvalho et al. 2011)
12.	ZnO	85–93.7	Soybean oil coconut oil	(Sachdeva and Saroj 2013, Yan et al. 2010)
13.	ZnO	96.88	Rapeseed oil	(Kim et al. 2013)
14.	TiO_2–ZnO and ZnO	98	Palm oil	(Madhuvilakku and Piraman 2013
15.	Mn Dopped ZnO	97	Mahua oil	(Baskrar et al. 2017
16.	Li/TiO_2	98	Canola oil	(Alsharifi et al. 2017)
17.	Li/ZnO	96.3	Soybean oil	(Xie et al. 2007)
18.	Magnetic $MgO/MgFe_2O_4$	82.4–91.2	Vegetable oil	(Alaei et al. 2018)
19.	Magnetic CaO/Fe_3O_4	69.7	Palm oil	(Mortadha et al. 2017)
20.	Magnetic Fe_3O_4	94	Vegetable oils	(Xie and Ma 2010)
21.	Magnetic $Cs/Al/Fe_3O_4$	94.8	Sunflower oil	(Feyzi et al. 2013)
22.	Magnetic $CsH_2PW_{12}O_{40}/Fe–SiO_2$	81		(Feyzi et al. 2014)
23.	Magnetic Fe_3O_4/MCM-41 composite-supported sodium silicate	99.2	Soybean oil	(Xie et al. 2018)

Bet-Moushoul et al. (2016) used five different composites of calcium oxide including commercial CaO, eggshell, mussel shell, calcite and dolomite-based nanocatalyst supported on gold nanoparticles (AuNPs) for the biodiesel synthesis from the sunflower oil. This study reported 94–97% and 89–91% of FAME yield.

Compared to the traditional solid basic catalyst CaO, the use of synthesized mesoporous 30KF/m-CaAl4(700)-700-3 catalyst showed higher reaction rate and the turnover frequency value (TOF) in the initial 1 hour with a yield of about 98% in 5 hours and near 100% selectivity for FAME. Calcium leaching amount with 30KF/m-CaAl4(700)-700-3 is lower than that with CaO and meets the EN14214 standard (Tao et al. 2012). A 100% yield of fatty acid monoalkyl esters (FAMEs) was achieved by use of synthesized 20KF/m-Mg2Fe-M composite (KF-loaded mesoporous Mg-Fe bi-metal oxides) under mild reaction conditions, in 1 hour reaction duration, extraordinarily low leaching amounts of K and Mg as well as good recyclability (Tao et al. 2013). This is due to the presence of larger amounts of medium and strong basic sites and uniform pore channels that favor the dispersion of active species and transportation of rapid molecules in mesoporous materials (Hua et al. 2011, Shi 2013). Therefore, this engineered mesostructured provides an efficient catalytic platform for a green biodiesel production process.

A 97.7 ± 2.14% yield of methyl ester was achieved by Boz et al. (2009) when they used the calcinated KF-impregnated nanoparticles of γ-Al_2O_3 as a heterogeneous catalyst for the transesterification of vegetable oil with methanol for the synthesis of biodiesel. This relatively high conversion was due to the relatively high basicity of the catalyst surface and the high surface-to-volume ratio of the nanoparticles of γ-Al_2O_3 (Boz et al. 2009).

In the supercritical/subcritical temperature, nano-MgO had higher catalytic activity. Nano-MgO improves the transesterification reaction of soybean oil with supercritical/subcritical methanol. During the transesterification reaction, a 99% yield was obtained at the stirring rate of 1,000 rpm with 3 wt% nano-MgO (Wang and Yang 2007).

The catalytic activity of three morphologically different nanocrystalline MgO materials, MgO (nanosheets MgO (I)), conventionally prepared MgO (MgO (II)) and aerogel prepared MgO (MgO (III)) were investigated during the transesterification of vegetable oils such as sunflower and rapeseed oils at low temperature. The performances of the nanocatalyst depend on the nature of the treated oil (Verziu et al. 2008). The use of $MgO/MgAl_2O_4$ nanocatalyst during biodiesel production from sunflower oil yielded 95% biodiesel (Rahmani Vahid and Haghighi 2016).

Immobilization of lipase enhances the efficiency of the enzyme as well as its reaction. Easy recovery of immobilized lipase from the transesterification reaction meditated by immobilized lipase for biodiesel production facilitates its repeated use (Dizge and Keskinler 2008, Dizge et al. 2009, Yagiz et al. 2007). Facile and fast separation of immobilized enzymes from the reaction mixture and magnetic nanoparticles are employed as carriers for enzyme immobilization (Shih-Hung et al. 2008, Guo and Sun 2004, Deng et al. 2003). Xie and Ma (2010) immobilized lipase onto magnetic Fe_3O_4 nanoparticles by using 1-ethyl-3-(3-dimethylaminopropyl) carbodiimide (EDAC) as an activating agent and used it as a biocatalyst to catalyze the transesterification of vegetable oils with methanol. This reaction attained the maximal conversion to methyl esters of 94%. This lipase catalyst can be used for three times without any loss of the activity (Xie and Ma 2010).

Palladium supported on carbon catalysts are the most active and selective catalysts, e.g., stearic acid carbon-supported palladium catalyst convert stearic acid completely with > 98% selectivity toward deoxygenated C17 products (Snåre et al. 2006, Simakova et al. 2010). In order to maximize the oxidative stability in soybean oil biodiesel, *in situ* generated palladium nanoparticles in imidazolium-based ionic liquids were used (Carvalho et al. 2011). This catalytic system partially hydrogenates biodiesel into mono-hydrogenated compounds and thus avoided the formation of saturated compounds with higher selectivity. The recovery and reuse of the ionic phase containing the catalyst are possible up to three times without significant loss in its catalytic performance.

Zinc oxide (ZnO) is one of the best transition metal oxide and can be used as nanoparticles for biodiesel production from crude algae oil, corn oil from DDGs, crude palm oil, crude soybean oil, crude coconut oil, waste cooking oil, food-grade soybean oil and food-grade soybean oil (Yan et al. 2010, Viswanatha et al. 2012). The ZnO nanoparticles is an efficient, heterogeneous, reusable and eco-friendly catalyst. The high catalytic activity, long catalyst life, and low leaching properties demonstrate these modified ZnO nanoparticles have potential in a commercial biodiesel production process (Sachdeva and Saroj 2013, Yan et al. 2010). Kim et al. (2013) used zinc nitrate as a source of zinc for the formation of zinc oxide nanoparticles and synthesized the biodiesel by using supercritical methanol method from the transesterification of rapeseed oil.

The use of mixed oxides of TiO_2-ZnO and ZnO nanocatalysts (synthesized by glycerol-nitrate combustion route) as an active and stable catalyst for the biodiesel production from the palm oil showed good catalytic performance (98% transesterification conversion at the optimum reaction) over the ZnO catalyst that could be a potential candidate for the large-scale biodiesel production (Madhuvilakku and Piraman 2013).

Manganese-doped zinc oxide heterogeneous nanocatalyst, when used for the production of biodiesel from Mahua oil, resulted in biodiesel yield of 97% (Baskrar et al. 2017). Lithium-ion is known to induce surface reactivity changes on the nanoparticles (Berger et al. 2007). The insertion of Li-ion improved the catalyst efficiency of TiO_2 without any alteration in structure. When Li/TiO_2 was used as a heterogeneous catalyst (prepared via impregnation method) for biodiesel production from canola oil resulted in 98% of transesterification yield (Alsharifi et al. 2017). Another heterogeneous catalyst Li/ZnO catalysts also prepared by using an impregnation method followed by calcination when used for soybean oil transesterification showed 96.3% soybean oil conversion (Xie et al. 2007).

Cs/Al/Fe_3O_4 magnetic nanocatalysts synthesized by the novel synthesis method showed high catalytic activity for biodiesel production and yielded 94.8% biodiesel under the optimal conditions by the transesterification of sunflower oil (Feyzi et al. 2013). Feyzi et al. (2014) further used the magnetic $CsH_2PW_{12}O_{40}$/Fe–SiO_2 nanocatalysts for biodiesel production from the sunflower oil. A combination of sol-gel and impregnation methods were used for the preparation of the magnetic $CsH_2PW_{12}O_{40}$/Fe–SiO_2 nanocatalysts. After investigations of catalytic performance under the effects of different $H_3PW_{12}O_{40}$/(Fe-SiO_2) weight percentage, loading of Cs as a promotor and calcination conditions, they reported air atmosphere at 600°C for 6 hours with a heating rate of 4°C min^{-1} is the best calcination condition. Thus, it recorded 81% of biodiesel yield under the optimal conditions (Feyzi et al. 2014).

The catalytic transesterification of palm seed oil resulted in 69.7% biodiesel yield and the specifications of generated biodiesel were according to ATSM, when nanocatalyst of CaO supported by Fe_3O_4 magnetic particles (prepared by a chemical precipitation method was used as a heterogeneous catalyst (Mortadha et al. 2017).

Alaei et al. (2018) utilized the magnetic and reusable MgO/$MgFe_2O_4$ nanocatalyst (prepared by combustion method) for biodiesel production from sunflower oil. Under optimum conditions, 82.4–91.2% biodiesel yield was recovered.

Xie et al. (2018) prepared Fe_3O_4/MCM-41/ECH/Na_2SiO_3 magnetic nanocatalyst. They first prepared the magnetic Fe_3O_4/MCM-41 and then sodium silicate was covalently bridged onto the magnetic supports by using epichlorohydrin(ECH) as a cross-linking reagent. When the prepared magnetic heterogeneous catalysts Fe_3O_4/MCM-41/ECH/Na_2SiO_3 was used for transesterification of soybean oil, a 99.2% biodiesel yield was achieved.

2.2 Use of Nanocatalyst in Second-Ggeneration Biorefinery

As described in Section 1, biofuel production from non-food parts of current crops or other non-food sources, such as straw, bark, wood chips from forest residues, used cooking oils, as well as other crops that are not used for food purposes (non-food crops), such as switchgrass, lemongrass, etc., fall under the category of the second-generation. Various nanocatalyst used in this category includes CaO, MgO, ZnO, magnetic nanoparticles, etc. (Table 2).

Kalanakoppal Venkatesh et al. (2018) used non-edible *Butea monosperma* oil (BMO) for the production of biodiesel by using CaO nanoparticle as the catalyst. A yield of 96.2% of *Butea monosperma* methyl ester (BMME or biodiesel) was obtained (Kalanakoppal Venkatesh et al. 2018).

A series of alkali metal ion (Li, Na and K) that impregnated CaO catalysts and their catalytic activities toward transesterification of used cottonseed oil were compared. Calcium oxide impregnated with Li_2CO_3 nanocatalyst was found to have the highest basic strength (Kumar and Ali 2010). Kaur and Ali (2011) got the > 99% of biodiesel conversion after transesterification of Karanja and jatropha oils to fatty acid methyl esters using Li-ion impregnated CaO as a heterogeneous catalyst.

There was 96.8% of biodiesel yield which was obtained when KF/CaO nanocatalyst (prepared by using the impregnation method) was used for the conversion of Chinese tallow seed oil to biodiesel under the optimal conditions. Thus, showed the potential applications of catalyst in the biodiesel industry (Wen et al. 2010).

A comparison between nano-CaO synthesis using the sol-gel method and nano MgO synthesis using sol-gel self-combustion showed that nano CaO provides better efficacy, reaction duration, repeatability, used catalyst weight percentage, methanol amount and biodiesel yield than the MgO. Due to the weaker basic affinity, nano-MgO is not capable of catalyzing the transesterification by itself but when combined

Table 2. Role of nanotechnology in second-generation biorefinery.

S. No.	Nanocatalyst Used	Yield (%)	Plant Source	Reference
1.	CaO	96.2	*Butea monosperma* oil	(Venkatesh et al. 2018)
2.	CaO+MgO	98.95	Used cooking oil	(Tahvildari et al. 2015)
3.	Li^+/CaO	99.0	Karanja and jatropha oils	(Kaur and Ali 2011)
4.	Lipase immobilized on magnetic nanoparticles	80	Used cooking oil	(Yu et al. 2013)
5.	KF/CaO	96.8	Chinese tallow seed oil	(Wen et al. 2010)
6.	Magnetics KF/CaO-Fe_3O_4	95	Chinese tallow seed oil	(Hu et al. 2011)
7.	ZnO	93.7	Crude algae oil, corn oil from DDGs, crude palm oil, waste cooking oil	(Yan et al. 2010)
8.	Sodium Impregnated Zinc Oxide	–	Used cottonseed oil, mutton fat, Karanja oil, and jatropha oil	(Ali et al. 2014)
9.	Ni-doped ZnO	95.20	Castor oil	(Baskar et al. 2018)
10.	Li_2CO_3	98	Cotton seed oil	(Kumar and Ali 2010)
11.	Nano IRON	–	Castor oil	(Rengasamy et al. 2014, Rengasamy et al. 2016)
12.	Nano-Mg-Al Hydrotalcite	–	Neem oil	(Manivannan and Karthikeyan 2013)
13.	Nano-Zn-Mg-Al Hydrotalcite	90.5	Neem oil	(Karthikeyan and Manivannan 2014)
14.	Copper-doped zinc oxide	97.18	Neem oil	(Gurunathan and Ravi 2015)
15.	CuO impregnated nanoFe_3O_4	90	Used cooking oil	(Tamilmagan et al. 2015)
16.	Sonochemical (nano La_2O_3-S) Hydrothermal (nano La_2O_3-H)	97.6 and 90.3	*Jatropha curcas* oil	(Zhou et al. 2015)

with nano-CaO it becomes a proper base for the catalyst. This combination resulted in a significant increase in the transesterification reaction yield which further lead to an increase in biodiesel production mass yield from the recycled cooking oil (Tahvildari et al. 2015).

The use of copper-doped zinc oxide heterogeneous nanocatalyst for transesterification reaction for biodiesel production from neem oil yielded 97.18% biodiesel. This yield was declined to 73.95% during the use of recycled nanocatalyst in the sixth cycle. The heterogeneous solid copper-doped zinc oxide nanocatalyst was synthesized by chemical co-precipitation (Gurunathan and Ravi 2015). Higher biodiesel yield (95.20%) was achieved when doped ZnO nanocatalyst was used as a heterogeneous catalyst for biodiesel production from castor oil. The higher reuse efficiency for 3 cycles was also recorded (Baskar et al. 2018).

Manivannan and Karthikeyan (2013) used Mg-Al nano-hydrotalcites as solid base catalysts for the transesterification of neem oil. The activity of the nanocatalyst was mainly associated with calcination temperature and the Mg/Al molar ratio. In another research (Karthikeyan and Manivannan 2014) on biodiesel production from neem oil, they used Zn-Mg-Al hydrotalcites as solid base catalysts and achieved a maximum ester conversion of 90.5%. For transesterification of a variety of feedstock (virgin cottonseed oil, used cottonseed oil, mutton fat, Karanja oil and jatropha oil), Ali et al. (2014) used the sodium impregnated zinc oxide catalysts (prepared by a wet impregnation method). The catalytic activity of nanocatalyst was found to be depended on the impregnated amount sodium, the calcination temperature, methanol to oil molar ratio) reaction temperature and the amount of free fatty acid contents in the feedstock. Biodiesel produced from castor oil using nanosized iron catalyst was according to the standard of biodiesel specification ASTM D6751 and the physicochemical properties of castor oil biodiesel were similar to the normal diesel (Rengasamy et al. 2014, Rengasamy et al. 2016).

Zhou et al. (2016) performed the comparative study of La_2O_3 and two nano-La_2O_3 as a heterogeneous catalyst for biodiesel synthesis by transesterification of *Jatropha curcas* L. oil. As expected, the activity of nano-La_2O_3 was significantly higher than that of conventional La_2O_3. This higher activity was due to the high base strength, large base amount, small particle size and large BET surface area on nanocatalyst. Between the two nanocatalysts, nano La_2O_3 catalysts were prepared through sonochemical (nano La_2O_3-S) and hydrothermal (nano La_2O_3-H) methods. Due to its simple preparation procedure and short preparation time, nano-La_2O_3-S was selected for further optimization. The FAME content and yield obtained were successively 97.6% and 90.3% under optimal conditions (Zhou et al. 2015).

The use of magnetic nanocatalysts is more advantageous over the non-magnetic catalysts in the separation process. These magnetic nanocatalysts possessed a unique porous structure with a ferromagnetic property. The use of magnetic nanoparticles for the production of biodiesel from the waste of the pure vegetable oils as feedstocks yielded 80% of biodiesel. The immobilized lipase can easily be recovered by a magnetic field for repeated use (Yu et al. 2013). Hu et al. (2011) prepared the nano-magnetic catalyst KF/CaO-Fe_3O_4 by a facile impregnation method. This nano-magnetic catalyst was used for biodiesel production from the Stillingia oil and reported a 95% yield of desired fatty acid methyl esters under optimal conditions. The use of α-Fe_2O_3 and CuO impregnated Fe_3O_4 nanoparticles when used for the transesterification of used cooking oil that yielded 90% of biodiesel of international standard. These nanocatalysts were efficient and reusable in the optimum conditions (Tamilmagan et al. 2015).

3. Nanotechnology in Biogas Production

Biogas is a renewable source of energy. It is a mixture of different gases such as methane (CH_4) and carbon dioxide (CO_2) and may have small amounts of hydrogen sulfide (H_2S), moisture and siloxanes. Biogas is produced from anaerobic digestion of organic wastes such as plant raw materials, agricultural waste, manure, municipal waste, plant material, sewage, green waste or food waste by the methanogenic bacteria. The activity of these methanogenic bacteria is affected by many metal ions (iron, cobalt and nickel) in trace amounts. These metal ion when used as nanoparticle can enhance or inhibit the anaerobic reaction and hence the biogas production (Feng et al. 2014, Feng et al. 2010, Malik et al. 2018). Therefore, in order to find out the effect of these nanoparticles, many experiments were conducted by the researchers (Table 3). Nanoparticles used for these studies include metal ion, metal oxides and magnetite.

A comparative analysis on the effect of bulk-sized and nanosized CuO and ZnO particles on biogas and methane production during anaerobic digestion of cattle manure for a period of 14 days at 36°C showed the importance of metal oxide particle size on the inhibitory effect of ZnO and CuO biogas and methane production (Risco et al. 2011). This inhibition is further confirmed by many researchers (Otero-González et al. 2014, Gonzalez-Estrella et al. 2013, Mu and Chen 2011).

Similarly, another study by Mu et al. (2011) showed that among four metal oxide nanoparticles (nano-TiO_2, nano-Al_2O_3, nano-SiO_2 and nano-ZnO) nano-ZnO showed

Table 3. Role of nanotechnology in biogas production.

S. No.	Nanoparticles Used	Feedstock	Effect	Reference
1.	Zero-valent iron	Fresh raw manure	Positive	(Karri et al. 2005)
2.	Zero-valent iron	Sewage sludge	Positive	(Su et al. 2013)
3.	Zero-valent iron	Digested sludge	Positive	(Yang et al. 2013)
4.	Zero-valent iron	Biomass from brewery wastewater	Positive	(Carpenter et al. 2015)
5.	Zero-valent iron and magnetite (Fe_3O_4)	Dewatered sludge	Positive	(Suanon et al. 2016)
6.	Zero-valent iron and magnetite (Fe_3O_4)	Manure	Positive	(Abdelsalam et al. 2017b)
7.	Magnetite	Food wastes	Positive	(Dalla Vecchia et al. 2016)
8.	Co, Ni, Fe and Fe_3O_4	Cattle dung slurry	Positive	(Abdelsalam et al. 2016)
9.	Co and Ni	Raw manure	Positive	(Abdelsalam et al. 2017a)
10.	Pristine iron and iron nanoparticles-coated zeolite	Domestic sludge	Positive	(Amen et al. 2017)
11.	Fe_2O_3	Sludge from wastewater treatment	Positive	(Casals et al. 2014)
12.	CuO and ZnO	Cattle manure	Negative	(Risco et al. 2011)
13.	CuO and ZnO	Wastewater	Negative	(Otero-González et al. 2014)
14.	TiO_2, Al_2O_3, SiO_2 and ZnO	Municipal sludge	TiO_2, Al_2O_3, SiO_2-No effect ZnO-Negative	(Mu et al. 2011)
15.	Cerium dioxide (CeO_2), titanium dioxide (TiO_2), silver (Ag) and gold (Au)	Wastewater	CeO_2-Negative TiO_2, Ag, Au-No Effect	(García et al. 2012)

inhibitory effect on methane generation and this inhibitory effect of nano-ZnO is also dosage-dependent (Mu et al. 2011). TiO_2, Al_2O_3, nano-SiO_2 and CeO_2 did not show any effect on biogas production (Mu et al. 2011, García et al. 2012).

A comparative study on the effect of nanoparticles (NPs) of trace metals such as Co, Ni, Fe and Fe_3O_4 on biogas and methane production from anaerobic digestion of livestock manure resulted in conclusion that the Ni-NPs yielded the highest biogas and methane production compared to Co, Fe and Fe_3O_4 NPs (Abdelsalam et al. 2016). The spherical shape of Co and Ni improved the biogas and methane production during the anaerobic digestion of slurry (Abdelsalam et al. 2017a). Gold nanoparticles did not show any effect on the biogas production (García et al. 2012) but the silver nanoparticle were both inhibiting as well as promoting effect (García et al. 2012, Yang et al. 2012, Yang et al. 2012).

Karri et al. (2005) and Su et al. (2013) reported enhancing methane production by the use of nanoscale zero-valent ion. This increase in the methane production is due to the reduction of H_2S in biogas by nanoscale zero-valent (Su et al. 2013, Karri et al. 2005, Yang et al. 2013, Suanon et al. 2016, Abdelsalam et al. 2017b, Carpenter 2015).

Suanon et al. (2016)—when used two iron nanoparticles, i.e., nanoscale zero-valent iron (nZVI) and magnetite (Fe_3O_4) to improve biogas production-showed that proper use of iron nanoparticles improve the biogas yield and it also regulates and controls the mobilization of metals during anaerobic digestion. Besides this, iron nanoparticles also promote the immobilization of phosphorus within the solid digestate. Abdelsalam et al. (2017a) improved the production of biogas and methane by using Fe_3O_4 magnetic nanoparticles during the digestion of raw manure. Enhanced production of biogas and methane using Fe_3O_4 as well as magnetite nanoparticles were further confirmed by many researchers (Yang et al. 2015, Dalla Vecchia et al. 2016, Casals et al. 2014).

The addition of pristine iron nanoparticles and iron nanoparticles-coated zeolite stimulated methane content up to 88% and 74% and thus showed the high efficiency and good performance of the iron zeolite system (Amen et al. 2017).

4. Nanotechnology in Bioethanol Biorefinery

Bioethanol is a promising substitute for fossil fuels. It is alcohol produced by the microbial fermentation of sugar-bearing or starch-bearing plants such as corn, sugarcane, sweet sorghum or lignocellulosic biomass (Anyanwu et al. 2018). The main composition of these plants and plant parts includes cellulose and hemicellulose, polymeric structures of carbohydrates and lignin (a complex organic polymer composed mainly by phenolic compounds) (Antunes et al. 2014).

For the production of the ethanol, the plant material is pretreated for the breakdown of the cellulose and hemicellulose fraction into fermentable monomers. This is done by the enzymatic hydrolysis. This enzymatic hydrolysis resulted in the production of monomeric glucose (Rai et al. 2016). According to Antunes et al. (2014), this monomer formation step consumes 18% of the total costs involved in the process of bioethanol production. Therefore, there is a need for the development of the advanced strategies to reduce the production cost recovery and recycling of enzymes. This can be achieved by the use of the immobilization of various enzymes such as cellulases and hemicellulases on a different nanomaterial, which can provide a way to recycle reuse of the enzymes (Table 4).

These immobilized enzymes are used in both first and second-generation biorefinery. The immobilization of enzymes on magnetic nanomaterials is a widely used method in bioethanol production due to its easy recovery by using a magnetic field and reuse efficiency for several cycles (Rai et al. 2016, Rai et al. 2017). Covalent binding or physical adsorption methods were mainly used for enzyme immobilization on nanoparticles (Abraham et al. 2014) (Table 4).

Cellulase is the most critical enzyme used during bioethanol production from the plant biomass. Due to endoglucanase, exoglucanase and β-glucosidase activity

Table 4. Role of nanotechnology in bioethanol production.

S. No.	Immobilization Base	Enzyme	Reuse Efficiency	Reference
1.	Magnetic poly[2-hydroxyethyl methacrylate–N-methacryloyl-(L)-phenylalanine	*Bacillus licheniformis* α-amylase	No loss after 10 cycles	(Uygun et al. 2012)
2.	Magnetic Fe_2O_3	Porcine pancreatic α-amylase	83% after 8 cycles	(Khan et al. 2012)
3.	Enzyme coating on polymer nanofibers	β-Glucosidase	91% after 20 days of incubation	(Lee et al. 2010)
4.	Iron oxide magnetic nanoparticles	β-Glucosidase (*Aspergillus niger*)	50% enzyme activity up to the 16th cycle	(Verma et al. 2013)
5.	Magnetic nanoparticle solid core supports bound	β-Glucosidase (*Aspergillus niger*)	60% after 10 cycles	(Park et al. 2018)
6.	Magnetic nanoparticle	Cellulase	70% after 5 runs	(Abraham et al. 2014)
7.	Non-porous, silica nanoparticle	Cellulase	–	(Lupoi and Smith 2011)
8.	Non-porous, silica nanoparticle	Cellulase	–	(Chang et al. 2011)
9.	*Saccharomyces cerevisiae* cells entrapped in matrix of alginate + magnetic nanoparticles and covalently immobilized on magnetite-containing chitosan and cellulose-coated magnetic nanoparticle	Cellulase	Continuous active for 42 days without loss of activity	(Ivanova et al. 2011)
10.	Magnetic (Fe_3O_4)	Cellulase	10% after 6 cycles	(Jordan et al. 2011)
11.	TiO_2	Cellulase	–	(Ahmad and Sardar 2014)
12.	MnO_2	Cellulase	50% after 5 cycles	(Cherian et al. 2015)
13.	Iron Oxide filled Magnetic Carbon Nanotube	Amyloglucosidase	40% after 10 cycles	(Goh et al. 2012)

cellulose catalyze the multi-step hydrolysis of cellulose to glucose. In contrast, cellulase is exceptionally sensitive to many environmental factors. Thus, posing serious questions about their industrial use (Shuttleworth et al. 2014). Therefore, researchers are continuously working to improve enzyme recyclability and efficiency. Immobilization of the cellulase is known to permit higher relentlessness and also upgrade the catalytic activity (Tebeka et al. 2009, Zhang et al. 2006, Lupoi and Smith 2011, Khoshnevisan et al. 2017) (Table 4).

Mesoporous silica nanoparticles were used by Chang et al. (2011) and Lupoi and Smith (2011) for the immobilization of cellulase enzyme through both physical adsorption and covalent bonding. After immobilization, immobilized cellulase provided 80% glucose yields that were recorded by Chang et al. (2011) and the yields nearly doubled in the case of the immobilization experiment done by the Lupoi and Smith (2011). For continuous ethanol fermentation processes, *Saccharomyces cerevisiae* cells were entrapped in an alginate matrix with incorporated magnetic nanoparticles (Ivanova et al. 2011). Jordan et al. (2011) used carbodiimide as a linking polymer for enzyme immobilization on magnetic Fe_3O_4 nanoparticles. Cellulase can be immobilized in the metal oxide nanoparticle such as TiO_2, MnO_2, SiO_2, etc. (Ahmad and Sardar 2014, Cherian et al. 2015).

Ahmad and Sardar (2014) compared the activity and stability of the immobilized cellulase (derived from *Saccharomyces cerevisiae*) on TiO_2 nanoparticles by two different approaches; physical adsorption and covalent coupling concluded covalent coupling increased the stability and activity of the immobilized enzyme. Cherian et al. (2015) showed that the immobilization of cellulase on MnO_2 nanoparticle enhanced the hydrolysis of agricultural waste and hence the ethanol production.

Uygun et al. (2012) used immobilized α-amylase with magnetic-nano-poly(HEMA-MAPA) nanoparticles which were prepared by the emulsion polymerization of HEMA and MAPA. At the end of the cycle, adsorbed α-amylase was desorbed with a 95% recovery. Porcine pancreatic α-amylase, when immobilized onto Fe_2O_3-NPs by a simple adsorption mechanism, was found to be stable against various types of physical and chemical denaturants and 83% of residual activity was shown by the immobilized α-amylase after the 8th consecutive use (Khan et al. 2012).

The β-Glucosidase is another enzyme that used cellulosic ethanol production and is responsible for the conversion of cellobiose to glucose. The β-Glucosidase immobilized by using polymer magnetic nanofibers and polymer nanofibers by entrapment method (Lee et al. 2010). The repeated use and separation by applying a magnetic field provide stability to entrapped β-glucosidase on magnetic nanofibers. Another method for immobilization of β-Glucosidase was described by Verma et al. (2013). They immobilized the β-Glucosidase (BGL) from *Aspergillus niger* to functionalize magnetic nanoparticles by covalent binding and recovered 50% enzyme activity up to the 16th cycle.

Park et al. (2018) developed and evaluated the β-Glucosidase immobilized magnetic nanoparticles as recoverable biocatalysts. The developed immobilized β-Glucosidase showed high thermal stability and they retained 60% of immobilized enzyme activity after 10 recycling steps.

Goh et al. (2012) by immobilizing amyloglucosidase (AMG) onto the magnetic single-walled carbon nanotube (mSWCNT) demonstrated the recovery of the magnetic single-walled carbon nanotube-amyloglucosidase (mSWCNT-AMG) complex several times onto the magnet. The immobilized enzyme retains its activity for at least one month when stored at 4°C for one month.

5. Nanobiotechnology in Microalgal Biorefinery

High productivity, fast growth rates, higher efficiency to capture CO_2 during photosynthetic growth and adaptability to various environmental conditions such as saline or contaminated water without much need of nitrogen fertilizers, harvesting throughout the year and little problems regarding land are the reasons what make microalgae a promising non-edible feedstock for the bio-based industry (Praveenkumar et al. 2014, Mata et al. 2010, Slade and Bauen 2013, Perrine et al. 2012, Christenson and Sims 2011, Rangabhashiyam et al. 2017, Pragya et al. 2013). Being rich in oil, minerals, carbohydrate and protein fractions, these can be used for the production of many other products such as chemicals, fuels, feed, biogas and other value-added products (Chew et al. 2017, González-Delgado and Kafarov 2011, Rizwan et al. 2018, 't Lam et al. 2017, Brasil et al. 2017). Although microalgal biorefineries have great potential in the field of biofuel production, certain technical and economic challenges still needed to be addressed. Therefore, to overcome these hurdles, extensive research efforts are in a continuous process. Nanotechnology, i.e., use of multifunctional nanoparticles have proposed a key solution to these technical glitches to improve biomass productivity, lipid extraction yield and biodiesel productivity as well as to enhance the biodiesel production (Lee et al. 2015a, Wang et al. 2015, Seo et al. 2017). Various kinds of nanoparticles such as magnetic nanoparticles, aminoclay nanoparticles, etc., are being used in harvesting, lipid extraction and lipid-to-diesel conversion in microalgal biorefineries (Bui et al. 2018, Lee et al. 2015).

The use of nanoparticles during the harvesting process lowered the energy consumption, environmental toxicity, processing cost and enhanced the final microalgal concentration and reliability (Lee et al. 2015b). Magnetic nanoparticles are mainly used during this process due to its fast-magnetic separation, harvesting efficiency and low contamination. Naked or the surface-functionalized magnetic particles are widely used nanoparticle during the harvesting process (Wang et al. 2015). Magnetic naked Fe_3O_4 nanoparticle, when used for the harvesting of *Botryococcus braunii*, *Chlorella ellipsoidea* and *Nannochloropsis maritima*, showed 95–99% of efficiency (Xu et al. 2011, Hu et al. 2013) (Table 5.1). To increase recovery efficiency, functionalizing the surface of the naked particles with cationic groups is an effective method. This is due to the negative surface charge of algal cells due to the presence of proteins, lipids and sugars on the surface of algal cells. Therefore, a positive charge on the surface of the particles improves separation. Besides this, surface-functionalized magnetite particles are energy efficient and rapid. Thus, when oxide magnetic particles, nanoscale zero-valent iron, naked Fe_3O_4 and $BrFe_{12}O_{19}$ nanoparticles were functionalized with various positively charged materials, such as diallyldimethylammonium chloride (PDDA), polyethylenimine (PEI), chitosan and (3-aminopropyl) triethoxysilane (APTES), aminoclay, etc., electrophoretic mobility of the particles as well as the isoelectric point also increased

Table 5.1. Nanotechnology-assisted microalgae harvesting.

S. No.	Nanoparticle	Efficiency (%)	Algae	Reference
1.	Naked Fe_3O_4 nanoparticle	99.9	*Botryococcus braunii*	(Xu et al. 2011)
		98.9	*Chlorella ellipsoidea*	
2.	Naked Fe_3O_4 nanoparticle	95	*Nannochloropsis maritima*	(Hu et al. 2013)
3.	Bare iron oxide nanoparticles (IONPs)	–	*Chlorella* sp. and *Nannochloropsis* sp.	(Toh et al. 2014a
4.	Surface functionalized IONPs (SF-IONPs)	98.89		
5.	Poly (diallyldimethylammonium chloride) (PDDA)-coated Fe_3O_4	99	*Chlorella* sp.	(Lim et al. 2012)
6.	Poly (diallyldimethylammonium chloride) (PDDA)-coated Fe_3O_4	90	*Chlamydomonas reinhardtii and Chlorella vulgaris*	(Toh et al. 2012)
7.	Chitosan (ChiL)-coated Fe_3O_4	99	*Chlorella* sp. KR-1	(Lee et al. 2013)
8.	Magnetic beads (MBs) carrying (DEAE–diethylaminoethyl and PEI–polyethylenimine)	99	*C. vulgaris*	(Prochazkova et al. 2013)
9.	Poly(diallyldimethylammonium chloride) (PDDA) and chitosan (ChiL)-coated Fe_3O_4	PDDA-98.21 ChiL-22.93	*Chlorella* sp.	(Toh et al. 2014b)
10.	chitosan (ChiL)-Coated Fe_3O_4	99	*Chlorella* sp.	(Toh et al. 2014)
11.	Polyethylenimine (PEI)-Coated Fe_3O_4	97	*Chlorella ellipsoidea*	(Hu et al. 2014)
12.	Polyethylenimine (PEI) Coated Nanoparticles	85	*Scenedesmus dimorphus*	(Ge et al. 2015)
13.	Polyacrylamide (CPAM) coated Fe_3O_4 nanocomposites	95	*Botryococcus braunii* and *Chlorella ellipsoidea*	(Wang et al. 2014)
14.	(3-aminopropyl)triethoxysilane (APTES)-functionalized $BaFe_{12}O_{19}$	85	*Chlorella* sp. KR-1	(Seo et al. 2014)
15.	Aminoclays with Mg^{2+} or Fe^{3+}, placed in metal centers by sol-gel reaction with 3-aminopropyltriethoxysilane (APTES)	93-99	*Chlorella vulgaris* (UTEX-265)	(Farooq et al. 2013)
16.	Aluminum chloride hexahydrate surface-modified nanoclay	100	*Chlorella* sp. KR-1	(Lee et al. 2013)
17.	Mg-aminoclay	100	*Chlorella* sp.	(Lee et al. 2014)
18.	Magnesium aminoclay [MgAC] and cerium aminoclay [CeAC]	100	Blue-green microalgae	(Ji et al. 2016)

Table 5.1 contd. ...

...Table 5.1 contd.

S. No.	Nanoparticle	Efficiency (%)	Algae	Reference
19.	Magnesium aminoclay/nanoscale zerovalent iron (nZVI)	100	*Chlorella* sp.	(Lee et al. 2014)
20.	Magnesium aminoclay-Fe_3O_4 (MgAC-Fe_3O_4) hybrid composites	100	*Chlorella* sp. KR-1, *Scenedesmus obliquus* and mixed microalgae	(Kim et al. 2018)
21.	Aminoclay-conjugated TiO_2	95	*Chlorella* sp.	(Lee et al. 2014)
22.	PVP/Fe_3O_4 composites	99	*Chlorella* sp.	(Seo et al. 2015)
23.	Triazabicyclodecene (TBD)-functionalized Fe_3O_4@silica core-shell	–	*Chlorella vulgaris*	(Chiang et al. 2015)

due to the additional functional groups present on the surface of the coated particles (Lim et al. 2012, Toh et al. 2012, 2014b, 2014, 2014a, Prochazkova et al. 2013, Ge et al. 2015, Hu et al. 2014, Wang et al. 2015, Seo et al. 2014, Farooq et al. 2013, Lee et al. 2013, 2014, Lee et al. 2014, Kim et al. 2018) (Table 5.1). In addition to increasing the harvesting efficiency, these nanoparticles also increased the overall cost and therefore the use of multifunctional nanoparticles is an alternative to reduce cost as these can be used in various downstream stages including harvesting, cell disruption, lipid extraction and oil conversion. These multifunctional nanoparticles include aminoclay-conjugated TiO_2, PVP/Fe_3O_4 composites, Triazabicyclodecene (TBD)-functionalized Fe_3O_4@silica core-shell nanoparticles (TBD-Fe_3O_4@Silica NPs) (Lee et al. 2014, Seo et al. 2015, Chiang et al. 2015).

Lipid extraction from the harvested microalgae is not so easy due to the presence of a hard algal cell wall (Halim et al. 2012). Many methods have been used for lipid extraction but none of them is fully satisfactory. Therefore, researchers are now utilizing the nanoparticles for better results. Many forms of nanoparticles are being utilized for the lipid extraction which includes dielectric, hard, spinose, magnetic and enzymatic nanomaterials. The use of the aminoclays during lipid extraction enhances the lipid extraction by destabilization of the cell wall from wet microalgal biomass. Various aminoclays used in the lipid extraction includes APTES clay, Al-APTES clay, Ca-APTES clay, Mg-N_3 clay (Lee et al. 2013), Fe-APTES clay, Mn-APTES clay and Cu-APTES (Lee et al. 2013). The use of surfactant-functionalized or enzyme-functionalized nanoparticles has been also used to enhance the lipid extraction processes (Tran et al. 2013, Tran et al. 2012) (Table 5.2).

The conversion of extracted microalgal lipid, oils to the fatty acid methyl ester (FAME) or diesel-range alkanes is the final step in biodiesel production in a microalgal biorefinery. This process is catalyzed by various homogeneous and heterogeneous catalysts. Due to various environmental, economic and recycling problems, a conventional homogeneous catalyst is not very useful. To solve these problems, heterogenous nanocatalysts are now being used for the production of biodiesel from the algal oil (Sani et al. 2013, Seo et al. 2017, Singh and Gaurav 2018). There are two kinds of nanocatalyst that are widely being used and these

Table 5.2. Nanotechnology-assisted lipid extraction.

S. No.	Nanoparticle	Conversion of Oil to Biodiesel (%)	Algae	Reference
1.	Mg–APTES clay, Al–APTES clay, Ca–APTES clay, and Mg–N_3 clay	100	*Chlorella* sp.	(Lee et al. 2013)
2.	Fe-APTES clay, Mn-APTES clay, and Cu-APTES	100	*Chlorella* sp. KR-1,3	(Lee et al. 2013)
3.	Aminoclay-conjugated TiO_2	–		(Lee et al. 2014)
4.	Fe_3O_4@SiO_2 nanoparticles using dimethyloctadecyl[3-(trimethoxysilyl) propyl] ammonium chloride as a linker	97.3	*C. vulgaris* ESP-31	(Tran et al. 2012)
5.	Lipase immobilized onto alkyl-grafted Fe_3O_4–SiO_2	97.6	*Chlorella vulgaris* ESP-31	(Tran et al. 2013)
6.	Immobilized cellulase into an electrospun polyacrylonitrile (PAN) nanofibrous membrane	–	*Chlorella pyrenoidsa*	(Fu et al. 2010)

include acid nanocatalyst and the base nanocatalyst (Carrero et al. 2011, Chiang et al. 2015, Seo et al. 2017, Kandel et al. 2014, Velasquez-Orta et al. 2013, Teo et al. 2014, Umdu et al. 2009, Park et al. 2015)

6. Conclusion and Future Aspects

A combination of biorefineries processes with the nanotechnologies is a significant technical advancement in the economics of biorefineries in terms of yield and product range. But there are certain steps in various biorefineries that still require scientific as well as technological advancements. Nanotechnology can significantly and effectively contribute to addressing these limitations. For this, nanotechnologists are required to design more active and selective nanocatalysts, more versatile nano-based catalytic systems as well as greener routes to nanoparticles production and utilization.

7. Disclosure of Potential Conflicts of Interest

Authors declare no potential conflicts of interest.

References

Abdelsalam, E., M. Samer, Y.A. Attia, M.A. Abdel-Hadi, H.E. Hassan and Y. Badr. 2016. Comparison of nanoparticles effects on biogas and methane production from anaerobic digestion of cattle dung slurry. Renewable Energy 87: 592–598.

Abdelsalam, E., M. Samer, Y.A. Attia, M.A. Abdel-Hadi, H.E. Hassan and Y. Badr. 2017a. Effects of Co and Ni nanoparticles on biogas and methane production from anaerobic digestion of slurry. Energy Conversion and Management 141: 108–119.

Abdelsalam, E., M. Samer, Y.A. Attia, M.A. Abdel-Hadi, H.E. Hassan and Y. Badr. 2017b. Influence of zero valent iron nanoparticles and magnetic iron oxide nanoparticles on biogas and methane production from anaerobic digestion of manure. Energy 120: 842–853.

Abraham, R.E., M.L. Verma, C.J. Barrow and M. Puri. 2014. Suitability of magnetic nanoparticle immobilised cellulases in enhancing enzymatic saccharification of pretreated hemp biomass. Biotechnology for Biofuels 7(1): 1–12.

Ahmad, R. and M. Sardar. 2014. Immobilization of cellulase on TiO_2 nanoparticles by physical and covalent methods: A comparative study. Indian Journal of Biochemistry & Biophysics 51(4, August): 314–320.

Alaei, S., M. Haghighi, J. Toghiani and B. Rahmani Vahid. 2018. Magnetic and reusable MgO/ $MgFe_2O_4$ nanocatalyst for biodiesel production from sunflower oil: Influence of fuel ratio in combustion synthesis on catalytic properties and performance. Industrial Crops and Products 117(September): 322–332.

Ali, A., P. Khullar and D. Kumar. 2014. Sodium impregnated zinc oxide as a solid catalyst for biodiesel preparation from a variety of triglycerides. Energy Sources, Part A: Recovery, Utilization, and Environmental Effects 36(18): 1999–2008.

Ali, S., I. Khan, S.A. Khan, M. Sohail, R. Ahmed, A. ur Rehman, M.S. Ansari and M.A. Morsy. 2017. Electrocatalytic performance of Ni@Pt core–shell nanoparticles supported on carbon nanotubes for methanol oxidation reaction. Journal of Electroanalytical Chemistry 795(April): 17–25.

Alsharifi, M., H. Znad, S. Hena and M. Ang. 2017. Biodiesel production from canola oil using novel Li/TiO_2 as a heterogeneous catalyst prepared via impregnation method. Renewable Energy 114: 1077–1089.

Amen, T.W.M., O. Eljamal, A.M.E. Khalil and N. Matsunaga. 2017. Biochemical methane potential enhancement of domestic sludge digestion by adding pristine iron nanoparticles and iron nanoparticles coated zeolite compositions. Journal of Environmental Chemical Engineering 5(5): 5002–5013.

Anon. 2017. The Future of the Western Cape Agricultural Sector in the Context of the 4th Industrial Revolution, Review: Biorefinery and Biofuels.

Antunes, F.A.F., A.K. Chandel, T.S.S. Milessi, J.C. Santos, C.A. Rosa and S.S. Da Silva. 2014. Bioethanol production from sugarcane bagasse by a novel Brazilian pentose fermenting yeast Scheffersomyces shehatae UFMG-HM 52.2: Evaluation of fermentation medium. International Journal of Chemical Engineering 2014.

Anyanwu, R.C., C. Rodriguez, A. Durrant and A.G. Olabi. 2018. Micro-macroalgae properties and applications. In Reference Module in Materials Science and Materials Engineering. Elsevier.

Astefanei, A., O. Núñez and M.T. Galceran. 2015. Characterisation and determination of fullerenes: A critical review. Analytica Chimica Acta 882: 1–21.

Balu, Alina M., B. Baruwati, E. Serrano, J. Cot, J. Garcia-Martinez, R.S. Varma and R. Luque. 2011. Magnetically separable nanocomposites with photocatalytic activity under visible light for the selective transformation of biomass-derived platform molecules. Green Chemistry 13(10): 2750–2758.

Balu, Alina Mariana, V. Budarin, P.S. Shuttleworth, L.A. Pfaltzgraff, K. Waldron, R. Luque and J.H. Clark. 2012. Valorisation of orange peel residues: waste to biochemicals and nanoporous materials. ChemSusChem. 5(9, September): 1694–1697.

Barrak, H., T. Saied, P. Chevallier, G. Laroche, A. M'nif and A.H. Hamzaoui. 2016. Synthesis, characterization, and functionalization of ZnO nanoparticles by N-(Trimethoxysilylpropyl) Ethylenediamine Triacetic Acid (TMSEDTA): Investigation of the interactions between phloroglucinol and ZnO@TMSEDTA. Arabian Journal of Chemistry 12(8): 4340–4347.

Baskar, G., I. Aberna Ebenezer Selvakumari and R. Aiswarya. 2018. Biodiesel production from castor oil using heterogeneous Ni Doped ZnO nanocatalyst. Bioresource Technology 250(December, 2017): 793–798.

Baskrar, G., A. Gurugulldevi, T. Nishanthini, R. Aiswarya and K. Tamilarasan. 2017. Optimization and kinetics of biodiesel production from mahua oilusing manganese doped zinc oxide nanocatalyst. Renewable Energy 103: 641–647.

Bell, G., M. Schuck, Stephen Jungmeier, Gerfried Wellisch, C. Felby, H. Jørgensen, M. Stichnothe, Heinz Clancy, S. De Bari, Isabella Kimura, R. van Ree and Jong De. 2014. IEA Bioenergy Task42 Biorefining. IEA Bioenergy Task42.

Benedé, J. 2014. Towards the automatic resolution of architectural variability in software product line architectures through model transformations. CEUR Workshop Proceedings 1258(12): 69–74.

Berger, T., J. Schuh, M. Sterrer, O. Diwald and E. Knözinger. 2007. Lithium ion induced surface reactivity changes on MgO nanoparticles. Journal of Catalysis 247(1): 61–67.

Berntsson, T., B. Sanden, L. Olsson and A. Asblad. 2014. What is a biorefinery? Systems Perspectives on Biorefineries 2008: 16–25.

Bet-Moushoul, E., K. Farhadi, Y. Mansourpanah, A.M. Nikbakht, R. Molaei and M. Forough. 2016. Application of CaO-based/Au nanoparticles as heterogeneous nanocatalysts in biodiesel production. Fuel 164: 119–127.

Bhatia, S. 2016. Natural polymer drug delivery systems: nanoparticles, plants, and algae. pp. 1–225. *In*: Natural Polymer Drug Delivery Systems. 1st ed. Switzerland: Springer International Publishing.

Bibalan, S.F. and S.M. Sadrameli. 2012. Kinetic modeling of sunflower oil methanolysis considering effects of interfacial area of reaction system. Iranian Journal of Chemical Engineering 9(1): 50–59.

Botero, C.D., D.L. Restrepo and C.A. Cardona. 2017. A comprehensive review on the implementation of the biorefinery concept in biodiesel production plants. Biofuel Research Journal 4(3): 691–703.

Boz, N., N. Degirmenbasi and D.M. Kalyon. 2009. Conversion of biomass to fuel: Transesterification of vegetable oil to biodiesel using KF loaded nano-γ-Al_2O_3 as catalyst. Applied Catalysis B: Environmental 89(3-4): 590–596.

Brasil, B.S.A., F.C.P. Silva and F.G. Siqueira. 2017. Microalgae biorefineries: The Brazilian scenario in perspective. New Biotechnology 39(October 25): 90–98.

Bridgwater, T. 2006. Biomass for energy. Journal of the Science of Food and Agriculture 86(12): 1755–1768.

Bui, V.K.H., D. Park and Y.-C. Lee. 2018. Aminoclays for biological and environmental applications: An updated review. Chemical Engineering Journal 336: 757–772.

Büscher, K., C.A. Helm, C. Gross, G. Glöckl, E. Romanus and W. Weitschies. 2004. Nanoparticle composition of a ferrofluid and its effects on the magnetic properties. Langmuir 20(6)(March 1): 2435–2444.

Campani, V., S. Giarra and G. De Rosa. 2018. Lipid-based core-shell nanoparticles: Evolution and potentialities in drug delivery. OpenNano 3(September 2017): 5–17.

Canakci, M. and J. Van Gerpen. 2003. A pilot plant to produce biodiesel from high free fatty acid feedstocks. Transactions of the Asae 46(4): 945–954.

Carpenter, A.W., S.N. Laughton and M.R. Wiesner. 2015. Enhanced biogas production from nanoscale zero valent iron-amended anaerobic bioreactors. Environmental Engineering Science 32(8): 647–655.

Carrero, A., G. Vicente, R. Rodríguez, M. Linares and G.L. Del Peso. 2011. Hierarchical zeolites as catalysts for biodiesel production from nannochloropsis microalga oil. Catalysis Today 167(1): 148–153.

Carvalho, M.S., R.A. Lacerda, J.P.B. Leão, J.D. Scholten, B.A.D. Neto and P.A.Z. Suarez. 2011. *In situ* generated palladium nanoparticles in imidazolium-based ionic liquids: A versatile medium for an efficient and selective partial biodiesel hydrogenation. Catalysis Science and Technology 1(3): 480–488.

Casals, E., R. Barrena, A. García, E. González, L. Delgado, M. Busquets-Fité, X. Font, K. Kvashnina, A. Sánchez and P. Víctor. 2014. Programmed iron oxide nanoparticles disintegration in anaerobic digesters boosts biogas production. Small 10(14): 2801–2808.

Chang, R.H.Y., J. Jang and K.C.W. Wu. 2011. Cellulase immobilized mesoporous silica nanocatalysts for efficient cellulose-to-glucose conversion. Green Chemistry 13(10): 2844–2850.

Cherian, E., M. Dharmendirakumar and G. Baskar. 2015. Immobilization of cellulase onto MnO_2 nanoparticles for bioethanol production by enhanced hydrolysis of agricultural waste. Cuihua Xuebao/Chinese Journal of Catalysis 36(8): 1223–1229.

Chew, K.W., J.Y. Yap, P.L. Show, N.H. Suan, J.C. Juan, T.C. Ling, D.J. Lee and J.S. Chang. 2017. Microalgae biorefinery: High value products perspectives. Bioresource Technology 229: 53–62.

Chiang, Y.D., S. Dutta, C.T. Chen, Y.T. Huang, K.S. Lin, J.C.S. Wu, N. Suzuki, Y. Yamauchi and K.C.W. Wu. 2015. Functionalized Fe_3O_4@silica core-shell nanoparticles as microalgae harvester and catalyst for biodiesel production. ChemSusChem 8(5): 789–794.

Chisti, Y. 2008. Biodiesel from microalgae beats bioethanol. Trends in Biotechnology 26(3): 126–131.

Chitra, P., P. Venkatachalam and A. Sampathrajan. 2005. Optimisation of experimental conditions for biodiesel production from alkali-catalysed transesterification of jatropha curcus oil. Energy for Sustainable Development 9(3): 13–18.

Christenson, L. and R. Sims. 2011. Production and harvesting of microalgae for wastewater treatment, biofuels, and bioproducts. Biotechnology Advances 29(6): 686–702.

Christian, P., F. Von Der Kammer, M. Baalousha and T. Hofmann. 2008. Nanoparticles: Structure, properties, preparation and behaviour in environmental media. Ecotoxicology 17(5): 326–343.

Dalla Vecchia, C., A. Mattioli, D. Bolzonella and E. Palma. 2016. Impact of magnetite nanoparticles supplementation on the anaerobic digestion of food wastes: Batch and continuous-flow investigations. Chemical Engineering Transactions 49: 1–6.

Degirmenbasi, N., S. Coskun, N. Boz and D.M. Kalyon. 2015. Biodiesel synthesis from canola oil via heterogeneous catalysis using functionalized CaO nanoparticles. Fuel 153: 620–627.

Deng, J., Y. Peng, C. He, X. Long, P. Li and A.S.C. Chan. 2003. Magnetic and conducting Fe_3O_4-polypyrrole nanoparticles with core-shell structure. Polymer International 52(7): 1182–1187.

Dizge, N. and B. Keskinler. 2008. Enzymatic production of biodiesel from canola oil using immobilized lipase. Biomass and Bioenergy 32(12): 1274–1278.

Dizge, N., C. Aydiner, D.Y. Imer, M. Bayramoglu, A. Tanriseven and B. Keskinler. 2009. Biodiesel production from sunflower, soybean, and waste cooking oils by transesterification using lipase immobilized onto a novel microporous polymer. Bioresource Technology 100(6): 1983–1991.

Eijck, V. and H. Henny. 2008. Prospects for jatropha biofuels in Tanzania: An analysis with strategic niche management. Energy Policy 36(1): 311–325.

Eijck, V., B. Batidzirai and A. Faaij. 2014. Current and future economic performance of first and second generation biofuels in developing countries. Applied Energy 135: 115–141.

Eijck, V.J., R. Henny, A. Balkema and A. Faaij. 2014. Global experience with jatropha cultivation for bioenergy: An assessment of socio-economic and environmental aspects. Renewable and Sustainable Energy Reviews 32(C): 869–889.

Farooq, W., Y.C. Lee, J.I. Han, C.H. Darpito, M. Choi and J.W. Yang. 2013. Efficient microalgae harvesting by organo-building blocks of nanoclays. Green Chemistry 15(3): 749–755.

Feng, X.M., A. Karlsson, B.H. Svensson and S. Bertilsson. 2010. Impact of trace element addition on biogas production from food industrial waste-linking process to microbial communities. FEMS Microbiology Ecology 74(1): 226–240.

Feng, Y., Y. Zhang, X. Quan and S. Chen. 2014. Enhanced anaerobic digestion of waste activated sludge digestion by the addition of zero valent iron. Water Research 52: 242–250.

Feynman, R. 1960. There's (still) plenty of room at the bottom. Engineering and Science, 22–36.

Feyzi, M., A. Hassankhani and H.R. Rafiee. 2013. Preparation and characterization of Cs/Al/Fe_3O_4 nanocatalysts for biodiesel production. Energy Conversion and Management 71: 62–68.

Feyzi, M., N. Leila and Z. Mohammad. 2014. Preparation and characterization of magnetic $CsH_2PW_{12}O_{40}$/Fe-SiO_2 nanocatalysts for biodiesel production. Materials Research Bulletin 60: 412–420.

FitzPatrick, M., P. Champagne, M.F. Cunningham and R. Whitney. 2010. A biorefinery processing perspective: Treatment of lignocellulosic materials for the production of value-added products. Bioresource Technology 101(23): 8915–8922.

Freedman, B., R.O. Butterfield and E.H. Pryde. 1986. Transesterification kinetics of soybean oil. J. Am. Oil Chem. 63(10): 1375–1380.

Frenkel, A.I., A. Yevick, C. Cooper and R. Vasic. 2011. Modeling the structure and composition of nanoparticles by extended X-Ray absorptionfine-structure spectroscopy. Annual Review of Analytical Chemistry 4(1): 23–39.

Fu, C.C., T.C. Hung, J.Y. Chen, C.H. Su and W.T. Wu. 2010. Hydrolysis of microalgae cell walls for production of reducing sugar and lipid extraction. Bioresource Technology 101(22): 8750–8754.

Ganesh, M., P. Hemalatha, M.M. Peng and H.T. Jang. 2017. One pot synthesized Li, Zr doped porous silica nanoparticle for low temperature CO_2 adsorption. Arabian Journal of Chemistry 10: S1501–S1505.

García, A., L. Delgado, J.A. Torà, E. Casals, E. González, V. Puntes, X. Font, J. Carrera and A. Sánchez. 2012. Effect of cerium dioxide, titanium dioxide, silver, and gold nanoparticles on the activity of microbial communities intended in wastewater treatment. Journal of Hazardous Materials 199-200: 64–72.

Ge, S., M. Agbakpe, W. Zhang and L. Kuang. 2015. Heteroaggregation between PEI-coated magnetic nanoparticles and algae: Effect of particle size on algal harvesting efficiency. ACS Applied Materials and Interfaces 7(11): 6102–6108.

Goh, W.J., V.S. Makam, J. Hu, L. Kang, M. Zheng, S.L. Yoong, C.N.B. Udalagama and G. Pastorin. 2012. Iron oxide filled magnetic carbon nanotube-enzyme conjugates for recycling of amyloglucosidase: Toward useful applications in biofuel production process. Langmuir 28(49): 16864–16873.

Gomez, L.D., C.G. Steele-King and S.J. McQueen-Mason. 2008. Sustainable liquid biofuels from biomass: The writing's on the walls. The New Phytologist 178(3): 473–485.

González-Delgado, A.-D. and V. Kafarov. 2011. Microalgae based biorefinery: Issues to consider. CTyF-Ciencia, Tecnologia y Futuro 4(4): 5–22.

Gonzalez-Estrella, J., R. Sierra-Alvarez and J.A. Field. 2013. Toxicity assessment of inorganic nanoparticles to acetoclastic and hydrogenotrophic methanogenic activity in anaerobic granular sludge. Journal of Hazardous Materials 260: 278–285.

Greenwell, H.C., L.M.L. Laurens, R.J. Shields, R.W. Lovitt and K.J. Flynn. 2010. Placing microalgae on the biofuels priority list: A review of the technological challenges. Journal of the Royal Society, Interface-the Royal Society 7(46): 703–26.

Guo, Z. and Y. Sun. 2004. Characteristics of immobilized lipase on hydrophobic superparamagnetic microspheres to catalyze esterification. Biotechnology Progress 20(2): 500–506.

Gurunathan, B. and A. Ravi. 2015. Process optimization and kinetics of biodiesel production from neem oil using copper doped zinc oxide heterogeneous nanocatalyst. Bioresource Technology 190: 424–428.

Hahn-Hagerdal, B., M. Galbe, M.F. Gorwa-Grauslund, G. Liden and G. Zacchi. 2006. Bio-ethanol-the fuel of tomorrow from the residues of today. Trends in Biotechnology 24(12, December): 549–556.

Halim, R., M.K. Danquah and P.A. Webley. 2012. Extraction of oil from microalgae for biodiesel production: A review. Biotechnology Advances 30(3): 709–732.

Hassan, N., K.N. Ismail, K.H. Ku Hamid and A. Hadi. 2018. CaO nanocatalyst for transesterification reaction of palm oil to biodiesel: Effect of precursor's concentration on the catalyst behavior. IOP Conference Series: Materials Science and Engineering 358: 012059.

Helwani, Z., M.R. Othman, N. Aziz, W.J.N. Fernando and J. Kim. 2009. Technologies for production of biodiesel focusing on green catalytic techniques: A review. Fuel Processing Technology 90(12): 1502–1514.

Hu, S., Y. Guan, Y. Wang and H. Han. 2011. Nano-magnetic catalyst KF/CaO-Fe_3O_4 for biodiesel production. Applied Energy 88(8): 2685–2690.

Hu, Y.R., F. Wang, S.K. Wang, C.Z. Liu and C. Guo. 2013. Efficient harvesting of marine microalgae nannochloropsis maritima using magnetic nanoparticles. Bioresource Technology 138: 387–390.

Hu, Y.R., C. Guo, F. Wang, S.K. Wang, F. Pan and C.Z. Liu. 2014. Improvement of microalgae harvesting by magnetic nanocomposites coated with polyethylenimine. Chemical Engineering Journal 242: 341–347.

IEA Bioenergy Task 42 Biorefinery [online]. 2013. IEA Bioenergy.

Ivanova, V., P. Petrova and J. Hristov. 2011. Application in the ethanol fermentation of immobilized yeast cells in matrix of alginate/magnetic nanoparticles, on chitosan-magnetite microparticles and cellulose-coated magnetic nanoparticles. J. Int. Rev. Chem. Eng. 3(2): 289–299.

Ji, H.M., H.U. Lee, E.J. Kim, S. Seo, B. Kim, G.W. Lee, Y.K. Oh, J.Y. Kim, Y.S. Huh, H.A. Song and Y. Lee. 2016. Efficient harvesting of wet blue-green microalgal biomass by two-aminoclay [AC]-mixture systems. Bioresource Technology 211: 313–318.

Jong, W. and J. Borm. 2008. Drug delivery and nanoparticles: Applications and hazards. International Journal of Nanomedicine 3(2): 133–49.

Jordan, J., C.S.S.R. Kumar and C. Theegala. 2011. Preparation and characterization of cellulase-bound magnetite nanoparticles. Journal of Molecular Catalysis B: Enzymatic 68(2): 139–146.

Kamata, K. and K. Sugahara. 2017. Base catalysis by mono- and polyoxometalates. Catalysts 7(11): 345.

Kamm, B. and M. Kamm. 2004a. Biorefinery systems. Chemical and Biochemical Engineering Quarterly 18(1): 1–6.

Kamm, B. and M. Kamm. 2004b. Principles of biorefineries. Applied Microbiology and Biotechnology 64(2): 137–145.

Kandel, K., J.W. Anderegg, N.C. Nelson, U. Chaudhary and I.I. Slowing. .2014. Supported iron nanoparticles for the hydrodeoxygenation of microalgal oil to green diesel. Journal of Catalysis 314: 142–148.

Karri, S., R. Sierra-Alvarez and J.A. Field. 2005. Zero valent iron as an electron-donor for methanogenesis and sulfate reduction in anaerobic sludge. Biotechnology and Bioengineering 92(7): 810–819.

Karthikeyan, C. and R. Manivannan. 2014. Environmentally benign neem biodiesel synthesis using nano-Zn-Mg-Al hydrotalcite as solid base catalysts. Journal of Catalysts 2014(3): 1–6.

Kaur, M. and A. Ali. 2011. Lithium ion impregnated calcium oxide as nano catalyst for the biodiesel production from Karanja and Jatropha oils. Renewable Energy 36(11): 2866–2871.

Khan, I., A. Abdalla and A. Qurashi. 2017. Synthesis of hierarchical WO_3 and Bi_2O_3/WO_3 nanocomposite for solar-driven water splitting applications. International Journal of Hydrogen Energy 42(5): 3431–3439.

Khan, I., K. Saeed and I. Khan. 2017. Nanoparticles: Properties, applications and toxicities. Arabian Journal of Chemistry 12(7): 908–931.

Khan, M.J., Q. Husain and A. Azam. 2012. Immobilization of porcine pancreatic α-amylase on magnetic Fe_2O_3 nanoparticles: Applications to the hydrolysis of starch. Biotechnology and Bioprocess Engineering 17(2): 377–384.

Khoshnevisan, K., F. Vakhshiteh, M. Barkhi, H. Baharifar, E. Poor-Akbar, N. Zari, H. Stamatis and A.K. Bordbar. 2017. Immobilization of cellulase enzyme onto magnetic nanoparticles: Applications and recent advances. Molecular Catalysis 442: 66–73.

Kim, B., V.K.H. Bui, W. Farooq, S.G. Jeon, Y.K. Oh and Y.C. Lee. 2018. Magnesium aminoclay-Fe_3O_4(MgAC-Fe_3O_4) hybrid composites for harvesting of mixed microalgae. Energies 11(6): 1–10.

Kim, M., H.S. Lee, S.J. Yoo, Y.S. Youn, Y.H. Shin and Y.W. Lee. 2013. Simultaneous synthesis of biodiesel and zinc oxide nanoparticles using supercritical methanol. Fuel 109: 279–284.

Kleinert, M. and T. Barth. 2008. Towards a lignincellulosic biorefinery: Direct one-step conversion of lignin to hydrogen-enriched biofuel. Energy and Fuels 22(2): 1371–1379.

Kobayashi, H., T. Komanoya, K. Hara and A. Fukuoka. 2010. Water-tolerant mesoporous-carbon-supported ruthenium catalysts for the hydrolysis of cellulose to glucose. ChemSusChem 3(4): 440–443.

Kumar, B., K. Jalodia, P. Kumar and H.K. Gautam. 2017. Recent advances in nanoparticle-mediated drug delivery. Journal of Drug Delivery Science and Technology 41: 260–268.

Kumar, D. and A. Ali. 2010. Nanocrystalline lithium ion impregnated calcium oxide as heterogeneous catalyst for transesterification of high moisture containing cotton seed oil. Energy and Fuels 24(3): 2091–2097.

Laurent, S., D. Forge, M. Port, A. Roch, C. Robic, L. Vander Elst and R.N. Muller. 2010. Magnetic iron oxide nanoparticles: Synthesis, stabilization, vectorization, physicochemical characterizations, and biological applications. Chemical Reviews 110(4)(April 14): 2574.

Lee, J.E., N. Lee, T. Kim, J. Kim and T. Hyeon. 2011. Multifunctional mesoporous silica nanocomposite nanoparticles for theranostic applications. Accounts of Chemical Research 44(10)(October 18): 893–902.

Lee, K., S.Y. Lee, J.G. Na, S.G. Jeon, R. Praveenkumar, D.M. Kim, W.S. Chang and Y.K. Oh. 2013. Magnetophoretic harvesting of oleaginous Chlorella sp. by using biocompatible chitosan/magnetic nanoparticle composites. Bioresource Technology 149: 575–578.

Lee, K., J.G. Na, J.Y. Seo, T.S. Shim, B. Kim, R. Praveenkumar, J.Y. Park, Y.K. Oh and S.G. Jeon. 2015. Magnetic-nanoflocculant-assisted water-nonpolar solvent interface sieve for microalgae harvesting. ACS Applied Materials and Interfaces 7(33): 18336–18343.

Lee, S.M., L.H. Jin, J.H. Kim, S.O. Han, H. Bin Na, T. Hyeon, Y.M. Koo, J. Kim and J.H. Lee. 2010. β-Glucosidase coating on polymer nanofibers for improved cellulosic ethanol production. Bioprocess and Biosystems Engineering 33(1): 141–147.

Lee, Y.C., Y.S. Huh, W. Farooq, J. Chung, J.I. Han, H.J. Shin, S.H. Jeong, J.S. Lee, Y.K. Oh and J.Y. Park. 2013. Lipid extractions from docosahexaenoic acid (DHA)-rich and oleaginous Chlorella sp. biomasses by organic-nanoclays. Bioresource Technology 137: 74–81.

Lee, Y.C., Y.S. Huh, W. Farooq, J.I. Han, Y.K. Oh and J.Y. Park. 2013. Oil extraction by aminoparticle-based H_2O_2 activation via wet microalgae harvesting. RSC Advances 3(31): 12802–12809.

Lee, Y.C., B. Kim, W. Farooq, J. Chung, J.I. Han, H.J. Shin, S.H. Jeong, J.Y. Park, J.S. Lee and Y.K. Oh. 2013. Harvesting of oleaginous Chlorella sp. by organoclays. Bioresource Technology 132: 440–445.

Lee, Y.C., H.U. Lee, K. Lee, B. Kim, S.Y. Lee, M.H. Choi, W. Farooq, J.S. Choi, J.Y. Park, J. Lee, Y. Oh and Y.S. Huh. 2014. Aminoclay-conjugated TiO_2 synthesis for Simultaneous harvesting and wet-disruption of oleaginous Chlorella sp. Chemical Engineering Journal 245: 143–149.

Lee, Y.C., K. Lee, Y. Hwang, H.R. Andersen, B. Kim, S.Y. Lee, M.H. Choi, J.Y. Park, O.Y. Kwan and S.H. Yun. 2014. Aminoclay-templated nanoscale zero-valent iron (NZVI) synthesis for efficient harvesting of oleaginous microalga, Chlorella sp. KR-1. RSC Advances 4(8): 4122–4127.

Lee, Y.C., S.Y. Oh, H.U. Lee, B. Kim, S.Y. Lee, M.H. Choi, G.W. Lee, J.Y. Park, Y.K. Oh, T. Ryu, Y.K. Han, K.S. Chung and Y.S. Huh. 2014. Aminoclay-induced humic acid flocculation for efficient harvesting of oleaginous Chlorella sp. Bioresource Technology 153: 365–369.

Lee, Y.C., K. Lee and Y.K. Oh. 2015. Recent nanoparticle engineering advances in microalgal cultivation and harvesting processes of biodiesel production: A review. Bioresource Technology 184: 63–72.

Lim, J.K., D.C.J. Chieh, S.A. Jalak, P.Y. Toh, N.H.M. Yasin, B.W. Ng and A.L. Ahmad. 2012. Rapid magnetophoretic separation of microalgae. Small 8(11): 1683–1692.

Liu, C.-C., W.-C. Lu and T.-J. Liu. 2012. Transesterification of soybean oil using CsF/CaO catalysts. Energy & Fuels 26(9)(September 20): 5400–5407.

Liu, X., H. He, Y. Wang, S. Zhu and X. Piao. 2008. Transesterification of soybean oil to biodiesel using CaO as a solid base catalyst. Fuel 87(2): 216–221.

Lupoi, J.S. and E.A. Smith. 2011. Evaluation of nanoparticle-immobilized cellulase for improved ethanol yield in simultaneous saccharification and fermentation reactions. Biotechnology and Bioengineering 108(12): 2835–2843.

Madhuvilakku, R. and S. Piraman. 2013. Biodiesel synthesis by TiO_2-ZnO mixed oxide nanocatalyst catalyzed palm oil transesterification process. Bioresource Technology 150: 55–59.

Malik, K., S. Majeed, X. Leng, F. Yusuf Hafeez, I. Saif, X. Li, Y. Zafar, S. Faisal and S. Zhao. 2018. A review on nanoparticles as boon for biogas producers—nano fuels and biosensing monitoring. Applied Sciences 9(1): 59.

Malik, M.A., P. O'Brien and N. Revaprasadu. 2002. A smple route to the synthesis of core/shell nanoparticles of chalcogenides. Chemistry of Materials 14(5)(May 1): 2004–2010.

Manivannan, R. and C. Karthikeyan. 2013. Synthesis of biodiesel from neem oil using Mg-Al nano hydrotalcite. In Advances in Nanoscience and Nanotechnology 678: 268–272. Advanced Materials Research. Trans Tech Publications.

Mansha, M., A. Qurashi, N. Ullah, F.O. Bakare, I. Khan and Z.H. Yamani. 2016. Synthesis of In_2O_3/graphene heterostructure and their hydrogen gas sensing properties. Ceramics International 42(9): 11490–11495.

Mansha, M., I. Khan, N. Ullah and A. Qurashi. 2017. Synthesis, characterization and visible-light-driven photoelectrochemical hydrogen evolution reaction of carbazole-containing conjugated polymers. International Journal of Hydrogen Energy 42(16): 10952–10961.

Marchetti, J.M., V.U. Miguel and A.F. Errazu. 2007. Possible methods for biodiesel production. Renewable and Sustainable Energy Reviews 11(6): 1300–1311.

Mata, T.M., A.A. Martis and N.S. Caetno. 2010. Microalgae for biodiesel production and other applications: A review. Renewable and Sustainable Energy Reviews 14: 217–32.

Math, M.C., S.P. Kumar and S.V. Chetty. 2010. Technologies for biodiesel production from used cooking oil—a review. Energy for Sustainable Development 14(4): 339–345.

Mohan, D., C.U. Pittman and P.H. Steele. 2006. Pyrolysis of wood/biomass for bio-oil: A critical review. Energy and Fuels 20(3): 848–889.

Mori, T. and T. Hegmann. 2016. Determining the composition of gold nanoparticles: A compilation of shapes, sizes, and calculations using geometric considerations. Journal of Nanoparticle Research: An Interdisciplinary Forum for Nanoscale Science and Technology 18(10): 295.

Mortadha, A.A., I.A. Al-Hydary and T.A. Al-Hattab. 2017. Nano-magnetic catalyst CaO-Fe_3O_4 for biodiesel production from date palm seed oil. Bulletin of Chemical Reaction Engineering & Catalysis 13(3): 460–468.

Mu, H. and Y. Chen. 2011. Long-term effect of ZnO nanoparticles on waste activated sludge anaerobic digestion. Water Research 45(17): 5612–5620.

Mu, H., Y. Chen and N. Xiao. 2011. Effects of metal oxide nanoparticles (TiO_2, Al_2O_3, SiO_2 and ZnO) on waste activated sludge anaerobic digestion. Bioresource Technology 102(22): 10305–10311.

Naik, S.N., V.V. Goud, P.K. Rout and A.K. Dalai. 2010. Production of first and second generation biofuels: A comprehensive review. Renewable and Sustainable Energy Reviews 14(2): 578–597.

Nanotechnology 101 [online]. 2008. nano.gov.

Nizami, A.-S. and M. Rehan. 2018. Towards nanotechnology-based biofuel industry. Biofuel Research Journal 5(2): 798–799.

Oda, M., M. Kaieda, S. Hama, H. Yamaji, A. Kondo, E. Izumoto and H. Fukuda. 2005. Facilitatory effect of immobilized lipase-producing rhizopus oryzae cells on acyl migration in biodiesel-fuel production. Biochemical Engineering Journal 23(1): 45–51.

Otero-González, L., J.A. Field and R. Sierra-Alvarez. 2014. Inhibition of anaerobic wastewater treatment after long-term exposure to low levels of CuO nanoparticles. Water Research 58: 160–168.

Park, H.J., A.J. Driscoll and P.A. Johnson. 2018. The development and evaluation of β-glucosidase immobilized magnetic nanoparticles as recoverable biocatalysts. Biochemical Engineering Journal 133: 66–73.

Park, J.-Y., M.S. Park, Y.-C. Lee and J.-W. Yang. 2015. Advances in direct transesterification of algal oils from wet biomass. Bioresource Technology 184(May): 267–275.

Perrine, Z., S. Negi and R.T. Sayre. 2012. Optimization of photosynthetic light energy utilization by microalgae. Algal Research 1(2): 134–142.

Pragya, N., K.K. Pandey and P.K. Sahoo. 2013. A review on harvesting, oil extraction and biofuels production technologies from microalgae. Renewable and Sustainable Energy Reviews 24: 159–171.

Praveenkumar, R., B. Kim, E. Choi, K. Lee, J.Y. Park, J.S. Lee, Y.C. Lee and Y.K. Oh. 2014. Improved Biomass and lipid production in a mixotrophic culture of chlorella sp. KR-1 with addition of coal-fired flue-gas. Bioresource Technology 171: 500–505.

Prochazkova, G., N. Podolova, I. Safarik, V. Zachleder and T. Branyik. 2013. Physicochemical approach to freshwater microalgae harvesting with magnetic particles. Colloids and Surfaces B: Biointerfaces 112: 213–218.

Rahmani Vahid, B. and M. Haghighi. 2016. Urea-nitrate combustion synthesis of MgO/$MgAl_2O_4$ nanocatalyst used in biodiesel production from sunflower oil: Influence of fuel ratio on catalytic properties and performance. Energy Conversion and Management 126: 362–372.

Rai, M., J.C.D. Santos, M.F. Soler, F.P.R. Marcelino, L.P. Brumano, A.P. Ingle, G. Swapnil, A. Gade and S.S.D. Silva. 2016. Strategic role of nanotechnology for production of bioethanol and biodiesel. Nanotechnology Reviews 6(5): 383–404.

Rai, Mahendra, A.P. Ingle, S. Gaikwad, K.J. Dussa and S. Silve. 2017. Nanotechnology for bioenergy and biofuel production. pp. 153–171. *In*: Rai, S. and S.S. Silva (eds.). Nanotechnology for Bioenergy and Biofuel Production. 1st ed. Springer International Publishing AG 2017.

Rajiv, P., R. Sivaraj, R.S.V. Priya and P. Vanathi. 2014. Synthesis and characterization parthenium mediated zinc oxide nanoparticles and assessing its medicinal properties. In International Conference on Advances in Agricultural, Biological & Environmental Sciences (AABES-2014) Oct 15–16, 2014 Dubai (UAE), 15–17. Dubai.

Ramacharyulu, P.V.R.K., R. Muhammad, J. Praveen Kumar, G.K. Prasad and P. Mohanty. 2015. Iron phthalocyanine modified mesoporous titania nanoparticles for photocatalytic activity and CO_2 capture applications. Physical Chemistry Chemical Physics 17(39): 26456–26462.

Rangabhashiyam, S., B. Behera, N. Aly and P. Balasubramanian. 2017. Biodiesel from microalgae as a promising strategy for renewable bioenergy production—a review. Journal of Environmental and Biotechnology Research 6(4): 260–269.

Ranganathan, S.V., S.L. Narasimhan and K. Muthukumar. 2008. An overview of enzymatic production of biodiesel. Bioresource Technology 99(10): 3975–3981.

Rawal, I. and A. Kaur. 2013. Synthesis of mesoporous polypyrrole nanowires/nanoparticles for ammonia gas sensing application. Sensors and Actuators, A: Physical 203: 92–102.

Reddy, C., V. Reddy, R. Oshel and J.G. Verkade. 2006. Room-temperature conversion of soybean oil and poultry fat to biodiesel catalyzed by nanocrystalline calcium oxides. Energy and Fuels 20(3): 1310–1314.

Rengasamy, M., S. Mohanraj, S.H. Vardhan and R. Balaji. 2014. Transesterification of castor oil using nano-sized iron catalyst for the production of biodiesel. Journal of Chemical and Pharmaceutical Sciences 2: 108–112.

Rengasamy, M., K. Anbalagan, S. Kodhaiyolii and V. Pugalenthi. 2016. Castor leaf mediated synthesis of iron nanoparticles for evaluating catalytic effects in transesterification of castor oil. RSC Advances 6(11): 9261–9269.

Risco, M.L., K. Orupõld and H.C. Dubourguier. 2011. Particle-size effect of CuO and ZnO on biogas and methane production during anaerobic digestion. Journal of Hazardous Materials 189(1-2): 603–608.

Rizvi, S.A.A. and A.M. Saleh. 2017. Applications of nanoparticle systems in drug delivery technology. Saudi Pharmaceutical Journal 26(1): 64–70.

Rizwan, M., G. Mujtaba, S.A. Memon, K. Lee and N. Rashid. 2018. Exploring the potential of microalgae for new biotechnology applications and beyond: A review. Renewable and Sustainable Energy Reviews 92(May): 394–404.

Rosarin, F.S. and S. Mirunalini. 2011. Nobel metallic nanoparticles with novel biomedical properties. Journal of Bioanalysis & Biomedicine 03(04): 85–91.

Sachdeva, H. and R. Saroj. 2013. ZnO nanoparticles as an efficient, heterogeneous, reusable, and ecofriendly catalyst for four-component one-pot green synthesis of pyranopyrazole derivatives in water. The Scientific World Journal 2013: 680671.

Sani, Y.M., W.M.A.W. Daud and A.R. Abdul Aziz. 2013. Solid acid-catalyzed biodiesel production from microalgal oil—the dual advantage. Journal of Environmental Chemical Engineering 1(3): 113–121.

Sarkar, J. and S. Bhattacharyya. 2012. Operating characteristics of transcritical CO_2 heat pump for simultaneous water cooling and heating. Archives of Thermodynamics 33(4): 23–40.

Sekhon, B. 2014. Nanotechnology in agri-food production: An overview. Nanotechnology, Science and Applications 7: 31–53.

Seo, J.Y., K. Lee, S.Y. Lee, S.G. Jeon, J.G. Na, Y.K. Oh and S. Bin Park. 2014. Effect of barium ferrite particle size on detachment efficiency in magnetophoretic harvesting of oleaginous chlorella sp. Bioresource Technology 152: 562–566.

Seo, J.Y., K. Lee, R. Praveenkumar, B. Kim, S.Y. Lee, Y.K. Oh and S. Bin Park. 2015. Tri-functionality of Fe_3O_4-embedded carbon microparticles in microalgae harvesting. Chemical Engineering Journal 280: 206–214.

Seo, J.Y., M.G. Kim, K. Lee, Y.-C. Lee, J.-G. Na, S.G. Jeon, S. Bin Park and Y.-K. Oh. 2017. Multifunctional nanoparticle applications to microalgal biorefinery. pp. 59–87. *In*: Rai, M. and S.S. da Silva (eds.). Nanotechnology for Bioenergy and Biofuel Production. Cham: Springer International Publishing. https://doi.org/10.1007/978-3-319-45459-7_4.

Serrano, E., G. Rus and J. García-Martínez. 2009. Nanotechnology for sustainable energy. Renewable and Sustainable Energy Reviews 13(9): 2373–2384.

Shaalan, M., M. Saleh, M. El-Mahdy and M. El-Matbouli. 2016. Recent progress in applications of nanoparticles in fish medicine: A review. Nanomedicine: Nanotechnology, Biology, and Medicine 12(3): 701–710.

Shih-Hung, H., L. Min-Hung and C. Dong-Hwang. 2008. Direct binding and ccharacterization of lipase onto magnetic nanoparticles. Biotechnology Progress 19(3): 1095–1100.

Shin, W.K., J. Cho, A.G. Kannan, Y.S. Lee and D.W. Kim. 2016. Cross-linked composite gel polymer electrolyte using mesoporous methacrylate-functionalized SiO_2 nanoparticles for lithium-ion polymer batteries. Scientific Reports 6(March): 1–10.

Shuttleworth, P.S., M. De Bruyn, H.L. Parker, A.J. Hunt, V.L. Budarin, A.S. Matharu and J.H. Clark. 2014. Applications of nanoparticles in biomass conversion to chemicals and fuels. Green Chemistry 16(2): 573–584.

Sigmund, W., J. Yuh, H. Park, V. Maneeratana, G. Pyrgiotakis, A. Daga, J. Taylor and J.C. Nino. 2006. Processing and structure relationships in electrospinning of ceramic fiber systems. Journal of the American Ceramic Society 89(2): 395–407.

Simakova, I., O. Simakova, P. Mäki-Arvela and D.Y. Murzin. 2010. Decarboxylation of fatty acids over Pd supported on mesoporous carbon. Catalysis Today 150(1-2): 28–31.

Singh, A. and K. Gaurav. 2018. Advancement in catalysts for transesterification in the production of biodiesel: A review 7: 1148–1158.

Slade, R. and A. Bauen. 2013. Micro-algae cultivation for biofuels: Cost, energy balance, environmental impacts and future prospects. Biomass and Bioenergy 53(0): 29–38.

Smith, V.H., B.S.M. Sturm, F.J. deNoyelles and S.A. Billings. 2010. The ecology of algal biodiesel production. Trends in Ecology and Evolution 25(5): 301–309.

Snåre, M., I. Kubičková, P. Mäki-Arvela, K. Eränen and D.Y. Murzin. 2006. Heterogeneous catalytic deoxygenation of stearic acid for production of biodiesel. Industrial and Engineering Chemistry Research 45(16): 5708–5715.

Stöcker, M. 2008. Biofuels and biomass-to-liquid fuels in the biorefinery: Catalytic conversion of lignocellulosic biomass using porous materials. Angewandte Chemie—International Edition 47(48): 9200–9211.

Su, L., X. Shi, G. Guo, A. Zhao and Y. Zhao. 2013. Stabilization of sewage sludge in the presence of Nanoscale Zero-Valent Iron (NZVI): Abatement of odor and improvement of biogas production. Journal of Material Cycles and Waste Management 15(4): 461–468.

Suanon, F., Q. Sun, D. Mama, J. Li, B. Dimon and C.P. Yu. 2016. Effect of nanoscale zero-valent iron and magnetite (Fe_3O_4) on the fate of metals during anaerobic digestion of sludge. Water Research 88: 897–903.

Sun, Y. and J. Cheng. 2002. Hydrolysis of lignocellulosic materials for ethanol production: A review. Bioresource Technology 83: 1–11.

't Lam, G.P., M.H. Vermuë, M.H.M. Eppink, R.H. Wijffels and C. van den Berg. 2017. Multi-product microalgae biorefineries: From concept towards reality. Trends in Biotechnology 36(2): 216–227.

Tahvildari, K., Y.N. Anaraki, R. Fazaeli, S. Mirpanji and E. Delrish. 2015. The study of CaO and MgO heterogenic nano-catalyst coupling on transesterification reaction efficacy in the production of biodiesel from recycled cooking oil. Journal of Environmental Health Science and Engineering 13(1).

Tamilmagan, A., Maheswari Priyabijesh and A. Gopal. 2015. Biodiesel production from waste cooking oil using green synthesized nano Fe_2O_3 and CuO impregnated nano Fe_3O_4. International Journal of ChemTech Research 8(5): 90–96.

Tao, G., Z. Hua, Z. Gao, Y. Chen, L. Wang, Q. He, H. Chen and J. Shi. 2012. Synthesis and catalytic activity of mesostructured KF/CaxAl 2O(X+3) for the Transesterification reaction to produce biodiesel. RSC Advances 2(32): 12337–12345.

Tao, G., Z. Hua, Z. Gao, Y. Zhu, Y. Chen, Z. Shu, L. Zhang and J. Shi. 2013. KF-loaded mesoporous Mg-Fe Bi-metal oxides: High performance transesterification catalysts for biodiesel production. Chemical Communications 49(73): 8006–8008.

Tebeka, I.R.M., A.G.L. Silva and D.F.S. Petri. 2009. Hydrolytic activity of free and immobilized cellulase. Langmuir: The ACS Journal of Surfaces and Colloids 25(3 February): 1582–1587.

Teo, S.H., Y.H. Taufiq-Yap and F.L. Ng. 2014. Alumina supported/unsupported mixed oxides of Ca and Mg as heterogeneous catalysts for transesterification of nannochloropsis sp. microalga's oil. Energy Conversion and Management 88: 1193–1199.

The Royal Society. 2008. Sustainable Biofuels: Prospects and Challenges. The Royal Society: Sustainable Biofuels: Prospects and Challenges.

Thomas, S.C., Harshita, P.K. Mishra and S. Talegaonka. 2015. Ceramic nanoparticles: fabrication methods and applications in drug delivery. Current Pharmaceutical Design 1(42): 6165–6188.

Toda, M., A. Takagaki, M. Okamura, J.N. Kondo, S. Hayashi, K. Domen and M. Hara. 2005. Green chemistry: Biodiesel made with sugar catalyst. Nature 438(7065 November): 178.

Toh, P.Y., S.P. Yeap, L.P. Kong, B.W. Ng, D.J.C. Chan, A.L. Ahmad and J.K. Lim. 2012. Magnetophoretic removal of microalgae from fishpond water: Feasibility of high gradient and low gradient magnetic separation. Chemical Engineering Journal 211-212: 22–30.

Toh, P.Y., B.W. Ng, A.L. Ahmad, D.C.J. Chieh and J. Lim. 2014a. The role of particle-to-cell interactions in dictating nanoparticle aided magnetophoretic separation of microalgal cells. Nanoscale 6(21): 12838–12848.

Toh, P.Y., B.W. Ng, A.L. Ahmad, D.C.J. Chieh and J.K. Lim. 2014b. Magnetophoretic separation of chlorella sp.: role of cationic polymer binder. Process Safety and Environmental Protection 92(6): 515–521.

Toh, P.Y., B.W. Ng, C.H. Chong, A.L. Ahmad, J.W. Yang, C.J. Chieh Derek and J. Lim. 2014. Magnetophoretic separation of microalgae: the role of nanoparticles and polymer binder in harvesting biofuel. RSC Advances 4(8): 4114–4121.

Tran, D.T., K.L. Yeh, C.L. Chen and J.S. Chang. 2012. Enzymatic transesterification of microalgal oil from chlorella vulgaris ESP-31 for biodiesel synthesis using immobilized burkholderia lipase. Bioresource Technology 108: 119–127.

Tran, D.T., B.H. Le, D.J. Lee, C.L. Chen, H.Y. Wang and J.S. Chang. 2013. Microalgae harvesting and subsequent biodiesel conversion. Bioresource Technology 140: 179–186.

Ullah, H., I. Khan, Z.H. Yamani and A. Qurashi. 2017. Sonochemical-driven ultrafast facile synthesis of SnO_2 nanoparticles: Growth mechanism structural electrical and hydrogen gas sensing properties. Ultrasonics Sonochemistry 34: 484–490.

Umdu, E.S., M. Tuncer and E. Seker. 2009. Transesterification of Nannochloropsis oculata microalga's lipid to biodiesel on Al_2O_3 supported CaO and MgO catalysts. Bioresource Technology 100(11): 2828–2831.

Uygun, D.A., N. Ozutuk, A. Akgol and A. Denizli. 2012. Novel magnetic nanoparticles for the hydrolysis of starch with bacillus licheniformis a-amylase. Journal of Applied Polymer Science 123: 2574–581.

van Ree, R., B. Annevelink, A. René van Ree, E. de Jong, J. Reijnders and K. Kwant. 2007. Status Report Biorefinery, 2007.

Velasquez-Orta, S.B., J.G.M. Lee and A.P. Harvey. 2013. Evaluation of FAME production from wet marine and freshwater microalgae by *in situ* transesterification. Biochemical Engineering Journal 76: 83–89.

Venkatesh, Y.K., R. Mahadevaiah, L.H. Shankaraiah, R. Suresh and A.R. Basanagouda. 2018. Preparation of a CaO nanocatalyst and its application for biodiesel production using butea monosperma oil: An optimization study. JAOCS, Journal of the American Oil Chemists' Society 95(5): 635–649.

Verma, M.L., R. Chaudhary, T. Tsuzuki, C.J. Barrow and M. Puri. 2013. Immobilization of β-glucosidase on a magnetic nanoparticle improves thermostability: Application in cellobiose hydrolysis. Bioresource Technology 135: 2–6.

Verziu, M., B. Cojocaru, J. Hu, R. Richards, C. Ciuculescu, P. Filip and V.I. Parvulescu. 2008. Sunflower and rapeseed oil transesterification to biodiesel over different nanocrystalline MgO catalysts. Green Chemistry 10(4): 373–381.

Viswanatha, R., T.G. Venkatesh, C.C. Vidyasagar and Y. Arthoba Nayaka. 2012. Preparation and characterization of ZnO and Mg-ZnO nanoparticle. Archives of Applied Science Research 4(1): 480–486.

Wang, L. and J. Yang. 2007. Transesterification of soybean oil with nano-MgO or not in supercritical and subcritical methanol. Fuel 86(3): 328–333.

Wang, S.K., F. Wang, Y.R. Hu, A.R. Stiles, C. Guo and C.Z. Liu. 2014. Magnetic flocculant for high efficiency harvesting of microalgal cells. ACS Applied Materials and Interfaces 6(1): 109–115.

Wang, S.K., A.R. Stiles, C. Guo and C.Z. Liu. 2015. Harvesting microalgae by magnetic separation: A review. Algal Research 9: 178–185.

Wen, L., Y. Wang, D. Lu, S. Hu and H. Han. 2010. Preparation of KF/CaO nanocatalyst and its application in biodiesel production from chinese tallow seed oil. Fuel 89(9): 2267–2271.

West, J.L. and N.J. Halas. 2000. A gold nanoparticle bioconjugate-based colorimetric assay, 215–217.

Xie, W., Y.Z. Zhenqiang and C. Hong. 2007. Catalytic properties of lithium-doped ZnO catalysts used for biodiesel preparations. Industrial and Engineering Chemistry Research 46(24): 7942–7949.

Xie, W. and N. Ma. 2010. Enzymatic transesterification of soybean oil by using immobilized lipase on magnetic nano-particles. Biomass and Bioenergy 34(6): 890–896.

Xie, W., Y. Han and H. Wang. 2018. Magnetic Fe_3O_4/MCM-41 composite-supported sodium silicate as heterogeneous catalysts for biodiesel production. Renewable Energy 125: 675–681.

Xu, L., C. Guo, F. Wang, S. Zheng and C.Z. Liu. 2011. A simple and rapid harvesting method for microalgae by *in situ* magnetic separation. Bioresource Technology 102(21): 10047–10051.

Xue, Y., X. Qiu, Z. Liu and Y. Li. 2018. Facile and Efficient synthesis of silver nanoparticles based on biorefinery wood lignin and its application as the optical sensor research-article. ACS Sustainable Chemistry and Engineering 6(6): 7695–7703.

Yagiz, F., D. Kazan and A.N. Akin. 2007. Biodiesel production from waste oils by using lipase immobilized on hydrotalcite and zeolites. Chemical Engineering Journal 134(1-3): 262–267.

Yan, S., S. Mohan, C. Dimaggio, M. Kim, K.Y.S. Ng and S.O. Salley. 2010. Long term activity of modified ZnO nanoparticles for transesterification. Fuel 89(10): 2844–2852.

Yang, Y., M. Xu, J.D. Wall and Z. Hu. 2012. Nanosilver impact on methanogenesis and biogas production from municipal solid waste. Waste Management 32(5): 816–825.

Yang, Y., Q. Chen, J.D. Wall and Z. Hu. 2012. Potential nanosilver impact on anaerobic digestion at moderate silver concentrations. Water Research 46(4): 1176–1184.

Yang, Y., J. Guo and Z. Hu. 2013. Impact of Nano Zero Valent Iron (NZVI) on methanogenic activity and population dynamics in anaerobic digestion. Water Research 47(17): 6790–6800.

Yang, Z., X. Shi, C. Wang, L. Wang and R. Guo. 2015. Magnetite nanoparticles facilitate methane production from ethanol via acting as electron acceptors. Scientific Reports 5: 1–8.

Yu, C.Y., L.Y. Huang, I.C. Kuan and S.L. Lee. 2013. Optimized production of biodiesel from waste cooking oil by lipase immobilized on magnetic nanoparticles. International Journal of Molecular Sciences 14(12): 24074–24086.

Yulianti, C.K., D. Hartanto, T.E. Purbaningtias, Y. Chisaki, A.A. Jalil, C.K.N.L.C.K. Hitam and D. Prasetyoko. 2014. Synthesis of CaOZnO nanoparticles catalyst and its application in transesterification of refined palm oil. Bulletin of Chemical Reaction Engineering and Catalysis 9(2): 100–110.

Zhang, P., Y.H. Himmel, M.E. Mielenz and R. Jonathan. 2006. Outlook for cellulase improvement: Screening and selection strategies. Biotechnology Advances 24(5): 452–481.

Zhao, L., Z. Qiu and S.M. Stagg-Williams. 2013. Transesterification of canola oil catalyzed by nanopowder calcium oxide. Fuel Processing Technology 114: 154–162.

Zhou, Q., H. Zhang, F. Chang, H. Li, H. Pan, W. Xue, D.Y. Hu and S. Yang. 2015. Nano La_2O_3 as a heterogeneous catalyst for biodiesel synthesis by transesterification of *Jatropha curcas* L. oil. Journal of Industrial and Engineering Chemistry 31: 385–392.

Zhu, L. 2015. Biorefinery as a promising approach to promote microalgae industry: An innovative framework. Renewable and Sustainable Energy Reviews 41(January 1): 1376–1384.

Index

P

T - #0104 - 111024 - C89 - 234/156/17 - PB - 9780367546335 - Gloss Lamination